$155.40

Selected Topics on
Electron Physics

PHYSICS OF ATOMS AND MOLECULES

A Chronological Listing of Volumes in this series appears at the back of this volume.

Selected Topics on Electron Physics

Edited by

D. Murray Campbell
University of Edinburgh
Edinburgh, Scotland

and

Hans Kleinpoppen
University of Stirling
Stirling, Scotland

Plenum Press • New York and London

Library of Congress Cataloging-in-Publication Data

Selected topics on electron physics / edited by D. Murray Campbell and
Hans Kleinpoppen.
      p.   cm. -- (Physics of atoms and molecules)
    "Proceedings of the Peter Farago Symposium on Electron Physics,
held as a satellite symposium to the fifth European Conference on
Atomic and Molecular Physics, held March 31-April 1, 1995, in
Edinburgh, Scotland"--T.p. verso.
    Includes bibliographical references and index.
    ISBN 0-306-45484-X
    1. Electrons. 2. Nuclear reactions.   I. Campbell, Murray.
II. Kleinpoppen, H. (Hans) III. Peter Farago Symposium on Electron
Physics (1995 : Edinburgh, Scotland) IV. European Conference on
Atomic and Molecular Physics (5th : 1995 : Edinburgh, Scotland)
V. Series.
QC793.5.E62S464 1996
539.7'2112--dc20   96-43717
                                                                  CIP

Proceedings of the Peter Farago Symposium on Electron Physics, held as a satellite symposium to the
Fifth European Conference on Atomic and Molecular Physics,
March 31 – April 1, 1995, at The Royal Society, in Edinburgh, Scotland

ISBN 0-306-45484-X

© 1996 Plenum Press, New York
A Division of Plenum Publishing Corporation
233 Spring Street, New York, N. Y. 10013

10 9 8 7 6 5 4 3 2 1

Printed in the United States of America

Peter Farago

Prof. H. Kleinpoppen
Unit of Atomic & Molecular Physics
University of Stirling
STIRLING FK9 4LA

Dr D. M. Campbell
Physics Department
University of Edinburgh
James Clerk Maxwell Buildings
King's Buildings
Mayfield Road
EDINBURGH   EH9 3JZ

Prof. R.M. Sillitto
Physics Department
University of Edinburgh
James Clerk-Maxwell Building
King's Buildings
Mayfield Road
EDINBURGH EH9 3JZ

A.J. Murray, N.J.Bowering and F.H.Read
Schuster Laboratory
The University
MANCHESTER   M13 9PL

Prof Dr G Baum and Prof Dr W Raith
Fakultät für Physik Universität Bielefeld
D-33501 Bielefeld
Germany

Prof E Reichert
Institüt für Physik
Johannes-Gutenberg Universität
Universität Mainz
D-55099 MAINZ
GERMANY

Prof. Albert Crowe and Igor Bray
Department of Physics
The University
Newcastle upon Tyne   NE1 7RU

Prof G.F. Hanne
Westfälische Wilhelms-Universität
Physikalisches Institut
Lehrstuhl f. Experimentelle Atomphysik
Wilhelm-Klemm-Strasse 10
D-48149 Münster
GERMANY

Prof S Chwirot, Dariusz Dziczek, Rajesh Srivastava
and Romans S Dygdala
Institut of Physics
Nicholas Copernicus University
ul.Grudziadzka 5-7
PL 87-100 TORUN
Poland

Prof. M.C. Standage
Dean
Faculty of Science & Technology
Griffith University
Nathan Brisbane
AUSTRALIA 4111

Prof. J.F. Williams
Department of Physics
The University of Western Australia
Nedlands
Western Australia 6009

Prof. P.G. Burke, CBE, FRS. and M.P. Scott
Department of Applied Mathematics &
Theoretical Physics
The Queen's University of Belfast
BELFAST BT7 1NN

Dr B. Lohmann
Institut für Theoretische Physik I
Westfalische-Wilhelms-Universität Münster
Wilhelm-Klemm-Str. 9
D-48149 Münster
Germany

H. Ast, C.T. Whelan, S Keller, J.Rasch. H.R.J.Walters
and R.M. Dreizler
Institut für Theoretische Physik
der J W Goethe-Universitat
Robert-Meyer-Str. 8-10
D-60054 Frankfurt am. Maine
Germany

Dr Klaus Bartschat
Department of Physics & Astronomy
Drake University
Des Moines
IA 50311
USA

Professor Joachim Kessler
Lehrstuhl für Experimentelle Atomphysik
Westfälische Wilhelms-Universität Münster
Wilhelm-Klemm-Strasse 10
D-48149 MUNSTER Germany

T.J. Gay, M.E. Johnston, K.W. Trantham and G.A. Gallup
Physics Department
University of Nebraska
Lincoln, Nebraska 68588-011
USA

Prof. K. Blum
Westfälische Wilhelms-Universität
Institut für Theoretische Physik I
Wilhelm-Klemm-Str. 9
48149 Münster
Germany

Dr W.R. Newell
Physics Department
University College London
Gower Street
London WC1E 6BT

Dr Norbert Böwering
Molekül- und Oberflächenphysik
Universität Bielefeld
Postfach 10 01 31
D-33501 Bielefeld
Germany

Prof. H.C. Siegmann
Laboratorium für Festkörperphysik
ETH-Hönggerberg
CH-8093 Zürich
Switzerland

H.Steidl and G. Baum
Fakultät für Physik
Universität Bielefeld
Postfach 100131
D33501 Bielefeld, Germany

Dr W. Nakel
Universität Tübingen
Physikalisches Institut
Auf der Morgenstelle 14
D-72076 Tübingen
Morganstelle, Germany

Prof. Dr. U. Heinzmann
Universität Bielefeld
Molekül- und Oberfläckenphysik
Postfach 100131
D-33501 Bielefeld
Germany

Dr A.V. Solov'yov
A.I. Ioffe Physical-Technical Institute
Politechnichnicheskaja 26
194021 St Petersburg
Russia

Dr Z. Roller-Lutz
Fakultät für Physik
Universitä Bielefeld
Postfach 100131
D-33501 Bilefeld
Germany

Dr B. Siegmann
Fakultät für Physik
Universität Bielefeld
D-33615 Bielefeld
Germany

Dr Bengt Skogvall and Gebhard von Oppen
Institut für Strahlungs- und Kernphysik
Technische Universität Berlin
Hardenbergstr. 36
D-10623 Berlin
Germany

Profs. U. Werner and H.O. Lutz
Fakultät für Physik
Universität Bielefeld
Postfach 10 01 31
D-33501 Bielefeld
Germany

Dr J.B. West
Daresbury Laboratory
Warrington
Cheshire   WA4 4AD

Dr M Ya Amusia
Institut for Fysik og Astronomi
Aarhus Universitet
DK-8000 Aarhus
Denmark

Dr G.C. King and G. Dawber
Department of Physics & Astronomy
University of Manchester
MANCHESTER   M13 9PL

Dr N.A. Cherepkov
Johannes Gutenberg-Universität
Fachbereich Physik (18)
Institut für Physik
D-55099 Mainz
Germany

Prof. F.H.M. Faisal and A.Becker
Fakultät für Physik
Universität Bielefeld
D-33615 Bielefeld
Germany

Prof. G.W.F. Drake
Department of Physics
University of Windsor
Windsor, Ontario NB9 3PA
Canada

Dr M. Brieger
Institut für Technische Physik DLR
Pfaffenwaldring 38-4o
D-70569 Stuttgart
Germany

Dr N.J. Mason
Department of Physics & Astronomy
University College London
Gower Street
London   WC1E 6BT

Dr A.J. Duncan
Atomic Physics Laboratory
University of Stirling
Stirling FK9 4LA, Scotland

PREFACE

In the spring of 1970 Peter Farago organised a three-day conference on Polarised Electron Beams at Carberry Tower, near Edinburgh. Although the development of the gallium arsenide source, which was to revolutionise the world of experimental polarised electron physics, was still some years in the future, the meeting provided an important forum for the exchange of ideas among theoreticians and experimentalists engaged in both high and low energy electron collision studies.

As soon as the decision had been taken to hold the 5th European Conference on Atomic and Molecular Physics in Edinburgh in 1995, it occurred to the editors of the present volume that it would be highly appropriate to mark the twenty-fifth anniversary of the Carberry Tower Conference by organising an ECAMP satellite meeting in honour of Peter Farago. The opportunity to pay tribute to Peter's many important contributions in the broad field of electron physics attracted colleagues from all over the world to the symposium, which was held in the rooms of the Royal Society of Edinburgh on 31st March and 1st April 1995. Peter himself, now Professor Emeritus at the University of Edinburgh, was present throughout the meeting. We were particularly happy to welcome back to Edinburgh many participants in the original Carberry Tower conference; these included Professor P. G. Burke, Professor J. Kessler, Professor E. Reichert and Professor H. C. Siegmann, whose review papers had been highlights of the 1970 meeting.

The range of scientific topics covered by the present symposium spans the eclectic research interests and activities of Peter Farago. Many of these are linked to effects in particle, atomic, and molecular physics, and in spin-related investigations in surface and solid state physics. The spectroscopic investigation of hydrogenic ions, in which Peter Farago collaborated with Canadian colleagues, has shown steady progress over two decades. The continuous development of research in the varied fields described in this volume provides a fascinating insight into the way in which experimental physics complements theoretical attempts to understand and explain some of the fundamental processes of nature.

The programme of papers present at the symposium was limited by the time available. However, a considerable number of distinguished physicists agreed to contribute further papers, which have been included in this publication. Richard Sillitto delivered a moving personal memoir of Peter at the symposium dinner, and he has kindly allowed its reproduction here. To all contributors we offer our grateful thanks for ensuring that the volume is a worthy tribute to Peter Farago.

We are also grateful to the Royal Society of Edinburgh and the Universities of Edinburgh and Stirling for financial support. Finally, we wish to join with all Peter's friends and colleagues in congratulating him on his many achievements in research, thanking him for his dedication and inspiration as a teacher, and wishing him many further happy and productive years.

D.M. Campbell
H. Kleinpoppen

PETER FARAGO: EARLY DAYS IN EDINBURGH

R. Sillitto

Physics Department
University of Edinburgh
James Clerk Maxwell Building
King's Buildings
Edinburgh   EH9 3JZ

It's both a pleasure and a privilege to respond to the Editor's invitation to contribute a brief note about how Peter Farago came to be in Edinburgh and his early days here.

Peter was born in 1918, the son of a schoolmaster from whom he certainly appears to have inherited his father's love of teaching and care for education. He studied for his degree in physics at the University of Budapest and Eötvös College, which was an institute modelled on the École Normale Supérieure in Paris and a similar institute in Italy. In those days he came under the influence of a most distinguished physicist, Zoltan Bay, who was to become to Peter in turn teacher, colleague, and lifelong friend. In 1946 Peter joined the Tungsram Research Laboratory, where he worked under Bay. Subsequently he became Professor of Physics at Eötvös College and Reader in Physics at the University of Budapest, in charge of the Chair which had been vacated when Georg Békésy, the future Nobel Laureate, moved to Harvard. Later, Peter became the Head of a Section of the Central Research Institute of Physics of the Hungarian Academy of Sciences, working in close cooperation with L. Jánossy who had just returned to Hungary from the Dublin Institute of Advanced Studies.

Peter, and his first wife Elizabeth and son "little Peter," left Hungary early in 1957 - Peter to travel to Leipzig where he had been invited to lecture, and then on to East Berlin under the auspices of the Hungarian Academy, while Elizabeth and "little Peter" set out privately on the same day to Paris to visit a relative of Elizabeth's; their visa applications were handled by different branches of the same Ministry, and the fact that the whole family had left the country wasn't realised until they were well away. Arrived in East Berlin, Peter crossed to the West and showed to British officials an invitation he had to visit London to lecture at Imperial College, and his onward journey to London was therefore made under British government auspices. The whole family was soon reunited in London. Being an *afficionado* of espionage fiction, I can't resist adding that, in London Peter found himself living in the same street as a chemical engineer from London

University, called Dr Peter Farago! After spending a few months at Imperial College, sponsored by Denis Gabor, Peter applied for and was appointed to a Senior Lectureship here in Edinburgh. Before he arrived I'd hunted in *Physics Abstracts* for some clues as to his scientific interests, and was intrigued to find that he (with Bay) had looked into some processes I was interested in - the so-called "photon coincidences" that are found in photoelectron coincidence experiments in coherent light fields. I remarked to our Head of department, Norman Feather, that our newly appointed colleague had some interests in common with mine, to which he replied drily: "I think you'll find he has a remarkable range of interests."

Arrived in Scotland, Peter settled into what must have seemed a rather strange department, with a fair quota of undoubtedly strange colleagues. He acquired a couple of research students, with whom he started to plan and set up an experiment to measure the g-factor anomaly of "free electrons," by a new and very elegant method which promised a degree of precision several orders of magnitude better than any previous measurements. When the project was advanced to a point at which he felt the research students could be left to get it up and running, Peter accepted an invitation to spend a year at Maryland in USA, complete with family. Left on their own, the two post-grads had, I suppose, a reasonable quota of mishaps, one of them verging on the catastrophic. For a time, friendly enquiries from the rest of us as to how the project was getting on received rather cool and tight-lipped responses. But progress was made, and after Peter's return the experiment was completed successfully; unfortunately by then another very elegant and much more expensive experiment by Crane and his school at Ann Arbor had produced a result which extended the precision by another order of magnitude. By us in Edinburgh, this was considered to be bad luck. I expect a different view was held in Ann Arbor!

Around this time, say 1961, Norman Feather must have decided that it would do me good to be exposed to Peter's influence full time, because he told me that because of pressure of space elsewhere Peter would be sharing my office. So for two or three years we sat and contemplated one another and shared our thoughts on the office blackboard. These thoughts were often concerned with the lack of evidence for the existence of "photons," and with the kind of evidence on the fluctuations of optical electromagnetic fields that could be obtained from photoelectron coincidences; and - Peter being Peter - with what could be learned about light-fields from their interaction with "free" electrons. We decided we should go down to NPL and there discuss possible experiments, and explore the possibility of borrowing some equipment. Peter wondered whether his stateless status would be an impediment to such a visit - it turned out that it wouldn't - and it must have been at that time that I ventured to ask him what his feelings were when he had left his own country, very probably for life. "I can accept being a foreigner in a foreign country," he replied, "but what I could not face was the thought of spending the rest of my life feeling like a foreigner in my own country."

One evening around this time, Peter and Elizabeth were entertaining in their Edinburgh flat some Hungarian friends who had left Hungary about the same time as they had and for similar reasons. They had settled in New York, but were passing through Edinburgh, and had dropped in. There was a lot of talk about, and it was really quite late when Peter glanced at the clock and said: "Isn't it great to know that at this time of night you won't hear a knock on the door, and open it to find uniformed police there!" Just then the doorbell rang, and when Peter opened the door he found himself confronted by uniformed police. "Sorry to disturb you sir, but we're wondering whether the owner of motor vehicle ZYX098 is here - it's parked in a rather exposed place, and it would be safer overnight if it could be moved a little." Yes - life was different here!

The Faragos settled in Edinburgh for good, with breaks for short sabbatical leaves overseas at the University of Munich (then in West Germany), the University of Windsor (Ontario), the FOM Institute of Atomic and Molecular Physics in Amsterdam, and the University of West Australia in Perth. Peter became a valued member of the community of British physicists; he has been a Fellow of the Institute of Physics since 1957, serving on some of its committees and for two years acting as Deputy Editor of the *Journal of Physics B*; he was elected a Fellow of the Royal Society of Edinburgh in 1961, and was appointed to a Personal Chair at the University of Edinburgh in 1967.

Much of Peter's work has been in fields close to the theme of this volume; Feather's remarks about the breadth of his interests has certainly been justified. And for those of us who have been able to share his interests at close quarters, that has been a delight, as has been his company, and his collegiality.

R.M. Sillitto

Richard Sillitto addressing the participants at the symposium dinner

Peter Farago's response to the participants at the symposium dinner

CONTENTS

ELECTRON-ATOM COLLISIONS

ELECTRON-MOLECULE COLLISIONS

PRODUCTION OF POLARIZED ELECTRONS AND ELECTRON SPIN FILTERS

ELECTRON-SOLID STATE, SURFACE AND CLUSTER INTERACTIONS

ION-ATOM AND MOLECULE COLLISIONS

PHOTOIONIZATION OF ATOMS, MOLECULES AND SOLIDS

SPECTROSCOPY

Participants at the symposium in honour of Peter Farago in the main meeting hall of the Royal Society of Edinburgh

(e,2e) COINCIDENCE MEASUREMENTS AT LOW ENERGY

A.J.Murray[†], N.J.Bowring and F.H.Read

Schuster Laboratory,
Manchester University,
Manchester, M13 9PL,
United Kingdom.
[†]email: A.Murray@fs3.ph.man.ac.uk

INTRODUCTION

Electron-atom ionisation event processes can be classified in three regions as defined by the excess incident electron energy above the ionisation threshold. The definition and range of these regions is necessarily somewhat arbitrary, however in this paper these regions are considered to be (figure 1):

- *The threshold region*, ranging from 0eV to around 4eV above threshold.

- *The intermediate energy region*, ranging from around 4eV to 100eV above threshold.

- *The high energy region*, where the incident electron energy exceeds 100eV above threshold.

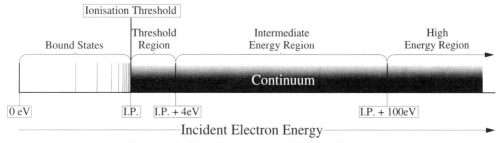

Figure 1. Regions of Ionisation as defined in this paper

Only the threshold and intermediate energy ionisation regions are considered here, since these are still largely unexplored experimentally and present challenging difficulties to theoretical models. Indeed, in the high energy region experimental results are now accurately modelled using either the Distorted Wave Born Approximation (DWBA) or in some cases using a Second Born approximation (for recent reviews see McCarthy and Weigold[1] and Coplan *et al*[2]).

Early experiments conducted in the threshold region have been successfully modelled using Wannier theory[3], although there is still contention as to the range of validity of this model with excess energy. In the Wannier model the ionisation process is considered to be dominated by correlations between the outgoing electrons, which have time to exchange energy and angular momentum through the Coulombic interaction. The ionic core plays a predominantly spectator role in this model, and the electrons are predicted to emerge asymptotically at a mutual angle of approximately π radians. As the excess energy reduces towards zero, the Wannier model predicts that the differential cross section (DCS) reduces to a delta function around a mutual angle of π radians, since the electrons have infinite time to communicate their mutual Coulombic repulsion to each other. Fully quantum mechanical DWBA models are also currently being developed near threshold[4], allowing inclusion of more complex interactions with the core. These models have proven successful when compared with (e,2e) experimental observations 2eV above the ionisation threshold for asymmetric and symmetric scattering geometries[5]. In this paper only the Wannier picture is considered in the threshold region.

By contrast with threshold and high energy experiments, recent and ongoing experiments conducted in the intermediate energy region[6-9] are not modelled satisfactorily using existing theoretical models. At intermediate energies the ionisation process involves all the complexities of exchange and capture, distortions in the incoming and outgoing channels, and short and long range correlations. None of these contributions can be ignored or can be considered to individually dominate the ionisation process. Experimental results in this region thus challenge theoretical understanding of ionisation, negating any complacency enjoyed by the success of existing models outside this region. Understanding intermediate energy ionisation presents significant theoretical challenges, since success will inevitably meld threshold and high energy models together allowing a uniform understanding of ionisation over all impact energies.

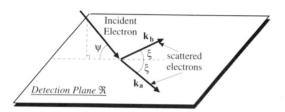

Figure 2. The experimental geometry for ionisation in the intermediate energy regime

In this paper selected results collected at incident electron energies 20eV and 40eV above ionisation are presented for a helium target, where the ejected and scattered electrons are detected with equal energies ($|\mathbf{k_a}|^2 = |\mathbf{k_b}|^2$) at equal angles ξ in the detection plane $\Re$ (figure 2). These results are compared with a Wannier model fit to experimental results obtained 1eV above the ionisation threshold by the Paris group[10] in coplanar geometry ($\psi = 0°$) and by the Manchester group[11] in the perpendicular plane geometry ($\psi = 90°$). An irreducible tensor parameterisation of the (e,2e) DCS as proposed by Klar and Fehr[12] is used to place these results on a common basis, allowing differences between the results in each energy regime to be highlighted.

Quantum mechanical calculations[13] using the recently introduced BBK final state wavefunction are presented and compared to the experimental results 40eV above the ionisation threshold. These results highlight both the favourable and the unfavourable aspects of present theory to explain experimental observations. At this energy a very deep minimum is observed in the DCS for an incident electron angle $\psi = 67.5°$, and a new suggestion as to a possible mechanism for this deep minimum is introduced based upon a semi-classical explanation of the generalised Ramsauer-Townsend minima observed in elastic scattering of electrons from atoms[14,15].

The intermediate energy experimental results presented here were collected using a fully computer controlled and computer optimised (e,2e) spectrometer developed at Manchester[16]. This spectrometer is unique since the spectrometer tuning is updated automatically under compute control throughout data collection using a Simplex optimisation technique. This experimental control is briefly reviewed. Essential for data collection over the exceptionally wide range of scattering geometries accessed by the spectrometer is the accurate alignment of all optics within the spectrometer over the full range of accessed geometries, and a new techniques allowing alignment to within $\pm 0.2°$ exploiting solid state laser diodes is described.

THE EXPERIMENTAL APPARATUS

The (e,2e) spectrometer is designed to measure angular correlations between electrons emerging from electron impact ionisation in the range from 1 to 100eV, while accessing as wide a range of detection geometries as mechanically possible from the coplanar to the perpendicular plane geometry. Figure 3 is a schematic of the apparatus configured in the perpendicular plane. The spectrometer is mounted vertically from a stainless steel flange supported in a cylindrical vacuum chamber pumped by a 500 l/s Balzers turbo-molecular pump. External magnetic fields are reduced to less than 5mG at the interaction region by μ-metal shields internally and externally, internal magnetic fields being avoided by manufacturing all components from non-magnetic materials. All electron optical elements are constructed from molybdenum, while the spectrometer mountings are 310 stainless steel.

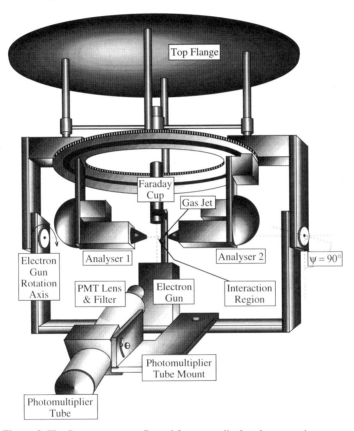

Figure 3. The Spectrometer configured for perpendicular plane experiments

The unselected electron gun of resolution approximately 0.6eV incorporates two triple aperture electrostatic lenses with intermediate beam and pencil angle defining apertures. The resulting electrons are selectable from 20 to 300eV at up to 4μA of current which is focused to a 1mm diameter beam at the interaction region, with zero beam angle and a pencil angle of approximately 2°. To facilitate electron beam placement and focusing, photons originating in the interaction region are accurately focused onto a photomultiplier tube via a lens, an aperture and an optical filter passing radiation at 450nm originating from direct helium excitation below threshold. The interaction volume is accurately focused onto the photocathode by back-focusing light from a source located at the photocathode onto the interaction region. The gun, photomultiplier tube, atomic beam source and Faraday cup assembly are mounted to allow rotation from the coplanar ($\psi = 0°$) to the perpendicular plane geometry ($\psi = 90°$).

Two identical hemispherical deflection analysers rotate in the horizontal detection plane $\Re$. The analyser input lenses are triple cylinder lenses with acceptance half angles approximately 3° over the range of energies studied. Following energy selection in the hemispheres the electrons are detected by channel electron multipliers. Internal collisions between the analysers and electron gun are prevented by position sensing optical interrupters connected to external control electronics.

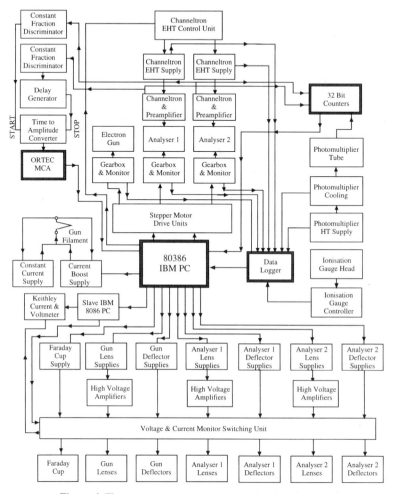

Figure 4. The spectrometer computer control and optimisation

Unique to this apparatus is the computer interface which automatically controls and optimises the spectrometer (figure 4). Electrostatic voltages to the electron gun and analysers are computer optimised to maintain focusing onto the interaction region during data collection, whereas the analyser and gun angles are sequentially adjusted to accumulate coincidence data over the full range of angles (ψ, ξ). At the heart of the system is an IBM 80386 PC which receives detailed information about the spectrometer status while controlling the spectrometer and accumulating coincidence data. The computer adjusts the spectrometer lens and deflector voltages, optimising these to either electron beam current, photomultiplier tube counts or analyser counts at regular intervals using a modified simplex technique based upon the method of Nelder and Mead[17]. The spectrometer runs unattended for continuous periods in excess of ten days, the resulting data being found to be more reliable and consistent than is possible with intermittent manual optimisation. Full details of the computer control and optimisation may be found in reference [16].

Imperative for experimental success is the accurate angular alignment of the analysers and electron gun. As the angles ψ, ξ_a and ξ_b are varied through all possible ranges the interaction region must remain stationary in space. In practice this is difficult to achieve, and so newly developed compact visible laser diodes (VLD's) have been employed in the alignment procedure. The VLD's produce 0.5mm diameter pencil laser beams that are directed through the centres of the analyser turntable axis, electron gun rotation axis, analyser input lenses and gun axis as defined by paired 0.5mm diameter axially spaced apertures. The crossing point of the turntable and gun rotation axis laser beams guarantees the true centre of rotation, which by definition is the centre of the stationary interaction region (see figure 5). Linear translators on the analyser and electron gun mounts are then adjusted so that the laser beams passing through the centres of the analysers and electron gun are brought to pass through this region for all angles (ψ,ξ). Finally the alignment photomultiplier tube is back-focused onto this crossing point and the gas jet is adjusted to centre the atomic beam. Projection of the laser beams passing through the analysers and electron gun onto a distant calibrated screen allows the individual rotation angles to be assessed to $\pm 0.2°$, ensuring accurate alignment of the spectrometer throughout the full range of experimental geometries.

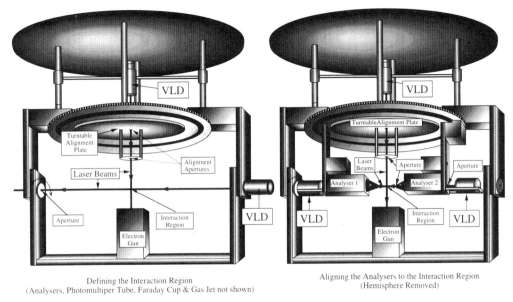

Defining the Interaction Region
(Analysers, Photomultiper Tube, Faraday Cup & Gas Jet not shown)

Aligning the Analysers to the Interaction Region
(Hemisphere Removed)

Figure 5. Aligning the spectrometer using Visible Laser Diodes (VLD's)

THRESHOLD EXPERIMENTS AND THE WANNIER THEORY

Experiments have been conducted for a number of years to evaluate the importance of electron-electron correlations near threshold. These include (e,2e) experiments and photo-double ionisation experiments, where angular and energy partitioning correlations between electrons emerging from the interaction can be studied[3,10,11,18]. Theoretical classical studies of these correlations were introduced by Wannier[19] in 1953; subsequent semi-classical studies by Peterkop[20] and Rau[21] confirmed the general results. Since that time numerous theoretical and experimental studies[22-25] have provided a detailed picture of near threshold ionisation, although only recently have attempts been made to evaluate a rigorous asymptotic wave function[4]. These new quantum mechanical calculations have recently been applied to experimental data obtained 2eV above threshold by the Kaiserslautern group, showing favourable comparisons between theory and experiment[5].

Experimental data at E = 1eV excess energy have been obtained ionising helium in coplanar geometry by the Paris group[10] and in the perpendicular plane by the Manchester group[11]. These results have been combined to evaluate partial wave amplitudes and phases for L ≤ 2 for the (e,2e) DCS in the Wannier model developed by Fournier-Lagarde et al [26] and Selles et al [27]. In this model the differential cross section may be written

$$DCS = \left(\frac{1}{4}|f_0|^2 + \frac{3}{4}|f_1|^2\right) \exp\left[\frac{-4\gamma^2 \ln 2}{\theta_0^2 \sqrt{E}}\right] \qquad (1)$$

where $\gamma = \pi - \theta_{12}$, $\theta_{12} = \cos^{-1}[\cos\theta_1 \cos\theta_2 + \sin\theta_1 \sin\theta_2 \cos(\varphi)]$ and f_0 and f_1 are singlet and triplet complex scattering amplitudes respectively. Limiting the partial wave expansion to $L \leq 2$ and normalising to the s wave amplitude yields $f_0 = f_0^S + f_0^P + f_0^D$ and $f_1 = f_1^P + f_1^D$ where :

$$f_0^S = E^\alpha \qquad \alpha = 0.1865 \qquad f_1^P = E^\alpha P_1(\cos\theta_1 - \cos\theta_2) = 0 \;\; if \;\; \theta_1 = \theta_2$$

$$f_0^P = E^\alpha P_0(\cos\theta_1 + \cos\theta_2)e^{i\theta_{P0}} \qquad f_1^D = \frac{3}{2}E^{\alpha - 1/4}d_1 e^{i\theta_{d_1}}\left(\cos^2\theta_1 - \cos^2\theta_2\right) = 0 \;\; if \;\; \theta_1 = \theta_2$$

$$f_0^D = E^\alpha\left[D_0 e^{i\theta_{d0}} + E^{-1/2}D_0' e^{i\theta_{d'0}}\left(1 + \frac{\gamma^2}{8}\right)\right]\left(\frac{3}{2}[\cos^2\theta_1 + \cos^2\theta_2] - 1\right)$$

$$- E^\alpha\left[D_0 e^{i\theta_{d0}} - E^{-1/2}D_0' e^{i\theta_{d'0}}\left(1 - \frac{\gamma^2}{8}\right)\right](3\cos\theta_1\cos\theta_2 - \cos\theta_{12})$$

Hawley-Jones et al [11] applied a least squares fit to the experimental data to determine the *energy independent* parameters in the above amplitudes. These were determined to be

$\theta_0(rad)$	P_0	$\theta_{p_0}(rad)$	D_0	$\theta_{d_0}(rad)$	D_0'	$\theta_{d'_0}(rad)$	P_1	D_1	$\theta_{d_1}(rad)$
1.30 ± 0.04	1.6 ± 0.3	$5.22^{+.04}_{-.07}$	$0.84^{+.07}_{-.04}$	$4.13^{+.04}_{-.11}$	$2.13^{+.34}_{-.28}$	$2.13^{+.05}_{-.07}$	$0.25^{+.09}_{-.04}$	$0.86^{+.07}_{-.06}$	3.1 ± 0.8

Hence the DCS can be obtained in symmetric geometry within the constraints of the Wannier model up to L=2 for all angles (ψ, ξ) by applying the transforms[7] :

$$\cos\theta_1 = \cos\theta_2 = \cos\xi \cos\psi \qquad\qquad \cot\frac{\phi}{2} = \cot\xi \sin\psi$$

Note that for symmetric geometry only *singlet* amplitudes contribute to the DCS.

Figure 6 shows a three dimensional surface representation of the differential cross section at E = 2.0eV, 1.0eV and 0.5eV respectively where the DCS magnitude derived from equation 1 is mapped onto the radial vector amplitude. It can be seen that the Wannier model predicts a lobe structure centred around $\theta_{12} = \pi$ and that the lobes narrow as the excess energy decreases, as expected from the higher degree of correlation between the outgoing electrons. Note also that the Wannier picture only predicts a two lobe structure, and therefore is invalid for higher energies where more complex structure is observed.

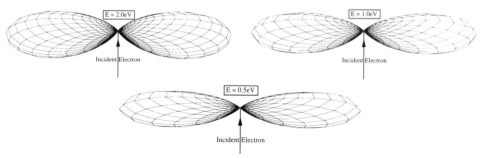

Fig 6 The 3D DCS Surface as derived from the fitted Wannier parameters at 2eV, 1eV and 0.5eV excess energy

6

INTERMEDIATE ENERGY EXPERIMENTS

Experiments in the intermediate energy regime have only recently been extensively conducted, principally in coplanar geometry at Kaiserslautern[9] and over the full range of geometries from coplanar geometry to the perpendicular plane geometry in Manchester[6-8]. The latter experiments conducted in symmetric geometry usually with equal energy sharing between the outgoing electrons, allow full 3 dimensional (e,2e) differential cross section surfaces to be deduced since the point $(\psi, \xi) = (\psi, 90°)$ is a common normalisation point throughout all gun angles ψ at any given energy.

Figures 7 shows results obtained 20eV and 40eV above the ionisation threshold for helium plotted on linear and logarithmic scales respectively. Notable in these results is the large difference in forward to backscatter peak ratios in coplanar geometry. At the lower energy the forward scatter peak is smaller than the backscatter peak with a ratio of 0.8:1, whereas at the higher energy the forward scatter peak dominates the backscatter peak with a ratio exceeding 7:1. As the gun angle ψ increases from coplanar geometry ($\psi = 0°$) to the perpendicular plane geometry ($\psi = 0°$) the forward peak is seen to evolve into the lower angle perpendicular plane triple peak, whereas the backscatter peak evolves into the central and higher perpendicular plane peaks. If ψ were to increase further from 90° to 180° reflection symmetry in the scattering plane $\Re$ (see figure 2) would reverse this trend.

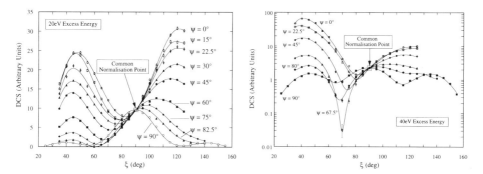

Figure 7. The DCS measured at 20eV and 40eV excess energy

At the lower excess energy the forward and backscatter peaks and inter-peak minima evolve monotonically between $\psi = 0°$ and $\psi = 90°$. This trend is found to be true for all energies where the forward/backward coplanar peak heights are less than unity. By contrast, for all energies where the forward/backward coplanar peak heights are greater than unity (30eV excess energy and above) the peaks are found to reduce monotonically with increasing gun angle but the inter-peak minima moves through a minimum at around $\psi = 67.5°$, increasing from this point as the gun angle ψ further increases through to 90°. For the energies studied the inter-peak minima is found to be deepest at an excess energy of 40eV, where deconvolution of the experimental resolution from the data suggests the cross section at this point approaches zero[7]. An explanation of this phenomena is suggested below in terms of a semiclassical model developed to explain the deep minima found in elastic scattering from atomic targets, commonly known as the generalised Ramsauer-Townsend minima.

The first fully quantum mechanical theoretical analysis of these results using the recently introduced Brauner, Briggs and Klar (BBK)[28] final state wavefunction has been proposed by Berekdar and Briggs[13]. This wavefunction has the advantage that it is exact in the asymptotic limit. Four separate mechanisms were considered to evaluate the full angular differential cross section:

- *Single binary collisions* between the projectile & target electrons.
- *Double binary collisions* between the projectile electron, ion core & the target electron.
- *Coulombic interactions* between the outgoing electrons of low relative velocity.
- *Interference terms* arising from the coherent sum of the scattering amplitudes.

The addition of the fourth term in the model was found to be necessary to approximate the very small differential cross section observed at $(\psi,\xi)= (67.5°,65°)$ for 40eV excess energy, where inclusion of the scattering amplitude from interaction with the *non-ionised* electron was found to be essential for the minima to appear.

Less favourably, the model does not predict the complete DCS over all angles (ψ,ξ), the results at $\psi = 45°$ predicting a *maxima* in the cross section at $\xi = 70°$ rather than a minima. This indicates that even with the best wavefunctions currently available theory cannot explain all details of the experimental observations in this energy regime (see figure 8).

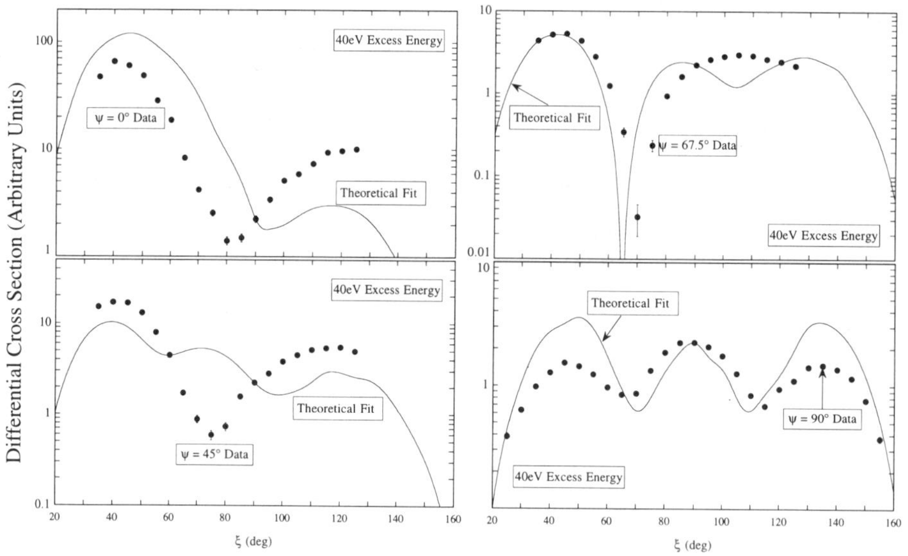

Figure 8. Theoretical calculation of the DCS at 40eV excess energy compared to experimental data

The existence of the deep minimum shown in Fig. 8 is of particular interest since a sharp dip has also been observed by Rösel *et al* [9] in coplanar symmetric geometry ($\psi = 0°$) at an incident energy of 500eV for a scattering angle $\xi \cong 90°$. Shallower minima have also been seen in (e,2e) differential cross sections for a variety of other conditions. The minimum in figure 8 and that of Rösel *et al* [9] are sufficiently sharp and deep to indicate the presence of some form of destructive interference, which in turn implies that the yield in the region of the dips must be explicable in terms of only one or two major collision mechanisms. This is surprising in view of the complexity usually assumed for the (e,2e) process in this energy region.

An analogous, long-standing problem of interpretation has existed for the deep minima seen in differential *elastic* electron-atom scattering cross sections at incident energies of the order of a few hundred electron volts, referred to as 'generalised Ramsauer-Townsend minima'. A convincing semi-classical interpretation of these minima has recently been developed (see for example Egelhoff[14], Burgdörfer *et al* [15]) in terms of the interference between two different semi-classical electron paths that sometimes exist for a give scattering angle. In one of the paths the impact parameter is small and the inner field of the atom is strong enough to deflect the incident electron through a total angle of more than 180°, causing the electron trajectory to 'loop' around the atom thus allowing interference with a second path that corresponds to the same scattering angle without looping (see figure 9). This model also indicates that elastic scattering under the conditions that yield generalised Ramsauer-Townsend minima is due predominantly to incident electrons that have a narrow range of impact parameters, which tends to restrict the participating angular momenta to a single dominant value (l=3). This therefore provides a correspondence between the semi-classical model and the accurate quantum mechanical calculations, which show (without offering a physical interpretation) that there is one dominant partial wave and that the angles at which the generalised Ramsauer-Townsend minima occur are near to the zeros of the third order Legendre polynomials.

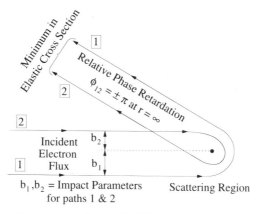

Figure 9. Semiclassical interpretation of generalised Ramsauer-Townsend minima in elastic scattering

It is possible that an analogous semi-classical model could help to explain the existence of the deep minima observed in (e,2e) differential cross sections. Figure 10 shows schematically how two paths might contribute to the minimum seen in figure 8. Here a 'binary' collision between the incident electron and a target electron is followed by a deflection of one of the outgoing electrons in the Coulomb field of the ion core. In fact the outgoing electrons are easily deflected through large angles under these conditions, as can be seen from the expression for the deflection angle θ of an electron of energy E and asymptotic impact parameter b in the field of a singly charged ion:

$$\tan(\frac{\theta}{2}) = \frac{1}{2Eb} \quad (in \;\; atomic \;\; units)$$

As an example for an energy of 20eV (equivalent to the energy shared at the asymptotic limit for the electrons in figure 8) and an impact parameter of 0.5 Å this gives $\theta = 71°$. It should also be noted that the outgoing electrons are more easily deflected than the incident electron because of their lower energy.

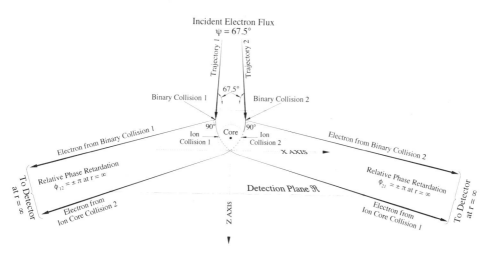

Figure 10. Path contributions to (e,2e) differential cross sections that may give rise to deep minima

Implicit in this interpretation of the deep minimum is the assumption that the mechanism illustrated in figure 10 (a binary collision followed by the interaction of an outgoing electron with the ion core) is the dominant mechanism for this region of incident energy and scattering angle. This has not been recognised before. In theoretical models used to fit the deep minima and other features of the differential cross sections (Zhang *et al* [29],

Whelan and Walters[30], and Berakdar and Briggs[13]) the emphasis has been on interactions that are important in the *incident* channel, rather than the effect of interferences and electron-ion interactions in the exit channel.

The semi-classical model of the (e,2e) process also provides a tentative explanation for the fact that backward scattering dominates over forward scattering for excess energies less than about 30eV (as discussed above). At these low energies the outgoing electrons have a high probability of bring deflected through angles in the region of 90° or more by the positive ion. Therefore after an initial binary collision, in which the two outgoing electrons start by having velocity components in the forward direction, the final direction of these electrons are more likely to be backward than forward.

PARAMETERISATION INTO IRREDUCIBLE TENSORIAL COMPONENTS

The (e,2e) experimental data presented here allows parameterisation of the measured differential cross section into a set $\Im_3$ of irreducible tensorial angular components. This parameterisation, used extensively in the nuclear physics field to derive angular momentum information from correlated particle decay, was first suggested for (e,2e) measurements by Klar and Fehr (1994)[12] but could not be exploited until the wide ranging angular measurements as presented here were obtained.

As the complete set of Legendre polynomials $\Im_2 = \{ P_l(\hat{k}_a, \hat{k}_b), l = 0...\infty \}$ form a natural expansion set for an angular correlated function of *two* vectors $\mathbf{k}_a$ and $\mathbf{k}_b$ (figure 11), so the irreducible set $\Im_3 = \{ I_{l_a l_b l_0}(\hat{k}_a, \hat{k}_b, \hat{k}_0), l_a, l_b, l_0 = 0...\infty, \Delta(l_a, l_b, l_0) \neq 0 \}$ forms the natural expansion set for a function of *three* vectors (see figure 12).

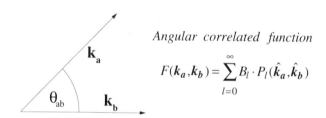

Angular correlated function

$$F(\mathbf{k}_a, \mathbf{k}_b) = \sum_{l=0}^{\infty} B_l \cdot P_l(\hat{k}_a, \hat{k}_b)$$

Figure 11. Angular correlated function of two vectors expanded into irreducible set of Legendre polynomials

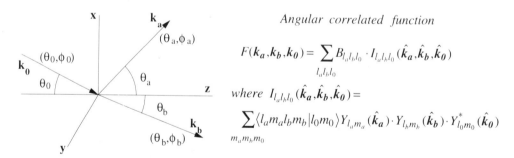

Angular correlated function

$$F(\mathbf{k}_a, \mathbf{k}_b, \mathbf{k}_0) = \sum_{l_a l_b l_0} B_{l_a l_b l_0} \cdot I_{l_a l_b l_0}(\hat{k}_a, \hat{k}_b, \hat{k}_0)$$

where $I_{l_a l_b l_0}(\hat{k}_a, \hat{k}_b, \hat{k}_0) =$

$$\sum_{m_a m_b m_0} \langle l_a m_a l_b m_b | l_0 m_0 \rangle Y_{l_a m_a}(\hat{k}_a) \cdot Y_{l_b m_b}(\hat{k}_b) \cdot Y^*_{l_0 m_0}(\hat{k}_0)$$

Figure 12. Angular correlated function of three vectors expanded into the irreducible set I_{lalbl0}

For symmetric (e,2e) experiments the angular correlated DCS function is conveniently expanded into the irreducible set based on the incident electron momentum vector $\mathbf{k_0}$ and the two outgoing electron momentum vectors $\mathbf{k}_a$ and $\mathbf{k}_b$. Letting the incident electron beam define the z-axis, the functions $I_{l_a l_b l_0}(\hat{k}_a, \hat{k}_b, \hat{k}_0)$ reduce to :

$$I_{l_a l_b l_0}(\hat{k}_a, \hat{k}_b, \hat{k}_0) = \sum_m (-1)^m \langle l_a m \ l_b m \ | l_0 0 \rangle \left[\frac{(l_a - |m|)!(l_b - |m|)!}{(l_a + |m|)!(l_b + |m|)!} \right]^{1/2} \cdot P_{l_a}^{|m|}(\theta_a) \cdot P_{l_b}^{|m|}(\theta_b) \cdot e^{im(\phi_a - \phi_b)} \quad (2)$$

10

Reflection symmetry restricts the parameters such that $(l_a + l_b + l_0)$ is even, whereas the symmetric geometry $(\phi_a = \pi - \phi_b, \theta_a = \theta_b, \xi_a = \xi_b)$ couples some functions thereby reducing the number of independent terms in the expansion set[31].

There are advantages in parameterising the DCS in this way. Firstly the expansion is independent of any theoretical model that is used to represent the DCS, and so forms a common 'language' through which details of the DCS can be studied. As an example the model specific results presented in the Wannier picture near threshold (section 3) can be re-parameterised using the $\mathfrak{I}_3$ set, allowing comparison with the intermediate energy results similarly parameterised. Secondly, the significance of the contributing integers (l_a, l_b, l_0) to the DCS indicate the *degree* of correlation existing between the electrons (high correlation $\Rightarrow$ high integer value). As an example, if $l_0 = 0$ then $B_{l_a l_b l_0}$ measures the degree of correlation between the outgoing electrons *independently* of the incident electron.

Experimental results inherently contain integration over the finite solid angles of detectors, incident electron beam pencil angles and the gas beam. These can be estimated from the experiment but are difficult to deconvolve from the data without the full 3-D form of the DCS. Parameterisation into the set $\mathfrak{I}_3$ allows this deconvolution to be performed and the true DCS estimated on a relative and an absolute scale. Finally, noting that the coefficients $B_{l_a l_b l_0}$ in the expansion are independent of angle but are dependent on the excess energy E_{exc} shared between the outgoing electrons, these coefficients can be further parameterised in terms of this excess energy, allowing the complete DCS to be obtained over all E_{exc}, ϕ, θ_a and θ_b.

Figure 13 illustrates the result of least squares fitting this parameterisation to the threshold and intermediate energy data presented above. Three constraints were applied to the fit. The first ensured that the DCS was zero for electrons emerging in the same direction ($\xi_a = \xi_b = 0, \pi$), the second ensured that the DCS was non-negative everywhere and the third ensured that the DCS did not have more than one point of inflection in the inaccessible regions between the points ($\xi_a = \xi_b = 0, \pi$) and the experimental data. 44 basis functions were found necessary to ensure a good fit, the values of l_a and l_b for each l_0 being chosen to reflect the angular symmetry of the experiment.

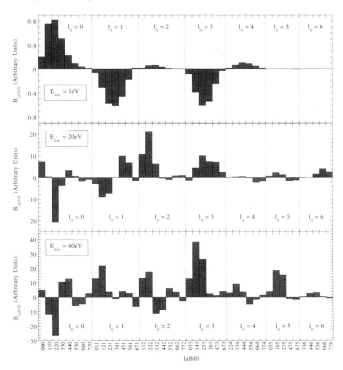

Figure 13. $B_{l_a l_b l_0}$ Parameters least squares fit to the 1eV Wannier fit & the intermediate energy data.

A number of features contrast the difference between the threshold result and the intermediate energy results. Firstly it is noted that at 1eV excess energy only l_0= 0 - 4 are significantly required to fit the Wannier model. This is a consequence at this energy of the dominance of the L = 0, 1 and 2 scattering amplitudes f_0 over the Gaussian correlation term in equation 1. As the excess energy decreases towards threshold, the influence of the Gaussian term increases the requirement for higher order terms in the parameterisation. Another striking difference between the results is the reversal in sign of the fitting parameters for the intermediate energy results compared to the threshold fit. The correlation terms no longer follow a set pattern dominated by the partial wave angular momenta as seen at the lower energy, indicating that at higher energies ionisation does not proceed in a clear way through well defined sharp angular momentum states.

Parameterisation of the intermediate energy data allows construction of three dimensional surface representations as applied to the Wannier model (section 3). Surface DCS representations are shown in figure 14 for E_{exc} = 20eV and 40eV contrasted with the 1eV Wannier model representation. The difference between the threshold and intermediate results is immediately obvious in the number and relative angles of the observed lobes. At 40eV excess energy the forward lobe/backward lobe volume ratio dominates, a trend that increases as the excess energy increases. Although the deep minima at ψ = 67.5° for E_{exc} = 40eV cannot easily be seen in this representation, these pictures are instructive since they model the *total* DCS surface in three dimensional space, including reflection symmetries and angular symmetries not obvious in the data presented graphically in figure 7.

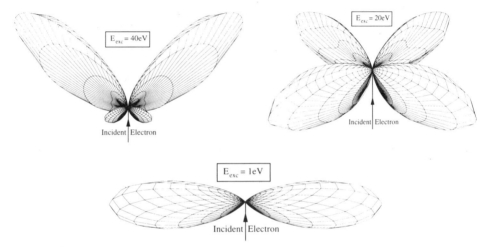

Figure 14. Normalised 3-D surface representations of the DCS for 40eV, 20eV and 1eV excess energy.

SUMMARY & CONCLUSIONS

Results have been presented from (e,2e) experiments ionising helium 1eV above threshold together with results in the intermediate energy regime 20eV and 40eV above threshold. The threshold results placed in the framework of the Wannier picture are evaluated using a new irreducible tensor parameterisation for the differential cross section, allowing comparison between these results and the parameterised intermediate energy results. Contrasts between the measured angular distributions in each of these regions has been highlighted both in conventional form and in a full three dimensional surface representation of the differential cross section.

For an excess energy 40eV above the ionisation threshold a deep minimum in the (e,2e) differential cross section is observed for equal energies and symmetric angles of the outgoing electrons. A suggestion based upon a semiclassical explanation of the Ramsauer-Townsend minima in elastic scattering from atoms has been proposed to explain this minimum, based upon the interference between the retardation phases of two possible paths to the detectors on either side of the ionic core. This interpretation also indicates that the dominant ionisation mechanism here is a binary collision followed by scattering of the outgoing electrons by the residual ion.

The intermediate energy results obtained over an exceptionally wide angular range presented here were collected with a unique fully computer controlled and computer optimised spectrometer. This spectrometer demonstrates superior operation compared to conventionally tuned instruments and achieves extensive operator independent data accumulation times. Details of the spectrometer and new alignment procedures have been reviewed.

Experiments conducted in the intermediate energy regime over a wide range of scattering angles provide the most complex and stringent tests of current ionisation theory, which at present fail to describe the experimental results. By contrast, theoretical models in the high energy regime and the threshold regime meet with far greater success. This indicates the direction of future experiments in the intermediate energy regime, which should extend the measurements up towards the high energy limit and down towards threshold. These future experiments will yield vital clues as to the complex mechanisms involved in the ionisation process.

REFERENCES

1. McCarthy I.E. and Weigold E. (1995) *Electron-Atom Collisions* (Cambridge University Press, UK)
2. Coplan M.A, Moore J.H. and Doering J.P. (1994), (e,2e) Spectroscopy, *Rev. Mod. Phys.* **66**:985
3. Read F.H. (1985), Threshold behavior of ionisation cross sections, *in Electron Impact Ionisation* (Eds T.D.Mark and G.H.Dunn, Springer-Verlag, New York), p42
4. Jones S. and Madison D.H. (1994), Asymptotically correct distorted wave calculations for low energy electron impact ionisation of helium, *J. Phys.B At. Mol. Opt. Phys* **27**:1423
5. Rösel T, Röder J, Frost L, Jung K, Ehrhardt H, Jones S. and Madison D.H. (1992), Absolute triple differential cross section for ionisation of helium near threshold, *Phys Rev A* **46**:2539.
6. Murray A.J. and Read F.H. (1992), Novel exploration of the helium (e,2e) Ionisation process, *Phys Rev Letters* **69**:2912
7. Murray A.J. and Read F.H. (1993), Evolution from the coplanar to the perpendicular plane geometry of helium (e,2e) differential cross sections symmetric in scattering angle and energy, *Phys Rev A* **47**:3724
8. Murray A.J, Woolf M.B.J. and Read F.H. (1992), Results from symmetric and non-symmetric energy sharing (e,2e) experiments in the perpendicular plane, *J.Phys.B At. Mol. Opt. Phys.* **25**:3021
9. Rösel, T, Dupré C, Röder J, Duguet A, Jung K, Lahmam-Bennani A. and Ehrhardt H. (1991), Coplanar symmetric (e,2e) cross sections on helium and neon, *J. Phys.B At. Mol. Opt. Phys.* **24**:3059.
10. Mazeau J, Huetz A and Selles P. (1986), in *Electronic and Atomic Collisions* (ed D.C. Lorents, W.E. Meyerhof and J.R. Peterson, Elsevier: Amsterdam) p141
11. Hawley Jones T.J, Read F.H, Cvejanovic S., Hammond P. and King G.C. (1992), Measurements in the perpendicular plane of angular correlations in near threshold electron impact ionisation of helium, *J. Phys.B At. Mol. Opt. Phys.* **25**:2393
12. Klar H. and Fehr M. (1992), Parameterisation of multiply differential cross sections, *Z. Phys. D.,* **23**:295
13. Berakdar J. and Briggs J. S. (1994), Interference effects in (e,2e) differential cross sections in doubly symmetric geometry, *J. Phys.B At. Mol. Opt. Phys.* **27**:4271.
14. Egelhoff W.F. (1993), Semiclassical explanation of the generalised Ramsauer-Townsend minima in electron-atom scattering, *Phys. Rev. Letters* **71**:2883
15. Burdörfer J, Reinhold C.O, Sterberg J. and Wang J. (1995) Semiclassical theory of elastic electron-atom scattering, *Phys. Rev. A* **51**:1248.
16. Murray A.J, Turton B.C.H. and Read F.H. (1992), Real time computer optimised electron coincidence spectrometer, *Rev. Sci. Inst.* **63**:3349
17. Nelder J.A. and Mead R.A. (1965), A simplex method for function minimisation, *Comput. J.,* **7**:308
18. Dawber G, Avaldi L, McConkey A.G, Rojas H, MacDonald M.A. and King G.C. (1995), Near threshold TDCS for photo-double ionisation of helium, *J. Phys. B. At. Mol. Opt. Phys.* **28**:L271
19. Wannier G.H. (1953), The threshold law for single ionisation of atoms or ions by electrons, *Phys. Rev.* **90**:817
20. Peterkop R. (1971), WKB approximation and threshold law for electron atom ionisation, *J. Phys. B. At. Mol. Phys.* **4**:513
21. Rau A.R.P. (1971), Two electrons in a coulomb potential. Double continuum wave functions and threshold law for electron atom ionisation, *Phys. Rev. A.* **4**:207
22. Crothers D.S.F. (1986), Quantal threshold ionisation, *J. Phys. B. At. Mol. Phys.* **19**:463
23. Altick P.L. (1985), Use of a long range correlation factor in describing triply differential electron impact ionisation cross sections, *J. Phys. B. At. Mol. Phys.* **18**:1841
24. Selles P., Mazeau J. and Huetz A. (1987), Wannier theory for P and D states of two electrons, *J. Phys. B. At. Mol. Phys.* **20**:5183

25. Cvejanovic S. and Read F.H. (1974) Studies of the threshold electron impact ionisation of helium, *J. Phys. B At. Mol. Phys.* **7**:1841
26. Fournier-Lagarde P, Mazeau J. and Huetz A. (1984), Electron impact ionisation of helium - a measurement of (e,2e) differential cross sections close to threshold, *J. Phys. B. At. Mol. Phys.* **17**:L591
27. Selles P, Huetz A. and Mazeau J. (1987), Analysis of e-e angular correlations in near threshold electron impact ionisation of helium, *J. Phys. B. At. Mol. Phys.* **20**:5195
28. Brauner M, Briggs J.S. and Klar H. (1989), 3 body dynamics in the reaction H(e,2e) H$^+$ at intermediate energies, *Z. Phys. D.* **11**:257
29. Zhang, X, Whelan, C. T. and Walters H. R. J. 1990, (e,2e) cross sections for ionization of helium in coplanar symmetric geometry, *J. Phys.B At. Mol. Opt. Phys.* **24**:3059.
30. Whelan C. T. and Walters H. R. J. 1990, A new version of the distorted wave impulse approximation - application to coplanar symmetric ionization, *J. Phys.B At. Mol. Opt. Phys.* **23**:2989.
31. Murray A.J, Read F.H. and Bowring N.J. (1994), Decomposition of experimentally determined atomic (e,2e) ionisation measurements *Phys Rev A* **49**:R3162

14

MEASUREMENT OF SPIN OBSERVABLES IN ELECTRON SCATTERING FROM ATOMS AND NUCLEONS

G. Baum and W. Raith

Universität Bielefeld
Fakultät für Physik
D-33501 Bielefeld

INTRODUCTION

Including spin in the investigation of scattering processes leads to several different spin observables depending on the polarizations of the beam and of the target, and on their relative orientation as well as their orientation with respect to the scattering plane. Experiments usually obtain these observables by measuring spin dependent asymmetries of the cross sections. We do not want to discuss the general aspects of these observables but rather want to concentrate on presentation of recent results which have been obtained by or with the involvement of our research groups.

The first topic relates very directly to the Peter Farago Symposium and new results nicely coincide in time with this occasion. It concerns elastic scattering of electrons from cesium atoms at low energies (2 eV - 20 eV) and the detection of a special interference between spin-orbit and exchange interaction in the collisions. The second topic involves a big jump in energy from 2 eV to 200 GeV and covers deep inelastic scattering of electrons (muons) from nucleons. These investigations are concerned with what at one time was called a "spin crisis" and can now be more aptly addressed as a spin problem connected with the proton (nucleon). As Peter Farago has always had a great interest in spin phenomena, this topic fits also very well into the frame of this symposium.

DETECTION OF SPIN-ORBIT INTERACTION IN THE ELASTIC SCATTERING OF ELECTRONS FROM ONE-ELECTRON ATOMS

Adding the two words "On the" to the title of this section gives exactly the title of a publication of Peter Farago.[1] With reference to a paper by Burke and Mitchell[2] he argued that the effect of spin-orbit interaction in the elastic scattering of electrons from one-electron atoms can be detected against the background of a dominant spin exchange effect by performing a single-scattering experiment, provided the target atoms are polarized at right angles to the scattering plane.

Of course at that time - polarized electron sources suitable for scattering experiments were still in their infant stages - it seemed very attractive to measure such a subtle effect by

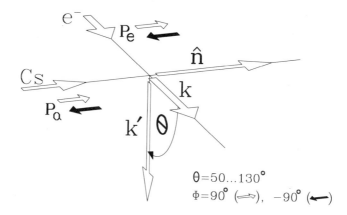

Figure 1. Scattering geometry.

using unpolarized electrons and just preparing a polarized atomic beam. In retrospect one can say that it is not that easy to spin-polarize cesium atoms with six-pole magnets (see for instance the discussion in reference[3]) and to obtain a sufficiently well defined polarization. Optical pumping, which is also mentioned in Faragos paper, had not yet had the benefits of tunable laser diodes. Now, about 20 years later, we have been able to see this effect of interference between spin-orbit and exchange for the first time.[4]

Spin Oberservables. The scattering geometry of our experiment with two highly spin polarized beams is shown in Figure 1. The polarized scattering cross section with spin directions perpendicular to the scattering plane can be expressed through the unpolarized cross section, σ_0, and through the different asymmetries (we basically follow the notation of Burke and Mitchell[2]): A_1 (special interference asymmetry), A_2 (spin orbit asymmetry), and A_{nn} (relativistic exchange asymmetry):

$$\sigma = \sigma_0 \, [1 + A_1 \, \vec{P}_a \cdot \hat{n} + A_2 \, \vec{P}_e \cdot \hat{n} - A_{nn} \, (\vec{P}_e \cdot \hat{n})(\vec{P}_a \cdot \hat{n})] \qquad (1)$$

The asymmetries can be isolated by observing event yields for different spin orientations and constructing from them either an unpolarized atomic beam ($\vec{P}_a = 0$), or an unpolarized electron beam ($\vec{P}_e = 0$), or evaluating suitably the spin antiparallel - parallel combinations (see below). $\vec{P}_a \, (\vec{P}_e)$ denotes the unit vector in direction of the atomic (electron) polarization.

Burke and Mitchell[2] have analyzed in a general way the scattering of electrons from Cs atoms, including spin-orbit effects. Six independent amplitudes (a_i, $i = 1$-6) are needed to completely describe the scattering. Constructing the collision matrix with the components of spin perpendicular to the scattering plane, the asymmetries introduced above can be expressed by the a_i's:

$$A_1 = 2\mathrm{Re} \, (a_1 a_2{}^* + a_3 a_4{}^*)/\sigma_0$$
$$A_2 = 2\mathrm{Re} \, (a_1 a_3{}^* + a_2 a_4{}^*)/\sigma_0$$
$$A_{nn} = 2\mathrm{Re} \, (-a_1 a_4{}^* - a_2 a_3{}^* + a_5 a_6{}^*)/\sigma_0 \qquad (2)$$

and the unpolarized cross section by:

$$\sigma_0 = \sum_{i=1}^{6} |a_i|^2.$$

A physical interpretation of the amplitudes in addition to their appearances in the above asymmetries can be given[2] for the following combinations, relevant to our spin-polarized electron atom scattering geometry: The amplitude $(a_1 + a_2 + a_3 + a_4)$ and $(a_1 - a_2 - a_3 + a_4)$ describes scattering where the spins both before and after the collision are aligned parallel and antiparallel to the normal of the scattering plane, respectively. The amplitude $(-a_5 + a_6)$ and $(a_5 + a_6)$ describes scattering where the change in the spin component normal to the scattering plane satisfies $\Delta M_s = \pm 2$ and $\Delta M_s = 0$, resepectively, with reorientation of the individual components, $\Delta m_i \neq 0$. The amplitude $(a_2 - a_3)$ is connected with spin-non-consering collisions, that is transitions from $S = 1$, $M_s = 0$ to $S = 0$, $M_s = 0$ states. The amplitude a_3 is the "standard" spin-orbit amplitude, responsible for the spin dependence of Mott scattering. The amplitude a_2 can be interpreted as describing the influence of "other" spin-orbit collision effects.[5] In the absence of spin-orbit interaction the amplitudes acquire the following values,[6] expressed through the customary triplet (t) and singlet (s) scattering amplitudes. $a_1 = (3/4)t + (1/4)s$; $a_4 = a_5 = a_6 = (1/4)t - (1/4)s = (1/2)g$; $a_2 = a_3 = 0$.

Looking at the expression for A_1 in equation (2) one sees the interference term $a_3 a_4^*$, an interference between spin-orbit a_3 and exchange a_4 (g) amplitude. The additional term $a_1 a_2^*$ is zero if the "other" spin-orbit effects (between valence-electron spin and scattered-electron orbit) are negligible. Thus A_1 can only be different from zero if spin-orbit and exchange effects are simultaneously present. The asymmetry A_2 contains the spin-orbit amplitudes a_3 and a_2 in connection with the main scattering amplitude a_1 and with exchange amplitude a_4, respectively. Thus, in the absence of spin-orbit interaction, A_2 vanishes ("spin-orbit asymmetry"). The asymmetry A_{nn} vanishes in the absence of exchange effects ("exchange asymmetry").

Prior to our experiment, other relevant experimental work on Cs had been performed by Klewer et al.[7] and by Gehenn and Reichert.[8] Klewer et al. measured the spin-orbit asymmetry, A_2, by scattering unpolarized and measuring the polarization of the scattered electrons at 13.5 eV, 20 eV and 25 e V over an angular range of 40° - 100°; Gehenn and Reichert measured the differential cross sections from 0.8 eV to 20 eV. Concerning the experimental work on A_1, there has been a search for a non-zero value in elastic scattering from spin-polarized sodium at 20 eV, 54.4 eV and 70 eV between 20° and 140°.[9] As expected, the observed asymmetry is less than 1%, at all energies and scattering angles studied, and is consistent with zero within the experimental errors.

Experiment. Our experimental arrangement can be seen in Figure 2. Noteworthy with regard to the electron beam is that it is produced by photoemission from a strained GaAs crystal. For the measurements presented here we typically obtained: $P_e = 0.7$, $I = 0.5\ \mu A$, and $\Delta E = 0.3$ eV as conditions in the scattering volume. The atomic beam has been described in detail.[10] We have $P_a = 0.9$ and $\rho = 5 \cdot 10^9/cm^3$. Continuous operation is possible over 200 h, after that the recirculating oven is empty. It can however be refilled without breaking the vacuum.

The hemispherical electron energy analyser has a radius of 3.3 cm. It is operated with a resolution of $\Delta E/E = 4.5\%$ and its acceptance solid angle is 12 msr. Above the threshold of 3.9 eV we can detect total ionization events through observation of the ions produced. We exploit this possibility for two purposes: a) calibration of the energy scale by observing the threshold with an accuracy of ± 0.1 eV; b) checking the collinearity of P_a with P_e (and with $\hat{n}$) by measuring the spin asymmetry of the total ionization cross section at 7.2 eV incident energy and comparing it with previously obtained values.[11] The experimental procedure to extract the different asymmetries of the differential cross sections is explained in Table 1.

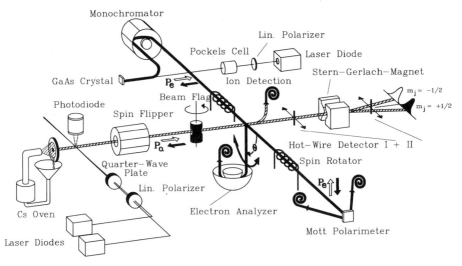

Figure 2. Experimental arrangement (schematic).

Results. In Figure 3 and 4 we show our results for 20 eV, 13.5 eV, 7 eV and 2 eV incident energies of the relative cross section σ_0, of the exchange asymmetry A_{nn}, of the spin-orbit asymmetry A_2, and of the asymmetry A_1, connected with special interference of spin-orbit and exchange amplitudes. The cross sections are compared with the data of Gehenn and Reichert,[8] normalized at one point. In all cases, for the cross section, the agreement in shape is quite satisfactory. In Figure 3, for 20 eV, the asymmetries A_{nn} and A_2 show pronounced structure at the cross section minimum near 67°. A_{nn} rises sharply from its lowest value (-0.1) to zero, whereas A_2 drops within 10° from its maximum value of 0.2 to its minimum value of -0.2. The spin-orbit asymmetry agrees nicely with the measurements of Klewer et al.[7] for P_2 ($A_2 \equiv P_2$), our measurement having a much smaller statistical error. The asymmetry A_1 is consistent with zero within the errors (note the enlarged scale for A_1). Figure 3 also displays our data obtained at 13.5 eV. The cross section has a rather smooth variation and exhibits only a shallow minimum. The asymmetries show also a smooth behavior with scattering angle, A_{nn} dropping nearly linear from 0 to ≈ -0.15, and A_2 being mostly positive with a maximum value of 0.1. Here the measurements of Klewer et al.[7] do not agree with our data, the discrepancies are around 5σ for individual points. Concerning A_1, there might be an indication of non-zero values in the range from 60° to 75°.

Table 1. Scheme of asymmetry determination.

Experiment	N_i	N_j	$\dfrac{N_i - N_j}{N_i + N_j}$	Asymmetry
$Cs^{\mathfrak{f}} + e^{\mathfrak{f}}$	$(N^{\uparrow\downarrow} + N^{\downarrow\uparrow})$	$(N^{\uparrow\uparrow} + N^{\downarrow\downarrow})$	Δ_{nn}	$A_{nn} = \dfrac{1}{P_e P_a} \cdot \Delta_{nn}$
$Cs + e^{\mathfrak{f}}$	$(N^{\downarrow\uparrow} + N^{\downarrow\downarrow})$	$(N^{\uparrow\uparrow} + N^{\uparrow\downarrow})$	Δ_2	$A_2 = \dfrac{1}{P_e} \cdot \Delta_2$
$Cs^{\mathfrak{f}} + e$	$(N^{\uparrow\uparrow} + N^{\downarrow\uparrow})$	$(N^{\uparrow\downarrow} + N^{\downarrow\downarrow})$	Δ_1	$A_1 = \dfrac{1}{P_a} \cdot \Delta_1$

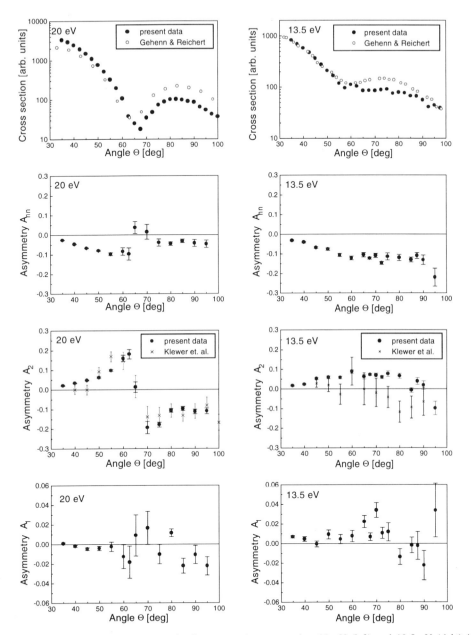

Figure 3. Cross section (relative) and spin asymmetries measured at 20 eV (left) and 13.5 eV (right) in comparison with previous experimental results (Ref. 7, 8), where available and as indicated.

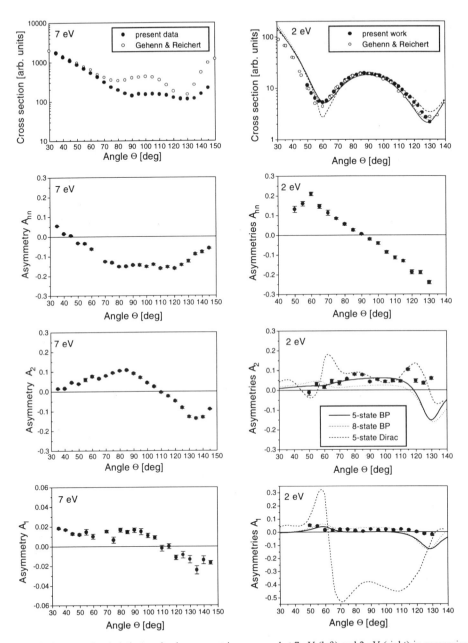

Figure 4. Cross section (relative) and spin asymmetries measured at 7 eV (left) and 2 eV (right) in comparison with previous experimental results (Ref. 8) and with theoretical calculations (Ref. 12 for BP and Ref. 13, 14 for Dirac), where available and as indicated.

20

The results of our investigations at 7 eV are shown in Figure 4. The angular range of the studies now extends from 35° to 145°. The exchange asymmetry A_{nn} has a zero crossing at 45°, acquires then relatively large values up to about ≈ -0.17, and tends towards zero for the backward angles. The spin-orbit asymmetry A_2 shows kind of sinusoidal variation with an amplitude of about 0.15, and zero crossing at 110°. The interference asymmetry A_1 has for the first time distinctly non-zero values of |0.02|. Apart from the forward scattering region the angular shape is similar to A_2. The false asymmetries were found to be below ±0.005. The results at 7 eV are described in more detail elsewhere.[4]

In Figure 4 we also show our data obtained at 2 eV. Here we performed exploratory studies in order to make a comparison with reliable theoretical calculations. The energy width for these first measurements was 0.3 eV. Measurements with reduced width are in preparation. The experimental data are compared with 3 different theoretical results: a) 5 state R-matrix calculation with semi-relativistic Breit-Pauli treatment;[12] b) 8 state R-matrix Breit-Pauli calculation,[12] and c) 5 state R-matrix calculation with full relativistic Dirac treatment.[13,14] The cross section shows good agreement between experiment and theory, the Dirac results having some small deviation in the vicinity of the two minima. For the asymmetry A_{nn} there are no theoretical data available yet. Our measurements show basically a linear decrease from + 0.2 to – 0.2 between 60° and 130°. The measured asymmetry A_2 has, throughout the angular region studied, small and positive values (≈ 0.05). From the theoretical curves the Breit-Pauli calculation seems to come closest. The Dirac results shows large deviation near the minima of the cross section. Very interesting is the behaviour of the asymmetry A_1. As measured, it has relatively large values of 0.05 at forward angles (50°). Note the change in scale of the ordinate compared to the previous diagrams of A_1. Whereas the Breit-Pauli treatments seem to be in accord with the experimental data up to 110°, the Dirac data show surprisingly large deviating values from 0.3 at 60° to -0.5 at 70°, followed by a very broad range (70° to 120°) of large negative values.

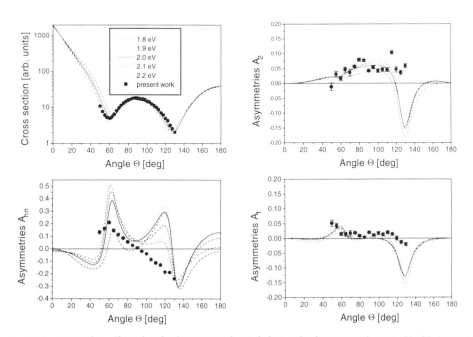

Figure 5. Comparison of our data for the cross section (relative) and spin asymmetries at 2 eV with 5 state BP calculations (Ref. 15) performed at 1.8 eV, 1.9 eV, 2.0e eV, 2.1 eV and 2.2 eV, as indicated.

In order to demonstrate that one does not expect a sensitive dependence of the cross section and of the spin observables on energy we show in Figure 5 our data at 2 eV, together with theoretical results of the 5 state Breit-Pauli calculation for 1.8 to 2.2 eV in steps of 0.1 eV.[15] None of the asymmetries changes rapidly with energy. This clearly demonstrates that a meaningful comparison can be made, even at the present experimental conditions of $\Delta E = 0.3 \pm 0.05$ eV.

Summary. With our highly polarized beams of electrons and cesium atoms we have obtained relative presice data ($\delta A \approx \pm 0.005$) at low energies for several spin observables. This allows a sensitive test of low energy electron-atom collision theories including the treatment of relativistic effect. First data at 2 eV show surprisingly poor agreement with calculations. For the future, the energy resolution of the experiment will be improved to $\Delta E = 0.1$ eV in order to eliminate any possible ambiguities. Besides elastic, we plan to perform superelastic and inelastic studies. In addition other spin observables are easily accessible, for instance some of the double spin asymmetries with both spin-directions oriented in the scattering plane. We acknowledge the work of B. Leuer, P. Baum, L. Grau, R. Niemeyer, M. Tondera who participate with us in these studies. We thank Prof. K. Bartschat for supplying data to us prior to their publication. This research is supported by the DFG (SFB 216).

SOME ASPECTS OF POLARIZED DEEP INELASTIC SCATTERING OF ELECTRONS FROM PROTONS

We will give an overview on results which the <u>S</u>pin-<u>M</u>uon-<u>C</u>ollaboration (SMC) has obtained in experiments concerning the proton at CERN since 1992. The SMC uses a 200 GeV polarized muon beam and scatters from polarized protons, as well as deuterons. Deep inelastic scattering can be interpreted in the parton model as elastic scattering from charged constituents (quarks) which are characterized kinematically by carrying a relative momentum fraction x of the proton. This is the Bjorken variable x which can vary between 0 and 1. In polarized scattering the spin-structure function g_1 contains information about the quark-spin distribution as function of x. The first moment of g_1 is the Ellis-Jaffe sum rule which is of special significance as it allows to obtain the quantity $\Delta\Sigma$, the contribution of quarks to the spin of the proton:

$$\Delta\Sigma = \Sigma_i\, \Delta q_i = \Delta u + \Delta d + \Delta s, \quad \text{with } \Delta q = \int_0^1 (q^+ - q^-)dx,$$

where q^+ (q^-) is the number density of quarks (including the sea quarks) of flavor i which have their spin parallel (antiparallel) to the spin of the parent proton. The quantity Δu (Δd, Δs), therefore, gives the net spin contribution of the individual u (d,s) quarks. Spin conservation requires for the spin of the proton

$$1/2\hbar = (1/2\Delta\Sigma + \Delta G + L_z)\,\hbar.$$

Besides contribution of the quarks ($\Delta\Sigma$) there can be contribution from the gluons (ΔG), which have spin 1, and from orbitial angular momentum (L_z) of quarks or gluons. ΔG is defined in a corresponding way to Δq, $\Delta G = \int_0^1 (G^+ - G^-)dx$ with obvious meaning of G^+ and G^-. The surprise in the results of the European Muon Collaboration (EMC) from their polarized measurements, carried out in 1984 and 1985 at CERN and published in 1988,[16] was that $\Delta\Sigma$ had a very small value and the quarks did not seem to contribute to the proton spin at all [$\Delta\Sigma = 0.12 \pm 0.16$].

Deep Inelastic Scattering (DIS). The Feynman diagram for deep inelastic scattering is well known. The muon of energy E scatters under an angle Θ and acquires a final energy E'. The virtual photon with energy $(E-E') = \nu$ and squared momentum transfer $Q^2 = 4EE'\sin^2\Theta/2$ gets absorbed by the proton, the final state of the proton is not observed (in-

22

clusive measurements). Besides the kinematic variables (E', Θ), or (Q^2, ν), the variables $x = Q^2/2Mν$ and $y = ν/E$ are used. The interaction is electromagnetic, connection to Quantum-Chromo-Dynamics (QCD) comes through the evolution of the structure functions with Q^2, in the unpolarized case known as scale breaking. The polarized evolution behavior has turned out to be particularly interesting. Questions like how the spin is transferred in the splitting processes q → q + G (Bremsstrahlung), G → q + q (pair creation), and G → G + G (triple gluon coupling), and how the spin structure function evolves with Q^2 have to be answered. Theoretical analyses have pointed out that $g_1(x)$ has an anomalous gluon contribution associated with it. As a consequence of this, the Ellis-Jaffe sum rule allows no statement about $\Delta\Sigma$, but only about the superposition $[\Delta\Sigma - (3\alpha_s/2\pi)\Delta G]$.

Spin Observables. Two spin observables can be measured in the high energy inclusive scattering, $A_\parallel$ and $A_\perp$. The incident lepton spin always has to be longitudinal to obtain any observable effects. Figure 6 shows the scattering geometry with scattering plane and polarization plane indicated. The polarized, double differential cross section, $d^2\sigma$, can be expressed by the unpolarized cross section, $d^2\sigma_0$, by the polarization direction of the proton target ($\vec{P}_p$), by the unit vector $\hat{k}$, parallel to the incoming momentum, by the vector normal to the scattering plane $\hat{n}$, and by the asymmetries $A_\parallel$ and $A_\perp$.

$$d^2\sigma = d^2\sigma_0 \{ 1 - [A_\parallel (\vec{P}_p \cdot \hat{k}) + A_\perp \; \hat{n} \cdot (\vec{P}_p \times \hat{k})] \}$$

$$A_\parallel = A = \frac{d^2\sigma^{\uparrow\downarrow} - d^2\sigma^{\uparrow\uparrow}}{d^2\sigma^{\uparrow\downarrow} + d^2\sigma^{\uparrow\uparrow}}$$

$$A_\perp = \frac{d^2\sigma^{\uparrow\rightarrow} - d^2\sigma^{\downarrow\rightarrow}}{d^2\sigma\uparrow\rightarrow + d^2\sigma^{\downarrow\rightarrow}}$$

(3)

The arrows ↑↓(↑↑) designate longitudinal spins which are antiparallel (parallel) to each other; the arrows ↑→(↓→) designate longitudinal lepton spin with helicity + (-) and transverse proton spin. For measuring $A_\parallel$ the proton unit polarization vector is oriented longitudinal: $(\vec{P}_p \cdot \hat{k})$ = cos ψ=1 and $(\vec{P}_p \times \hat{k})$ = sin ψ = 0. For measuring $A_\perp$ the proton unit polarization vector is oriented transverse, $(\vec{P}_p \cdot \hat{k})$ = cos ψ = 0 and $\hat{n}\cdot (\vec{P}_p \times \hat{k})$ = cos φ. The scattering asymmetry will be zero if scattering occurs perpendicular to the polarization plane (cos φ = 0).

To relate in a more direct way to their physical significance, the asymmetries $A_\parallel$ and $A_\perp$ are customarily expressed by corresponding virtual photon absorption asymmetries, A_1 and A_2. Here D, D', $\gamma = \sqrt{Q^2}/ν$, and y are known kinematical factors, D being the depolarization factor for the spin helicity transfer from incoming lepton to the virtual photons. $\gamma \leq 0.1$ for the conditions of the SMC experiment.

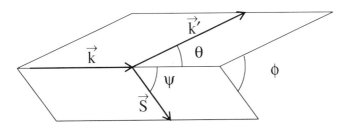

Figure 6. Scattering geometry, showing the scattering plane with momenta $\vec{k}$ and $\vec{k}'$ and the polarization plane with momentum $\vec{k}$ and proton spin $\vec{S}$ ($\vec{S} = \vec{P}_p$), forming an angle φ between them.

$$A_{\parallel} = D \left(A_1 + \gamma \frac{1-y}{1-y/2} A_2 \right)$$

$$A_{\perp} = D' \left(A_2 - \gamma (1-y/2) A_1 \right)$$

(4)

The polarized structure functions g_1, g_2 and the unpolarized structure functions F_1, F_2 appear in the theoretical expression for $d^2\sigma$. Using the above relations one obtains

$$A_1 = \frac{g_1 - \gamma^2 g_2}{F_1} \quad ; \qquad A_2 = \frac{\gamma (g_1 + g_2)}{F_1}$$

(5)

and in very good approximations

$$g_1 = \frac{A_{\parallel} F_2}{D \ 2x \ (1+R)} \quad ; \qquad g_2 = \frac{A_{\perp} F_2}{\gamma D' \ 2x \ (1+R)} - \frac{y}{2} g$$

(6)

Here the relation between the unpolarized structure functions is

$$F_1 = \frac{F_2}{2 \ x \ (1+R)} .$$

(7)

The asymmetries are the primary quantities measured. To obtain from them the spin structure functions according to equation (6), unpolarized data have to be used. $F_2(x)$ is taken from a NMC parametrization[17] and $R(x)$ from SLAC data.[18]

The transversity,[19] a spin dependent structure function relating to transverse spin distributions of quarks when the proton is polarized transversely, can not be obtained from a double-spin asymmetry measurement. To measure it, the polarization of struck quarks coming from transversely polarized protons has to be analyzed.

Interpretation. In the quark-parton model the spin structure function $g_1(x)$ can be interpreted in a simple way: Because of conservation of angular momentum J_z, a virtual photon of positive (negative) helicity can only be absorbed by a positive (negative) helicity quark, the absorption strength being proportional to the square of the quark charge. Thus $\sigma_{1/2}$ [J_z (photon) = 1, J_z (proton) = -1/2)] is proportional to the number of positive helicity quarks, $q^+(x)$, which have their spin in the same direction as the proton, and $\sigma_{3/2}$ [J_z (photon) = 1, J_z (proton) = 1/2)] is proportional to $q^-(x)$:

$$\sigma_{1/2} \propto \sum_i e_i^2 \ q_i^+(x) \quad ; \qquad \sigma_{3/2} \propto \sum_i e_i^2 \ q_i^-(x) .$$

(8)

With $A_1 = (\sigma_{1/2} - \sigma_{3/2}) / (\sigma_{1/2} + \sigma_{3/2})$ and $A_1 = g_1/F_1$ one obtains

$$g_1(x) = 1/2 \sum_i e_i^2 \left(q_i^+(x) - q_i^-(x) \right)$$

$$F_1(x) = 1/2 \sum_i e_i^2 \left(q_i^+(x) + q_i^-(x) \right)$$

(9)

One sees that the structure function $g_1(x)$ measures the difference between the positive and negative spin distributions, whereas the unpolarized structure function F_1 measures the sum. An interpretation of $g_2(x)$ on a partonic level cannot be given. Here, interference between

24

longitudinal and transverse photon absorption is involved and an insight into the significance of g_2 is complicated.[20] The spin content of the proton due to contribution of quarks can be obtained - at least in a first, naive way - from the Ellis-Jaffe sum rule for the proton, Γ_1^p,

$$\Gamma_1^p = \int_O^1 g_1^P(x)dx = \frac{1}{18} \left(4\Delta u + \Delta d + \Delta s \right) . \tag{10}$$

Re-writing leads to

$$
\begin{aligned}
\Gamma_1^{p(n)} &= \left(\genfrac{}{}{0pt}{}{+}{-}\right) \frac{1}{12} (\Delta u - \Delta d) + \frac{1}{36} \left(4\Delta u + \Delta d + 2\Delta s \right) + \frac{1}{9} (\Delta u + \Delta d + \Delta s) \\
&= \left(\genfrac{}{}{0pt}{}{+}{-}\right) \frac{1}{12} (F + D) + \frac{1}{36} \left(3F - D \right) + \frac{1}{9} (\Delta\Sigma) ,
\end{aligned}
$$

where correction factors from QCD-evolution in Q^2 have been omitted. F,D are axial current matrix elements between baryons. Their values are obtained from the neutron beta decay and from hyperon decay rates.

Experiment. The experiments which are being performed at CERN by the SMC use a muon beam. The advantage and strength of the SMC experiment is its high energy which allows to access the low-x region. With electrons, polarized DIS experiments have been carried out at SLAC by several different collaborations: E80/E130 until 1983, E142 in 1993, and E143 in 1994. The target material was either butanol or NH_3 (EMC, E143) for the protons and deuterated butanol or ND_3 (E143) for the deuterons. As neutron target, ^{3}He was taken (E142).

The 200 GeV muon beam at CERN obtains its polarization in a "natural" way. Positive pions are produced in proton-Be collisions. Out of the decay spectrum positive muons are selected from the high energy part of the spectrum. These muons are negatively polarized as can be seen by considering the spin directions of the decay in the CM-system: The two particles are leaving in opposite directions, with the muon-neutrino spin (negative helicity) being compensated by the muon spin, which therefore has a direction opposite to its momentum. The polarization has been measured in a separate experiment to be $P_{\mu^+} = -0.82\pm0.04$.[21] It has to be pointed out that the beam polarization is not easily reversible. A reversal would require a change of the beam line to transport a μ^- beam, for which variations are expected in intensity, location and phase space of the beam, introducing considerable systematic errors into the measurements.

The spectrometer of the CERN experiment has been developed over the years, with many improvements and changes in detector components. The present arrangement of the apparatus is seen in Figure 7. The basic method of detection and measurement of DIS events has remained unaltered, with definition of the incoming muons in momentum, time, and space in the first part of apparatus, followed by momentum and vertex position measurement of the scattered particles, and muon identification after a large absorber wall. Due to the large number of detection planes (150) the spectrometer acceptance is largely insensitive to variations of efficiencies of the individual detectors in time.

The target is polarized by using the method of dynamic nuclear polarization, involving low temperatures (0.5 K), high magnetic field (2.5 T), microwave irradiation at 70 GHz, and extended NMR measurements.[22] A schematic drawing is shown in Figure 8. The beam passes through the inner bore, which is kept as much as possible free of material other

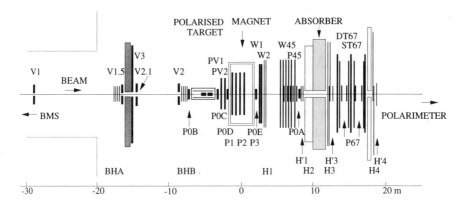

Figure 7. A schematic plane view the muon spectrometer. BMS: beam momentum spectrometer; V: veto counters for muon halo; BH: muon beam hodoscope; P: multiwire proportional chambers; W: drift chambers; H: hodoscope planes for trigger; H2: calorimeter; ST: streamer tubes.

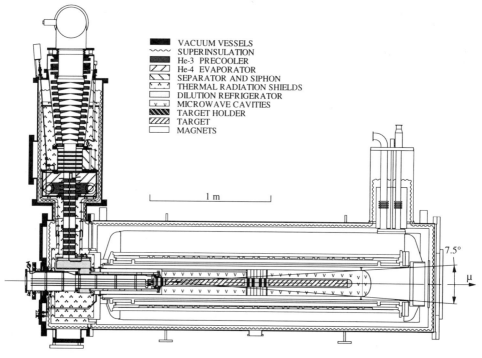

Figure 8. Schematic drawing of the SMC polarized target.

than that of the target proper. There are two target cells, an upstream and a downstream one, each 60 cm in length and 5 cm in diameter. The separation betweem them is 30 cm and they are polarized longitudinally in opposite directions. Together with the vertex resolution capability of ±3 cm, this allows spin asymmetry measurements with small systematic errors (see below), basically equivalent to an infinitely fast reversal of a single-cell target. Butanol has 10 polarizable protons (deuterons) per molecule. For the protons a polarization of $P_p = 0.86$ was obtained as average over a 150 days data taking period, for the deuterons $P_d = 0.5$.

As the observable asymmetries are quite small, it is important to reverse the spin directions of the target halves frequently in order to reduce systematic errors and minimize false asymmetries. The main method used for this purpose is a rotation of the solenoid field direction by 180°. This is achieved with the help of a superconducting dipole coil, wound on the solenoid, which allows to keep the magnetic field above 0.5 T during the rotation procedure. A rotation was performed every 5 hours during data taking, requiring 10 minutes time-out when the field was not longitudinal. To guard against possible systematic effects depending on the solenoid field direction, reversals were also performed every two weeks by changing sign of the polarization with an interchange of the microwave frequencies of the DNP for the target halves. The DIS event yields from the two target halves are determined by the following expression:

$$N_{u(d)} = n_{u(d)} \, a_{u(d)} \, \phi \, \sigma_o \, [1 - f P_\mu P_p \, DA_1] \,,$$

where the index u refers to the upstream and d to the downstream target cell, n is the number density of target nucleons, a the acceptance of the spectrometer, ϕ the beam flux, σ_o the unpolarized cross section, and f the fraction of events coming from polarized nucleons. By taking the ratio of the ratio N_u/N_d of events accumulated before a rotation to the same quantitity, N'_u/N'_d after a rotation,

$$\frac{N_u \, N'_d}{N_d \, N'_u} = (1 + 4 f P_\mu \, P_p \, D \, A_1) \,,$$

one sees that dependences on beam flux, target density, and spectrometer acceptance cancel. The latter two cancel only if they are constant in time. Relevant is the acceptance, for which fluctuations can be controlled at a level of 10^{-3} leading to false asymmetries $\Delta A_1 < 0.01$. The average size of the expression ($4 f P_\mu \, P_p \, D \, A_1$) for the measurements on the proton is about 0.02.

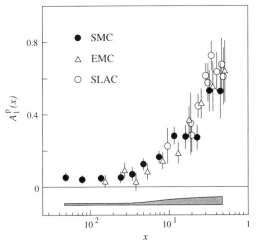

Figure 9. The virtual-photon proton cross section asymmetry A^p_1 as a function of the Bjorken scaling variable x. Only statistical errors are shown with the data points. The size of the of the systematic errors for the SMC points is indicated by the shaded area.

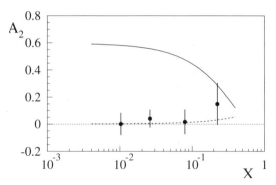

Figure 10. The asymmetry A_2 of the proton as a function of x. The solid line shows $\sqrt{R}$ from the SLAC parametrization, and the dashed line the results obtained with $\bar{g}_2 = 0$. The error bars represent statistical errors, only.

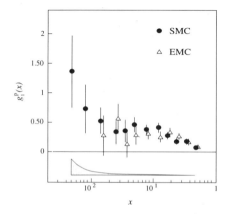

Figure 11. The spin dependent structure function $g_1^P(x)$ at the average Q^2 of each x bin. Only statistical errors are shown with the data points. The size of the systematic errors for the SMC data is indicated by the shaded area. The EMC points are reevaluated using the NMC F_2 parametrization.[17]

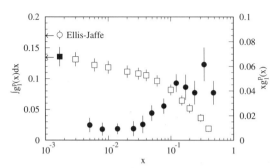

Figure 12. The solid circles (right-hand axis) show the structure function xg_1^P as a function of the Bjorken scaling variable x, at $Q_0^2 = 10$ GeV^2. The open boxes (left-hand axis) show $\int_{xm}^1 g_1^P(x)dx$, where x_m is the value of x at the lower edge of each bin. Only statistical errors are shown. The solid square shows our result for the first moment of $g_1^P(x)$, with statistical and systematic errors combined in quadrature. Also shown is the theoretical prediction by Ellis and Jaffe.

Results. Figure 9 shows the recent results of the SMC[23] for A_1^p, together with the earlier data of the EMC[16] and of the previous SLAC-E80, E130[24] collaborations. The comparison clearly shows that the new measurements confirm the results of the earlier experiments. The systematic errors have been significantly reduced and the kinematic region has been extended down to $x = 0.003$. The SMC also measured the spin asymmetry on a transversely polarized proton target.[25] The virtual-photon asymmetry A_2 (see Figure 10) is found to be significantly smaller than its positivity limit $\sqrt{R}$. The result also helped to reduce the uncertainty in the determination of the spin structure function g_1 from the measurements with longitudinal polarization. The spin structure function g_1^p was evaluated from the asymmetry A_1^p in each x bin, using the relation given before and taking unpolarized structure function data from NMC and SLAC. Figure 11 shows the results at the average Q^2 of each bin which varies from 1.3 GeV^2 at the lowest x value to 58 GeV^2 at the highest .x value. For evaluating the integral $\int g_1^p \, dx$ at a fixed Q_o^2, it is assumed that A_1 is independent of Q^2 and g_1^p is recalculated with F_2 and R taken at Q_o^2. This is shown in Figure 12 for $Q_o^2 = 10$ GeV^2. The final result for the first moment of g_1^p at $Q_o^2 = 10$ GeV^2 is

$$\Gamma_1^p \, (10 \text{ GeV}^2) = 0.136 \pm 0.011 \pm 0.011.$$

This value is about 2.5 standard deviations below the prediction of Ellis and Jaffe. With further evaluation one obtains $\Delta\Sigma = 0.22 \pm 0.15$ and $\Delta s = -0.12 \pm 0.06$. The SMC thus finds that only a small fraction of the nucleon spin is due to the helicity of the quarks, and that the strange sea is negatively polarized, in agreement with the original statements of the EMC.

Summary. All the recent results of the polarized DIS experiments [23,26,27,28] show a violation of the Ellis-Jaffe prediction for the first moment of the spin structure function for the proton as well as for the deuteron. This leads to similar statements about $\Delta\Sigma$ and Δs as was given above for the proton results of the SMC. An unexpectedly rich spectrum of spin effects has emerged, especially through the extensive theoretical treatments. More data about the spin structure of the deuterons will come in the near future from further measurements at CERN (SMC) and at SLAC (E154 and E155), as well as at DESY (Hermes). For the more distant future, there are several experiments under construction (RHIC[29]) or in planning.[30] It is hoped to obtain results for the gluon spin ΔG and for the individual quark polarization Δq_i. In discussion is also a measurement of the transverity function $h_1(x)$ which gives information about the spin distribution of quarks inside a transversely polarized proton. This research project is supported by the BMBF.

References

1. P.S. Farago, *J.Phys.B***7**, L28 (1974).
2. P.G. Burke and J.F.B. Mitchell, *J.Phys.B* **7**, 214 (1974).
3. G. Baum, M.S. Lubell, and W. Raith, *Phys.Rev.A* **5**, 1073 (1972).
4. B. Leuer et al., *Z.Phys.D* **33**, 39 (1995).
5. W. Raith et al. *in* Polarized Electron/Polarized Photon Physics, ed. H. Kleinpoppen, Plenum Press, New York and London, to be published.
6. P.S. Farago, *in* Electron and photon interactions with atoms, eds. H. Kleinpoppen and M.R.C. McDowell, Plenum Press, New York and London, 235 (1976).
7. M. Klewer, M.J.M. Beerlage, and M.J. Van der Wiel, *J.Phys.B* **12**, L525 (1979).
8. W. Gehenn and E. Reichert, *J.Phys.B* **10**, 3105 (1977).
9. J.J. McClelland et al., *J.Phys.B* **23**, L21 (1990).
10. G. Baum et al., *Z.Phys.D* **22**, 431 (1991).
11. G. Baum et al., *J.Phys.B* **26**, 331 (1993).
12. K. Bartschat, *J.Phys.B* **26**, 3595 (1993).
13. U. Thumm and D.W. Norcross, *Phys.Rev.A* **47**, 305 (1993).

14. U. Thumm, K. Bartschat, and D.W. Norcross, *J.Phys.B* **26**, 1587 (1993).
15. K. Bartschat, private communication.
16. EMC, J. Ashman et al., *Phys.Lett.B* **206**, 364 (1988).
17. NMC, P. Amaudruz et al., *Phys.Lett.B* **295**, 159 (1992).
18. L.W. Whitlow et al., *Phys.Lett.B* **250**, 193 (1990).
19. R.L. Jaffe and X. Ji, *Phys.Rev.Lett.* **67**, 552 (1991).
20. R.L. Jaffe, *Comments Nucl.Part.Phys.* **19**, 239 (1990).
21. SMC, B. Adeva et al., *Nucl.Instrum.Meth.A* **343**, 363 (1994).
22. SMC, B. Adeva et al., *Nucl.Instrum.MethodsA* **349**, 334 (1994).
23. SMC, D. Adams, et al., *Phys.Lett.B* **329**, 399 (1994).
24. SLAC E80, E130, G. Baum et al., *Phys.Rev.Lett.* **51**, 1135 (1983).
25. SMC, D. Adams et al., *Phys.Lett.B* **336**, 125 (1994).
26. SMC, D. Adams et al., submitted to *Phys.Lett.B* (CERN-PPE/95-97).
27. SLAC E143, K. Abe et al., *Phys.Rev.Lett.* **74**, 346 (1995).
28. SLAC E143, K. Abe et al., SLAC preprint SLAC-PUB-95-6734.
29. G. Bunce et al., *Particle World* **3**, 1 (1992).
30. HMC, G.K. Mallot et al., Letter of Intent, CERN/SPSLC 95-27.

THE ELECTRIC FORMFACTOR OF THE NEUTRON DETERMINED BY QUASIELASTIC SCATTERING OF LONGITUDINALLY POLARIZED ELECTRONS FROM ³He AND ²D.

Collaboration A3 at MAMI: H.G. Andresen[1], J. R. M. Annand[4], K. Aulenbacher[2], J. Becker[2], J. Blume-Werry[1], Th. Dombo[1], P. Drescher[2], J. E. Ducret[1], D. Eyl[1], H. Fischer[2], A. Frey[1], P. Grabmayr[3], S. Hall[4], P. Hartmann[1], T. Hehl[3], W. Heil[2], J. Hoffmann[2], J. D. Kellie[4], F. Klein[1], M. Leduc[5], M. Meierhoff[2] H. Möller[2], Ch. Nachtigall[2], M. Ostrick[1], E. W. Otten[2], R. O. Owens[4], S. Plützer[2], E. Reichert[2], D. Rohe[2], M. Schäfer[2], H. Schmieden[1], K.-H. Steffens[1], R. Surkau[2], Th. Walcher[1].

[1]Institut für Kernphysik, University of Mainz, Germany;
[2]Institut für Physik, University of Mainz, Germany;
[3]Physikalisches Institut, University of Tübingen, Germany;
[4]Kelvin Laboratory, Glasgow, Scotland;
[5]ENS, Paris, France;

presented by E. Reichert

INTRODUCTION

The recent availability of polarized electron beams at electron accelerator laboratories adds another technique to probing the electromagnetic structure of nuclear matter [1]. The present paper discusses the application of polarized electron scattering to the determination of the electric formfactor of the neutron. Its value at medium energies is only poorly known hitherto. Electron scattering is dominated by interaction with the neutron magnetic moment. Therefore the contribution of the charge distribution to the scattering cross section is scarcely detectable in case of unpolarized collision partners. The elastic scattering of longitudinally polarized electrons from a polarized neutron target offers a new and better chance to measure the neutron electric formfactor.

SCATTERING ASYMMETRIES AND FORMFACTORS

The kinematics of scattering longitudinally polarized electrons from a polarized neutron target are sketched in figure 1. It is assumed that the neutron polarization vector is in the scattering plane.

The elastic cross section has the form [1]:

$$\sigma = \sigma_{Mott} \cdot f_{rec}^{-1}[I_0 + (\Delta_x P_x^n + \Delta_z P_z^n)P^e]$$

(1)

with

$$I_0 = \frac{(G_E^n)^2 + \tau(G_M^n)^2}{1 + \tau} + 2\tau(G_M^n)^2 \cdot tan^2(\vartheta_e/2)$$

(2)

$$\Delta_x = -2\sqrt{\frac{\tau}{1 + \tau}} tan(\vartheta_e/2) \cdot G_E^n G_M^n$$

(3)

$$\Delta_z = -2\tau\sqrt{\frac{1}{1 + \tau} + tan^2(\vartheta_e/2)} \cdot tan(\vartheta_e/2) \cdot (G_M^n)^2$$

(4)

and

G_E^n	=	electric formfactor
G_M^n	=	magnetic formfactor
σ_{Mott}	=	Mott cross section
f_{rec}^{-1}	=	recoil correction
τ	=	$-\dfrac{Q^2}{4M^2}$
Q	=	4–momentum transfer
M	=	neutron mass
P^e	=	electron spinpolarization
P_x^n	=	x–component of neutron polarization
P_z^n	=	z–component of neutron polarization

G_E^n is small compared to G_M^n. So $(G_M^n)^2$ dominates the spinindependent part of the cross section in (2). The classical Rosenbluth analysis [2] does not work here in determining G_E^n. But the spindependent terms Δ_x in (3) and Δ_z in (4) offer a good chance to get G_E^n experimentally. The cross section of scattering longitudinally polarized electrons from neutrons that are polarized with respect to the direction of 3–momentum transfer $\vec{q}$ along direction x or z respectively (helicity system of the outgoing neutron) is:

$$\sigma_{x,z}^{\pm} = \sigma_{Mott} \cdot f_{rec}^{-1}[I_0 \pm \Delta_{x,z}P^n P^e]$$

(5)

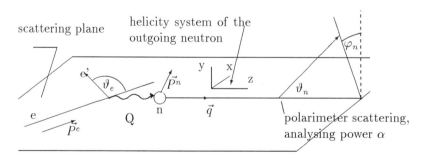

Figure 1: Kinematics of the elastic $\vec{e}$-$\vec{n}$–process. $\vec{P^e}$ = electron polarization; $\vec{P^n}$ = neutron polarization

Where the $\pm$–sign stands for scattering of positive or negative helicity electrons respectively.

In experiment one determines the two asymmetries A_x and A_z:

$$A_{x,z} = \frac{\sigma_{x,z}^+ - \sigma_{x,z}^-}{\sigma_{x,z}^+ + \sigma_{x,z}^-} = \frac{\Delta_{x,z}}{I_0} \cdot P^n P^e \qquad (6)$$

Inserting expressions (3) and (4) gives the ratio of asymmetries A_x and A_z:

$$\frac{A_x}{A_z} = \frac{\Delta_x}{\Delta_z} = \frac{1}{\sqrt{\tau + \tau(1+\tau)tan^2(\vartheta_e/2)}} \cdot \frac{G_E^n}{G_M^n} \qquad (7)$$

or

$$\frac{G_E^n}{G_M^n} = \sqrt{\tau + \tau(1+\tau)tan^2(\vartheta_e/2)} \cdot \frac{A_x}{A_z} \qquad (8)$$

So the measurement of the ratio A_x/A_z determines G_E^n in units of G_M^n. In deriving (7) and (8) respectively constant polarization values P^e, P^n have been assumed. In this case P^e and P^n drop out. So one has not to know the absolute polarization values of electron and target spin for the evaluation of G_E^n/G_M^e as long as the values do not change when switching from one asymmetry measurement to the other. This is a very helpful feature of the scheme, because absolute determination of polarization is often a cumbersome task.

In another scheme the necessity of having a polarized target may be avoided, if the spinpolarization of the recoil neutron is measured. Using (1) one gets the components of recoil polarization $\vec{P_{rec}^n}$:

$$P_{rec,x}^n = \frac{\Delta_x}{I_0} \cdot P^e \qquad (9)$$

and

$$P_{rec,z}^n = -\frac{\Delta_z}{I_0} \cdot P^e \qquad (10)$$

$P_{rec,x}$ may be determined by e.g. np–scattering with analysing power α as indicated in figure 1. The longitudinal component $P_{rec,z}^n$ has to be rotated through an angle of 90^0 e.g. by precession in a magnetic field before spinanalysis in np–scattering. The asymmetries $\mathcal{A}_{x,z}$ expected in a polarimeter np–scattering are given by

$$\mathcal{A}_{x,z} = \frac{I_{x,z}(\varphi_n) - I_{x,z}(\varphi_n + 180)}{I_{x,z}(\varphi_n) + I_{x,z}(\varphi_n + 180)} = \alpha \cdot P_{rec,x,z}^n \cdot cos\varphi_n \qquad (11)$$

Both schemes are used in current experiments at MAMI. There is no target of free neutrons that is dense enough. So in Mainz the quasielastic scattering from the neutron bound in ^3He and ^2D respectively is investigated instead. A polarized neutron target, which is needed in the first scheme, is approximated by a polarized ^3He–target, which represents to a good approximation a system of two paired unpolarized protons and a single polarized neutron [3]. One investigates the exclusive quasielastic reaction $^3\vec{H}e(\vec{e},e'n)$. In the second scheme the quasielastic reaction $^2D(\vec{e},e'\vec{n})$ is studied with polarization analysis of the recoil neutron.

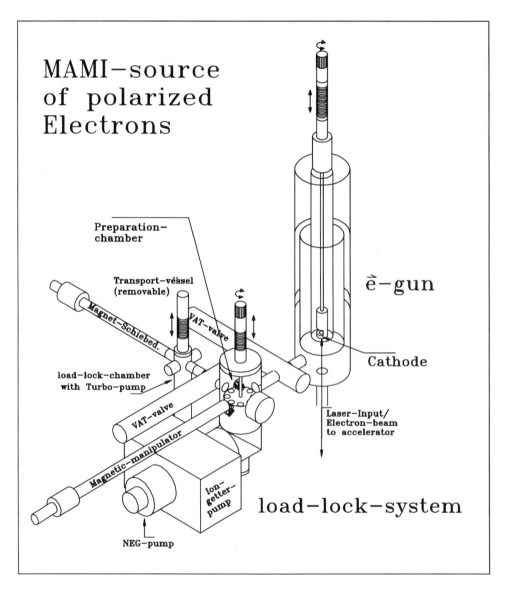

Figure 2: MAMI–source of polarized electrons

SPINPOLARIZED ELECTRON BEAM OF MAMI

The source of polarized electrons installed at MAMI is based on photoelectron emission from III–V–semiconductor cathodes [4]. Figure 2 sketches very schematically the electron gun and an UHV–system attached to it that allows to replace cathodes without breaking the gun vacuum [5]. UHV in the gun is necessary for avoiding excessive degradation of the cathode and is achieved here by use of ion– and NEG–pumps, that maintain a residual pressure of $2 \times 10^{-11} mb$. The cathode is illuminated by a circularly polarized light beam from a laser system not shown. Helicity of the light may be switched from positive to negative sign by switching the sign of $\lambda/4$–voltage of a pockels cell in the path of the light beam. In this way the helicity of the electron beam may be reversed quickly. This is important for correction of instrumental asymmetries in polarization experiments. Photocathodes made from bulk GaAs [6, 7], from InGaP [8, 9], and from strained GaAs [10] have been used up to now. At present a strained layer GaAsP–cathode fabricated by Mamaev, Yavich and coworkers [10] is installed. Using this cathode the MAMI source produces a beam with a degree of polarization of 75%. Emission current may be 20 μA. Because of the limited life time of a cathode it has to be replaced from time to time. At MAMI this is done every 21 hours. Replacement and readjustment of the beamline takes 3 hours so that a duty factor of 85% is achieved. The beam produced is d.c., so a greater portion of the current is lost in the chopper buncher system at injection into MAMI. Only 10% of the current emitted from the cathode is transmitted to a target at the exit of the accelerator. So the source has to deliver e.g. 20 μA polarized current for 2 μA at the target.

In a race track microtron the beam being accelerated has to pass through transverse magnetic fields. In such fields the spin vector of an electron precesses faster than its momentum because of its anomalous magnetic moment [11]. The orientation of the polarization vector relative to momentum varies accordingly during beam acceleration in MAMI and the final orientation may depend severely on the energy to which the system is tuned. Therefore one has to provide means that allow adjustment of the spinorientation of the final beam. In the MAMI scheme a spin rotator is incorporated in the beamline between source and injection that is sketched in figure 3 [12]. It allows the setting of any spinorientation of the beam at the target.

DETECTOR

Figure 4 shows the detector set up used for investigating the quasifree collisions $^3\vec{H}e(\vec{e},e'n)$ and $^2D(\vec{e},e'\vec{n})$. Scattering angles of electrons are measured by a 16×16 matrix of lead glass detectors. The energy resolution is sufficient for discrimination of π– production. An imaging air–Čerenkov–detector is used to reject events that stem from electrons scattered at the entrance and exit windows of the $^3\vec{H}e$–target cell and events produced by photons from π^0–decay. Scattering angles of recoil neutrons are measured by two walls of plastic scintillators, the neutron kinetic energies are determined from the neutron time of flight.

In the case of the $^2D(\vec{e},e'\vec{n})$ experiment the transverse spinpolarization of the recoil neutrons has to be analysed (equation 11). This is done by a technique proposed by Taddeucci and coworkers [13, 14]. The neutron detector–polarimeter consists of two walls of plastic scintillators bars with two veto detectors, all of them with phototubes on either end. So both, horizontal and vertical coordinates of the neutron interaction vertices may be determined. The first wall serves as the polarimeter target (second

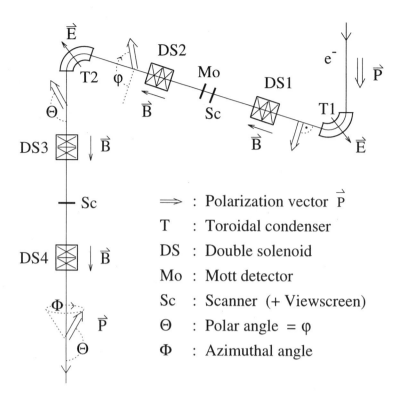

Figure 3: Spinrotator of MAMI [12]

scattering in figure 1) and exploits the analysing power of n–p–scattering in the plastic essentially. The veto detectors reject events produced by charged particles. We intend to install a magnet between target and first detector wall. It will be used to rotate the longitudinal component of the neutron polarization through 90^0 so that it may also be analysed by the two–wall–polarimeter.

TARGETS

A liquid deuterium target is used in case of the $^2D(\vec{e}, e'\vec{n})$ investigations. The polarized ^3He–target bases on a development started by G. K. Walters and coworkers in 1963 [15]. It produces spinpolarized ^3He–nuclei in atomic ground state via metastability exchange scattering with optically pumped and oriented metastable $(1s2s^3S_1)He$–atoms. Dense polarized ^3He–targets have been obtained in recent years by application of powerful lasers in the pumping process [16, 17, 18]. Figure 5 shows the ^3He–target developed in Mainz [18]. Metastable ^3He–atoms are produced by a weak r.f. discharge in an optical pumping cell at a gas pressure of roughly 1 mb. They are polarized by absorption of circularly polarized 1083 nm radiation from a LNA–laser. The nuclear spin polarization of the metastables is very effectively transfered to the ground state atoms in the cell via metastability exchange collisions. The scheme works only in a low pressure discharge. To get higher ^3He densities the polarized ^3He–gas is compressed by a Toepler pump [19] and fed to a target cell at pressures around 1 bar. Spinrelaxation times around 50 h are achieved in the target cell. Through a capillary system some He may leak back to the optical pumping stage where its nuclear polarization is refreshed again.

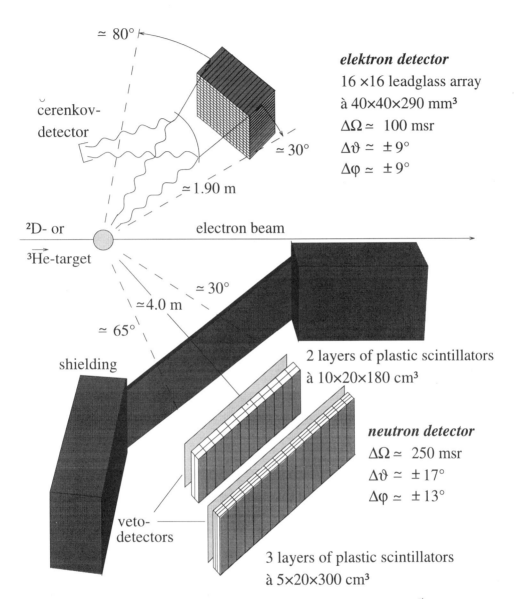

elektron detector

16 ×16 leadglass array

à 40×40×290 mm³

$\Delta\Omega \simeq$ 100 msr

$\Delta\vartheta \simeq \pm 9°$

$\Delta\varphi \simeq \pm 9°$

cerenkov-
detector

$\simeq 80°$

$\simeq 30°$

$\simeq 1.90$ m

²D- or

³He-target

electron beam

$\simeq 30°$

$\simeq 4.0$ m

$\simeq 65°$

shielding

2 layers of plastic scintillators

à 10×20×180 cm³

neutron detector

$\Delta\Omega \simeq$ 250 msr

$\Delta\vartheta \simeq \pm 17°$

$\Delta\varphi \simeq \pm 13°$

veto-
detectors

3 layers of plastic scintillators

à 5×20×300 cm³

Figure 4: The MAMI–detector for investigation of quasielastic $^{3}\vec{H}e(\vec{e},e'n)$ and $^{2}D(\vec{e},e'\vec{n})$ collisions.

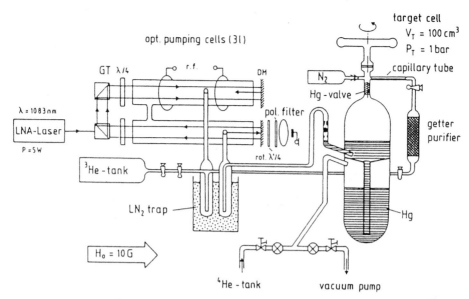

Figure 5: Polarized ^3He–target used in investigation of quasifree collisions $^3\vec{H}e(\vec{e}, e'n)$ at MAMI.

In equilibrium a nuclear spinpolarization of 60% is reached in the pumping cell while in the target cell at one bar the degree of polarization so far observed is just below 40%.

FIRST $^2D(\vec{e}, e'\vec{n})$ AND $^3\vec{H}e(\vec{e}, e'n)$ RESULTS

Polarization experiments at MAMI started in 1992/93 with the investigation of the polarization transfer to the proton in elastic $^1H(\vec{e}, e'\vec{p})$– and quasielastic $^2D(\vec{e}, e'\vec{p})$– collisions. The results are in accordance with transfer values that are calculated using proton formfactors published in literature [20, 21, 22].

A subset of the final detector sketched in figure 4 has been used to start pilot experiments $^3\vec{H}e(\vec{e}, e'n)$ and $^2D(\vec{e}, e'\vec{n})$ at a 4–momentum transfer of $-Q^2 = 8fm^{-2}$. In the investigation $^2D(\vec{e}, e'\vec{n})$ the neutron detector is run as a polarimeter [23]. The modularity of the plastic walls and the time difference in signal readout at the ends of the plastic bars allow full reconstruction of polar angles ϑ_n and azimuthal angles φ_n of neutrons after scattering in the first wall. In case of transversely polarized neutrons one expects a $\cos\varphi_n$ dependence of the scattering asymmetry $\mathcal{A}_x$ of equation (11) measured by the polarimeter. This is clearly observed in the experimental data of figure 6. The asymmetry $\mathcal{A}_x$ for events integrated over the ϑ_n–angular range $10^0 \leq \vartheta_n \leq 35^0$ is plotted as a function of azimuth φ_n. The cos–fit to the data points has an amplitude of $(2.53 \pm 0, 61)\%$. The analysing power of the polarimeter is not calibrated yet.

In the investigation $^3\vec{H}e(\vec{e}, e'n)$ the neutron detector operates as a time–of–flight device only. Again only a part of the final detector set up is used in an exploratory experiment. Asymmetries Δ_x, Δ_z of equation (1) were measured using four different target spin orientations A, B, C, and D [24]. In orientations A and B the target spin is nearly transverse with respect to momentum transfer pointing in directions +x and -x of figure 1 approximately, while in orientations C and D it points essentially along +z and -z respectively. The deviations from pure transverse and longitudinal alignments

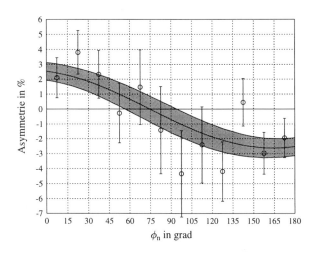

Figure 6: Asymmetry $\mathcal{A}_x$ detected by the neutron polarimeter as a function of azimuth angle φ_n in $^2D(\vec{e}, e'\vec{n})$.

of the target polarization vector are produced by the earth magnetic field that was not compensated and so adds to the guiding field H_0 of the ^3He–target sketched in figure 5. Asymmetries A_x and A_z of equations (7) and (10) respectively are evaluated from the data by taking

$$A_x = \frac{1}{2}(A_A - A_B); \qquad A_z = \frac{1}{2}(A_C - A_D) \tag{12}$$

The experimental data of table 1 give $A_x = 0.89 \pm 0.30$ and $A_z = -7.40 \pm 0.73$. Using relation (7) we get

$$\frac{A_x}{A_z} = -0.120 \pm 0.042 \pm 0.01 \tag{13}$$

Using the value of the magnetic formfactor G_M^n given by the empirical dipol fit

$$G_M^n = \mu_N \cdot (1 - Q^2/18.2)^{-2} \tag{14}$$

with $\quad \mu_N$ = neutron magnetic moment
we arrive at a final value of the electrical formfactor of the neutron

$$G_E^n(-Q^2 = 8fm^{-2}) = 0.035 \pm 0.012 \pm 0.005 \tag{15}$$

This value is in agreement with the work of Platchkov and coworkers [25]. In figure 7 the G_E^n–data of Platchkov et al. are shown. They are evaluated from elastic $^2D(e,e')$–data using the Paris potential. The solid line shown is the best fit using the parametrization

$$G_E^n(Q^2) = \frac{-a\mu_N\tau}{1 + b\tau} \cdot \frac{1}{(1 - Q^2/18.23)^2} \tag{16}$$

with a = 1.25 ±0.13 and b = 18.3 ±3.4 . The values are model dependent in so far as they depend on the deuteron potential assumed. Best fits obtained on the basis of

Table 1: Results of asymmetry measurements at $-Q^2 = 8\,fm^{-2}$ in exclusive collisions $^3\vec{H}e(\vec{e}, e'n)$.

orientation	$\theta(deg)$	$\phi(deg)$	asymmetry (%)		
A	88.2	-1.8	+0.44±0.42		
B	88.2	181.9	-1.35±0.44		
			$(A_A - A_B)/2$	=	$A_x = +0.89 \pm 0.30$
C	2.1	-56.2	-7.22±1.01		
D	177.7	-56.2	+762±1.07		
			$(A_C - A_D)/2$	=	$A_z = -7.40 \pm 0.73$
			ratio: A_x/A_z	=	$-0.120 \pm 0.042 \pm 0.01$

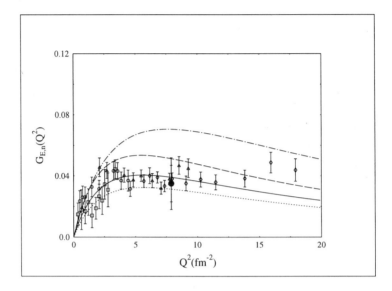

Figure 7: Comparison of the G_E^n value determined from $^3\vec{H}e(\vec{e}, e'n)$ collisions (full circle) with data from [25] (open symbols). The solid curve gives the two parameter best fit of relation (16) to the data points of [25] using the Paris potential. In addition the corresponding two parameter fits using the RSC (dotted), Argonne V14 (dashed), or Nijmegen (dash–dotted) potentials are shown.

RSC–, Argonne–V14–, or Nijmegen–potentials are also indicated in figure 7. Our fairly modelindependent data point at $-Q^2 = 8\,fm^{-2}$ supports the analysis of Platchkov and coworkers.

CONCLUSION

Experiments with spinpolarized electrons have been started successfully at the Mainz race track microtron MAMI. First exploratory investigations of the quasielastic collisions $^2D(\vec{e}, e'\vec{n})$ and $^3\vec{H}e(\vec{e}, e'n)$ demonstrate the potentialities one gets in using spinpolarized collision partners in the study of the electromagnetic structure of nucleons and nuclei.

ACKNOWLEDGEMENTS

This work is supported by the *Deutsche Forschungsgemeinschaft* in the *Sonderforschungsbereich 201, project A3.*

References

[1] T. W. Donnelly and A. S. Raskin, "Considerations of polarization in inclusive electron scattering from nuclei.," *Annals of Physics*, vol. 69, pp. 247–351, 1986.

[2] M. N. Rosenbluth, "High energy scattering of electrons on protons," *Phys. Rev.*, vol. 79, pp. 615–619, 1950.

[3] B. Blankleider and R. M. Woloshyn, "Quasi-elastic scattering of polarized electrons on polarized ^{3}He," *Phys. Rev. C*, vol. 29, pp. 538–552, 1984.

[4] D. T. Pierce, F. Meier, and P. Zürcher, "Negative electron affinity GaAs: A new source of spinpolarized electrons," *Appl. Phys. Letters*, vol. 26, p. 670, 1975.

[5] K. Aulenbacher, C. Nachtigall, H. G. Andresen, P. Drescher, H. Euteneuer, H. Fischer, D. v. Harrach, P. Hartmann, J. Hoffmann, P. Jennewein, K.-H. Kaiser, H. J. Kreidel, S. Plützer, E. Reichert, K.-H. Steffens, and M. Steigerwald, "The MAMI–source of polarized elctrons," vol. to be published, 1995.

[6] W. Hartmann, D. Conrath, W. Gasteyer, H.-J. Gessinger, W. Heil, H. Kessler, L. Koch, E. Reichert, H. G. Andresen, T. Kettner, B. Wagner, J. Ahrens, J. Jethwa, and F. P. Schäfer, "A source of polarized electrons based on photoemission of GaAsP," *Nucl. Instr. and Meth.*, vol. A286, pp. 1–8, 1990.

[7] H. G. Andresen, K. Aulenbacher, T. Dombo, P. Drescher, H. Euteneuer, H. Fischer, D. v. Harrach, P. Hartmann, P. Jennewein, K. H. Kaiser, S. Koebis, H. J. Kreidel, C. Nachtigall, S. Plützer, E. Reichert, K.-H. Steffens, and T. Weis, "Operating experience with the MAMI polarized electron source," in *Proceedings of the Workshop on Photocathodes for Polarized Electron Sources for Accelerators, Stanford, September 8 – 10, 1993; SLAC–report 432* (M. Chatwell, J. Clendenin, T. Maruyama, and D. Schultz, eds.), (Stanford, USA), pp. 2–12, SLAC, 1994.

[8] V. Alperovich, Y. Bolkhovityanov, A. Paulish, and A. Terekhov, "New material for photoemission electron source: semiconductor alloy InGaAsP grown on GaAs substrate," *Nucl. Instr. and Meth. in Phys. Res. A*, vol. 340, pp. 429 – 435, 1994.

[9] A. S. Terekhov, "New material for photoemission electron source: Semiconductor alloy InGaAsP grown on GaAs substrate," in *Proceedings of the Workshop on Photocathodes for Polarized Electron Sources for Accelerators, Stanford, September 8 – 10, 1993; SLAC–report 432* (M. Chatwell, J. Clendenin, T. Maruyama, and D. Schultz, eds.), (Stanford, USA), pp. 369–381, SLAC, 1994.

[10] Y. A. Mamaev, J. C. Gröbli, B. S. Yavich, Y. P. Yashin, F. Meier, A. V. Subashiev, N. N. Faleev, M. S. Galaktianov, D. Guarisco, I. V. Kochnev, S. A. Starovoitov,

A. Vaterlaus, E. Reichert, and S. Plützer, "Spinpolarized photoemission from In-GaAs, AlGaInAs, GaAs, and GaAsP," in *Proceedings of the Workshop on Photocathodes for Polarized Electron Sources for Accelerators, Stanford, September 8 – 10, 1993; SLAC–report 432* (M. Chatwell, J. Clendenin, T. Maruyama, and D. Schultz, eds.), (Stanford, USA), pp. 157–173, SLAC, 1994.

[11] V. Bargman, L. Michel, and V. L. Telegdi, "Precession of the polarization of particles moving in a homogeneous electromagnetic field," *Phys. Rev. Lett.*, vol. 2, pp. 435–437, 1959.

[12] K. H. Steffens, H. G. Andresen, J. Blume-Werry, F. Klein, K. Aulenbacher, and E. Reichert, "A spinrotator for producing a longitudinally polarized electron beam with MAMI.," *Nucl. Instr. Meth.*, vol. A325, pp. 378–383, 1993.

[13] T. N. Taddeucci *Nucl. Instr. Meth.*, vol. A241, p. 448, 1985.

[14] R. Madey, "Bates experiment E85–05," tech. rep., BATES, 1988.

[15] F. D. Colegrove, L. D. Schearer, and G. K. Walters, "Polarization of He^3 gas by optical pumping," *Phys. Rev.*, vol. 132, pp. 2561–2572, 1963.

[16] M. Leduc in *7th Int. Conf. on Polarization Phenomena in Nuclear Physics* (A. Boudard and Y. Terrien, eds.), Coll. de Physique 51 (1990), Suppl. C6-317, 1990.

[17] R. G. Milner, R. D. McKeown, and C. E. Woodward *Nucl. Instrum. Meth.*, vol. A274, p. 56, 1989.

[18] G. Eckert, W. Heil, M. Meyerhoff, E. W. Otten, R. Surkau, W. Werner, M. Leduc, P. J. Nacher, and L. D. Schearer, "A dense polarized ^3He target based on compression of optically pumped gas," *Nucl. Instr. Meth.*, vol. A320, pp. 53–65, 1992.

[19] R. S. Timsit, J. M. Daniels, E. I. Dennig, A. K. Kiang, and A. D. May, "An experiment to compress polarized ^{3}He gas," *Can. J. Phys.*, vol. 49, pp. 508–516, 1971.

[20] D. Eyl, *Messung des Spintransfers in elastischer und quasielastischer Streuung polarisierter Elektronen in den Reaktionen $H(\vec{e}, e'\vec{p})$ und $D(\vec{e}, e'\vec{p})$.* PhD–thesis, Institut für Kernphysik der Joh. Gutenberg Universität Mainz, 1993.

[21] F. Klein, "The electric formfactor of the neutron," in *Proceedings of the 14th International Conference on Few Body Problems in Physics, Williamsburg, Virginia, USA, May 26–31, 1994*, 1994.

[22] D. Eyl, A. Frey, H. G. Andresen, J. R. M. Annand, K. Aulenbacher, J. Becker, J. Blume-Werry, T. Dombo, P. Drescher, H. Fischer, P. Grabmayr, S. Hall, P. Hartmann, T. Hehl, W. Heil, J. Hoffmann, J. D. Kellie, F. Klein, M. Meyerhoff, C. Nachtigall, M. Ostrick, E. W. Otten, R. O. Owens, S. Plützer, E. Reichert, R. Rieger, H. Schmieden, R. Sprengard, K.-H. Steffens, and T. Walcher, "First measurement of the polarization transfer on the proton in the reactions $H(\vec{e}, e'\vec{p})$ and $D(\vec{e}, e'\vec{p})$," *Z. Phys.A*, vol. 352, pp. 211–214, 1995.

[23] A. Frey, *Messung der Neutron–Rückstoßpolarisation in der Reaktion $D(\vec{e}, e'\vec{n})$ zur Bestimmung des elektrischen Formfaktors des Neutrons.* PhD–thesis, Institut für Kernphysik der Joh. Gutenberg Universität Mainz, 1994.

[24] M. Meyerhoff, D. Eyl, A. Frey, H. G. Andresen, J. R. M. Annand, K. Aulenbacher, J. Becker, J. Blume-Werry, T. Dombo, P. Drescher, J. E. Ducret, H. Fischer, P. Grabmayr, S. Hall, P. Hartmann, T. Hehl, W. Heil, J. Hoffmann, J. D. Kellie, F. Klein, M. Leduc, H. Möller, C. Nachtigall, M. Ostrick, E. W. Otten, R. O. Owens, S. Plützer, E. Reichert, D. Rohe, M. Schäfer, L. D. Schearer, H. Schmieden, K.-H. Steffens, R. Surkau, and T. Walcher, "First measurement of the electric formfactor of the neutron in the exclusive quasielastic scattering of polarized electrons from polarized ^{3}He," *Phys. Lett. B*, vol. 327, pp. 201–207, 1994.

[25] S. Platchkov, A. Amroun, S. Auffret, J. M. Cavedon, P. Dreux, J. Duclos, B. Frois, D. Goutte, H. Hachemi, J. Martino, X. H. Phau, and I. Sick, "The deuteron $A(q^2)$ structure function and the neutron electric formfactor," *Nucl. Phys. A*, vol. 510, pp. 740–758, 1990.

IS SINGLE ELECTRON EXCITATION IN HELIUM NOW FULLY UNDERSTOOD?

Albert Crowe[1] and Igor Bray[2]

[1]Department of Physics
University of Newcastle upon Tyne
Newcastle upon Tyne
NE1 7RU
United Kingdom

[2]Electronic Structure of Materials Centre
The Flinders University of South Australia
GPO Box 2100
Adelaide 5001
Australia

INTRODUCTION

Electron impact excitation of atomic and molecular systems has been a subject for both experimental and theoretical study since Franck and Hertz demonstrated the loss of energy of electrons passing through mercury vapour in 1914. Experimentally the great majority of studies on short-lived excited states have been carried out by observation of either the scattered electron or the photon from decay of the excited state. In principle these latter measurements can yield the energy dependence of the cross section for exciting a particular state of the atom. In practice there are substantial difficulties in extracting this information from the spectrum line intensity which is actually measured. Particular problems are associated with cascade from more highly excited states and with the polarisation of the radiation. Major reviews of this type of work have been given by Moisewitsch and Smith[1] and Heddle and Gallagher.[2] Interestingly the latter omits comparison with theory on the basis that 'it does not appear to us that detailed calculations have much to offer at the present time'.

Measurement of the angular distribution of scattered electrons following excitation of a particular state leads to an angular differential cross section. Even for the n=3 states of helium, limited energy resolution prevents isolation of some individual states. An example of this work in helium is that of Trajmar *et al*.[3]

A major disadvantage of both methods is the integration over direction of the undetected particle. The pioneering electron-photon angular and polarisation correlation experiments of Kleinpoppen and co-workers[4,5] and the corresponding analyses of Fano and Macek,[6] Macek and Jaecks[7] have provided detailed means of study of electron impact excitation processes. Both techniques use coincidence methods to observe the scattered electron and decay photon from a single atom. The information obtainable from these experiments is discussed later. It should also be noted that the method makes use of the higher resolution of the photon channel to isolate states, such as some helium n=3 states, not separable by electron spectroscopy. Hence this method can also be used as a means of determination of the differential cross sections for these states.[8]

Theoretical methods of predicting the results of electron impact excitation processes have traditionally fallen into three groups depending on the incident electron energy. At low energies a number of methods have been used to solve a set of coupled integrodifferential equations. These methods include the linear algebraic equations method,[9] variational methods[10] and the R-matrix method.[11] At high energies Born-type calculations can be used to good effect. For intermediate energies, above the ionisation threshold, most methods rely on attempts to extend either the low energy methods upwards[12] or the high energy methods downwards.[13] While these techniques have met with some success, usually over limited angular ranges, they fail to produce consistent agreement with reliable experiments over a range of states and kinematics.

Recent correlation experiments for 1^1S-3^1D excitation in helium have shown the problem for non-dipole processes. Figure 1 highlights the inadequacy existing prior to mid-1994 by comparing available theory and experiment for the circular polarisation Stokes parameter P_3 (defined later) for the 3^1D state of helium at 40eV and small scattering angles. There is excellent agreement between the experiments. The different theories, on the other hand, do not reproduce the experimental results and strongly disagree with each other.

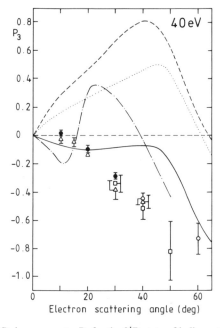

Figure 1. Variation of the Stokes parameter P_3 for the 3^1D state of helium at 40eV – Status of comparison between theory and experiment – mid 1994. *Circles* (open and closed) – Donnelly and Crowe[14] and private communication; *squares* – Beijers *et al*;[15] *triangles* – Mikoza *et al*;[16] *solid line* – Bartschat and Madison[17] – DWBA-EP; *dashed line* – Bartschat and Madison[17] – DWBA-GP; *dotted line* – Cartwright and Csanak[18] – FOMBT; *dash-dot line* – Mansky and Flannery[19] – DMET.

Here we show how comparison between recent correlation experiments in Newcastle and the Convergent Close Coupling (CCC) calculations at Flinders have transformed the situation. Excitation of the 3^1D and 3^3D states are chosen because of the severity of the test they provide for theory.

THE POLARISATION CORRELATION METHOD
– ANALYSIS APPLIED TO HELIUM D STATES

The method is now well established and only a brief outline of its application to D states in helium is given here. A full discussion is given by Andersen *et al.*[20] Figure 2 gives the angular part of the charge cloud density for the five magnetic substates $|d_M\rangle$ of the exited D state. In the natural co-ordinate frame (z perpendicular to the xy scattering plane) only the M = 0, ±2 states have positive reflection symmetry. The excited state can therefore be described in terms of the states $|LM\rangle$ as

$$\psi(3^1 D) = a_{+2}(E,\varphi)|2+2\rangle + a_0(E,\varphi)|2\ 0\rangle + a_{-2}(E,\varphi)|2-2\rangle$$

where the scattering amplitudes can be written $a_2 = \alpha_2 e^{i\beta_2}$, $a_0 = \alpha_0$, $a_{-2} = \alpha_{-2} e^{i\beta_{-2}}$.

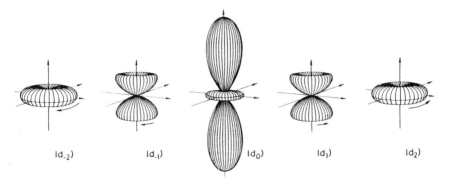

$|d_{-2}\rangle$ $|d_{-1}\rangle$ $|d_0\rangle$ $|d_1\rangle$ $|d_2\rangle$

Figure 2. Angular part of the charge cloud density for the five magnetic substates of a d-state, quantized along the vertical z-axis.

Rather than determine the magnitudes and relative phases of the complex scattering amplitudes, it is convenient to express the results of these correlation experiments in terms of the shape and dynamics of the excited state. These in turn can be related to the polarisation of the D-P radiation observed.

In the experiment, the polarisation state of the radiation is determined as follows. For radiation emitted perpendicular to the x-y scattering plane, Stokes parameters P_1, P_2, P_3 are measured where

$$I_z P_1 = I_z(0°) - I_z(90°)$$

$$I_z P_2 = I_z(45°) - I_z(135°)$$

$$I_z P_3 = I_z(RHC) - I_z(LHC)$$

$I_z(\alpha^\circ)$ is the intensity of photons transmitted by a linear polariser with its transmission axis at an angle α to the electron beam direction, I_z is the total photon intensity in the z direction and *RHC* and *LHC* refer to the handedness of the circular polarisation.

The polarisation analysis is completed by a linear polarisation measurement in the scattering plane defined as

$$I_y P_4 = I_y(0^\circ) - I_y(90^\circ)$$

i.e. analogous to P_1 but for radiation emitted in the y direction.

The data are presented in terms of the following parameters which are related to the measured Stokes parameters as shown for singlet D states.[20] The expectation value of the angular momentum (which can only be transferred perpendicular to the scattering plane), $L_\perp$, where

$$L_\perp = 2(\alpha_2^2 - \alpha_{-2}^2) = -2P_3 I_z$$

where

$$I_z = \frac{2(1 + P_4)}{4 - (1 - P_1)(1 - P_4)}$$

The alignment angle γ and the degree of linear polarisation P_ℓ are defined in terms of P_1 and P_2 by

$$P_1 + iP_2 = P_\ell e^{2i\gamma}$$

and the density matrix element ρ_{00} by

$$\rho_{00} = \alpha_0^2 = 3(1 - I_z)/2$$

The degree of polarisation P is given by

$$P = \sqrt{P_\ell^2 + P_3^2}$$

For the triplet D state, fine structure depolarisation of the radiation must be taken into account in determining the excited state parameters. The following relations are obtained (Crowe *et al*[21])

$$P_\ell = \sqrt{\overline{P}_1^2 + \overline{P}_2^2}$$

where the reduced polarisations $\overline{P}_i$ relating to the excited state produced in the collision are given in terms of the measured polarisations P_i by

$$\overline{P}_i = P_i \frac{I_z}{\frac{2}{3}(G_2 - 1) + I_z} \qquad \text{for } i = 1,2$$

$$L_\perp = -\frac{2}{G_1} P_3 I_z$$

$$\rho_{00} = \frac{1}{2} + \frac{1}{G_2}\left(1 - \frac{3}{2} I_z\right)$$

Here the degree of polarisation P is given by

$$P = \sqrt{\overline{P}_\ell^2 + \overline{P}_3^2}$$

where
$$\overline{P}_3 = (G_2/G_1)P_3 \frac{I_z}{\frac{2}{3}(G_2 - 1) + I_z}$$

$G_1 = 43/54$ and $G_2 = 71/150$ characterise the depolarisation. γ is defined as for the singlet D-state.

THE EXPERIMENT

Figure 3 shows a schematic diagram of the present apparatus emphasising the arrangement for measurement of the Stokes parameters. Briefly a beam of electrons moving in the $+x$ direction crosses a beam of helium atoms. Electrons scattered into a small solid angle enter an electrostatic hemispherical analyser and those which have excited n=3 states are transmitted and detected. The emitted photons are polarisation analysed and detected using the systems shown both in and perpendicular to the scattering plane.

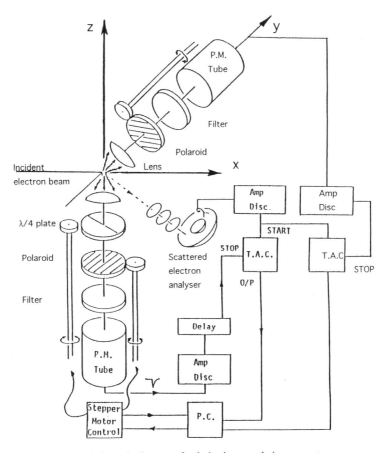

Figure 3. Schematic diagram of polarisation correlation apparatus

Each consists of a lens with its focus at the interaction region followed by a rotatable polaroid, interference filter and fast photomultiplier. For P_3 measurements, a quarter wave plate is inserted before the polariser. Pulses from the electron analyser start the ramps of two time-to-amplitude converters. One is stopped by pulses from the photon detector perpendicular to the scattering plane and the other by pulses from the in-plane photon detector. Polaroid rotation and data accumulation are controlled by a personal computer.

THE CCC METHOD

Details of the CCC method for electron-helium scattering have been given by Fursa and Bray.[22] The basic ideas are very simple. The method uses the close-coupling formalism, with the target states being obtained by diagonalising the target Hamiltonian in an explicitly antisymmetrized two-electron basis constructed from orthogonal Laguerre functions. All of the resulting states are square-integrable, and so may be included in the close-coupling formalism. With increase of the number of Laguerre functions, the negative energy states converge to the true discrete eigenstates, while the positive energy states provide a quadrature rule for the integration over the true continuum. Such a formulation recasts the solution of the Schrödinger equation, which describes the scattering, to simply establishing convergence, to a desired accuracy, in the T-matrix of interest with increasing number of states N.

Due to computational resources being limited we have to be judicious in the way we choose our target states. This has to be done in consideration of the scattering process we wish to describe. Here we are concerned with discrete state excitation. These are very well modelled by the frozen-core approximation, where we constrain one of the target electrons to be the $1s$ orbital of the He$^+$ ion. Even the ground state is reasonably well modelled by this approximation, with the error in the corresponding energy being of order 3%. The usage of this model allows us to test for convergence in only one-electron space, and in effect has reduced the electron-helium scattering problem to a three-body problem. Thus, the frozen-core model provides for an ideal start to the study of electron-helium scattering. Though we have developed technology for studying say one-electron ionisation with excitation or even (e,3e) processes, our first application has been to simple discrete excitation processes within the frozen-core model.

COMPARISONS OF CCC PREDICTIONS WITH EXPERIMENT

Figure 4 shows a comparison between the measured and calculated Stokes parameters P_1–P_4, the excited state parameters γ, P_ℓ, $L_\perp$, ρ_{00} and the degree of polarisation P for the 3^1D state of helium excited by 40 eV electrons. It can be seen that the experimental data for all parameters are well reproduced by the 62-state CCC calculation of Bray et al.[23] Importance of the inclusion of 33 continuum states in this calculation is demonstrated by comparison with the 29 discrete state CC calculation. For all parameters and in particular P_3 and P_4 there is no agreement with experiment when the continuum is ignored. As an illustration of earlier theoretical calculations the Distorted Wave Born Approximation with Excited State potentials of Bartschat and Madison[17] is shown. Comparison with P_1, P_2 data confirms the conclusion from figure 1 of the inadequacy of this theoretical approach for all Stokes parameters.

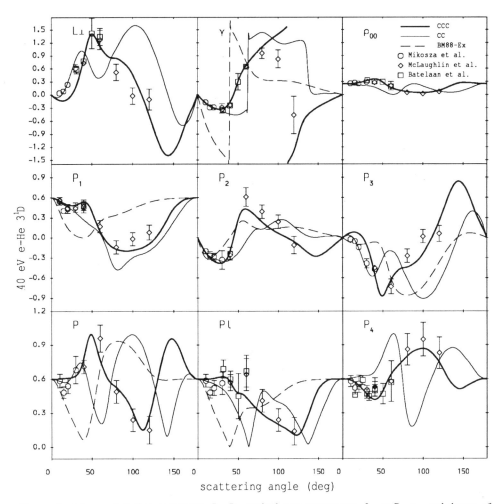

Figure 4. Measured Stokes parameters P_1–P_4, excited state parameters $L_\perp$, γ, P_ℓ, ρ_{00} and degree of polarisation P for the 3^1D state of helium excited by 40 eV electrons. The 62-state CCC and 29-state CC calculations are from Bray et al,[23] the BM-88-Ex calculations are Distorted Wave Born Approximation by Bartschat and Madison.[17] Measurements are from the Newcastle group,[14,24,25] Batelaan et al[26] and Mikoza et al.[16]

Figure 5 shows the corresponding situation for the 3^3D state of helium at the same incident electron energy, 40eV.[21] Again agreement between the experimental data and a 75-state CCC calculation is excellent over all scattering angles for all Stokes parameters and hence for all excited state parameters. For the sake of clarity, preliminary data for P_1, P_2, P_3 and γ by Beyer et al[27] for scattering angles up to 80° and data for P_3 and P_4 up to 120° by Batelaan et al,[26] have been omitted. They are in general agreement with the data from this laboratory. Neither the Distorted Wave Born[17] nor the First Order Many Body Theory calculation[28] reproduce the data. Detailed comparison with these experiments and calculations are given in reference 21.

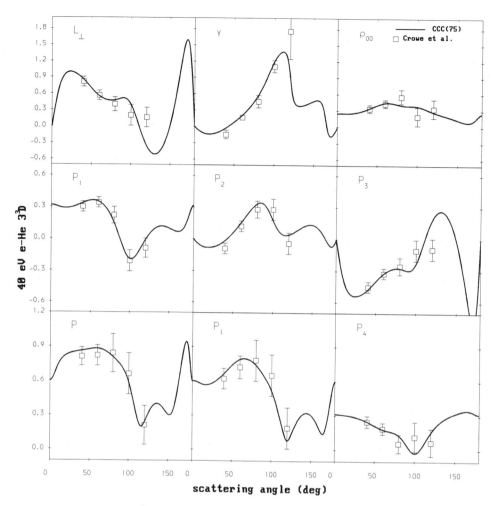

Figure 5. As figure 4 for the 3^3D state of helium. The experimental data and CCC results are from reference 21.

Given the excellent agreement between the CCC calculations and the correlation experiments for both singlet and triplet D states, it is interesting to look at longer standing problems relating to P-state correlation studies. An example of this was the disagreement for the charge cloud alignment γ between the angular correlation data of Neill and Crowe[29] and a 19-state R-matrix calculation of Fon *et al*.[30] Figure 6 shows the disagreement. Beyond a scattering angle of 60°, experiment shows the charge cloud continuing to rotate in the same direction whereas theory predicts that its rotation is in the opposite direction. A subsequent polarisation correlation experiment by Neill *et al*[31] confirmed the original experimental data. Figure 6 shows that the CCC calculation of Fursa and Bray[22] confirms the experimental data.

The longest standing discrepancy between theory and correlation experiments is for excitation of the 2^1P state of helium at 80 eV. There were major discrepancies between the original large angle angular correlation data of Hollywood *et al*[32] and various theories, for example, the Distorted Wave Born approximation of Beijers *et al*[33] and a 5-state R-matrix calculation of Fon *et al*.[34] Although the original data were supported by independent measurements of Slevin *et al*,[35] theory found support from data by Steph and Golden.[36]

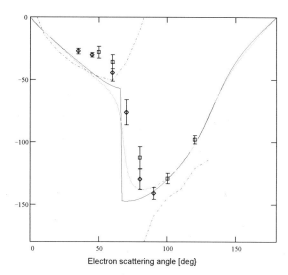

Figure 6. The alignment parameter γ for the 3^1P state of helium at 29.6eV. The experimental data are from Neill *et al*.[29,31] The solid line is the 75-state CCC of Fursa and Bray.[22] The dotted line is a 69-state CCC.[22] The dash-dot line is a 19-state R-matrix calculation of Fon *et al*.[30] Note that γ can only be defined within an arbitrary 180° range and hence the apparent discontinuity in the data of Fon *et al*.

The discrepancy was most pronounced in terms of a λ-parameter (mathematically equivalent to $(1 + P_1)/2$). Figure 7 highlights the discrepancies. For consistency the situation is presented in terms of the P_ℓ, γ and $L_\perp$ parameters. P_ℓ shows the discrepancies most clearly but smaller effects can be seen in γ and $L_\perp$. It can be seen that the 75-state CCC calculation[22] agrees with the data of Hollywood *et al* and Slevin *et al*.

CONCLUSIONS

The polarisation correlation measurements presented here for excitation of both singlet and triplet D states in helium provide the most stringent test of theory for any single electron excitation process in helium available at this time. Unlike all previous calculations the CCC method accurately reproduces the experimental data over a wide range of incident electron energies and scattering angles. Moreover it has been shown capable of resolving disagreements between different experiments and between theory and experiment for P excitation. As would be expected, the CCC method accurately predicts differential and integrated cross sections for S, P and D states. In the sense that these processes can be accurately predicted theoretically, these processes can be considered understood.

It is significant that, unlike in the case presented in figure 7, the corresponding CCC calculations by Bray and Stelbovics[37] for the 2p excitation in atomic hydrogen leave discrepancies between theories and experiment unresolved. Having demonstrated convergence in the much simpler problem, they argued that there is not much more the theorists could do, and requested further experimental investigation. The successes of the CCC theory when applied to similar measurements (many spin-resolved) for the sodium[38] and helium[22] targets suggest to us that the understanding of electron-atom scattering theory is in fact satisfactory in the case of light quasi one- and two-electron targets.

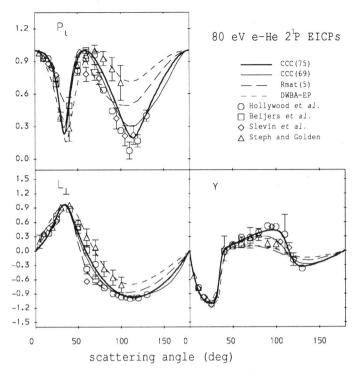

Figure 7. The parameters P_ℓ, γ, $L_\perp$ for the 2^1P state of helium. The CCC calculations are from Fursa and Bray,[22] 5-state R-matrix from Fon *et al*,[34] DWBA-EP and experimental data from Beijers *et al*.[33] Other experimental data are due to Hollywood *et al*,[32] Slevin *et al*[35] and Steph and Golden.[36]

From the theoretical standpoint, we wish to be able to provide cross sections which are of sufficient accuracy for practical applications. The general qualitative and often quantitative agreement with the highly detailed experimental data allows us to be confident in the fundamentals and practicality of the CCC theory. Its primary strength over other techniques is the ability to provide an accurate treatment of the target continuum, which includes the ability to generate the various differential ionisation cross sections. The results presented have provided the foundation for studying cases where more than one-electron excitation processes are important, and hence the frozen-core approximation is inappropriate.

ACKNOWLEDGMENTS

This work is supported by the United Kingdom EPSRC and the Australian Research Council.

REFERENCES

1. B.L. Moisewitsch and S.J. Smith, *Rev. Mod. Phys.* 40:238 (1968).
2. D.W.O. Heddle and J.W. Gallagher, *Rev. Mod. Phys.* 61:221 (1989).
3. S. Trajmar, D.F. Register, D.C. Cartwright and G. Csanak, *J. Phys. B* 25:4889 (1992).
4. M. Eminyan, K.B. McAdam, J. Slevin, M.C. Standage and H. Kleinpoppen, *J. Phys. B* 7:1519 (1974).

5. M.C. Standage and H. Kleinpoppen, *Phys. Rev. Lett.* 36:577 (1976).

6. U. Fano and J.H. Macek, *Rev. Mod. Phys.* 45:553 (1973).

7. J. Macek and D.H. Jaecks, *Phys. Rev. A* 4:2288 (1971).

8. H. Batelaan, F. Vercoulen, J. van Eck and H.G.M. Heideman, *J. Phys. B* 25:4911 (1992).

9. B.I. Schneider and L.A. Collins, *Comput. Phys. Rep.* 10:49 (1989).

10. K. Takatsuka and V. McKoy, *Phys. Rev. A* 30:1734 (1984).

11. P.G. Burke and W.D. Robb, *Adv. At. Mol. Phys.* 11:143 (1975).

12. P.G. Burke and H.M. Shey, *Phys. Rev.* 126:147 (1962).

13. H.R.J. Walters, *Phys. Rep.* 116:1 (1984).

14. B.P. Donnelly and A. Crowe, *J. Phys. B* 21:L637 (1988).

15. J.P.M. Beijers, S.J. Doornenbal, J. van Eck and H.G.M. Heideman, *J. Phys. B* 20:6617 (1987).

16. A.G. Mikoza, R. Hippler, J.B. Wang and J.F. Williams, *Z. Phys. D* 30:129 (1994).

17. K. Bartschat and D.H. Madison, *J. Phys. B* 21:153 (1988).

18. D.C. Cartwright and G. Csanak, *J. Phys. B* 20:L583 (1987).

19. E.J. Mansky and M.R. Flannery, *J. Phys. B* 24:L551 (1991).

20. N. Andersen, J.W. Gallagher and I.V. Hertel, *Phys. Rep.* 165:1 (1988).

21. A. Crowe, B.P. Donnelly, D.T. McLaughlin, I. Bray and D.V. Fursa, *J. Phys. B* 27:L795 (1994).

22. D.V. Fursa and I. Bray, *Phys. Rev. A* (in press).

23. I. Bray, D.V. Fursa and I.E. McCarthy, *Phys. Rev. A* 51:500 (1995).

24. D.T. McLaughlin, B.P. Donnelly and A. Crowe, *Z. Phys. D* 29:259 (1994).

25. D.T. McLaughlin, A.W. Baerveldt, D.G. McDonald, A. Crowe, J. van Eck and H.G.M. Heideman, *J. Phys. B* 27:L19 (1994).

26. H. Batelaan, J. van Eck and H.G.M. Heideman, *J. Phys. B* 24:L397 (1991).

27. H.J. Beyer, H.A. Silim and H. Kleinpoppen, *Proc. 16th Int. Conf. on Physics of Electronic and Atomic Collisions (New York)* ed. A. Dalgarno et al (New York: AIP) Abstracts p165.

28. G. Csanak and D.C. Cartwright, *J. Phys. B* 20:L603 (1987).

29. P.A. Neill and A. Crowe, *J. Phys. B* 21:1879 (1988).

30. W.C. Fon, K.A. Berrington and A.E. Kingston, *J. Phys. B* 23:4347 (1990).

31. P.A. Neill, B.P. Donnelly and A. Crowe, *J. Phys. B* 22:1417 (1989).

32. M.T. Hollywood, A. Crowe and J.F. Williams, *J. Phys. B* 12:819 (1979).

33. J.P. Beijers, D.H. Madison, J. van Eck and H.G.M. Heideman, *J. Phys. B* 20:167 (1987).

34. W.C. Fon, K.A. Berrington and A.E. Kingston, *J. Phys. B* 13:2309 (1980).

35. J. Slevin, H.Q. Porter, M. Eminyan, A. Defrance and G. Vassilev, *J. Phys. B* 13:3009 (1980).

36. N.C. Steph and D.E. Golden, *Phys. Rev. A* 21:1848 (1980).

37. I. Bray and A.T. Stelbovics, *Phys. Rev. A* 46:6995 (1992).

38. I. Bray, *Phys. Rev. A* 49:1066 (1994).

(e, eγ) AND (e,2e) EXPERIMENTS INVOLVING POLARIZED ELECTRONS

G. F. Hanne

Physikalisches Institut
Universität Münster
Wilhelm-Klemm-Str. 10
D-48149 Münster

INTRODUCTION

Recent experimental work on (e,eγ) and (e,2e) coincidences involving polarized electrons is reviewed. The origin of the observed spin asymmetries are exchange effects, spin-orbit interaction of the continuum electrons and spin-orbit coupling within the target in conjunction with orbital orientation and exchange. These mechanisms are explained in some detail. Experimental and theoretical investigations have been performed for excitation and ionization of outer-shell electrons as well as for ionization of inner-shell electrons.

DISCUSSION OF SPIN EFFECTS

Excitation and ionization of atomic targets by electron impact are important processes in gas discharges, plasmas, etc. In the past a variety of investigations were performed to study these processes in some detail. Recent experimental progress has made feasible (e,eγ) and (e,2e) experiments involving polarized electrons. Spin effects in inelastic electron-atom collisions can be explored on the most fundamental level by means of these types of experiments. Sources of polarized electrons with a polarization of $P \geq 0,7$ and currents of up to a μamp have been developed[1] with which such investigations can be performed succesfully.

What is the nature of the spin effects that influence inelastic collisions of electrons with atoms? We can distinguish three different situations:

(i) The spin-orbit interaction of the entire system (continuum electrons plus target electrons) is so small that it does not influence the collision process. In that case spin effects are only caused by exchange, and LS coupling holds for the entire system. If the collisions lead to transitions between certain target states that belong to a fine-structure multiplet the (small) spin-orbit splitting is accounted for by an angular-momentum coupling procedure.

(ii) We assume that the spin-orbit interaction of the continuum electron is still negligible. A weak violation of LS coupling in the target can be accounted for by a description in the intermediate-coupling scheme, where the different fine-structure states have approximately the same radial wave functions. Spin effects are still caused only by exchange, but LS coupling is violated for the entire system. An example of such a situation are the $2p^6$–$2p^53s$ transitions in Ne ($Z = 10$) where an intermediate-coupling scheme applies. In certain situations, e.g. at high energies and small scattering angles, exchange is negligible and, thus, the scattering is not spin dependent. In that case no effect from the violation of LS coupling is observed. But this, of course, does not mean that LS coupling holds in this case; the target states must always be described in the intermediate-coupling scheme.

(iii) The LS coupling in the target is strongly violated, so that in the description of the intermediate-coupling scheme different radial wave functions must be used. In that case the spin-orbit interaction of the continuum electrons is likely to be important as well. This could be an adequate description for the heavy targets xenon and mercury.

Case (i) results in certain relationships for spin-dependent observables that are only valid in LS coupling. Thus a violation of LS coupling results in a violation of such relationships which can be observed experimentally. However, it is, in general, not possible to distinguish case (ii) from case (iii).

We will not give here a complete overview over all the possible types of experiments involving polarization of electrons and atoms. However, we can state that it is not feasible to date to measure the electron polarization after scattering in a coincidence experiment, because the efficiency of such detectors is too small to give reasonable high coincidence signals. Thus generalized T and U parameters and the polarization function S_P cannot be determined. Hence we are left with observables that are connected with the polarizations of the initial electron and atom beam.

Two types of spin asymmetries can be evaluated from a measurement of the coincidence cross sections. If the polarized electrons are scattered by polarized atoms a spin antiparallel-parallel asymmetry

$$A = \frac{\sigma(\uparrow\downarrow) - \sigma(\uparrow\uparrow)}{\sigma(\uparrow\downarrow) + \sigma(\uparrow\uparrow)} \tag{1}$$

of the coincidence signal will be found. However, this asymmetry may depend on the orientation of the initial polarization vectors with respect to the scattering plane if spin-orbit effects play a significant role. In that case one may also obtain a spin up-down asymmetry

$$S_A = \frac{\sigma(\uparrow) - \sigma(\downarrow)}{\sigma(\uparrow) + \sigma(\downarrow)} \tag{2}$$

in the coincidence signal for scattering of polarized electrons from unpolarized atoms. In (2) the directions up and down refer to the normal on the scattering plane.

In an (e,eγ) experiment we can measure spin asymmetries for various positions of the photon polarization analyzer, whereas in a (e,2e) experiment no further polarization analyzing element is included.

ELECTRON-PHOTON COINCIDENCES

In inelastic collisions, the magnetic sublevels of the target may show an anisotropic population. This will, in general, result in an anisotropic shape and in an orientation of the charge-cloud distribution of the excited target electron. If we detect scattered electrons and emitted photons in coincidence, we will select excited atoms that have scattered the electrons into a particular direction.

Such an ensemble of excited atoms will emit polarized light, where the linear polarization components will give a picture of the shape of the charge cloud, i. e. the alignment, whereas the circular polarization components are a measure of the target orientation after scattering. We are particularly interested in how the incident electron's spin affects the resulting charge-cloud distribution.

In Figure 1 is shown a scheme of our apparatus. We can map the excited target by means of a photon detector that looks perpendicular to the scattering plane and another detector - not shown here - that looks parallel to the scattering plane.

I should mention that one can, of course, measure the polarization of the emitted light without detecting the scattered electrons. Such experiments with polarized electrons have been proposed about 25 years ago by Farago and Wykes[2] in Edinburgh, and we use them frequently for calibration purposes.[3,4]

Here I want to review electron-photon coincidence studies. We did such experiments with mercury[5] and xenon atoms[6] and as an example some typical results for mercury are shown.

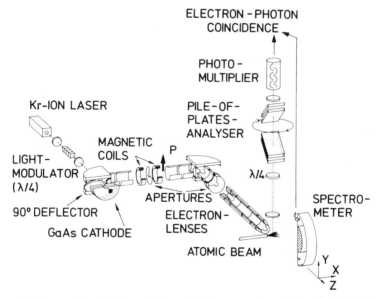

Figure 1. Scheme of our electron-photon coincidence experiment with polarized electrons

The first result (Figure 2) shows how the shape of the charge cloud depends on the initial spin orientation. We studied the excitation of the $6\,^3P_1$ state of mercury at 8 eV.[5] The excitation energy of this famous Franck-Hertz transition is about 4.9 eV, thus the electrons are scatterd with a final energy of about 3.1 eV. The scattering angle is 20°.

It is clearly seen that the shape of the resulting charge cloud of the excited target electron depends strongly on the spin orientation of the incident electron. The alignment angle through which the charge cloud is tilted from the incident electron beam direction is also strongly dependent on the initial spin orientation. The size of the charge-cloud distribution is a measure of the excitation cross section. These distributions indicate therefore the magnitude of the cross sections for electrons with spin up and spin down. The average of the two charge-cloud distributions gives the distribution that one would observe for unpolarized electrons.

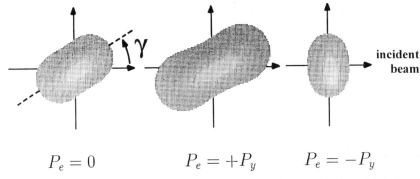

$$P_e = 0 \qquad\qquad P_e = +P_y \qquad P_e = -P_y$$

Figure 2. Charge-cloud distributions for Hg atoms (after electrom-impact excitation of the 6 $^3P_1$ state with polarized electrons at $E = 8$ eV and $\theta = 20°$). View perpendicular to the scattering plane

The agreement between experimental data and data from an R-matrix calculation by Bartschat *et al*[7] is quite good, as can be seen from another parameter of the same experiment. From measurements of the circular polarization we can extract the orientation of the target electron, and this orientation is plotted in Figure 3 as a function of scattering angle for different spin orientations of the incident electrons.

For s - p transitions there exists an propensity rule that states that the orientation of a charge cloud is positive for scattering of unpolarized electrons by small angles. This is obviously not true for spin-resolved collisions where the electron spin is initially down. A careful analysis of our data by Bartschat and Andersen *et al*[8] shows that in the case of the 6 $^3P_1$ excitation of mercury spin flips are very likely for spin-down electrons, but those spin-down electrons whose spin is not flipped tend to transfer a negative angular momentum to the atom! This is also indicated in Figure 3 by a Fano-Kohmoto plot for the collision system presented here. Unfortunately, only a few data points are available for the spin-resolved orientation parameter. Thus we plan to revive that investigation.

First results for similar experiments for xenon have been obtained recently.[6] These investigations are still going on.

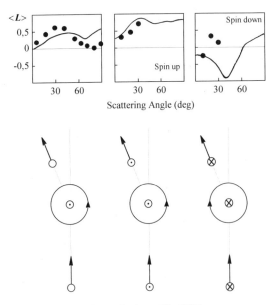

Figure 3. Spin-resolved orientation for excitation of Hg (6 $^3P_1$ state, parameters see Figure 2)

TRIPLE DIFFERENTIAL IONIZATION CROSS SECTIONS:
(e, 2e) COINCIDENCES

The dynamics of electron-impact ionization of atomic targets has focused much interest on understanding the final-state correlations of this three-body problem where two continuum electrons are emitted after scattering from a positively charged core. Spin exchange effects for integral ionization cross sections have been measured for one-electron atoms and for metastable helium atoms for more than 15 years.[9] Ionization processes can also be studied by socalled (e,2e) experiments, where scattered and ejected electron are detected in coincidence. Although there have been numerous studies of the (e,2e) problem, only little attention has been paid to spin effects in these collisions. However, triple differential cross sections may, in general, depend on the spin orientation of the incident electrons.

We can distinguish two situations. If we scatter from polarized targets, such as lithium, the cross section for electron and target spin parallel may differ from that for antiparallel spins due to exchange effects or - in other words - due to the fact that the PAULI principle requires spatial wave functions of different symmetry for the two spin orientations.

Results from an experiment of this type have been obtained recently by the Bielefeld group.[10] Large spin antiparallel-parallel asymmetries for ionization of lithium atoms are observed and are predicted by distorted wave calculations.[11] If scattered and ejected electrons have the same energy and symmetric angles, the spatial wave function of the final state is symmetric, and thus the Pauli principle states that the total spin must be antisymmetric. This results in an asymmetry of unity for this situation, which is, however, not obtained in this experiment[10] due to imperfect angular and energy resolution. A convolution of the theoretical results from Zhang et al[11] from Belfast that includes the experimental resolution shows fair agreement with the experimental data.

For the case of atomic excitation it has been shown in experimental[12] and theoretical[13-15] investigations that spin up-down asymmetries exists for electron-impact excitation of the bound states of rare-gas atoms with a np^6 configuration in the ground state. The prediction that spin up-down asymmetries may also be observed in (e,2e) coincidence studies[16] has stimulated very recent experimental[17,18] and theoretical[19,20] investigations. From an experimental point of view, (e,2e) studies with xenon and krypton targets are of particular interest, since their fine-structure splittings of 1.31 eV and 0.67 eV, respectively, can be resolved without major difficulties.

A simple picture shown in Figure 4 illustrates the mechanism that produces spin up-down asymmetries in electron-impact excitation of heavy rare-gas atoms. In the excitation, a vacancy is produced in the closed p^6 shell, e.g., a $np^6 \rightarrow np^5(n + 1)l$ transition. It is well established that the COULOMB interaction may produce an oriented ionic $np^5\ ^2P$ core; i.e., the cross sections for exciting the $m_l = \pm 1$ magnetic sublevels of the ionic ^2P core are different for a quantization axis orientation perpendicular to the scattering plane.[21] This orbital orientation effects a spin orientation of the ionic core, if its final J state is resolved, e.g., in the $^2P_{1/2}$ configuration the projections of target spin $\langle s_c \rangle$ and orbital angular momentum $\langle l_c \rangle$ are opposite. Let us, for the sake of simplicity, assume that in a collision the $^2P_{1/2}$ core is completely orientated, say with $m_l = +1$. The spin of this state is therefore down, which means that the spin $\langle s_e \rangle$ of the excited $(n + 1)l$ electron is up, because in the initial np^6 configuration the spins compensated each other. It is obvious that in such a situation a spin up-down asymmetry may be observed since both direct and exchange scattering are possible for an incident spin-up electron while only direct scattering is possible for an incident spin-down electron.

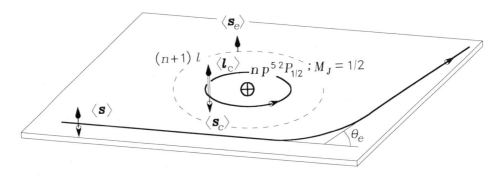

Figure 4. Simple picture of spin effects in np^5 ($^2P_{1/2}$)$(n + 1)l$ excitation of heavy rare gases

Figure 5 demonstrates the size of the effects you can expect from (e,2e) experiments with xenon. Remember that the effect does not depend on the strength of the spin-orbit interaction within the target. You merely need to resolve the fine-structure splitting. First calculations by Jones *et al*[19] from the University of Missouri predict large spin effects for which an example is shown here. Spin-resolved cross sections and spin up-down asymmetries are plotted for electron-impact ionization of xenon at 2 eV above threshold. The solid circles are relative measurements of Rösel *et al*[22] from Kaiserslautern for unpolarized electrons. The agreement between experimental data and theory for unpolarized electrons is very good, so it were an interesting experiment to measure the spin resolved cross sections for which large spin effects are predicted.

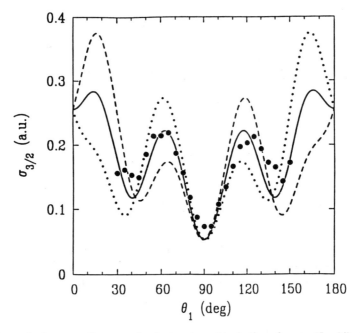

Figure 5. Spin-resolved cross sections $\sigma_{3/2}$ for electron impact ionization of xenon ($J = 3/2$ core) at 2 eV above threshold. The angle θ_1 is measured counterclockwise from the forward beam direction. Theory[17] for σ: Solid line, statistical average (cross section for unpolarized electrons); dotted line, spin-up cross sections; dashed line, spin-down cross sections. • Experimental results for unpolarized electrons from Rösel *et al.*[22]

First experiments have been performed by a group in Australia[17] at 147 eV incident energy and by our group[18] at 40 eV. Spin effects are clearly seen, in agreement with predictions, but they are somewhat smaller than shown in Figure 5.

A quite different experiment is running in Tübingen.[23] Spin up-down asymmetries for excitation of inner-shell electrons are clearly obseved and predicted[24,25] for electrons that are emitted into large angles (recoil peak) due to the spin-orbit interaction that these electrons experience in the field of the nuclei of Ag atoms.

It is interesting to note that the fine-structure effect[8] may contribute as well for ionization of electrons from subshells with $l > 0$. This has been predicted recently[26] and is the subject of a new experimental investigation.[27]

Acknowledgements. The experimental work performed in Münster was supported by the Deutsche Forschungsgemeinschaft in Sonderforschungsbereich 216 *Polarization and Correlation in Atomic Collision Complexes*.

REFERENCES

1. "Workshop on Photocathodes for Polarized Electron Sources for Accelerators," M. Chatwell, J. Cledenin, T. Maruyama, and D. Schultz, eds., SLAC Report 432 (1994)
2. P. S. Farago and J. Wykes, *J. Phys. B* 2: 747 (1969)
3. T. J. Gay, *J. Phys. B* 16: L553 (1983)
4. M. Uhrig, A. Beck, J. Goeke, F. Eschen, M. Sohn, G. F. Hanne, K. Jost, and J. Kessler, *Rev. Sci. Instrum.* 60: 872 (1989)
5. M. Sohn and G. F. Hanne, *J. Phys. B* 25: 4627 (1992)
6. M. Uhrig, G. F. Hanne, and J. Kessler, *J. Phys. B* 27: 4009 (1994)
7. K. Bartschat, private communication (1991)
8. N. Andersen, K. Bartschat, J. T. Broad, G. F. Hanne, and M. Uhrig, to be published (1995)
9. see *e. g.* J Kessler, *Adv. At. Mol. Phys.* 27: 81 (1991)
10. G. Baum, W. Blask, P. Freienstein, L. Frost,, S. Hesse, W. Raith, P. Rappolt, and M. Streun, *Phys. Rev. Lett.* 69: 3037 (1992)
11. X. Zhang, T. Colm, T. Whelan, and H. R. J. Walters, *J. Phys. B* 25: L457 (1992)
12. M. Dümmler, G. F. Hanne, and J. Kessler, *J. Phys. B.* 28: in press (1995)
13. G. F. Hanne, *Phys. Rep.* 95: 95 (1983)
14. K. Bartschat and D. H. Madison, *J. Phys. B* 20: 5839 (1987)
15. N. T. Padial, G. D. Meneses, F. J. da Paixão, G. Csanak, and D. C. Cartwright, *Phys. Rev.* A23: 2194 (1990)
16. G. F. Hanne, *in*: "Correlations and Polarization in Electronic and Atomic Collisions and (e,2e) Reactions," P. J. O. Teubner and E. Weigold, eds., IOP Conference Series Number 122, Bristol (1992)
17. T. Simon, Thesis (Universität Münster) (1995)
18. J. Lower, private communication (1995)
19. S. Jones, D. H. Madison, and G. F. Hanne, *Phys. Rev. Lett.* 72: 2554 (1994)
20. S. Mazevet, private communication (1995)
21. N. Andersen, J. W. Gallagher, and I. V. Hertel, *Phys. Rep.* 165: 1 (1988)
22. T. Rösel, R. Bär, K. Jung, and H. Ehrhardt, *in*: "Invited Papers and Progress Reports, European Conference on (e,2e) Collisions and Related Problems," H. Ehrhardt, ed., Universität Kaiserslautern (1989)
23. H.-Th. Prinz, K.-H. Besch, and W. Nakel, *Phys. Rev. Lett.* 74: 243 (1995)
24. R. Tenzer and N. Grün, *Physics Letters* A194: 300 (1994)
25 D. H. Jakubaßa-Amundsen, *J. Phys. B* 28: 259 (1995)
26. S. Keller, private communication (1995)
27. W. Nakel, private communication (!995)

ELECTRON - PHOTON POLARISATION CORRELATION EXPERIMENT ON ELECTRON IMPACT EXCITATION OF THE FIRST 4^1P_1 STATE OF Ca ATOMS

Stanisław Chwirot,[1] Dariusz Dziczek,[1] Rajesh Srivastava,[2] and Roman S. Dygdała [1]

[1] Institute of Physics
Nicholas Copernicus University
ul. Grudziądzka 5-7
PL 87-100 Toruń, Poland
[2] Department of Physics
University of Roorkee
Roorkee 247667, India

Introduction

Excitation processes of metal atoms are of great interest for astrophysics and plasma physics. Ca is for instance an element contributing many lines to spectra of solar - type stars and also the lines of other metal atoms are often used for diagnostic purposes [1,2]. Alkaline earth atoms are often seen as relatively simple multielectron systems since for most purposes they can be considered composed of two valence electrons outside an inert inner electron core.

Electron impact excitation of alkaline earth atoms have only in recent years become a target of coincidence experiments. The fundamental review of Andersen et al.[3] summarising the work done in the field up to 1985 mentioned only two such experiment involving excitation of Ba atoms [4,5]. It took about ten years to start active research on these collision systems and the first reports on coincidence studies on electron impact excitation of $^1P_1$ state of Ca were published by Kleinpoppen's group in 1988 [6]. Since then the field has developed at a good speed and a series of new data has been accumulated both from theoretical and experimental studies (see for instance [7,8,9]). Most of the coincidence data published until now, are however either of a preliminary character or reported by one group only for particular energies and/or scattering angles. The data are very interesting and indicate that the expected spin dependent effects can indeed be observed for these moderately heavy atoms. There is a little doubt that in the coming years more experimental material will be needed to provide more complete information on the excitation parameters for electron - alkaline earth atom systems and thus to clarify discrepancies between theoretical and experimental data and to stimulate further progress in the field of collision studies.

All the above mentioned reasons made us to concentrate the work of our newly established coincidence laboratory on studies of electron impact excitation of alkaline earth atoms. This paper gives us opportunity to join the family of the coincidence groups and to present the current status of the experiment and our first data on the electron excitation of the first 4s4p $^1P_1$ state of Ca obtained using coherence analysis approach.

Experimental Aspects

A geometry of the experimental set-up used in this study is typical for coherence analysis experiments (Figure1.). The scattering plane is defined by the momentum of the incident electron beam and the momenta of the scattered electrons. Photons of a wavelength 422.6 nm arising from the decay of the 4^1P_1 state are detected in a direction perpendicular to the scattering plane.

The experiment is housed in a cylindrical vacuum chamber made of stainless steel (dimensions: 67 cm in diameter by 56 cm in height). The tank is evacuated by an oil diffusion pump with a pumping speed of 2000 dcm^3·s^{-1}. A typical operating pressure is of order of 5x10^{-7} mbar with Ca atomic beam flowing into the chamber and the cold trap filled with liquid nitrogen. The vertical component of magnetic field is compensated using a pair of Helmholtz coils placed outside the tank and a μ-metal shield is used to prevent a field produced by the oven heaters from entering the chamber. The residual magnetic field present in the interaction region can still affect electron trajectories at low energies and for that reason the measurements were carried out at the electron incidence energy of 100 eV. In a near future the compensation with the Helmholtz coils will be replaced by a set of two μ-metal liners placed inside the tank making it essentially a field-free region.

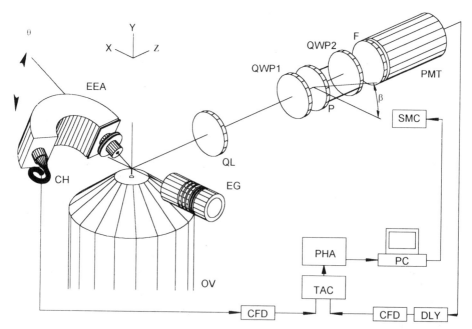

Figure 1. A schematic diagram of the apparatus and measurement geometry (CH - channeltron, EEA - electron energy analyser, EG - electron gun, OV - atomic beam oven, QL - quartz lens, F - filter, P - linear polariser, QWP - retarding plates, SMC - stepper motor controller, CFD - constant fraction discriminator, DLY - delay, TAC - time-to-amplitude converter, PHA - pulse-height analyser, PC - computer).

A beam of Ca atoms is produced by a resistively heated stainless steel oven and collimated by a nozzle and an aperture of the magnetic shield. The main body of the oven is typically heated to 620°C and the nozzle is kept at a temperature of 700°C. The temperatures remain constant within 1°C during the measurements. The atomic density at the interaction region located about 8 mm above the nozzle was estimated to be approximately 3 x 10^{12} atoms·cm^{-3}. Liquid nitrogen trap placed over the oven prevents a contamination of the chamber with Ca atoms and reduces the pressure of pump oil vapours. Two bifilar sets of Thermocoax (Philips) wires are used to heat both the oven's main body and the nozzle. Such a geometry minimises the magnetic field of heating currents leaking into the interaction region. The oven is separated from the rest of the system by a water-cooled heat-sink, absorbing its thermal radiation, and a magnetic shield attenuating residuals of its magnetic field.

The electron beam is produced by a commercial electron gun (EG-402EL, Comstock) providing stable currents of about 2 μA. The scattered electrons are firstly selected for the scattering angle and the energy loss of 2.9 eV using the hemispherical 160° electrostatic electron energy analyser (AC-901, Comstock) and then detected with the channeltron (X959BL, Philips). Three-element einzel lenses are used both at the output of the electron gun and at the input of the electron energy analyser. The combined electron energy resolution of the electron beam and the spectrometer is 0.5 eV which is adequate to resolve the electrons scattered elastically and those resulting from inelastic collisions and excitation of the 4 $^1P_1$ state (see Figure 2.). The acceptance angle of the electron spectrometer is estimated to be approximately 3°.

Photons emitted in a direction perpendicular to the scattering plane are collected by a quartz lens positioned in such a way that the interaction region is at its focus. The acceptance angle of the lens was estimated to be approximately 15°. The fluorescence light collected by the lens passes through a polarisation analyser and appropriate interference filter (custom made), and then it is detected by a photomultiplier (XP2020, Philips).

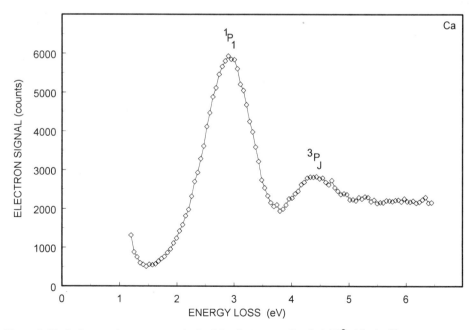

Figure 2. Typical energy loss spectrum obtained for electrons scattered at 25° at the incidence energy of 100 eV.

The polarimeter is composed of the $\lambda/4$ plate (mica, Spindler & Hoyer) and the linear polariser (Spindler & Hoyer, extinction ratio better than 99.99 %) coupled with another $\lambda/4$ plate so that the light emerging from the polarising optics is always circularly polarised to ensure independence of the measured photon fluxes on a polarisation of the detected light. The linear polariser and the second retarder are rotated together and the accuracy of positioning them with respect to axes of the first retarder is approximately $0.6°$. The alignment of the axes of all elements of the polarimeter with respect to the scattering plane and to each other is tested by measuring the integrated (non-coincidence) Stokes parameters. Due to the axial symmetry of the non-coincidence measurement configuration one of the Stokes parameters should be equal zero, and placing a retarder axis at a right angle with respect to the direction of the incoming electron beam should not influence the measured degree of the linear polarisation.

The pulses from the electron and photon detectors are amplified (preamplifier model 9301, EG&G Ortec) and fed into constant fraction discriminators (model 935, EG&G Ortec). The electron pulses start the time-to-amplitude converter (type 1701, Polon) and the photon pulses, suitably delayed, are used to stop it. The output of the converter is monitored continuously with a multichannel analyser operated in a pulse-height-analysis mode. Fig.3 shows a typical coincidence spectrum recorded during one measurement cycle with the excitation of the 4 $^1P_1$ state of Ca by electrons scattered at angle of $25°$.

The measurements typically took two to four days of integration preceded by two or three days of stabilisation of the system with all the elements working continuously. The experimental procedure involves the accumulation of true coincidence counts at each setting of the linear polariser's angle required for a determination of the Stokes parameters. Measurements of the linear and circular polarisation are made successively without or with the first $\lambda/4$ retarder. Typically the polariser scans the set of the angles required in 15 min. steps in a repetitious manner to minimise the effect of possible long-term instabilities of measurement conditions (like drifts in the efficiency of detectors used).

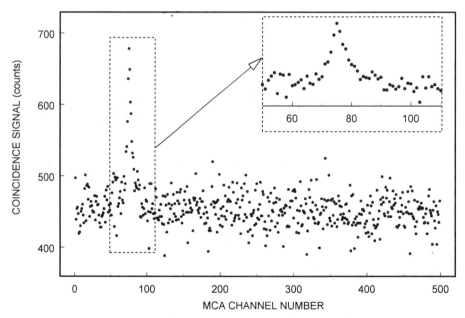

Figure 3. An example of the electron-photon coincidence signal from the $^4P_1$ excitation of Ca at the incidence energy of 100 eV with electrons scattered at $25°$.

The rate of real coincidences is scaled to the electron count rate to account for variations in electron current and in atomic beam concentration. A care was taken to reduce possible sources of systematic errors and the procedures involved all the typical precautions like shielding the possible sources of stray electric fields and coating all the surfaces in vicinity of the interaction region with colloidal graphite. The operation of the system and the data acquisition are controlled by a suitably interfaced personal computer.

It has been observed that in our experimental conditions the background of false coincidences is not uniform but changes exponentially with the delay time[10]. The effect was corrected for by a least squares routine analysing the exponential characteristics of the background in regions with false coincidences only and then interpolating the true levels of background in the channels of interest. The polarisation of the 422.6 nm line of Ca was measured in a broad energy range by Ehlers and Gallagher[11]. It changes quite rapidly for energies between 60 and 150 eV and thus can be used as a test of the beam energy. One finds from their data that for the beam energy of 100 eV the linear polarisation degree is approximately -14.8 ± 0.4 % for directly measured results, and approximately -16.5 ± 0.4 % when results are corrected for the imperfection of their polarisation analyser and some geometrical factors. Our measurements yield -15.5 %, which is in a good agreement taking into account 1% accuracy of our result.

As shown in reference [11] the finite acceptance angle of a photon detector and the divergence of the electron beam may result in values of a measured apparent polarisation lower than the true one. In our case the maximum correction for polarisation parameters was found to be about 2 % i.e. much less than experimental errors of the data.

Results and Discussion

The measured values of the Stokes parameters at incident energy of 100 eV and the relevant values of the so-called electron impact coherence parameters are shown in Table 1. The same quantities are graphically presented in Figure 4 together with the theoretical results of the relativistic distorted wave calculation (as reported by Srivastava et al. [8] but obtained at 100 eV).

Table 1. Values of the measured Stokes parameters and of the calculated electron impact excitation parameters obtained for different scattering angles at the incidence energy of 100 eV. The errors represent one standard deviation.

SCATTERING ANGLE	10°	15°	20°	25°
P_1	0.0247	0.2576	0.3168	0.5628
σ	0.0681	0.1181	0.0957	0.1062
P_2	-0.1606	-0.4397	-0.4767	-0.4257
σ	0.0764	0.0972	0.0952	0.1187
P_3	-0.6411	-0.4658	-0.5386	-0.5490
σ	0.0919	0.0993	0.1077	0.1662
P_l	0.1625	0.5096	0.5724	0.7057
σ	0.0859	0.1436	0.1323	0.1563
P	0.6614	0.6904	0.7859	0.8941
σ	0.1102	0.1730	0.1701	0.2254
γ (°)	-40.63	-29.82	-28.20	-18.55
σ(°)	5.47	0.83	0.58	0.45

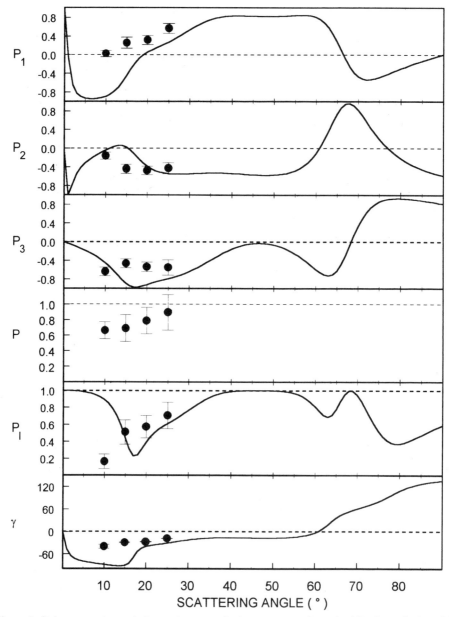

Figure 4. Stokes parameters and electron impact excitation parameters determined for the excitation of the $^4P_1$ state of Ca at the electron incidence energy of 100 eV and compared with the theoretical calculations (relativistic distorted wave calculation as described by Srivastava et al.[8]). The error bars represent one standard deviation.

It is surprising that for the low scattering angles and high incidence energy of the electrons the total polarisation measured is less than one at least for the two lowest angles. In the absence of spin dependent interactions the process under study should be completely coherent and one would not expect such interactions to be important at our experimental conditions. Such a conclusion is additionally supported by the results obtained at lower incidence energies for a similar transition in Mg [12] and for Ba [13]. On the other hand, total polarisations measured by Kleinpoppen's group at lower incident energies for similar transitions in Ca [14] also deviated from one beyond the experimental error bars between 15° and 25° and even more significant effect was obtained by them for the excitation of Sr at electron impact energy of 45 eV [15]. At this stage, however, we believe that any definite conclusions regarding a possible incoherence of the excitation process would be premature. Despite all the precautions it is still possible that the measurement as such introduces some kind of systematic error to the data although the non-coincidence polarisation value is consistent with the results of the careful work of Ehlers and Gallagher[11].

A general agreement between the measured Stokes parameters and the theoretical data is rather good. Both P_1 and P_2 change in a way predicted by the theory although in both cases one can see a shift of the two sets of the data with experimental values shifted by about 5° towards lower scattering angles with respected to the theoretical curves. A different situation was found for the P_3 parameter which show only little variations in the range of scattering angles studied in this work. This observation may suggest that perhaps the retarder we use is not as good as specified in the catalogue and introduces a degree of elliptical polarisation thus lowering both the measured values of the circular polarisation and the total polarisation degree.

On the other hand, however, the agreement between the experiment and the theory is for our data much better than for the experiments carried on a lower incident energy of 45 eV (see [8]). In the light of these lower energy data it will be very interesting to see the results of measurement in the region of slightly larger scattering angles between 30° and 60°. In this region the theory and experiment clearly diverge at energy of 45 eV for P2 and P3 and whatever the outcome of our measurements at 100 eV the results will undoubtedly help to find a source of the discrepancy. Preparations for such measurements are currently in progress and we expect to start them after completing the changes in the design of the vacuum chamber necessary to open this range of angles to the measurements.

Acknowledgements

This work was supported by a grant from Committee of Scientific Research (KBN, No 2 P03B 068 09). One of us (S.C.) thanks H. Kleinpoppen and J. Beyer for helpful discussions and advise regarding experiments with calcium.

References

1. G.Smith, Oscillator strengths for neutral calcium lines of 2.9 eV excitation,
   J. Phys. B: At. Mol. Opt. Phys. **21**:2872 (1988).
2. L. H. Aller, "Astrophysics", 2nd edition, Ronald, New York (1963).
3. N. Andersen, J. W. Gallagher and I. V. Hertel, Collisional alignment and orientation of atomic outer shells. I. Direct excitation by electron and atom impact, *Phys. Rep.* **165**:1 (1988).
4. D. F. Register, S. Trajmar, S. W. Jensen, R. T. Poe, Electron scattering by laser-excited barium atoms, *Phys. Rev. Lett.* **41**:749 (1978).
5. D. F. Register, S. Trajmar, G. Csnak, S. W. Jensen, M. A. Fineman, R. T. Poe, Investigation of superelastic electron scattering by laser-excited Ba, experimental procedures and results, *Phys. Rev. A.* **28**:151 (1983).
6. M. A. K. El-Fayoumi, H. J. Beyer, F. Shahin, Y. A. Eid and H. Kleinpoppen, Electron-photon coincidence study of the 4^1P state of calcium, *in:* "11th ICAP Book of Abstracts", XI-2, Paris (1988).
7. R. K. Houghton, M. J. Brunger, G. Shen and P. J. O. Teubner, Electron impact excitation of the 3^3P state in magnesium, *J. Phys. B: At. Mol. Opt. Phys.* **27**:3573 (1994).
8. R. Srivastava, T. Zuo, R. P. McEachran and A. D. Stauffer, Excitation of the $^{1,3}P_1$ states of calcium, strontium and barium in the relativistic distorted-wave approximation, *J. Phys. B: At. Mol. Opt. Phys.* **25**:3709 (1992).

9. R. Srivastava, R. P. McEachran and A. D. Stauffer, Excitation of the D states of cadmium and barium by electron impact, *J. Phys. B: At. Mol. Opt. Phys.* **25**:4033 (1992).
10. Chr. Holzapfel, On statistics of time-to-amplitude converter systems in photon counting devices, *Rev. Sci. Instrum.* **45**:894 (1974).
11. V. J. Ehlers and A. Gallagher, Electron excitation of the calcium 4227-Å resonance line, *Phys. Rev. A.* **7**:1573 (1973).
12. M.J. Brunger, J.L. Riley, R.E. Scholten and P.J.O. Teubner, Electron - photon coincidence studies in magnesium. *J. Phys. B: At. Mol. Opt. Phys.* **22**:1431 (1989).
13. P.W. Zetner, Electron scattering from laser-excited barium. Proc. of XVIII ICPEAC, Aarhus 1993, *in*: "AIP Conference Proceedings 295", p. 306, AIP Press, New York (1993).
14. E.I.M. Zohny, M.A.K. El Fayoumi, H. Hamdy, H.-J. Beyer, Y.A. Eid, F. Shahin and H. Kleinpoppen: Electron-photon polarisation correlation measurements for the $^4P_1$ state of Ca, *in*: "XVI ICPEAC Book of Abstracts", p. 173, New York (1989).
15. H. Hamdy, H.-J. Beyer, K. R. Mahmoud, E.I.M. Zohny, G. Hassan and H. Kleinpoppen, Electron-photon polarisation correlation measurements for the first excited $^1P_1$ states of alakli-earth metals, *in*: "XVII ICPEAC Book of Abstracts", p. 132, Griffith University Press, Brisbane (1991).

THE STEPWISE EXCITATION ELECTRON-PHOTON COINCIDENCE TECHNIQUE

A T Masters, W R MacGillivray and M C Standage

Laser Atomic Physics Laboratory
School of Science
Griffith University
Nathan, Queensland
Australia 4111

INTRODUCTION

Following the initial development in the 1970's of the electron-photon coincidence technique by Kleinpoppen and coworkers in which angular correlation[1-3] and polarized-photon correlation[4] techniques were first applied to the study of electron-atom collisions, coincidence techniques have become a widely used and very sensitive method for investigating atomic collision dynamics. The introduction of laser techniques to the study of atomic collision processes followed the development of the tunable dye laser in the 1970's. Stepwise excitation techniques were developed by several groups in which laser and electron excitation procecesses were used to excite target atoms in a number of non-coincidence investigations of atomic collision processes (see ref.5 for a review of this work).

In 1989, the electron-photon coincidence and stepwise excitation techniques were brought together in the first stepwise electron-photon coincidence experiment which was reported by Murray et al[6]. In a subsequent series of papers, these authors have presented an extensive theoretical and experimental study of the technique. In this paper, the stepwise electron-photon coincidence technique is critically reviewed.

THEORY

Figure 1 shows a typical Type 1[5] stepwise excitation scheme in which a target atom is first excited by electron impact from the ground state lg> to a first excited state le>. Laser excitation further excites the atom from state le> to the higher lying state li>. The stepwise excited atom may decay to state le> or lf> emitting a photon. The stepwise excitation signal is usually obtained by detecting photons emitted as the atom decays to state lf> in delayed coincidence with electrons inelastically scattered from state le>. State le> may be connected by an optically allowed transition to the ground state lg>, or it may be a metastable state. To date, stepwise coincidence experiments have only been performed for states in

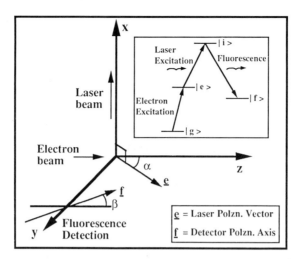

Figure 1. Schematic diagram of the experimental geometry of the stepwise coincidence experiment. $\underline{e}$ and $\underline{f}$ are respectively the polarization vectors of laser light and the photon detector. The insert shows the stepwise electron and laser excitation scheme.

which the transition $|e>->|g>$ is optically allowed. However, non-coincidence stepwise experiments have been performed on metastable states (see ref. 5). It should be noted that even if $|e>$ is a metastable state, then it is in principle possible to obtain a coincidence signal using the stepwise excitation method provided the metastable state is connected to higher lying states by optically allowed transitions.

As in a conventional coincidence experiment, the directions of the incident and scattered electrons determine the scattering plane in a stepwise excitation coincidence experiment. Figure 1 also shows the experimental geometry used in the experiment described in this paper. Linearly polarized laser light traverses the interaction region along the x-axis and fluorescence photons emitted by the stepwise excited atom are detected perpendicular to the (x-z) scattering plane along the y-axis. The laser light is linearly polarized at angle α to the z-axis. Both circular and linear analysis of the fluorescent light is used to fully determine the Stokes parameters of the fluorescent light.

The theoretical description of the stepwise excitation coincidence technique has been fully described elsewhere[5,7] and here only the main features are briefly recalled. The fluorescent intensity, I, emitted by a stepwise excited atom is represented by

$$I = \sum_{m'n'} \rho^i_{m'n'} F_{n'm'} \tag{1}$$

where $\rho^i_{m'n'}$ is the density matrix element associated with the upper excited magnetic substates $|m'>$ and $|n'>$ of state $|i>$. F is an emission operator which represents the fluorescence decay and has matrix elements, such that

$$F_{n'm'} = \sum_{q} <n'|\mathbf{f}.\mathbf{P}|q> \times <q|\mathbf{f}^*.\mathbf{P}|m'> \tag{2}$$

where $\mathbf{f}$ is the polarization vector of the photon detector, $\mathbf{P}$ is the electric dipole moment operator and $|q>$ are substates of the final state $|f>$. The laser excitation step can be formally represented by the expression

$$\rho^{i}_{m'n'} = \sum_{mn} A_{m'n'mn} \rho^{e}_{mn} \tag{3}$$

where ρ^{e}_{mn} are density matrix elements of the electron-impact excited state $|e>$.

The combination of Eqn.1 and Eqn.3 gives

$$I = \sum_{\substack{m' \; n' \\ m \; n}} F_{n'm'} A_{m'n'mn} \rho^{e}_{mn} \tag{4}$$

For weak optical excitation, it can be shown[5] that the term $A_{m'n'mn}$ can be represented by the expression

$$A_{m'n'mn} = \frac{1}{\Gamma_{i}\Gamma_{e}} < m'|e.P|m > < n|e^{*}.P|n' > \tag{5}$$

where e is the polarization vector of the laser light and Γ_{e} and Γ_{i} are spontaneous emission decay rates for the states $|e>$ and $|i>$.

For strong optical excitation, a nonperturbative theoretical description is required. The use of rate equations is generally not appropriate for treating the laser-atom interactions involved in stepwise excitation processes because rate equations deal only with atomic populations and do not treat atomic coherences. Several techniques are available which do treat coherences, such as the semiclassical density matrix and Atomic operator techniques[7,8].

A considerable advantage is gained if the weak optical excitation treatment can be applied, because analytical expressions can be derived which link the Stokes parameter measurements made on the stepwise coincidence signals with the atomic collision parameters describing the collisionally excited state. For the stepwise coincidence experiment discussed here, the electron impact excited transition was the $6^{1}S_{0}$ - $6^{1}P_{1}$ (185nm) transition of Hg. The laser excited transition was the $6^{1}P_{1}$ - $6^{1}D_{2}$ (579nm) transition and fluorescence was detected from the $6^{1}D_{2}$ - $6^{3}P_{1}$ (313nm) transition. The latter transition is optically allowed due to the breakdown of LS coupling in Hg. The Stokes parameters for this stepwise excitation scheme are given by[5,9]:

$$IP_{1\alpha} = \lambda\left[\frac{27}{2} - 6\cos^{2}\alpha\right] + \frac{9}{2}\cos 2\alpha + \frac{9}{2}(1-\lambda)\cos\delta$$

$$IP_{2\alpha} = -3(5 - \cos 2\alpha)[\lambda(1-\lambda)]^{\frac{1}{2}}\cos\chi \tag{6}$$

$$IP_{3\alpha} = 9(3 + \cos 2\alpha)[\lambda(1-\lambda)]^{\frac{1}{2}}\sin\phi$$

$$I = \lambda(6\cos^{2}\alpha + \frac{1}{2}) + 5(\cos^{2}\alpha + \frac{17}{2}) + \frac{1}{2}(1-\lambda)(8\sin^{2}\alpha - 9)\cos\delta$$

$P_{1\alpha}, P_{2\alpha}, P_{3\alpha}$ are the Stokes parameters for a particular laser polarization angle α and I is the total stepwise coincidence intensity. Five atomic collision parameters are required to characterize the density matrix that represents the electron-impact of an ^{1}S - ^{1}P transition. In the collision frame, the atomic collision parameters are: $\lambda = \rho_{00}/\sigma$, $\cos\delta = \rho_{1-1}/\rho_{11}$, $\cos\chi = \mathrm{Re}\,\rho_{10}/(\rho_{00}\rho_{11})^{\frac{1}{2}}$, $\sin\phi = \mathrm{Im}\,\rho_{10}/(\rho_{00}\rho_{11})^{\frac{1}{2}}$ and $\sigma = \rho_{00} + 2\rho_{11}$. Electron-impact excited populations and Zeeman coherences are respectively represented by diagonal, ρ_{mm}, and off-diagonal, ρ_{mn}, density matrix elements.

The laser polarization at angle α to the z axis provides an additional parameter which enables a full determination of the collision parameters for an S - P transition to be made from the Stokes parameters measured perpendicular to the scattering plane. The values of α used in the experiment were 0^0 and 90^0.

It might be thought that the conditions under which the analytical expressions for the stepwise coincidence Stokes parameters, given in Eqn.6, can be used are restricted to when the laser intensity is very low, or the decay time of the collisionally excited state is so short that to a good approximation only one laser photon interacts with the atom during the stepwise excitation process. However, quantum electrodynamical calculations indicate that at least for some cases, the weak excitation approach remains valid for quite high laser intensities. The calculations were performed using the Atomic operator method for the laser excited 579nm transition in Hg used in the stepwise excitation coincidence experiment. Details of the calculation are given in ref.7 and will not be repeated here. It was assumed that π - excitation ($\Delta m_j = 0$) of the transition in Hg occurred with single mode laser light. All the associated decay rates and branching ratios were included in the calculation, which is free of approximations except for the usual rotating wave and electric dipole approximations. The calculation assumes that the excitation process is coherent and symmetries between the 6^1P_1 matrix elements have been used which arise because of the reflection symmetries associated with the scattering plane. The initial values used in the calculation of the resulting four independent 6^1P_1 density matrix elements were:

$$\rho_{00} = 0.5, \quad \rho_{11} = 0.25, \quad \mathrm{Re}\,\rho_{10} = 0.25, \quad \mathrm{Im}\,\rho_{10} = 0.25$$

Figure 2 shows the time evolution of density matrix elements for a laser intensity of 500mW/mm^2, which corresponds to the intensity of the laser beam at the centre of the interaction region used in the experiment. Such an intensity is well above what would normally be regarded as corresponding to weak excitation conditions. Figure 2 shows the behaviour of the population ρ_{00} of the $m_j = 0$ sublevel. The curve shows a secondary peak that reaches a maximum 5-6ns after the initial excitation of the atom due to Rabi nutational cycling of the population. The Rabi frequency is 160MHz for the ($m_j=0 \rightarrow m_j=0$) transition. Also shown is the time dependent behaviour of the ρ_{00}^{μ} sublevel population of the laser excited 6^1D_2 state. The time evolution of the 6^1D_2 state is also driven by the laser induced Rabi nutational cycling. It is out of phase with the time evolution of the 6^1P_1 state in that when the population of the lower state reaches a minimum, the upper state population reaches a maximum. The decay time of the upper excited state population is significantly shorter than the natural lifetime of 10.8ns of this state. This is due to the laser interaction which couples the longer lifetime upper state to the short lived (1.3ns) lower state.

Figures 2 also shows the time dependent behaviour of the natural frame parameters P_{lin}, γ, $L_\perp$ and ρ_{00} which respectively describe the ratio of the major to the minor axis of the charge cloud, its tilt with respect to the incident electron direction, the angular momentum imparted to the atom in the collision, and the spinflip cross section. It will be noticed that during the initial decay of the 6^1P_1 state, these parameters remain constant until the population approaches zero, at which time significant excursions from the initial values occur. As the maximum of the secondary peak in the population occurs, most of the parameters return to approximately their initial values. It should be noted that after 10-15ns the magnitudes of the density matrix elements become negligible and do not contribute significantly to the stepwise coincidence signal.

The significance of these variations in the natural frame parameters can be determined by calculating the average values of the natural frame parameters over the duration of the secondary peak. Because the contribution to the coincidence signal is proportional to the

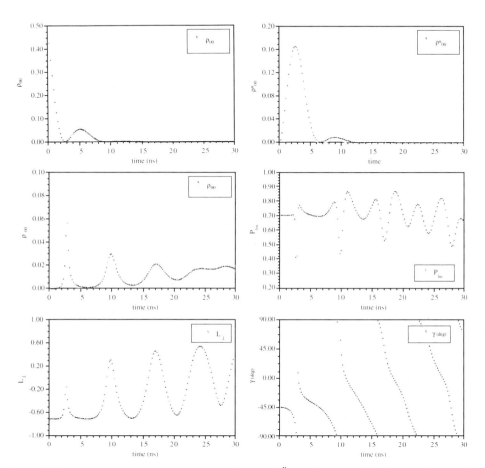

Figure 2. The time evolution of the lower (ρ_{00}) and upper (ρ_{00}^u) excited state sublevel populations, and the natural frame parameters.

differential cross-section at any instance in time, the averaging was carried out by performing an integration over time for the duration of the secondary peak weighted by the differential cross-section. The resulting values, shown in Table 1, indicate that the overall effect is small. A similiar calculation for the initial decay shows a smaller effect compared to that for the secondary peak. Moreover, integration of the coincidence signal over time, which is the usual manner of analysing coincidence data, further reduces any deviations from that expected if weak excitation conditions applied due to inclusion of signal corresponding to the initial decay. Calculations presented in ref.7 show that the variation of the stepwise coincidence Stokes parameters for the range of laser intensities 0 - 500mW/mm^2 is a few percent. It is clear in this example, that in spite of the effects of the strong laser excitation, such as enhanced decay rates, and transient changes to the shape of the atomic charge cloud, the Stokes parameters are little changed compared to that expected for the weak excitation case. This may be attributed to the very rapid decay rate of the lower excited state and hence the limited number of Rabi nutational cycles that can occur in any excitation process.

Table 1: Averaged natural frame parameters (for the secondary peak) for the 6^1P_1 state of Hg as a function of Rabi frequency. Initial values used in the calculation are also shown.

Rabi Frequency	$<\rho_{00}>$	$<L_\perp>$	$<P_{Lin}>$	$<\gamma>$
160 MHz	0.003	-0.669	0.714	-41.0^0
120 MHz	0.004	-0.672	0.513	-28.9^0
80 MHz	0.007	-0.678	0.694	-38.3^0
40 MHz	0.019	-0.679	0.649	-33.7^0
Initial values	0.0	-0.707	0.707	-45.0^0

EXPERIMENT

Figure 3 shows a diagram of the stepwise excitation coincidence experiment. A full description of the apparatus has been presented in ref.9. Briefly, the experiment is based on a conventional crossed beam configuration in which mutually orthogonal electron and laser beams intersect with an atomic beam which passes through the interaction region perpendicular to the laser beam. The collimation factor of the atomic beam was 5:1 which resulted in a 1/e Doppler width of 300 MHz. The estimated atom density in the interaction region was 2-5 x 10^{16} m^{-3} at a tank operating pressure of about 2 x 10^{-6}mbar.

Fluorescence photons were collected by an f1.0 lens with a solid angle of 0.51 sr. A narrow bandwidth optical filter was used to select the required wavelength corresponding to the stepwise excited fluorescence channel and a cooled photomultiplier tube was used to reduce the dark current. A combination of linear and circular polarizers was used to measure the Stokes parameters of the detected radiation. Inelastically scattered electrons were detected using a cylindrical mirror analyser fitted with an aperture lens system that imaged the interaction region into the analyser. An actively stabilized Spectra Physics 380D single-mode ring dye laser provided laser light for the experiments. The Doppler width of 300MHz for the atomic beam is the limiting factor which determines the spectral resolution, however the overall spectral resolution is sufficient to fully resolve the hyperfine structure of the 6^1D_2 -6^1P_1 (579 nm) line of Hg[9]. As a result, the stepwise coincidence experiments yield data for

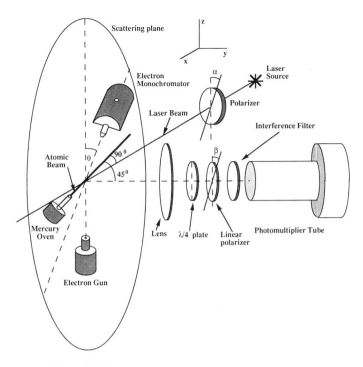

Figure 3. Schematic diagram of experimental apparatus.

which the depolarizing effects of the hyperfine structure are experimentally eliminated, thereby removing the need to resort to theoretical corrections to allow for such effects.

Standard coincidence counting techniques were used to obtain the signals shown in Figure 4. Most of the sources of systematic errors in a stepwise excitation coincidence experiment are similar to those that arise in a conventional coincidence experiment, although the manner in which they manifest themselves may be somewhat different. One such systematic error is the finite volume effect, which is discussed in ref.9. For the stepwise case, it is found from both theoretical considerations and experiment that the P_{190} Stokes parameter is the parameter which is most sensitive to the finite volume effect. This is because when the laser light is linearly polarized at right angles to the nominal scattering plane for a P_{190} measurement, rotation of the effective scattering plane about the z-axis due to the finite volume effect renders the laser light no longer perpendicular to the effective scattering plane. The effect of laser intensity on stepwise coincidence signals has already been discussed above and is an additional source of possible systematic error that needs to be evaluated by calculations and experimental verification.

Radiation trapping manifests itself in a somewhat different manner in a stepwise coincidence experiment as compared to a conventional coincidence experiment. Figure 4a shows a stepwise signal obtained at a low atomic density in the interaction region, whereas Figure 4b shows a signal obtained under the same experimental conditions except that the atomic density was approximately twenty times higher. An explanation of trapping effects in a stepwise coincidence experiment is based on two trapping mechanisms[11,12]. The first of these mechanisms involves an indirect trapping process in which a stepwise excitation process takes place which is followed by the atom returning to the ground state via state |e> (see Figure 1). A fluorescence photon from the |e> ->|g> resonance transition is emitted and may be recaptured by another ground state atom. If this atom is in the laser beam, another

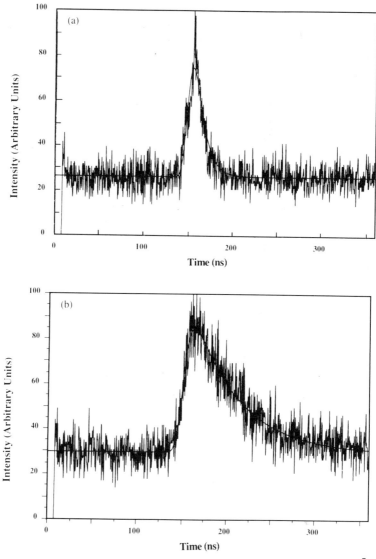

Figure 4. Stepwise coincidence signals (a) Tank pressure (Hg partial vapour pressure), 6.0 x 10^{-7} Torr, decay time, 9ns. (b) Tank pressure, 1.1 x 10^{-5} Torr, decay time, 52ns.

stepwise excitation can take place with the subsequent emission of a photon via the $|i\rangle \rightarrow |f\rangle$ transition, or further trapping events may occur. The time delay associated with a single indirect trapping is characterized by the time for the second maximum to appear in the population of state $|e\rangle$, which is determined by the inverse of the Rabi nutational frequency. The second trapping mechanism is conventional trapping in which, following electron impact excitation to $|e\rangle$, the atom decays directly back to the ground state emitting a photon which can be recaptured by an adjacent ground state atom. For the Hg stepwise excitation experiment discussed here, the direct trapping mechanism has a delay time of 1.3 ns associated with it, whereas the indirect trapping mechanism has a delay time which will depend on the laser intensity, but is typically 6-10 ns. Clearly, various combinations of direct and indirect trapping events can also occur.

A theory has been developed by Masters et al 1995[12] which uses a multipole treatment of the radiation trapping process to describe radiation trapping for stepwise coincidence signals. The theory of conventional radiation trapping in terms of state multipoles has been given by a number of authors and was recently extended to conventional coincidence experiments[13]. The time evolution of all state multipoles is exponential, with a decay rate given by $\Gamma (1 - A_k n)$, where for a J=1 -> J=0 transition, $A_0=1$, $A_0=0.5$ and $A_2=0.7$, and Γ is the untrapped decay rate. n is the probability of absorption of the resonance line radiation. The work of Masters et al 1995 gives the following theoretical expressions for the stepwise coincidence Stokes parameters for the 6^1S -> 6^1P_1 -> 6^1D_2 -> 6^3P_1 stepwise excitation scheme:

$$P_{10} = \frac{21 - 3(7 - 12\lambda)C_2 / C_0}{47 + (7 + 6\lambda)C_2 / C_0}$$

$$P_{190} = \frac{27}{26C_0 / C_2 + 1}(2\lambda - 1)$$

$$P_{20} = -\frac{36C_2 / C_0}{47 + (7 + 6\lambda)C_2 / C_0}\sqrt{\lambda(1 - \lambda)} \cos \chi$$

$$P_{30} = \frac{108\sqrt{\lambda(1 - \lambda)} \sin \phi}{47C_0 / C_1 + (7 + 6\lambda)C_2 / C_1}$$

(7)

where $C_k = [\Gamma_i^R \Gamma_e^R (1 - A_k n)^2]^{-1}$. The decay rates Γ_e^R and Γ_i^R are respectively the decay rates for states |e> and |i> in the presence of the laser excitation. These expressions were derived under the assumption that $\cos \delta = -1$ and in the absence of radiation trapping (i.e. when $n=0$) reduce to the expressions given in Eqn.6. A striking feature of the theoretical results is that they predict the slope of the P_{10} Stokes parameter as a function of the absorption probability, n, can be either positive or negative depending on the value of the atomic collision parameter λ. The value of λ at which the slope of P_{10} is zero is 0.78. The behaviour of the other Stokes parameters as a function of n is predicted to be monatonically decreasing with n. The theory also predicts an approximately linear dependence of all the Stokes parameters on the atom density in the interaction region, particularly for small values of n.

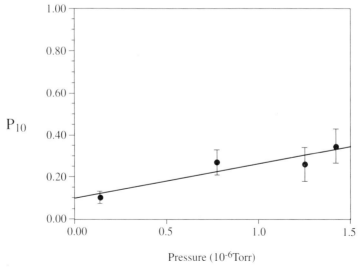

Figure 5. Stokes parameter P_{10} data plotted against Hg partial vapour tank pressure.

Figure 5 shows recently obtained data for P_{10} as a function of Hg partial vapour tank pressure. The atom density in the interaction region is proportional to the tank pressure and measurements of the decay rates of stepwise coincidence signals as a function of tank pressure show an approximately linear relationship between the absorption probability n and the tank pressure at lower values of n[11,12]. It will be noted that the P_{10} data exhibits a positive slope with tank pressure. The value obtained for the λ atomic collision parameter is 0.14.

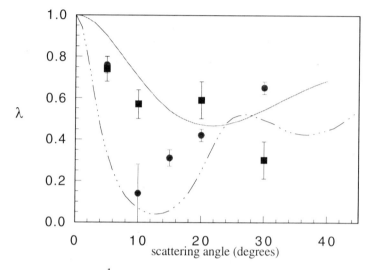

Figure 6. λ parameter data for the 6^1P_1 state of Hg vs electron scattering angle. ■ 16 eV ● 50 eV incident electron energy. —— 16 eV —— – – 50 eV DWBA calculation [14].

Figure 6 shows typical atomic collision data for the 6^1P_1 state of Hg obtained from the stepwise coincidence experiment. Measurements of the λ parameter are shown versus electron scattering angle for 16 eV and 50 eV incident electron energy. The 16 eV data has been previously reported in ref. 9. The theoretical curves are obtained from a Distorted Wave Born Approximation calculation (DWBA)[14]. Although the agreement between theory and experiment could not be described as good, the theory does describe the general features of the λ parameter data.

Table 2 shows complete sets of atomic collision parameter data for two incident electron energies (16 eV, 50 eV) and two electron scattering angles (10°, 20°). Also shown are the corresponding DWBA theoretical results for the natural frame parameters. Several conclusions can be drawn from a comparison of the experimental and theoretical data.

Although the agreement between theory and experiment is not good, it is better for the higher energy data, particularly with respect to the shape of the atomic charge cloud, as measured by P_{lin}, and its tilt, as given by γ, in relation to the incident electron beam. The measured angular momentum transferred to the atom, $L_\perp$, is significantly smaller than the theoretical values and a significant loss of coherence is evident, as measured by $|\mu|$ and P_{Tot} being less than unity, although the loss of coherence is less at the higher energy. The better agreement between theory and experiment at the higher energy is not surprising in view of the fact that the DWBA theory is an intermediate energy theory. However, the results do suggest that further development of theory is required to more accurately model collision processes involving heavy target atoms such as Hg.

Table 2. Collision frame and natural frame atomic collision parameters for the Hg 6^1P_1 state. *Indicates an assumed value. ( ) DWBA theoretical values. See refs 3 & 5 for full definitions of atomic collision parameters.

Incident Energy	16 eV		50 eV	
Scattering angle	10°	20°	10°	20°
Atomic Collision Parameters				
λ	0.57±0.07	0.59±0.09	0.14±0.14	0.42±0.04
$\cos\delta$	-0.96±0.57	-0.99±0.70	-1.00*	-1.00*
$\cos\chi$	-0.13±0.13	0.10±0.16	0.31±0.43	-0.42±0.02
$\sin\phi$	0.02±0.04	-0.21±0.08	-0.07±0.04	-0.32±0.02
$|\mu|$	0.13±0.06	0.23±0.05	0.32±0.21	0.52±0.02
ρ_{00}	0.01±0.12	0.00±0.15	0.00*	0.00*
	(0.00)	(0.00)	(0.00)	(0.00)
P_{Lin}	0.19±0.13	0.20±0.10	0.75±0.14	0.44±0.01
	(0.92)	(0.65)	(0.96)	(0.67)
$L_\perp$	-0.02±0.04	0.20±0.08	0.05±0.03	0.32±0.02
	(0.40)	(0.76)	(0.26)	(0.75)
γ	$21°^{+15°}_{-21°}$	$-14°^{+29°}_{-19°}$	-81.77±14.75	55.31±18.23
	(-31.90)	(-47.56)	(-76.63)	(69.58)
P_{Tot}	0.19±0.1	0.30±0.08	0.75±0.14	0.54±0.02
	(0.99)	(0.99)	(1.00)	(1.00)

CONCLUSIONS

The stepwise coincidence method combines the capacity of the electron-photon coincidence method to completely characterize an inelastic collision process with the spectral resolution only available from laser spectroscopic techniques. There are a number of advantages that the stepwise coincidence technique has to offer. It offers an alternative to conventional coincidence techniques for investigating collision processes that involve vacuum-ultraviolet transitions in atomic targets. For a number of atoms, the higher lying transitions that can be accessed by stepwise excitation techniques lie in the optical part of the spectrum which allows conventional optical components to be used to analyse fluorescence. The high spectral resolution of the stepwise excitation technique permits atomic fine and hyperfine structure to be resolved for many atomic transitions, thereby allowing the depolarizing effects of such structure on coincidence signals to be experimentally eliminated. The polarization of the laser light is an additional experimental parameter not available in a conventional coincidence experiment, which enables a full measurement of atomic collision parameters to be made for a simpler experimental geometry than is the case for conventional methods. Radiation trapping effects are reduced because these are confined to the small region illuminated by the laser beam.

The stepwise coincidence method also has several drawbacks when compared with the conventional coincidence method. It is inherently less efficient than the conventional coincidence method because the laser excitation process is typically only 20% efficient. A detailed understanding of the laser excited transition is required to enable atomic collision parameters to be accurately measured. The need to use a laser adds additional complexity and cost to undertaking stepwise coincidence experiments as compared to conventional coincidence experiments.

Stepwise coincidence methods can be applied to open up new regimes of investigation, such as the collisional excitation of metastable states and higher lying transitions. Many metastable states are connected to higher lying states by allowed transitions, so that stepwise coincidence techniques offer a way of measuring atomic alignment and orientation parameters for collisionally excited metastable states. Type II[5] stepwise excitation processes in which laser excitation of the target atom to a first excited state is followed by electron excitation can provide a method for investigating excited state-excited state transitions.

ACKNOWLEDGMENTS

The financial support of the Australian Research Council is gratefully acknowledged.

REFERENCES

1. M. Eminyan, K. MacAdam, J. Slevin and H. Kleinpoppen. Phys. Rev. Lett.31:411 (1973).
2. ------------. J.Phys.B 7:1519 (1974).
3. N. Andersen, J.W. Gallagher and I.V. Hertel. Phys. Rep.1:165 (1988).
4. M.C. Standage and H. Kleinpoppen. Phys. Rev. Lett. 36:577 (1976).
5. W.R. MacGillivray and M.C. Standage. Phys. Rep. 168:1 (1988).
6. A.J. Murray, W.R. MacGillivray and M.C. Standage. Phys. Rev. Lett. 62:411 (1989).
7. A.J. Murray, C.J. Webb, W.R. MacGillivray and M.C. Standage. J.Phys.B 23:3373 (1990).
8. P.M. Farrell, W.R. MacGillivray and M.C. Standage. Phys. Rev. A. 37:1519 (1988).
9. A.J. Murray, R. Pascual, W.R. MacGillivray and M.C. Standage. J.Phys.B. 25:1915 (1992).
10. C.J. Webb, W.R. MacGillivray and M.C. Standage. J. Phys.B. 18:1701 (1985).
11. A.J. Murray, W.R. MacGillivray and M.C. Standage. Phys. Rev. A. 44:3162 (1991).
12. A.T. Masters, A.J. Murray, R. Pascual and M.C. Standage. submitted to Phys. Rev. A. (1995).
13. J.F. Williams, A.G. Mikosza, J.B. Wang and A.B. Wedding. Phys. Rev. Lett. 69: 757 (1992).
14. D.H. Madison, private communication.

THE PHOTON-PHOTON CORRELATION METHOD

James F Williams

Department of Physics
The University of Western Australia
Nedlands 6907
Western Australia

1. INTRODUCTION

This paper concerns the development of the cascading photon-photon correlation method in atomic scattering to determine information about the electron charge cloud of excited states. A brief discussion is given of earlier studies of 'two photon' correlations in atomic physics, similar earlier studies in nuclear physics and their relevance to the present topic.

The atomic studies were made in three areas of considerable physics diversity but with common experimental methods. Some examples are given here to indicate the diversity of the measurements. First, the nature of simultaneous two-photon decay was studied by observation[1] of the angular distributions of the continuum radiation from the two-photon decay of the 2^2S state of He^+ and of atomic hydrogen[2]. Secondly, for cascading radiations, the lifetimes, g-values and branching ratios[3] of atoms and the determination of the absolute quantum efficiencies of photon detectors[4] have been studied. But the main interest has been in testing Bell's inequality[5] and the interpretation of quantum mechanics, for example, by studying the directional correlations of cascading photons[6], linear polarization correlations[7,8], circular polarisation correlations[7,9] and polarization correlations using time-varying polarisers[10,11] as well as the perturbation of a polarization correlation due to an external magnetic field[12]. Detailed accounts have been given of the use of the 'two photon' correlations method and its results in testing Bell's theorem and the EPR paradox[13] and the atomic lifetimes work[3].

The above studies have a number of experimental considerations similar to those used in scattering studies particularly the selection of transitions for coincidence detection of the emitted photons and the determination of the Stokes parameters. Such measurements are obviously limited to photon energies which can be detected by single particle counting photomultipliers and suitable wavelength filters and polarisers. But it is frequently the polarisers which cause concern because of the difficulty of determining the retardation of the retarders at VUV wavelengths. Occasionally the theoretical form of the two-photon angular and polarisation correlations has been given to relate the amplitude of the correlations with the angular momentum of the states in the cascades. The properties of some of the more important transitions are given elsewhere[4,12,14-24] but several experimental aspects deserve mention here. Mercury[25] has always been an attractive atom for study because of its ease of pumping and suitable energy levels. The zero nuclear spin isotopes are attractive because of their simplified spectra. Let (J_i, J, J_f) represent the cascade in which J_i, J and J_f angular momenta of the initial, intermediate and final states. The (1,1,0) cascade[25] of ^{198}Hg from the 9^1P_1 state emits 567.6 nm and 404.7 nm photons for which the final 6^3P_0 level is not the ground state of the atom, consequently resonance trapping can be avoided and high target densities can be used. The (1,1,0) cascade of the zero nuclear spin isotope ^{200}Hg from the 7^3S_1 state emits 435.8 nm and 253.7 nm photons which have been used[26] used to indicate the merits of atomic beams, rather than a static gas, for enabling two excitation steps from the ground state to the 7^3S_1 state at physically different locations so that there are essentially no rapidly decaying states other than the cascade states in the region viewed by the photon detectors. The (0,1,0) cascade of calcium[10,27], starting from the upper $4p^2\ ^1S_0$ state with photons of wavelengths 551.3 nm and 422.7 nm, similarly shows simplified spectra for a pure sample of the isotope with zero nuclear spin and even though the initial S state is of limited interest in scattering amplitude determinations the methods have wider applicability.

For molecules the two-photon method has been applied[28] to determine the lifetime of the $a^3\Sigma_g$ state of molecular hydrogen averaged over vibrational and rotational levels. The first photon came from a number of upper molecular bands and the second photon arose from transitions from the $a^3\Sigma_g$ state to $b^3\Sigma_u$ state. The method was also used to determine the lifetime of the $B^2\Sigma^+$ state of the CN radical obtained in a dissociation process[29] using the cascade $H^2\Pi_g$ - $B^2\Sigma^+$ - $X^2\Sigma^+$ by detecting the (0,0) band at 284.3 nm and the (0,1) band at 302.6 nm as the first photon and the (0,0), (0,1) and (1,1) bands in the 359 to 387 nm range as the second photon. An average lifetime of 61.1 ± 7.6 nsec was determined. These measurements are of interest because they indicate how the two-photon technique can be extended to radicals. They also indicated that the inherent response times of electron and photon detectors and the transit times of electrons through an electron energy analyser meant that shorter lifetimes could be measured using the two photon method rather than the electron-photon delayed coincidence method.

However nuclear physics studies provide a more direct link and relevant background to the present applications of the sequential cascading photon correlations to scattering studies. Hamilton[30] appears to have initiated the study of photon-photon angular and polarisation correlations for the case of gamma rays using time-dependent perturbation theory. Subsequently extensive treatments of the theory of angular correlations of nuclear emissions have been given[31,32]. Other studies are of interest because they concern physical phenomena which have not yet been pursued with the two-photon technique in atomic physics. Some examples are the determination of arbitrary multipole moments[33] triple gamma ray detection in which the first gamma provided a quantisation axis[34] and (p,γ,γ) measurements from oriented ^{12}C where the incident proton direction provides the quantisation direction for the gamma decays[35]. The 'triple correlation of gamma ray' type of measurements under special conditions can give a test of time-reversal invariance[36]. Also external static or dynamic fields with well-defined symmetry axes cause perturbations of the correlations which allow the determination of the nuclear 'g' factor or a determination of the electric quadrupole moment. The method gives some indication of the applicability of the two-photon correlation method for atom excitation processes in the presence of external fields.

Finally, a number of the essentials for atomic physics were provided by Fry[6] who gave a general formula for two-photon coincidence rates with consideration of the effects of solid angles, plane polarisation, hyperfine structure for nuclear spins from 0 to 9/2 and anisotropic initial state populations. The principle is straight forward. The first radiation in a direction with given spin and angular momentum selects an ensemble of the excited state. Succeeding radiations then show an angular or polarisation correlation, with respect to the initial ensemble, as a function of the angle between the two photons which reflects the angular momentum of the states in the cascade. However that work did not initiate subsequent work in scattering studies presumably because the relationship of the correlations to scattering amplitudes and phases was not indicated.

2.   ATOMIC CORRELATIONS

The particle correlations measurements of atomic physics have their foundations in the work of Fano[37]. He showed how a coincidence experiment, in which two or more of the scattered or emitted particles from a collision complex are detected in time coincidence, selects a given ensemble of atoms and specifies a unique quantum state of the excited atom, that is with specific amplitudes and phases. These quantities, or equivalently the multipole moments, the density matrix or the expectation values of components of angular momentum of the excited state, can be deduced from the polarization and directional correlations of the outgoing particles. After the theoretical developments of Rubin et al[38], Macek and Jaecks[39], Wykes[40] and Fano and Macek[41], the first successful experiments were carried out by Kleinpoppen and colleagues[42]. The scattered electron- radiated photon correlation measurements arose from the above work because it was able to determine excitation amplitudes and relative phases, and hence their interference or coherence parameters, rather than moduli-squared excitation amplitudes, ie cross sections, for a particular electron scattering angle. The essence of subsequent work can be deduced from the reviews, for example, of Andersen et al[43] and Bartschat[44], for example, and the recent successes of the convergent close coupling calculations[45]. A physical picture of the electronic charge cloud of the excited state has emerged describing the height, width, length and the angle of the axis of the charge cloud with respect to the quantisation axis (the alignment), and the internal motion (orientation) of the charge cloud. The topical review by Hippler[46] gives an excellent overview of basic ideas with particular application to few-electron atom-atom collisions.

Current interest is moving to higher angular momentum states, excited triplet states and the role of various spin effects, such as exchange, spin-orbit interaction in the excitation process, particularly to test theoretical treatment of target and continuum electron effects. The experimental limitations of the electron-photon technique become apparent when measurements are attempted for states with angular momentum greater than unity; then the separation of a given energy level becomes increasingly difficult and as the electron energy resolution is improved, the count rate decreases and the statistical accuracy decreases. For D states, the complete determination of the amplitudes and phases requires a triple (e,γ_1,γ_2) coincidence measurement with obvious difficulties. Subsets of this triple measurement are (e,γ_1) and (e,γ_2) measurements, which can only determine up to rank $K=2$ state multipoles because of detection of a single dipole photon, and (γ_1,γ_2) measurements with cylindrical symmetry average over the electron scattering angles and hence only allow determination of the zero component of the multipoles with rank up to four. The observation of the scattered energy-loss electron in coincidence with the second cascade photon has limitations arising from branching ratios and excited state lifetimes. Some of these difficulties for 3D state excitation of atomic hydrogen and helium have been addressed[47]. The development of the sequential cascading photon-photon correlation method arose from its attraction as a more accessible subset of the triple coincidence measurements and it has since been shown to provide data not accessible by other techniques. The current status and some recent results are now described.

(a) Atomic hydrogen

Triple coincidence measurements on an excited 3D state access up to rank 4 state multipoles and can provide a complete description of this excited state. Heck and Gauntlett[48] derived the triple (e,γ_1,γ_2) correlation expression where the sequential cascading Balmer-alpha photon and Lyman-alpha photon are detected in coincidence with the scattered n=3 energy loss electron. In the initial two-photon (γ_1,γ_2) studies in atomic hydrogen[49], angular correlation studies were made because the use of a reflection type polariser needed to analyse the n=2→1 (121.6 nm) photons could be avoided. The photon detectors for the n=3→2 (656.2 nm) and n=2→1 (121.6 nm) photons were located in the plane defined by the incident electron beam and the first of the cascading photons and the incident beam direction provided a quantisation axis. For that geometry the radiated photon-photon (normalised) intensity is given by

$$N(\beta,\theta) = <T(0)_{00}^+>[1 + \tfrac{1}{6}P_2(\cos(\theta - \beta))] + <T(2)_{00}^+> [\sqrt{5} + \frac{1}{12\sqrt{5}}P_2(\cos(\theta - \beta))]$$

$$+ <T(2)_{20}^+> [\frac{19\sqrt{14}}{100}P_2(\cos\beta) + \frac{37\sqrt{14}}{300}P_2(\cos\theta) - \frac{22\sqrt{5}}{300}\Lambda(2;\beta,\theta)]$$

$$+ <T(2)_{40}^+> \tfrac{9}{10}\Lambda(4;\beta,\theta) \qquad (1)$$

where $\Lambda(l;\beta,\theta)] = \dfrac{4p}{\sqrt{5(2l+1)}} \sum_m \begin{pmatrix} l & 2 & 2 \\ m & -m & 0 \end{pmatrix} Y_{lm}(\beta,0)Y_{2m}(\theta-\beta,0)$

and β and θ are the angles of the first and second photons with respect to the quantisation axis. The measurements of the sequential cascading photons and their angular correlations were difficult because of the nearly degenerate $3^2S_{1/2}$, 3^2P_j and 3^2D_j states whose decay radiations to the n=2 levels cannot be separated by convenient optical filters. However their lifetimes [$t(3^2S) = 158$, $t(3^2P) = 5.3$ and $t(3^2D) = 15.5$ nsec] are sufficiently different, and longer than instrumental timing resolutions of the order of 1 nsec, that they can in principle be separated under appropriate conditions. Fig. 1 shows the essential features of the sinusoidal $(\theta-\beta)$ correlation for an incident electron energy of 290 eV. Data are shown for $\beta=40°$ and $90°$, plotted as $\theta_{12}=(\beta-\theta)$ and normalised to the $\beta=40°$ and $\theta_{12}=180°$ datum point. For a fixed β the correlation is of the form $I(\beta,\theta) = a + b\cos\theta_{12}$ and the effect of a non-zero value of β is to reduce the amplitude of the sinusoid. The displacement of the correlations from the incident beam direction by the angle of the first photon detector is not shown because the data are plotted as a function of the angular difference of the two detectors.

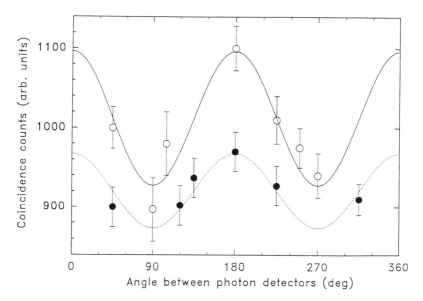

Fig. 1. The angular correlation of the (656.2 nm) photon - (121.6 nm) photon coincidence count rate as a function of the angle between the two photon detectors for B = 40^0 (open circles) and B=90^0 (closed circles). The lines-of-best-fit to the data are of the form (a + b cosθ).

A least squares regression of the data to the above equation determined the three $<T(2)>$ state multipoles which were normalised by setting $T(0)_{00}^+$ equal to the Born value at 290 eV of 0.0128 ± 26, to give the values $<T(2)_{00}> = 0.0037 \pm 15$, $<T(2)_{20}> = 0.0032 \pm 6$ and $<T(2)_{40}> = -0.0005 \pm 29$. The experimental uncertainties are one standard deviation. The multipoles are related to the cross sections σ_m for exciting the 3d magnetic sublevel m by

$<T(0)_{00}^+> = \sigma_{3s}; \quad <T(2)_{00}^+> = (1/\sqrt{5})[2\sigma_2+2\sigma_1+\sigma_0];$

$<T(2)_{20}^+> = (\sqrt{2}/\sqrt{7})[2\sigma_2-\sigma_1-\sigma_0]$ and $<T(2)_{40}^+> = (\sqrt{2}/\sqrt{35})[\sigma_2-4\sigma_1+3\sigma_0]$ (2)

from which the values are readily deduced for $\sigma_{3d_0} = -0.0005\pm20$, $\sigma_{3d_1} = 0.0010\pm20$ and $\sigma_{3d_2} = 0.0033\pm5$ a.u., using the normalisation $\sigma_{3s} = 0.0128\pm26$ a.u. The first Born values for the above quantities are $<T(2)_{k0}>$ are 0.00430, 0.00160, 0.000428 for k=0,2,4 and for the σ_{3dm} are 0.00138, 0.00129 and 0.00283 a.u. for m=0,1,2 respectively. The large experimental uncertainties are not unexpected for the incident electron energy of 290 eV where the cross sections are small, however they limit comparison with other theories. Nevertheless the data are adequate to indicate the potential of the two photon method for extensive measurements at lower energies where the data of Mahan et al[50] of the cross sections for the separated 3S, 3P and 3D states, obtained with a non-coincidence but time-modulated beam method, can be extended.

(b) Helium

The two-photon technique has been applied more successfully[51] to the 3^1D state of helium by observing the polarisation correlation of the first photon (667.8 nm) in coincidence with the unpolarised second photon (58.4 nm). This approach was indicated because the polarisation state of the first photon can be more readily determined than that of the second photon and the circular polarisation can be readily measured for a full determination of the Stokes parameters. For a geometry in which the first photon detector is located normal to the scattering plane and θ_{12} is equal to the difference between the azimuthal angles of the two photon detectors, the unpolarised coincidence intensity is given by

$I = \Sigma_k (c_k - d_k \cos2\theta_{12}) T(2)_{k0},$

the Stokes parameter P_i are given by $IP_2 = 0$, $IP_3 = 0$ and

$IP_1 = \Sigma_k (a_k - b_k \cos2\theta_{12}) T(2)_{k0}$

where a,b,c and d are constants for a given rank k multipole. The $T(2)_{k0}$ (k=0,2,4) multipoles are deduced from the sinusoidal correlation but a more accurate value of $<T(2)_{20}>$ can be deduced from the non-coincident photon signal. The non-coincident intensity $I^{(\gamma_1)}$ and the linear polarisation $P^{(\gamma_1)}$ are defined by

$I^{(\gamma_1)}P^{(\gamma_1)} = -\frac{3\sqrt{14}}{8}<T(2)_{20}>$ and $I^{(\gamma_1)} = \sqrt{5} <T(2)_{00}> - \frac{\sqrt{14}}{8}<T(2)_{20}>$

The data were normalised by measuring the ratio $\sigma(3^1S) / \sigma(3^1D) = 0.66$ and comparing it with the recommended value of 0.64 of de Heer et al[52]. Then the cross sections σ_m for exciting the 3d magnetic sublevel m are determined using the same relations as given for hydrogen. An example of the data obtainable from such work is shown in fig 2 which indicates the changing role of the angular momentum transfer in the collision process and the appropriateness of the theoretical models. The success of the convergent close coupling model[45] is readily attributed to inclusion of coupling to all states while other theories include coupling to fewer states. The calculations which predict larger values for σ_2 generally overestimate the contribution from that part of the electron charge cloud which is aligned perpendicular to the direction of the incident electron and which becomes excited in those collisions that involve large momentum transfers. Further details are given elsewhere[51]. In summary, these measured values of σ_m cannot be obtained from electron-photon correlations and indicate the value of the two-photon correlations method.

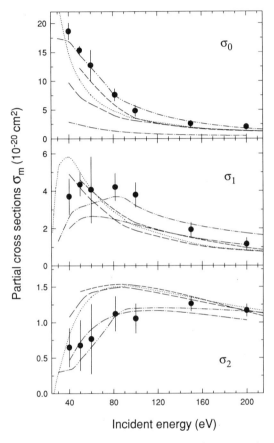

Fig. 2. The partial cross sections σ_m for excitation of the magnetic sublevels m of the 3^1D state of helium. The experimental data are shown as full circles, the calculated values are from convergent close coupling (dash-two dots), multichannel eikonal (dots), 10-state eikonal (long dash), 22-state diagonalisation, and distorted-wave Born approximation methods as discussed in detail elsewhere[51].

3. DISCUSSION AND CONCLUSION

The above studies indicate that the photon-photon technique offers some experimental advantages. First for every coincidence measurement the lifetime of the middle state can be automatically determined although this has usually not been done. For the, say, nine measurements required for an angular or polarisation correlation, an accurate lifetime should be determined. Secondly, a significant experimental advantage is that the solid angle of the photon detectors can be at least an order of magnitude larger than the electron detector acceptance solid angle. Of necessity the scattered electron energy loss analyser must have a solid angle small enough to specify the scattering angle with little ambiguity and this usually means within about one degree especially for small scattering where the scattering cross section is usually large and changing rapidly with angle. By contrast, integrating the photon polarisation over an acceptance angle as large as 60 degrees still allows useable Stokes parameters to be extracted from the signals. Thirdly there is always some difficulty in separating higher states by their electron energy loss but usually it will be easier to separate both photons with wavelength filters.

Finally, the quantum efficiencies of both photon detectors can be determined. This is most advantageous if absolute cross section values are of interest and are being determined in conjunction with other detectors. Consider the source of N atoms per second decaying in the cascade, the number of true coincidences per second is $N_c = \zeta \, \varepsilon_1 \, \varepsilon_2 \, N$ where ε_i are the effective quantum efficiencies of the ith detector and ζ is the maximum fraction of the photons which can be observed for the given solid angles of the detectors. The number of photons detected from the first transition is $N_1 = 0.5(1-\cos\theta_1)\varepsilon_1 N$ and the effective quantum efficiency of the second detector is $\varepsilon_2 = 0.5(1-\cos\theta_1)N_c / \zeta N_1$. Considering only photons originating from the upper state, the effective quantum efficiency of the first photon detector is given by $N_2 = 0.5 \, (1-\cos\theta_2)\varepsilon_2 N$. Then $\varepsilon_1 = 0.5(1-\cos\theta_2)N_c / \zeta N_2$. Here the quantum efficiency is the probability of obtaining an output electron pulse from the photomultiplier for each incident photon. Also ζ is given[6] by

$$\zeta = 0.25(1-\cos\theta_1)(1-\cos\theta_2) - (0.0625) \, A \, \sin^2\theta_1 \sin^2\theta_2 \cos\theta_1 \cos\theta_2$$

and the factor A, a constant for each transition is readily calculated[53] using an algebraic package such as Mathematica.

The use of an external field to mix even and odd parity states[54] is challenging. For photon-photon coincidences there is the considerable simplification that the energy and angle of scattering of the energy loss electron do not have to be determined. The interferences of the even and odd parity states can then be explored for $n \geq 3$ states.

Several recent studies have indicated excitation processes for possible two-photon measurements. The measurement[55] of the integrated (over scattering angle) Stokes' parameters for the excitation by spin-polarised electrons of the $2p^53p(^3D_3)$ state of neon indicate that spin-orbit interaction during the excitation can be neglected, just as for the excitation of the lightest inert gas atom, helium. Also Baerveldt et al[56] have determined the natural parameters for higher excited states of neon, such as the 3p'[3/2]2, 3p[3/2]2, 3p[5/2]2, 3p[5/2]3 states. ie for J=2 and 3 states. They showed that the target parameter L$_{perp}$ starts at a negative value as a function of electron scattering angle, changes sign between 40 and 70 degrees and that this behaviour is different from that for excitation of the He*(1snp), He*(1s3d) and Ne*($2p^53s$) states at relatively low collision energies. Whether the J=2 states are excited mainly through their singlet component without exchange or the J=3 state can only be excited by electron exchange could be further explored using the two photon polarisation correlation measurements for those states for which cascading photons can be found.

Finally, a theoretical description of the electron-impact excitation of the singlet-triplet mixed 1s4f state of helium has been given[57]. Using the first-order DWBA method for an incident electron energy of 100 eV it was found that the intensity and polarisation of the first photon from the 1s4f to 3^1D transition was time modulated due the singlet and triplet mixing in the 1s4f state and that quantum beats appeared in the time spectra of the cascade photon from the subsequent 3^1D - 2^1P transition. Detection of the first photon at 1870 nm is not possible and triple (e,γ_2,γ_3) will be quite difficult, so initial studies will concentrate on the two-photon (γ_2,γ_3) measurements as a means of deducing magnetic sublevels excitation cross sections and the higher rank, zero component multipoles.

ACKNOWLEDGEMENTS

This research was supported by the Australian Research Council and the University of Western Australia.

References

1. N. Lipeles, R. Novak, and N. Tolk, Phys. Rev. Lett. 15:690 (1965).
2. A.J. Duncan and H. Kleinpoppen, Ch. 7 in "Quantum Mechanics versus Local Realism" F. Selleri, ed., Plenum Press, New York (1988).
3. R.E. Imhof and F.H. Read, Rep. Prog. Phys. 40:1(1975)
4. F. Cristofori, P. Fenici, G.E. Figerio, and P.G. Sona, Phys. Lett. 6:171(1963).
5. J.S. Bell, Physics. 1:195 (1964).
6. E.S. Fry, Phys Rev. A8:1219 91973); Phys. Rev. Lett. 37: 465 (1976).
7. J.F. Clauser, Phys. Rev. Lett., 37: 1223 (1976); Nuovo Cimento B33: 740 (1976).
8. J.F. Clauser and A. Shimony, Rep. Prog. Phys. 41:1881 (1978).
9. F. Falciglia, A. Garuccio, G. Iaci, and L. Pappalardo, Lett. Nuovo Cim. 38: 52(1983)
10. A. Aspect, P. Grangier, and G. Roger, Phys. Rev. Lett. 49:91(1982).
11. A. Aspect, P. Dalibard and G. Roger: Phys. Rev. Lett., 49:1804 (1982).
12. A.M. Dumont, Thése de Doctorat, Paris, (1970).
13. D. Home and F Selleri, Rivista del Nuovo Cimento 14:1 (1991).
14. E. Brannen, F.R. Hunt, R.H. Adlington, and R.W. Nichols, Nature 175:8180 (1955).
15. R.D. Kaul, J. Opt. Soc. Am. 56:1262 (1966).
16. C. Camhy-Val and A.M. Dumont, Astron. and Astrophys. 6:27 (1970).
17. J. Paridié, Thése de Doctorat, Bordeaux, (1967).
18. C.A. Kocher and E.D. Commins, Phys. Rev. Lett. 18: 575 (1967).
19. C.A. Kocher, Ph.D.Thesis (Uni. of California, Berkeley) 56 (1967).
20. G.H. Nussbaum and F.M. Pipkin, Phys. Rev. Letters. 19: 1089 (1967).
21. C. Camhy-Val and A.M. Dumont, Compt. Rend. 267 B: 689 (1968).
22. C. Camhy-Val, A.M. Dumont, M. Dreux, and R. Vitry, Phys. Letters. 32A:233 (1970).

23. M. Popp, G. Schäfer, and E. Bodenstedt, Z. Physik. 240:71 (1970).

24. A.M. Dumont, C. Camhy-Val, and R. Vitry, Compt. Rend. 271B:1021 (1970).

25. R.A. Holt and F.M. Pipkin. University of Harvard Preprint (1974).

26. E.S. Fry and R.C. Thompson, Phys. Rev. Lett. 37:465 (1976).

27. S.J. Freedman and J.F. Clauser, Phys Rev. Lett 28: 938 (1972).

28. K.A. Mohamed, G.C. King, and F.H. Read, J. Phys. B9:3159 (1976).

29. K.A. Mohamed, G.C. King, and F.H. Read, J. Elect. Spect. Rel. Phen. 12:229 (1977).

30. D.R. Hamilton, Phys. Rev. 58:122 (1940).

31. S. Devons and L.J. Goldfarb in "Encyclopedia of Physics", S. Flugge, ed., (Springer, Berlin) 42:362 (1957).

32. H. Frauenfelder and R.M. Steffen, in "Alpha, Beta and Gamma Ray Spectroscopy", K. Seigbahn, ed., North Holland, Amsterdam, (1965).

33. D.L. Falkoff and G.E. Uhlenbeck, Phys. Rev. 79:323 (1950).

34. L.C. Biedenharn and M.E. Rose, Rev. Mod. Phys. 25:729 (1953).

35. L.C. Biedenharn, G.B. Arfken, and M. E. Rose, Phys Rev. 83:586 (1951).

36. V. de Sabbata, Il Nuovo Cimento, Ser 10, 21:659, 1058 (1961)

37. U. Fano, Rev. Mod. Phys. 29:74 (1957).

38. K. Rubin, B. Bederson, M Goldstein, and R.E. Collins, Phys. Rev.182:201 (1969).

39. J. Macek and D.H. Jaecks, Phys. Rev. A4:1288 (1991).

40. J. Wykes, J. Phys. B5:1126 (1972).

41. U. Fano and J.H. Macek, Rev. Mod. Phys. 45:553 (1973).

42. M. Eminyan, K.B. MacAdam, J. Slevin, M. Standage, and H. Kleinpoppen, J. Phys. B7:1519 (1974).

43. N. Andersen, J.W. Gallagher, and I.V. Hertel, Phys. Rep. 165:1(1988).

44. K. Bartschat, Comments At. Mol. Phys. 27:239 (1992); Phys. Rep. 180:1 (1989).

45. I. Bray and A.T. Stelbovics, Phys. Rev. A48:4787 (1992).

46. R. Hippler, J. Phys. B26:1 (1993).

47. S. Chwirot and J. Slevin, Comments At. Mol. Phys. 26:11 (1991).

48. E.L. Heck and J.P. Gauntlett, J. Phys. B19:3633 (1986).

49. J.F. Williams, M. Kumar, and A.T. Stelbovics, Phys. Rev. Lett. 70:1024(1993).

50. A.H. Mahan, A. Gallagher, and S.J. Smith, Phys. Rev. A13:156 (1976).

51. A.G. Mikosza, R. Hippler, J. Wang, and J.F. Williams, Phys. Rev. Lett. 71 235(1993).

52. F.J. deHeer, R.Hoekstra, A.E.Kingston, and H.P. Summers, Nucl. Fusion.3: 19(1992).

53. J. Wang and J.F. Williams, Comput. Phys. Commun. 75:275 (1993).

54. E.L. Heck and J.F. Williams, J. Phys. B20:2871 (1987).

55. J.E. Furst, W. Wijayaratna, D.H. Madison, and T.J. Gay, Phys. Rev. A47:3775 (1993).

56. A.W. Baerveldt, J. van Eck, and H.G.M. Heideman, J Phys B27:1857 (1994).

57. J. B. Wang, J.F. Williams, A.T. Stelbovics, J.E. Furst, and D.H. Madison, accepted Phys Rev A (1995)

ANALYSIS OF SCATTERING AND EXCITATION AMPLITUDES IN POLARIZED-ELECTRON-ATOM COLLISIONS FOR LIGHT ONE-ELECTRON ATOMS

Hans Kleinpoppen

Atomic Physics Laboratory
University of Stirling
Stirling, Scotland.

ABSTRACT

An experimental analysis is proposed to extract partial differential cross-sections for elastic scattering and for excitation of the ^2P states of light alkali metal atoms. Relevant data for the ratios of direct, exchange and interference cross sections to the differential cross section for the elastic case and the inelastic case of ^2P excitation of scattering partially spin polarized electrons on partially polarized light one-electron atoms are linked to the degrees of polarization of the electrons and atoms before and after the scattering process. An example of a five-state close coupling approximation is given for the above ratios of elastic electron-lithium scattering which illustrate the richness of the varieties of structure in these partial differential cross sections.

INTRODUCTION

As the techniques of producing and handling polarized electron and polarized atom beams are becoming more sophisticated it is possible to study atomic collision processes in greater depth and detail.[1] The application of electron-photon angular and polarization correlations allows for extracting information on state parameters, on magnetic sub-level cross sections and, in certain cases, on phases; these data are usually averaged over the spin states of the projectile and the target. However, recent theoretical framework has provided us with sufficient data predicting possible outcome of a collision process in detail. Such data decompose differential cross-sections into partial differential cross sections for excitation of different orbital angular momentum states of the target. These partial differential cross-sections are further broken up in terms of physically meaningful spin dependent processes in electron-atom collisions,[2] for instance for elastic scattering of polarized electrons on polarized light one-electron atoms (i.e. neglecting spin-orbit interactions in the collisional interaction), we have for both spin-polarizations parallel to the incident electron direction

(i)	Direct Scattering	$e\uparrow + A\downarrow \rightarrow e\uparrow + A\downarrow$	
		$e\downarrow + A\uparrow \rightarrow e\downarrow + A\uparrow$,	$\sigma^d = |f|^2$

(ii)	Exchange Scattering	$e\uparrow + A\downarrow \rightarrow e\downarrow + A\uparrow$	
		$e\downarrow + A\uparrow \rightarrow e\uparrow + A\downarrow$,	$\sigma^{ex} = |g|^2$

iii)	Interference Scattering	$e\uparrow + A\uparrow \rightarrow e\uparrow + A\uparrow$	
		$e\downarrow + A\downarrow \rightarrow e\downarrow + A\downarrow$.	$\sigma^{int} = |f\text{-}g|^2$.

This arrangement[3] makes the collision process quite transparent with respect to the spin states of the projectile and the target (for light spin ½ targets).

In this paper we take up the case study of light spin ½ atomic targets. The light alkali atoms are quite suitable for practical reasons.[2,4] In this case it is a reasonable approximation to neglect the spin-orbit interaction and hyperfine coupling during the short time of collision. It means that exchange degeneracy due to the indistinguishability of the target and projectile electrons plays by far the dominant role in altering spin polarizations of the projectile and the target for collisions at low energies.

The spin analysis of the scattered target atoms, when they decay to the ground state, is affected by the fine- and hyperfine-structure splittings, the finite width of the excited states and the rules governing these transitions. These complications, however, do not influence the scattered electrons since the collision time is too small to have set up magnetic and spin-orbit interactions in the atom or between the atom and the projectile electrons. For these considerations, our *spin analysis* is based on *spin-measurement on the scattered electrons*.

Elastic Collisions

If a beam of partially polarized electrons of polarization P_e is scattered on partially polarized spin half targets (polarization P_a parallel to P_e) of low atomic number the differential cross section for elastic scattering for $P_a = 0 = P_e$ is

$$\sigma(E,\theta) = \tfrac{1}{2}(\,|f(E,\theta)|^2 + |g(E,\theta)|^2 + |f(E,\theta) - g(E,\theta)|^2)$$

$$= \tfrac{1}{2}(\sigma^d + \sigma^{ex} + \sigma^{int}) \,. \qquad (1)$$

If

$$S = \sigma_e\uparrow + \sigma_e\downarrow = \sigma_a\uparrow + \sigma_a\downarrow$$

is the sum of the up and down spin-polarized electrons of atom cross-sections, and

$$M_e = \sigma_e\uparrow - \sigma_e\downarrow \,,\; M_a = \sigma_a\uparrow - \sigma_e\downarrow$$

are the relevant up-down asymmetries then obviously the respective polarizations after scattering are:

$$P_{\acute{e}} = \frac{\sigma_e\uparrow - \sigma_e\downarrow}{\sigma_e\uparrow + \sigma_e\downarrow} = \frac{M_e}{S}$$

and

$$P_{\acute{a}} = \frac{\sigma_a\uparrow - \sigma_a\downarrow}{\sigma_a\uparrow + \sigma_a\downarrow} = \frac{M_a}{S} \;;$$

one can calculate the polarization of the atoms and electrons after the scattering to:

$$P_a' = \frac{\sigma(P_e+P_a)(P_eP_a-1)+S(P_e+P_a) - SP_eP_aP_e'}{SP_eP_a} \qquad (2)$$

$$P_a' = \frac{\sigma(P_e+P_a)(P_eP_a-1)+S(P_e+P_a) - SP_eP_aP_a'}{SP_eP_a} \qquad (3)$$

These relations are, of course, not valid for $P_e = 0$ or $P_a = 0$ or both $P_e = 0 = P_a$. Obviously using eqs. (2) and (3) one can restrict oneself to one of the measurements for P_a' or P_e' after scattering.

The ratios of the partial differential cross-sections to the differential cross section turn

$$\frac{\sigma^d}{2\sigma} = \frac{1}{2} - \frac{S/\sigma}{4}\left[\frac{P_e'+P_a'}{P_e+P_a} - \frac{P_e'-P_a'}{P_e-P_a}\right]$$

$$= \frac{1}{4(P_a-P_e)}\left[\frac{S}{\sigma}(\frac{1}{P_a}-P_e')-\frac{1}{P_a}+P_a\right] \qquad (4)$$

$$= \frac{1}{(P_e-P_a)}\left[\frac{S}{\sigma}(\frac{1}{P_e}-P_a')-\frac{1}{P_e}+P_e\right]$$

$$\frac{\sigma^{ex}}{2\sigma} = \frac{1}{2} - \frac{S/\sigma}{4}\left[\frac{P_e'+P_a'}{P_e+P_a} + \frac{P_e'-P_a'}{P_e-P_a}\right]$$

$$= \frac{1}{2(P_e-P_a)}\left[\frac{S}{\sigma}(\frac{1}{P_e}-P_e')-\frac{1}{P_e}+P_e\right]$$

$$= \frac{1}{2(P_a-P_e)}\left[\frac{S}{\sigma}(\frac{1}{P_a}-P_a')-\frac{1}{P_a}+P_a\right] \qquad (5)$$

and

$$\frac{\sigma^{int}}{2\sigma} = \frac{S}{2\sigma}\left(\frac{P_e' + P_a'}{P_e + P_a}\right) = \frac{1}{2}\left[S/\sigma - \frac{(1-P_aP_e)}{P_eP_a}\right] = \frac{1}{2} - \frac{1}{2}\left[\frac{1-(S/\sigma)}{P_eP_a}\right].$$

Another useful quantity which gives the phase between f and g is

$$\frac{Re(fg^*)}{\sigma} = \frac{1}{P_aP_e}(1 - \frac{S}{\sigma}) \qquad (6)$$

If we put $f = |f|e^{i\gamma_f}$ and $g = |g|e_{i\gamma_g}$, then for elastic as well as inelastic scattering we obtain

$$\delta_{fg} = \gamma_g - \gamma_f = \cos^{-1}\left[\frac{1}{2P_eP_a}\left(\frac{1 - S/\sigma}{\sqrt{\dfrac{\sigma^d\sigma^{ex}}{2\sigma}}}\right)\right] \qquad (7)$$

Theoretical data for these quantities are expected to show an interesting interference structure in their angular distribution. Following the above scheme these quantities are measurable and may prove to be a substantial test ground for theoretical predictions.

Polarization Analysis of the Scattered Electrons and Recoiling Atoms

For partially polarized electrons and partially polarized atoms the scattering process is symbolically written as

$$e(\uparrow\downarrow) + A(^2S,\uparrow\downarrow) \rightarrow e(\uparrow\downarrow) + A^*(^2P,\uparrow\downarrow) \; ;$$

the arrows of different lengths indicate a measure of the partial polarization.

We characterise the excited states $^2P_{1/2,3/2}$ of the atom by the quantum numbers m_s and m_ℓ since the collision time ($\sim 10^{-6}$ns) is extremely short compared to the spin-orbit relaxation time ($\sim 10^{-3}$ns) for light alkalis even at low energy scattering. Therefore, spin-orbit coupled states need not be considered during the collisional excitation so that the excited atom is first described by the state vector $|s, \ell, m_\ell, m_s>$. On the other hand for all processes that occur after a time larger than $\sim 10^{-3}$ns emission of impact radiation during the lifetime of ~ 10 ns) the spin-orbit coupling causes fine structure splitting into $^2P_{1/2}$ and $^2P_{3/2}$ states. For this reason we have to work in a $|j, m_j>$ representation after the excitation. The state vector $|j, m_j>$ can be expressed as a superposition of states $|m_s, m_\ell >$ as

$$|j,m_j> \;=\; \sum_{m_\ell m_s} C(s)(\ell\, m_s m_\ell\, jm_j)\,|m_s, m_\ell > \; .$$

With these considerations and taking due care of all the composite interaction processes (i.e. direct, exchange and interference interactions) with their appropriate amplitudes f_m and g_m for the magnetic sublevels with $m_\ell = 0, \pm 1$ (i.e., $f_0, g_0, f_0\text{-}g_0$ and $f_1, g_1, f_1\text{-}g_1$) the differential cross-section results in

$$\sigma(S \rightarrow P) \;=\; [\sigma_o(S \rightarrow P) + 2\sigma_1(S \rightarrow P)] \tag{8}$$

In the following analysis all these quantities implicitly refer to the process $^2S \rightarrow {}^2P$ where

$$\sigma(S \rightarrow P) = 1/2 \left[(|f_o|^2 + |g_o|^2 + |f_o\text{-}g_o|^2) + 2(|f_1|^2 + |g_1|^2 + |f_1\text{-}g_1|^2) \right]\frac{k'}{k} \; ; \tag{9}$$

k is the wave vector before and k' after the scattering. Putting

$$
\begin{aligned}
\sigma^d &= (\sigma_o^d + 2\sigma_1^d) = (|f_o|^2 + 2|f_1|^2)\frac{k'}{k} \\[4pt]
\sigma^{ex} &= (\sigma_o^{ex} + 2\sigma_1^{ex}) = (|g_o|^2 + 2|g_1|^2)\frac{k'}{k} \\[4pt]
\sigma^{int} &= (\sigma_o^{int} + 2\sigma_1^{int}) = (|f_o - g_o|^2 + 2|f_1 - g_1|^2)\frac{k'}{k}
\end{aligned}
\tag{10}
$$

(8) becomes

$$2\sigma = \sigma^d + \sigma^{ex} + \sigma^{int} \tag{11}$$

The cross-section for electrons with spin-up after scattering is

$$4\sigma^\uparrow_e = (1+P_e)(1-P_a)\sigma^d + (1-P_a)(1-P_e)\sigma^{ex} + (1+P_e)(1+P_a)\sigma^{int} \tag{12}$$

and similarly for spin-down scattered electrons is

$$4\sigma^\downarrow_e = (1+P_a)(1-P_e)\sigma^d + (1+P_e)(1-P_a)\sigma^{ex} + (1-P_e)(1-P_a)\sigma^{int} \tag{13}$$

Corresponding relations for the scattered atoms are:

$$4\sigma_a^\uparrow = (1+P_a)(1-P_e)\sigma^d + (1+P_e)(1-P_a)\sigma^{ex} + (1+P_e)(1+P_a)\sigma^{int} \tag{14}$$

and

$$4\sigma_a^\downarrow = (1+P_e)(1-P_a)\sigma^d + (1+P_a)(1-P_e)\sigma^{ex} + (1-P_a)(1-P_e)\sigma^{int} \tag{15}$$

It is trivial to verify that

$$S = S_e = \sigma_e^\uparrow + \sigma_e^\downarrow = S_a = \sigma_a^\uparrow + \sigma_a^\downarrow = (\sigma - P_e P_a \eta) \tag{16}$$

where

$$2\eta = \sigma^d + \sigma^{ex} - \sigma = 2[Re(f_o^* g_o + Re(f_1^* g_1]\frac{k'}{k}, \tag{17}$$

$$\eta_o = \frac{k'}{k} Re(f_o^* g_o); \quad \eta_1 = Re(f_1^* g_1)\frac{k'}{k} \quad .$$

Note that for $P_e = 0$ or $P_a = 0$ or $P_e = P_a = 0$, we get $\sigma = S$ \hfill (18)

The spin up-down asymmetry for the scattered electrons and atom is given by

$$\sigma_e^\uparrow - \sigma_e^\downarrow = M_e = (P_e + P_a)\sigma - P_a\sigma^d - P_e\sigma^{ex} \tag{19}$$

$$\sigma_a^\uparrow - \sigma_a^\downarrow = M_a = (P_e + P_a)\sigma - P_e\sigma^d - P_a\sigma^{ex} \quad , \tag{20}$$

respectively.

Adding and subtracting (19) and (20) we get:

$$\sigma^d + \sigma^{ex} = 2\sigma - (\frac{M_e + M_a}{P_e + P_a}) \tag{21}$$

and

$$\sigma^d - \sigma^{ex} = \frac{M_e - M_a}{P_e - P_a} \tag{22}$$

Since, by definition, the polarizations after scattering are

$$P_e' = \frac{M_e}{S} \qquad P_a' = \frac{M_a}{S}$$

eqs. (21) and (22) give

$$\frac{\sigma^d}{2\sigma} = \frac{1}{2} - \frac{1}{2}(S/\sigma) \left[\frac{P_e P_a' - P_e' P_a}{P_e^2 - P_a^2} \right] \qquad (23)$$

and

$$\frac{\sigma^{ex}}{2\sigma} = \frac{1}{2} - \frac{1}{2}(S/\sigma) \left[\frac{P_e P_e' - P_a P_a'}{P_e^2 - P_a^2} \right] . \qquad (24)$$

Furthermore from (9) and (10), we obtain

$$\sigma^{int} = 2\sigma - (\sigma^d + \sigma^{ex}) = 2\sigma - 2\sigma + S \left(\frac{P_e' + P_a'}{P_e + P_a} \right) = S \left(\frac{P_e' + P_a'}{P_e + P_a} \right)$$

or

$$\frac{\sigma^{int}}{2\sigma} = (S/2\sigma) \left(\frac{P_e' + P_a'}{P_e + P_a} \right) \quad ; \qquad (25)$$

comparing (16) and (17) and using (25) we get

$$S = \sigma \frac{(P_e P_a - 1)(P_e + P_a)}{P_e P_a (P_e' + P_a') - (P_e + P_a)} \qquad (26)$$

or

$$P_a' = \frac{\sigma(P_e + P_a)(P_e P_a - 1) + S(P_e + P_a) - S P_e P_a P_e'}{S P_e P_a} \qquad (27)$$

and

$$P_e' = \frac{\sigma(P_e + P_a)(P_e P_a - 1) + S(P_e + P_a) - S P_e P_a P_a'}{S P_e P_a} \qquad (28)$$

Eqs. (27) and (28) are of considerable importance with regard to the experimental situation, e.g., if we start with both electrons and target atoms partially polarized, then the two polarizations after scattering are correlated so that spin-polarization measurements are only necessary on any one of the scattered particles. This gives us a greater degree of choice suitable for the experimental situation. For example if we eliminate P_a' in our expressions we get:

$$\frac{\sigma^d}{2\sigma} = \frac{1}{2} - \frac{1}{2}\left[\frac{(1 - P_e P_a) + (\frac{S}{\sigma})(P_e' P_a - 1)}{P_a(P_a - P_e)}\right] \tag{29}$$

$$\frac{\sigma^{ex}}{2\sigma} = \frac{1}{2} + \frac{1}{2}\left[\frac{(1 - P_e P_a) + (\frac{S}{\sigma})(P_e' P_e - 1)}{P_e(P_a - P_e)}\right] \tag{30}$$

$$\frac{\sigma^{int}}{2\sigma} = \frac{1}{2}\left[\frac{(\frac{S}{\sigma}) - (1 - P_e P_a)}{P_e P_a}\right] = \frac{1}{2} - \frac{1}{2}\left[\frac{1 - (\frac{S}{\sigma})}{P_e P_a}\right] \tag{31}$$

Similar relations can be developed by eliminating P_e' instead of P_a' . Note that (27) and (28) break down for $P_a = 0$ or $P_e = 0$ or both $P_a = P_e = 0$ and (23), (24) and (25) must be used, with the obvious disadvantage of having to measure both P_e' and P_a' after scattering.

It is therefore reasonable in practice to start with both electrons and atoms partially polarized and measure the polarization P_e' of the scattered electrons.

It is obvious that by measuring the spin polarizations P_e, P_a, P_e' or P_a' and unpolarized and polarized differential cross sections σ and S, the cross sections for direct, exchange and interference scattering (σ^d, σ^{ex}, σ^{int}) can be determined from (23), (24), and (25).

CONCLUDING REMARKS

We have attempted to present a theory of measurement to extract $\sigma^d/2\sigma$, $\sigma^{ex}/2\sigma$, $\sigma^{int}/2\sigma$, $Re(f^*g/\sigma)$ for elastic scattering and $\sigma^d/2\sigma$, $\sigma^{ex}/2\sigma$, $\sigma^{int}/2\sigma$ for $^2S \rightarrow {}^2P$ excitation of light spin half targets by low energy electrons. These dimensionless quantities can be compared with those predicted by different theoretical approaches and may prove important for assessment and evaluation of these approximation schemes in finest details. We have plotted examples of these numbers against scattering angles ($\theta_e = 0°$ to $180°$) at a few low energies in adjoining figures, using close coupling approximations including exchange interaction. As can be seen these data show pronounced structure. The complete determination of the scattering amplitudes f_o, g_o, f_1, g_1 is not, however, possible unless electron photon coincidence measurements are done with both electron and atom beams polarized as pointed out by Kleinpoppen[3] and recently by Hanne[4]. A detailed theory is being worked out to accomplish this task.

We finally draw attention to equivalent amplitude descriptions of the above scattering of electrons by one-electron atoms, in collisions of one-electron ions or atoms with targets of one-electron ions or atoms[5]. However, in such scattering further types of collision processes may particularly occur in connection with quasi-molecular interactions.

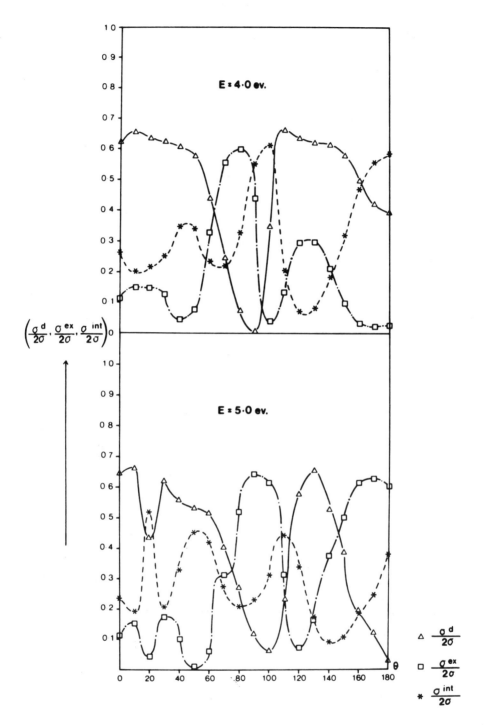

Fig. 1: Elastic Scattering of electrons by lithium atoms as a function of the scattering angle at collision energies E=4eV and E=5eV. These ratios of the partial to the double differential cross sections have been calculated by Khalid, S.M.[5] based upon a closed coupling code of the (2s-2p-3s-3p-3d)-approximation of Jacubovicz, H. and Moores, D.L.

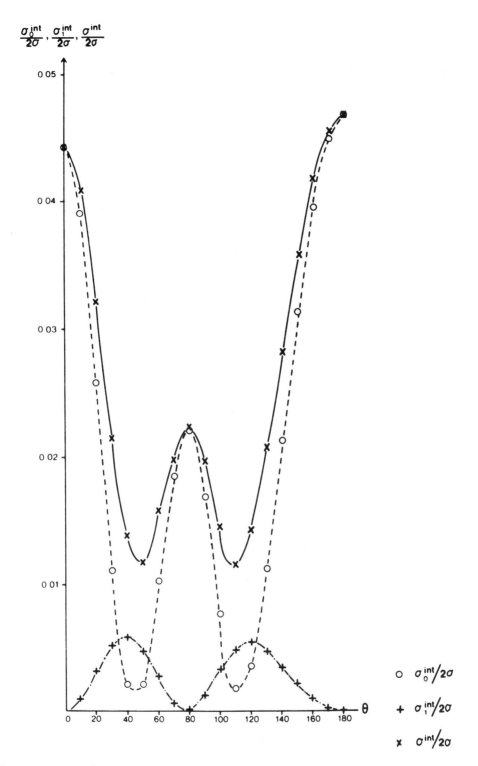

Fig. 2: Ratios of the interference excitation cross sections to the double differential cross section for excitation of the sodium, e-Na3s → 3p, at an energy of 2.2 eV, calculated[5] by four state exchange 3s-3p-3d-4s close coupling approximation of Jacubovicz, H. and Moores, D.L.

REFERENCES

1. See various papers on theoretical and experimental investigations on electron-atom collisions in these Proceedings.
2. N.F. Mott, H.S.W. Massey. "The Theory of Atomic Collisions," 3rd edition, Oxford, Clarendon Press, (1965).
3. H. Kleinpoppen. "Phys.Rev.," A3, 2015, (1971), and K.Blum and H. Kleinpoppen. "Int. Jour. Quantum Chemistry Symposium," **10**, 321 (1976).
4. G.F. Hanne. "Coherence and Correlation in Atomic Collisions," eds. H. Kleinpoppen and J.F. Williams, Plenum Press, New York, p593 (1980).
5. S.M. Khalid. "PhD Thesis", Stirling University (1981).
6. H.O. Lutz. "Correlation and Polarization of Electron and Atomic Collisions and (e,2e) Reactions," *in*: Int. Conf. Series No.122, Flinders University, Adelaide, eds. P.J.O. Teubner and E. Weigold, p343 (1991).

ELECTRON-ATOM SCATTERING AT INTERMEDIATE ENERGIES

P G Burke and M P Scott

Department of Applied Mathematics and Theoretical Physics
The Queen's University of Belfast
Belfast BT7 1NN
Northern Ireland

1. INTRODUCTION

Over the last few years there has been much interest, both theoretically and experimentally, in electron-atom scattering at intermediate energies. The theoretical work has been driven mainly by the many applications in astronomy, laser physics and plasma physics which require accurate data for their interpretation and by experimental advances which provide an important stimulus to theory. The increasing availabililty of more powerful computers has enabled the development of more sophisticated theoretical methods to study electron scattering in the intermediate energy regime, which faces particular difficulties owing to the infinite number of open channels which occur in this energy region. In order to obtain accurate results in the intermediate energy region, the theoretical approaches used must be able, in some way, to allow for loss of flux into these highly excited states and continuum states of the target atom or ion. Recent theoretical approaches which have been proposed include extensions of low-energy methods, the use of optical potentials and extensions of the Born approximation by the inclusion of higher-order terms.

Many of the extensions of low energy methods are based on augmenting the target state basis by the inclusion of pseudostates in addition to the target atomic or ionic states of interest. These pseudostates are not eigenstates of the target Hamiltonian but are chosen in some way to represent the higher lying bound states and continuum states of the target. Intermediate energy $e^- - H$ scattering has been studied using this approach by Burke and Webb[1], Burke and Mitchell[2] and Callaway and Wooten[3,4,5]. More recently Callaway and Oza[6] have used a Slater basis to represent the pseudostates to obtain accurate results for the $1s - 2s$ and $1s - 2p$ excitation. This process has also been sucessfully investigated by Burke et al.[7], Scott et al.[8] and Scholz et al.[9] using an intermediate energy R-matrix (IERM) basis for the pseudostates, and by

Bray and Stelbovics[10,11] using a Sturmian basis. The method adopted by Bray and co-workers, referred to as the converged close coupling (CCC) method, has been used very sucessfully in a number of calculations, most recently in electron scattering by atomic helium[12].

The use of optical potentials to account for the loss of flux into the infinity of open channels in the intermediate energy region has been studied by a number of authors. These include Bransden and Coleman[13] and Bransden *et al.*[14] who have studied electron scattering by atomic hydrogen using the optical potential calculated to second order, where $1s, 2s$ and $2p$ states were retained in P space. McCarthy and Stelbovics[15], Bray *et al.*[16] and McCarthy[17] have extended the optical potential beyond second order and also made allowance for exchange. This coupled channel optical model has been used sucessfully in $e^- -$Na scattering[18], but some discrepancies still remain when compared with the pseudostate methods for $e^- -$H scattering. Work on the use of optical potentials has also been carried out by Callaway *et al.*[19], who constructed an optical potential using a set of pseudostates. This method has been used in the study of $n = 1 \rightarrow 3$ and $n = 2 \rightarrow 3$ excitations in $e^- -$H scattering.

To conclude this brief review of intermediate energy methods, we mention the work of Byron *et al.*[20,21,22] who have sucessfully included higher order terms in the Born series using eikonal-Born-series (EBS) approaches. This method has proven particularly useful at the higher end of the intermediate energy region.

In this article we will concentrate on the recent theoretical developments based on the R-matrix approach[23]. In section 2 we present a summary of the theory which is illustrated by recent results presented in section 3. Concluding remarks and proposed direction of future work is discussed in section 4.

2. THEORY

Before discussing in detail the scattering of electrons by atoms and ions at intermediate energies we first consider the scattering process at low energies. That is, the process

$$e^- + A_i \rightarrow e^- + A_j \tag{1}$$

where A_i and A_j are the initial and final states of the target and where by low energies we mean that the velocity of the scattered electron is of the same order or less than that of the target electrons playing an active role in the collision.

Basic scattering equations

The Schrödinger equation describing the scattering of an electron by a target atom or ion is

$$H_{N+1} \Psi = E \Psi \tag{2}$$

where N is the number of electrons in the target atom or ion, E is the total energy of the system and the $(N + 1)$-electron Hamiltonian H_{N+1} is given by

$$H_{N+1} = \sum_{i=1}^{N+1} \left(-\frac{1}{2} \nabla_i^2 - \frac{Z}{r_i} \right) + \sum_{i>j=1}^{N+1} \frac{1}{r_{ij}}, \tag{3}$$

Z being the nuclear charge, and we have used atomic units throughout. We have written $r_{ij} = |\mathbf{r}_i - \mathbf{r}_j|$ where $\mathbf{r}_i$ and $\mathbf{r}_j$ are the spatial coordinates of the ith and jth electrons.

The solution of equation (2) has the following asymptotic form

$$\Psi_i \underset{r \to \infty}{\sim} \Phi_i \chi_{1/2m_i} e^{ik_i z} + \sum_j \Phi_j \chi_{1/2m_j} f_{ji}(\theta, \phi) \frac{e^{ik_j r}}{r} \quad (4)$$

where Φ_i are the N-electron target eigenfunctions with eigenenergies w_i which satisfy the equation

$$\langle \Phi_i | H_N | \Phi_j \rangle = w_i \delta_{ij} . \quad (5)$$

$\chi_{1/2m_i}$ and $\chi_{1/2m_j}$ are the spin eigenfunctions of the incident and scattered electrons, $f_{ij}(\theta, \phi)$ is the scattering amplitude and the wave numbers k_i and k_j are related to the total energy of the system by

$$E = w_i + \frac{1}{2} k_i^2 = w_j + \frac{1}{2} k_j^2 . \quad (6)$$

The outgoing wave in equation (4) contains contributions from all target states that are energetically allowed (i.e. $k_j^2 \geq 0$). We note that when the target is ionic, logarithmic phase factors have to be included in the exponential terms in equation (4).

The differential cross section for the transition from initial state i to the final state j can be obtained by calculating the incident and scattered flux in equation (4) to give the equation

$$\frac{d\sigma_{ji}}{d\Omega} = \frac{k_j}{k_i} \left| f_{ji}(\theta, \phi) \right|^2 . \quad (7)$$

The total cross section can then be obtained by averaging over initial spin states, summing over final spin states and integrating over all scattering angles.

Standard low energy R-matrix theory

The R-matrix theory proceeds by partitioning configuration space into two regions by a sphere of radius $r = a$, where r is the relative coordinate of the scattering electron and the centre of gravity of the target. In the internal region electron ($r \leq a$) exchange and correlation effects between the scattering electron and the target electrons are important and the corresponding $(N + 1)$-electron complex behaves very much like a bound state. In the external region ($r \geq a$) electron exchange between the scattered electron and the target can be neglected provided the radius a of the sphere is sufficiently large to completely envelope the target states of interest. The scattered electron then moves in the long-range potential of the target and the solution in this region can then be obtained by using an asymptotic expansion or perturbation approach.

In the internal region the solution of equation (2) at an energy E is expanded in terms of an energy independent R-matrix basis

$$\Psi_E = \sum_k A_{Ek} \psi_k , \quad (8)$$

where the basis states ψ_k are given by

$$\psi_k = A \sum_{ij} \Phi_i(1, ..., N) u_j(N + 1) a_{ijk} + \sum_i \chi_i(1, ..., N + 1) b_{ik} . \quad (9)$$

The first expansion includes all those target eigenstates and pseudostates Φ_i of the N-electron target of interest. The u_j are continuum functions representing the motion of the $(N + 1)$th electron. The operator A antisymmetrizes the first expansion in accordance with the Pauli exclusion principle. The second expansion in equation (9) consists of quadratically integrable functions, χ_i, which are included to allow for short range correlation effects between the scattered electron and the target. The coefficients a_{ijk} and b_{ik} are chosen so that the R-matrix basis diagonalises the $(N + 1)$-electron Hamiltonian matrix in the internal region.

Extension to intermediate energies

Equation (9) has proved to be a very satisfactory basis for low energy electron scattering calculations and detailed calculations have been carried out for many atomic and ionic targets[51]. However, difficulties arise at intermediate energies owing to the infinite number of open channels which are present in this energy regime. In the intermediate energy R-matrix approach (IERM)[7] both the outer valence electron of the target and the scattering electron are expanded in terms of the same continuum basis u_j. This gives rise to 'continuum-continuum' type terms in the basis states ψ_k. It is the inclusion of these terms, which are omitted in the standard low-energy theory, that enable the ψ_k basis states to represent the intermediate energy scattering wavefunction contained in the internal region. The expansion basis in the low-energy method, given by equation (9), lack this completeness at intermediate energies. As an example, we consider the basis states used in the IERM approach to electron scattering by atomic hydrogen. The two-electron R-matrix basis states, $\Theta_k^{LS\pi}$, are given by

$$
\Theta_k^{LS\pi}(r_1, r_2) = [1 + (-1)^S P_{12}] \sum_{n_1}^{n_{1max}} \sum_{n_2}^{n_{2max}} \sum_{l_1}^{l_{1max}} \sum_{l_2}^{l_{2max}} r_1^{-1} u_{n_1 l_1}(r_1) r_2^{-1} u_{n_2 l_2}(r_2)
$$
$$
* \mathcal{Y}_{l_1 l_2 L M_L}(\hat{r}_1, \hat{r}_2) a_{n_1 l_1 n_2 l_2 k}^{LS\pi} \tag{10}
$$

The angular functions $\mathcal{Y}_{l_1 l_2 L M_L}(\hat{r}_1, \hat{r}_2)$ are eigenfunctions of L^2, l_1^2 and l_2^2 . The one-electron radial basis functions, $u_{nl}(r)$, are made up of bound hydrogenic orbitals together with members of a complete set of continuum orbitals. n_{1max} is the maximum value of n_1 retained in expansion (10) for each angular momentum l_1 and n_{2max} is the maximum value of n_2 retained in expansion (10) for each angular momentum l_2. In the IERM method n_{1max} is set equal to n_{2max} and l_{1max} is set equal to l_{2max}. This works well for transitions involving the $n = 1$ and $n = 2$ states at intermediate energies. However, if we wish to study transitions involving the $n = 3$ states we need to increase the R-matrix boundary radius to completely envelope the $n = 3$ bound orbitals. It is then necessary to increase n_{1max} and n_{2max} and we find that the Hamiltonian matrix becomes very large (of the order of 5000) and the calculation computationally difficult. However, it has been found possible to reduce n_{2max}, thus reducing the overall size of the Hamiltonian matrix, without loss of accuracy in the cross section results[24]. This is the scheme adopted in the *reduced* IERM approach[25,26]. We note that if n_{2max} and l_{2max} are reduced so that the summation over n_2 and l_2 include only the bound hydrogenic orbitals then expansion (10) is equivalent to the standard low-energy expansion (9).

While this reduced approach now enables us to obtain physical observables for transitions involving up to the $n = 3$ states, we still encounter computational difficulties if we wish to study transitions involving more highly excited states. The radius of the internal region, and the consequent size of the basis necessary to obtain accurate results, becomes very large and a more efficient approach is then to divide the internal region into a number of sub-regions. As a result, work is currently ongoing on the development of a 2-dimensional R-matrix propagator[7,27]. In this method the (r_1, r_2) space is subdivided as illustrated in figure 1. The Schrödinger equation is then solved

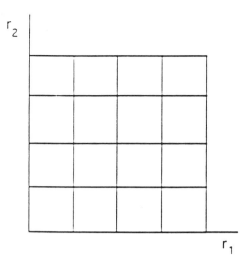

Figure 1. Subdivision of the (r_1, r_2) plane used in the 2-d R-matrix propagator.

independently in each sub-region using similar numerical one-electron basis functions as in the IERM approach given by expansion (10). It is then possible to propagate a global R-matrix from one sub-region to another until the final R-matrix is obtained on the R-matrix boundary. The asymptotic calculation is then carried out in the usual way. The size of the Hamiltonian matrices that need to be diagonalized in each sub-region is thus considerably reduced at the expense of introducing an additional propagation step in the calculation.

In principle, the 2-d R-matrix propagator can be extended to treat intermediate energy electron collisions with complex atoms where two electrons lie outside an N-electron core. This is achieved by diagonalizing the (N+2)-electron Hamiltonian in the innermost sub-region and then propagating the R-matrix into the outer sub-regions which contain only two outer electron wavefunctions.

Finally in this section, we mention a recent development based on the standard low energy R-matrix approach. Improved numerical techniques now allow us to include a much greater number of pseudostates in expansion (9) than was hitherto possible. This enables us to extend the work of Burke and Mitchell[2] to include pseudostates up to $n \simeq 8$ without encountering any numerical difficulties. In this work the extra pseudostates included in expansion (9) are of the form

$$\Phi_{nl} = \sum_{i=l+1}^{n_{max}} a_{nli}^{n_{max}} r^i \exp(-\alpha r) \tag{11}$$

The $a_{nli}^{n_{max}}$ are uniquely defined by orthogonality between the states and by the requirement that the pseudostates diagonalise the one-electron Hamiltonian. The range parameter α is chosen so that the pseudostates have approximately the same range as the highest lying physical state of interest. One advantage of this development is that the standard R-matrix programs for complex atoms and ions can be simply modified enabling cross sections for arbitrary atoms to be calculated at intermediate energies.

Results obtained using the various R-matrix approaches discussed in this section will be presented in section 3.

3. RESULTS

In this section we review recent work on electron scattering at intermediate energies using the R-matrix method discussed in section 2 above. In figure 2 we show the integrated $1s - 2s$ excitation cross section for electron scattering from atomic hydrogen from threshold to 4 Ryd. The IERM results[8] and the coupled pseudostate calculations[6] are in very good agreement with experiment[29,30] over the whole energy range while the 1s-2s-2p close coupling calculation[28] overestimates the cross section by more than a factor of two at low energies. This illustrates the importance of allowing for loss of flux into highly excited and continuum channels at these intermediate energies.

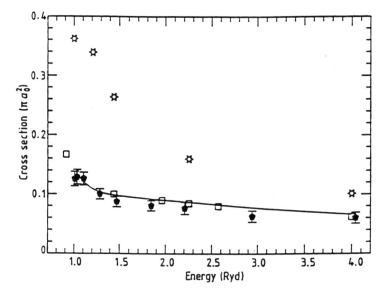

Figure 2. The $e^- - H$ $1s - 2s$ excitation cross section. Full curve: IERM results (ref. 8); open squares: coupled pseudostate (ref. 6); six-point stars: 1s-2s-2p close coupling (ref. 28); full pentagons: experiment (refs. 29 and 30) (from ref. 8).

Investigation of the angular distribution of the λ and R parameters for the $1s - 2p$ excitation obtained by measuring the scattered electron in coincidence with the decay Lyman-α photon provide a more stringent test of the theory. Comparison between theory and experiment for these observables enables a test of the magnitude and the relative phases of the excitation amplitudes to be made.

The angular distribution of these parameters at 54.42 eV (4 Ryd.) is presented in figure 3. At small scattering angles (less than 90^0) the theoretical results, with the exception of the 1s-2s-2p close coupling calculation, agree well with experiment. At larger angles, however, the theoretical results differ considerably from the experimental measurements. A recent calculation (not shown in this figure) by Bray and Stelbovics[11] using a Sturmian expansion, which is in close accord with the IERM results, also differs from experiment at large scattering angles. Clearly there is still need for further work in this area both theoretically and experimentally.

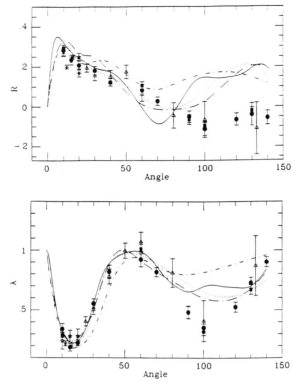

Figure 3. The $e^- - H$ R and λ parameters for $1s - 2p$ excitation at 54.42eV. Theory: full curve: IERM results (ref. 9); short dash-dot line: 1s-2s-2p close coupling (ref. 28); long dash-dot line: pseudostate (ref. 31); dotted line: pseudostate (ref. 32). Experiment: solid circles (ref. 33); solid squares (ref. 34); five-point open stars (ref. 35); triangles (ref. 36); seven-point stars (ref. 37) (from ref. 9).

In figure 4 we show results for the integrated excitation cross section for the $1s - 3l$ transitions for $e^- - H$ scattering in the energy range $1.0 - 2.5$ Ryd. In the reduced IERM calculation[25,26] an inner region boundary radius of 38 au was used to completely envelope the $n = 3$ hydrogenic orbitals. We see that both the reduced IERM results[25] and the coupled pseudostate results[38], which allow for loss of flux into the continuum channels, are in good agreement with the experimental measurements[39]. However, the 15-state close coupling results[40], where no allowance has been made for loss of flux into these continuum channels, tend to overestimate the cross section by about a factor of two.

While we have concentrated on examples which illustrate the importance of allowing for loss of flux into highly excited and ionization channels at intermediate energies, we note the importance of including these *extra* channels even at low scattering energies. Recent studies[43,46,52] on electron scattering from atomic hydrogen below the $n = 3$ threshold show that while the inclusion of the continuum channels has very little effect on the widths and positions of low lying resonances structures, there is still a discrepancy of around 10% in the excitation cross section between calculations which do not allow for loss of flux into highly excited and ionization channels[44,46] and those which do[43,45,52]. We note that the coupled pseudostate calculation of Callaway[45] are in good agreement with the experimental measurements of Williams[47].

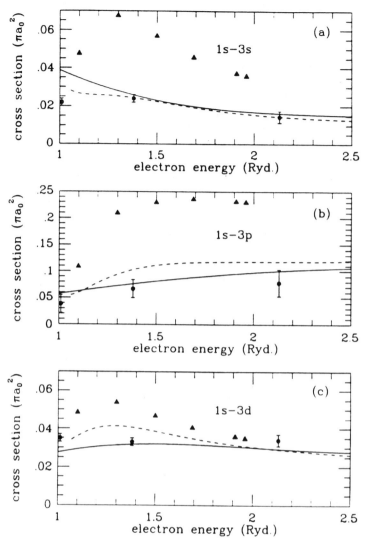

Figure 4. The $e^- - H$ $1s - 3l$ integrated excitation cross sections. Full curve: reduced IERM results (ref. 24); dashed curve: coupled pseudostate (ref. 38); solid trianges: 15-state close coupling (ref. 40); experiment (ref. 39) (from ref. 24).

The importance of the inclusion of highly excited and ionization channels in electron scattering from He$^+$ is illustrated in figure 5 where we consider as an example the $2s - 3p$ transition for $^1S^e$ scattering. Again we see that the 15-state close coupling results obtained using the same basis functions as in the calculation of Aggarwal *et al.*[41] overestimate the cross section when compared with the IERM results[42]. However, the difference is not as great as for electron scattering from atomic hydrogen in a comparable energy range.

The IERM results discussed in figures 2-5 have been obtained without the need to subdivide the (r_1, r_2) plane as described in section 2. However, if we are to proceed to transitions involving more highly excited states this subdivision becomes essential. In

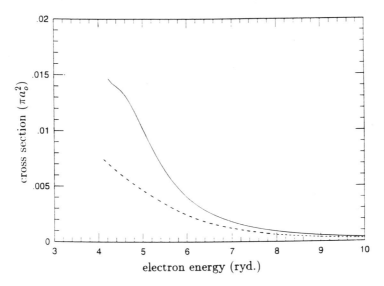

Figure 5. The e⁻ −He⁺ ¹Sᵉ scattering cross section for the 2s − 3p transition. Dashed curve: reduced IERM; full curve: 15-state close coupling (see ref. 42).

figure 6 we present some initial results for the 1s − 4s cross section for e⁻ −H ¹Sᵉ scattering at intermediate energies using the 2-d R-matrix propagator[48]. In this calculation the inner region boundary radius was taken as 60 *au* to completely envelope the $n = 4$ hydrogenic orbitals. Within this inner region the (r_1, r_2) plane was subdivided into 16 square blocks of length 15 *au* (see figure 1). These data are compared with the results of a 15-state close coupling calculation using the same approximation as in reference 40, where we again see a significant overestimate in the cross section by the 15-state close coupling results.

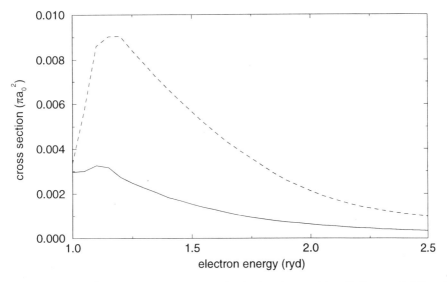

Figure 6. The e⁻ −H ¹Sᵉ scattering cross section for the 1s − 4s transition. Full curve: 2-d R-matrix propagator; dashed curve: 15-state close coupling (from ref. 48).

Finally in this section we consider results for e⁻−H scattering obtained using the R-matrix with pseudostates approach discussed in section 2 above. In figure 7 we compare these results for the $1s − 2s$ excitation cross section for $^1S^e$ scattering with data obtained using the 2-d R-matrix propagator and the IERM approach in the energy region just above the ionization threshold. The results in this energy region from all three calculations are in very good agreement. As an example of the comparison of these calculations at lower energies we consider in figure 8 the analysis of the first $^1S^e$ resonance above the $n = 2$ hydrogenic threshold. Whereas no minimum principle exists which is applicable to the cross section data in figure 7, we know that the eigenphase sum given in figure 8 forms a lower bound on the exact result[50]. From this we conclude that, while the results for the eigenphase sum from both the R-matrix pseudostate calculation and the 2-d R-matrix propagator are in good accord, the 2-d R-matrix propagator is the 'better' calculation in this energy region. This is to be expected as the 2-d R-matrix propagator includes more correlation functions in (r_1, r_2) space than the other methods . The position and width of the $^1S^e$ resonance are 0.8620 Ryd and 0.002 83 Ryd, respectively, using the 2-d R-matrix propagator and 0.8621 Ryd and 0.002 83 Ryd using the R-matrix with pseudostates approach.

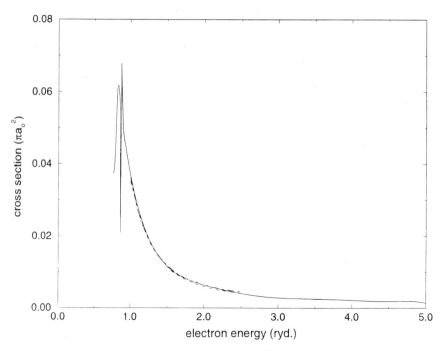

Figure 7. The e⁻−H $1s − 2s$ excitation cross section for $^1S^e$ scattering. Full curve: R-matrix with pseudostates; dashed curve: 2-d R-matrix propagator; dashed-dotted curve: reduced IERM (see ref. 49).

One of the advantages of this new R-matrix with pseudostate approach is that it uses existing computer codes which have been developed to consider electron scattering from complex atoms and ions. Hence it is fairly straightforward to extend the above procedure to consider electron scattering at intermediate energies from more complex targets and work is currently underway on electron scattering from helium and beryllium.

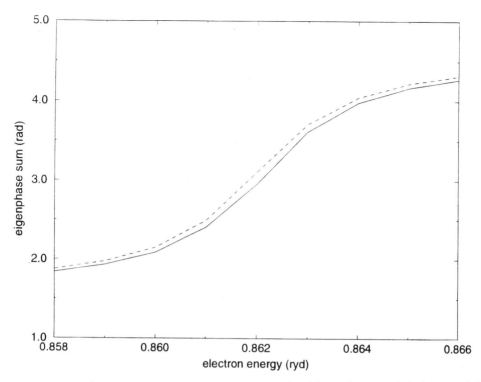

Figure 8. The $^1S^e$ eigenphase sum. Full curve: R-matrix with pseudostates; dashed curve: 2-d R-matrix propagator (see ref. 49).

4.CONCLUSIONS

Electron impact excitation of low lying states of simple atoms and ions can now be accurately calculated at intermediate energies. However, more stringent tests of both theory and experiment, such as evaluation of various coincidence parameters, still require further attention both theoretically and experimentally. New approaches are currently being developed to treat excitation of Rydberg states and ionization but the scattering of electrons by complex atoms at intermediate energies still requires further theoretical and computational development. There is an urgent need for more experimental studies of collision cross sections at intermediate energies for comparison and to provide a stimulus for future theoretical development.

5.ACKNOWLEDGEMENTS

Some of the recent results reported in this review were obtained in collaboration with Prof.K.Bartschat, Mr.B.R.Odgers and Mr.E.T.Hudson and we wish to acknowledge many discussions with these workers. We also wish to acknowledge support from the EPSRC, the NATO (grant no. CRG 900144) and the EC, supported by contract No ERB CHRX CT 920013.

REFERENCES

1. P.G.Burke and T.G.Webb J.Phys.B:At.Mol.Phys. **3**, L131 (1970)
2. P.G.Burke and J.F.B.Mitchell J.Phys.B:At.Mol.Phys. **6**, 665 (1973)
3. J.Callaway and J.W.Wooten Phys.Lett.A **45**, 85 (1973)
4. J.Callaway and J.W.Wooten Phys.Rev.A **9**, 1924 (1974)
5. J.Callaway and J.W.Wooten Phys.Rev.A **11**, 1118 (1975)
6. J.Callaway and D.H.Oza Phys.Rev.A **32**, 2628 (1985)
7. P.G.Burke, C.J.Noble and M.P.Scott Proc.Roy.Soc.A **410**, 289 (1987)
8. M.P.Scott, T.T.Scholz, H.R.J.Walters and P.G.Burke
J.Phys.B:At.Mol.Opt.Phys. **22**, 2097 (1989)
9. T.T.Scholz, H.R.J.Walters, P.G.Burke and M.P.Scott
J.Phys.B:At.Mol.Opt.Phys. **24**, 3055 (1991)
10. I.Bray and A.Stelbovics Phys.Rev.Lett.A **69**, 53 (1992)
11. I.Bray and A.Stelbovics Phys.Rev.A **46**, 6995 (1992)
12. I.Bray, D.Fursa and I.E.McCarthy Phys.Rev.A **51**, 500 (1995)
13. B.H.Bransden and J.P.Coleman J.Phys.B:At.Mol.Phys. **5**, 537 (1972)
14. B.H.Bransden, T.Scott, R.Shingal and J.Raychoudhury
J.Phys.B:At.Mol.Phys. **15**, 4605 (1982)
15. I.E.McCarthy and A.Stelbovics Phys.Rev.A **28**, 2693 (1983)
16. I.Bray, I.E.McCarthy, J.Mitroy and K.Ratnavelu Phys.Rev.A **39**, 4998 (1989)
17. I.E.McCarthy Comm.Atom.Molec.Phys. **24**, 343 (1990)
18. I.Bray, D.A.Konovalov and I.E.McCarthy Phys.Rev.A **44**, 7830 (1991)
19. J.Callaway, K.Unnikrishnan and D.H.Oza Comment.At.Mol.Phys.
13, 19 (1987)
20. F.W.Byron,Jr., C.J.Joachain and R.M.Potvliege J.Phys.B:At.Mol.Phys. **14**, L609 (1981)
21. F.W.Byron,Jr., C.J.Joachain and R.M.Potvliege J.Phys.B:At.Mol.Phys. **15**, 3915 (1982)
22. F.W.Byron,Jr., C.J.Joachain and R.M.Potvliege J.Phys.B:At.Mol.Phys. **18**, 1637 (1985)
23. P.G.Burke and W.D.Robb Adv.Atom.Molec.Phys. **11**, 143 (1975)
24. M.P.Scott and P.G.Burke J.Phys.B:At.Mol.Opt.Phys. **26**, L191 (1993)
25. M.P.Scott, B.R.Odgers and P.G.Burke J.Phys.B:At.Mol.Opt.Phys. **26**, L827 (1993)
26. B.R.Odgers, M.P.Scott and P.G.Burke J.Phys.B:At.Mol.Opt.Phys. **27**, 2577 (1994)
27. M.LeDourneuf, J.M.Launay and P.G.Burke J.Phys.B:At.Mol.Opt.Phys. **23**, L559 (1990)
28. A.E.Kingston, Y.C.Liew and P.G.Burke J.Phys.B:At.Mol.Phys. **15**, 2755 (1982)
29. W.E.Kauppila, W.R.Ott and W.L.Fite Phys.Rev.A **1**, 1099 (1970)
30. R.L.Long,Jr., D.M.Cox and S.J.Smith J.R.N.B.S.A. **72**, 521 (1968)
31. B.H.Bransden, I.E.McCarthy, J.D.Mitroy and A.Stelbovics Phys.Rev.A. **32**, 166 (1985)
32. W.L.van Wyngaarden and H.R.J.Walters J.Phys.B:At.Mol.Phys. **19**, 929 (1986)
33. J.F.Williams J.Phys.B:At.Mol.Phys. **14**, 1197 (1981)
34. J.F.Williams Aust.J.Phys. **39**, 621 (1986)
35. S.T.Hood, E.Weigold and A.J.Dixon J.Phys.B:At.Mol.Phys. **12**, 631 (1979)
36. E.Weigold, L.Frost and K.J.Nygaard Phys.Rev.A. **21**, 1950 (1980)

37. J.Slevin, M.Eminyan, J.M.Woolsey, G.Vassiler and H.Q.Porter
J.Phys.B:At.Mol.Phys. **13**, L341 (1980)

38. J.Callaway and K. Unnikrishnan J.Phys.B:At.Mol.Opt.Phys. **26**, L419 (1993)

39. A.H.Mahan, H.Gallaher and S.J.Smith Phys.Rev.A **43**, 156 (1976)

40. K.M.Aggarwal, K.A.Berrington, P.G.Burke, A.E.Kingston and A.Pathak
J.Phys.B:At.Mol.Opt.Phys. **24**, 1385 (1991)

41. K.M.Aggarwal, K.A.Berrington, A.E.Kingston and A.Pathak
J.Phys.B:At.Mol.Opt.Phys. **24**, 1757 (1991)

42. B.M.McLaughlin, M.P.Scott and P.G.Burke (1995) to be submitted

43. B.R.Odgers, M.P.Scott and P.G.Burke J.Phys.B:At.Mol.Opt.Phys. **28**, 2973 (1995)

44. A.Pathak, A.E.Kingston and K.A.Berrington J.Phys.B:At.Mol.Opt.Phys. **21**, 2939 (1988)

45. J.Callaway Phys.Rev.A **26**, 199 (1982)

46. W.C.Fon, K.Ratnavelu and K.M.Aggarwal Phys.Rev.A. **49**, 1786 (1994)

47. J.F.Williams J.Phys.B:At.Mol.Opt.Phys. **21**, 2107 (1988)

48. M.P.Scott, B.R.Odgers, P.G.Burke J.Phys.B:At.Mol.Opt.Phys. (1996) submitted

49. K.Bartschat, E.T.Hudson, M.P.Scott, P.G.Burke and V.M.Burke
J.Phys.B:At.Mol.Opt.Phys. **29**, 115 (1996)

50. W.A.McKinley and J.H.Macek Phys.Lett. **10**, 210 (1964)

51. P.G.Burke and K.A.Berrington Atomic and Molecular Process: An R-matrix Approach, IOP Publishing Ltd.,Bristol (1993)

52. D.A.Konovalov and I.E.McCarthy J.Phys.B:At.Mol.Opt.Phys. **27**, L741 (1994)

SPIN POLARIZATION OF AUGER ELECTRONS; RECENT DEVELOPMENTS

B Lohmann

Universität Münster
Institut für Theoretische Physik I
Wilhelm-Klemm-Str. 9
D-48149 Münster
GERMANY

Abstract. The theory of spin polarization of Auger electrons is reviewed. Special attention is paid to the in-scattering plane components of the spin polarization vector. A simpler experiment is discussed. Within the LS coupling scheme, theoretical predictions for the spin polarization vector for Auger transitions on argon and krypton are given. A degree of spin polarization of up to 50 % is predicted. Possible experimental set-ups are discussed.

INTRODUCTION

Nowadays, the observation of angular distribution and spin polarization has been recognized as a useful tool to obtain more refined information about the Auger emission process, i.e. to obtain information about the matrix elements and the scattering phases. A recent review on the present stage of research has been given by Mehlhorn[1].

So far, the number of theoretical and experimental investigations concerning the spin polarization of Auger electrons is still small (e.g. Lohmann *et al*[2] and refs. therein).

The main topic has been the investigation of the so-called "dynamic" spin polarization parameter which is connected with the component of the spin polarization vector perpendicular to the scattering plane. However, experiments[3,4] have reported an almost vanishing dynamic spin polarization.

Meanwhile, experiments are in progress to measure the so-called "intrinsic" in-scattering plane components of the spin polarization vector of the Auger electrons[5,6][†].

[†]The reader is also referred to the contribution by Heinzmann ibid.

A first experiment measuring the intrinsic spin polarization components on Ba has been recently carried out by Kuntze et al[7].

Theoretical predictions have been given by Kabachnik and Lee[8] and recently by Lohmann et al[2] who provided full parameter sets for Auger transitions on Ar, Kr, and Xe. In the following, the general theory is reviewed and some results are presented and discussed.

However, there is still need to guide experimentalists in future experiments. On the other hand, it is useful to investigate Auger transitions which we can understand more easily. We will discuss such an experiment in the following where we make use of the fact that the Auger lines to observe are isotropic with respect to their angular distribution, however, still remain spin polarized, provided the ionic hole state has been oriented.

GENERAL THEORY

The present understanding of the Auger emission is that of a two-step process (e.g. see the review by Mehlhorn[1]). Considering the case of a primary photoionization process, and using circularly polarized light, this can be written as

$$\gamma + A \longrightarrow e_e + A^+ \tag{1}$$
$$\searrow A^+ \longrightarrow A^{++} + e_{Auger} . \tag{2}$$

By taking the dipole approximation into account, observing the Auger electrons under a certain angle θ, and detecting its spin polarization vector components yields the general expression for the angular distribution

$$I(\theta) = \frac{I_0}{4\pi} \left(1 + \alpha_2 \mathcal{A}_{20} P_2(\cos\theta)\right) , \tag{3}$$

and for the cartesian components of the spin polarization vector we obtain

$$p_x = \frac{3/4\,\gamma_1 \mathcal{A}_{10}\,\sin 2\theta}{1 + \alpha_2 \mathcal{A}_{20} P_2(\cos\theta)} , \tag{4}$$

$$p_y = \frac{\xi_2 \mathcal{A}_{20}\,\frac{3}{2}\sin 2\theta}{1 + \alpha_2 \mathcal{A}_{20} P_2(\cos\theta)} , \tag{5}$$

and

$$p_z = \frac{\mathcal{A}_{10}(\beta_1 + \gamma_1 P_2(\cos\theta))}{1 + \alpha_2 \mathcal{A}_{20} P_2(\cos\theta)} . \tag{6}$$

We have chosen the incident beam axis as quantization axis. The scattering plane, as shown in figure 1, is defined by the quantization axis and the direction of Auger emission. The cartesian component of the spin polarization vector perpendicular to the scattering plane is denoted by p_y. One should note, that the structure of the equations is similar to the relations obtained for the emission of photoelectrons (e.g. see the book by Kessler[9]). Here, $\mathcal{A}_{10}$ and $\mathcal{A}_{20}$ describe the orientation and alignment of the initial state. The anisotropy and spin polarization parameters α_2, ξ_2, β_1, and γ_1 are describing the Auger emission dependent parameters of angular distribution α_2 and spin polarization, respectively. Explicit expressions for these parameters in terms of matrix elements may be found in the literature[10,11,2]. The two latter β_1 and γ_1 are denoted as the "intrinsic" and ξ_2 as "dynamic" parameters. $P_2(\cos\theta)$ denotes the second Legendre polynomial.

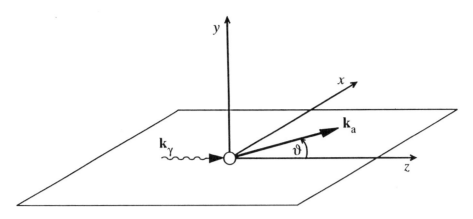

Figure 1: The coordinate frame and the reaction geometry determined by the incident photon beam axis (momentum $\mathbf{k}_\gamma$) and the direction of Auger emission (momentum $\mathbf{k}_a$).

Generally, the angle resolved intensity is a function of the total intensity, the anisotropy parameter and the alignment tensor, whereas the y-component of the spin polarization vector depends on the dynamic spin polarization parameter, the anisotropy coefficient, and the alignment tensor. The in-plane components of the spin polarization vector usually depend on all parameters, orientation and alignment, and on the anisotropy and spin polarization parameters, too. Thus, their experimental determination is not without problems in the most cases.

SOME RESULTS

The anisotropy and spin polarization parameters have been calculated by the author and co-workers for several Auger transitions of the rare gases[2]. The method of calculation has been extensively described in the cited paper. In short, the multiconfigurational Dirac-Fock program of Grant $et\ al$[12] is employed to calculate bound state wavefunctions. The continuum wavefunction is calculated within a model potential created from the final state wavefunctions of the doubly ionized atom.

A selection of data are shown in table 1 for the Xe $N_4O_{2,3}O_{2,3}$ Auger transitions. Experimental and theoretical data for the relative intensities are also given. They show relatively good agreement.

Concerning the orientation and alignment parameters $\mathcal{A}_{10}$ and $\mathcal{A}_{20}$ there are Hartree-Fock calculations by Berezhko $et\ al$[15] and by Kabachnik and Lee[8], both using the same model. Using their results the photon-energy dependence of the spin polarization parameters has been calculated. In figure 2 the P_z-component of the spin polarization vector has been plotted for the Xe $N_{4,5}O_{2,3}O_{2,3}\,^1S_0$ Auger decays, together with experimental data for the Xe $4d$ partial cross section [16,17]. As can be seen, the magnitude of the spin polarization can reach 50% in an area of photon energy where the cross section is high. However, for the general type of Auger transitions the experimental determination of the spin polarization vector is still a challenging task.

Table 1: The relative intensities, angular anisotropy, and spin polarization parameters for the Xe $N_4O_{2,3}O_{2,3}$ Auger transitions (data from Lohmann *et al*[2]). The Auger energies are given in a.u.'s.

Xe $N_4O_{2,3}O_{2,3}$						
Final state	Rel. intensity		Parameter			
	Theo.	Exp.†	α_2	β_1	γ_1	ξ_2
$^1S_0$	1.0	1.0	-1.0	-0.447	0.894	0
$^1D_2$	1.60	1.46	0.052	-0.429	-0.262	-0.073
$^3P_0$	0.01	0.03	-1.0	-0.447	0.894	0
$^3P_1$	0.45	0.35	-0.835	0.594	-0.731	-0.014
$^3P_2$	0.18	0.11	0.165	0.110	0.321	0.216

†Aksela *et al*[13]

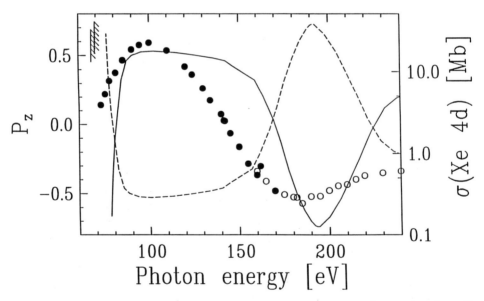

Figure 2: P_z dependent on photon energy, solid line: Xe $N_4O_{2,3}O_{2,3}{}^1S_0$, dashed line: Xe $N_5O_{2,3}O_{2,3}{}^1S_0$ under $\theta = 90^o$. The points indicate experimental values for the Xe $4d$ partial cross section, filled dots: Becker *et al*[16], open dots: Lindle *et al*[17].

A SIMPLER EXPERIMENT

Most recently, we therefore suggested a simpler experiment to measure the intrinsic spin polarization of Auger transitions[18]. A close inspection of equations (3–6) shows that there must be an initial orientation $\mathcal{A}_{10}$ to observe the "intrinsic" in-plane components of the spin polarization vector. However, there is no need for a non-zero alignment. Indeed Auger transitions with an oriented but non-aligned initial state can be found. Considering the L_2MM Auger transitions of the rare gases, we have an initial p-hole state with $J = 1/2$. Thus, an orientation can be created by circularly polarized light but no alignment can be found.

$$\mathcal{A}_{20} = 0 . \tag{7}$$

By this, equations (3–6) are significantly simplified. First, the intensity becomes isotropic

$$I(\theta) = \frac{I_0}{4\pi} . \tag{8}$$

The denominator of equations (4–6) equals one which yields

$$p_x = \frac{3}{4} \gamma_1 \mathcal{A}_{10} \sin 2\theta , \tag{9}$$

$$p_y = 0 , \tag{10}$$

and

$$p_z = \mathcal{A}_{10} \left(\beta_1 + \gamma_1 P_2(\cos\theta)\right) . \tag{11}$$

Thus, the spin polarization component p_y, perpendicular to the scattering plane, vanishes because it depends only on the alignment only, and we remain with the in-plane components of the spin polarization vector.

This is of physically significance, since we now have the most interesting feature that the emitted Auger electrons can be spin polarized though still remain isotropic with respect to their angular distribution. This aspect simplifies the experiment enormously because only the three parameters $\mathcal{A}_{10}$, β_1, and γ_1 have to be determined, whereas the spin polarization vector components p_x and p_z in such experiments generally depend on the anisotropy parameter and alignment, too. In the following, we illustrate our proposed experiment for some Auger transitions on Ar and Kr.

RESULTS

Spin polarization parameters

The Ar L_2MM multiplet has been extensively discussed by Lohmann and Larkins[18] under the aspect of a possible spin polarization experiment. As we have pointed out in the cited paper, large intrinsic spin polarization parameters have been found for most of the multiplet transitions. Data for β_1 and γ_1 for the Ar $L_2M_{2,3}M_{2,3}$ multiplet are shown in table 2.

The surprisingly constant value of $\beta_1 = -1/3$ for most of the transitions can be explained by closely inspecting the general expression for the parameter β_1, i.e. equation (5) of Lohmann et al[2]. We find that no interference terms between matrix elements with different angular momentum ℓ are possible. For the Auger transitions

Table 2: The relative intensities and spin polarization parameters β_1 and γ_1 for the Ar $L_2M_{2,3}M_{2,3}$ Auger transitions (data from Lohmann and Larkins[18]). The experimental Auger energies have been taken from Mehlhorn and Stalherm[19] and are given in a.u.'s.

Ar $L_2M_{2,3}M_{2,3}$				
Final state	Auger energies	Parameters		
		I_0	β_1	γ_1
$^1S_0$	7.4605	14.03	-0.3333	1.3333
$^1D_2$	7.5483	38.83	-0.3333	0.5754
$^3P_0$	7.6053	10.97	-0.3333	1.3333
$^3P_1$	7.6075	24.32	0.5555	-0.2222
$^3P_2$	7.6123	11.85	-0.3333	0.0898
		$\sum$ 100		

considered, the additional selection rule $j = j'$ applies. As a result, one ends up with

$$
\beta_1 = - \left[\sum_{\ell j} | \langle J_f(\ell_{1/2})j : 1/2 \| V \| 1/2 \rangle |^2 \right]^{-1}
$$

$$
\times 2 \sum_{\ell j} (2j+1) \begin{Bmatrix} \frac{1}{2} & \frac{1}{2} & 1 \\ j & j & 2 \end{Bmatrix} \begin{Bmatrix} \frac{1}{2} & \frac{1}{2} & 1 \\ j & j & \ell \end{Bmatrix} | \langle J_f(\ell_{1/2})j : 1/2 \| V \| 1/2 \rangle |^2 \, . \tag{12}
$$

Therefore, no interference occurs between partial waves with different angular and total angular momentum. As a matter of fact, the summation is over two partial waves, only and the product of Wigner 6j-coefficients and factors in the numerator turns out to be 1/3 for both terms. Thus, the parameter β_1 becomes independent of the matrix elements which finally yields the result. This is of physically importance since it shows an interesting feature of the Ar L_2MM spectrum. Although more than one partial wave is emitted, i.e. the Auger decay proceeds via two channels, the intrinsic spin polarization parameter β_1 remains independent of the matrix elements for most of the Auger transitions. This is due to the fact that the usually extended summation over the possible interfering partial waves (e.g. see equations (5) and (6) of Lohmann *et al*[2]) of the emitted Auger electron is enormously reduced for a $J = 1/2$ initial ionic state by selection rules.

The question occurs whether this is a general behaviour or only a special property of the Ar L_2MM spectrum. Most recently, we therefore investigated the spectra of the Kr $L_2M_xM_{4,5}$ (x =1,...,5) Auger transitions[20]. Due to the fact that a d-electron is involved in forming the final doubly ionized state, a different behaviour might occur. First results for the intrinsic spin polarization parameters and intensities for some selected transitions of the Kr $L_2M_{2,3}M_{4,5}$ Auger lines are shown in table 3. Our data show that the behaviour of the β_1-parameter is in fact a general one, and not restricted to the argon case.

Table 3: The Auger energies, relative intensities and intrinsic spin polarization parameters β_1 and γ_1 for selected Auger transitions of the Kr $L_2M_{2,3}M_{4,5}$ multiplet. The Auger energies are given in a.u.'s.

	Kr $L_2M_{2,3}M_{4,5}$			
Final	Auger		Present result	
state	energies	I_0†	β_1	γ_1
$^1F_3$	50.7487	46.85	-0.3333	0.2766
$^1P_1$	50.7640	23.57	-0.3333	-0.2121
$^1D_2$	51.1084	20.37	0.5513	-0.2243

†Note, that the whole multiplet has been normalized to 100%. The three singlet transitions cover $\sim 90\%$ of the total intensity.

Predictions for the spin polarization vector

The in-plane components P_x and P_z of the spin polarization vector are functions of the intrinsic spin polarization parameters β_1 and γ_1 and of the orientation parameter $\mathcal{A}_{10}$. Let us assume that the primary ionization process is by photoionization from the ground state, which is $^1S_0$ for the rare gases. Following the work of Bussert and Klar[21] and applying the LS coupling scheme approximation which has been investigated, by photoionization, for instance, by Berezhko et al[22], we obtain for the orientation of an intermediate ion state with a $np_{1/2}$ hole the general expression

$$\mathcal{A}_{10}(np_{1/2}) = \frac{|D(s)|^2 - \frac{1}{2}|D(d)|^2}{|D(s)|^2 + |D(d)|^2} , \tag{13}$$

where $D(\ell)$ denotes the dipole matrix element, and ℓ describes the angular momentum of the photoelectron. In LS coupling only two partial waves (s and d) are emitted. The above equation yields upper and lower bounds for the orientation

$$-0.5 \leq \mathcal{A}_{10}(2p_{1/2}) \leq 1.0 . \tag{14}$$

Usually, out of the region of the Cooper minimum and well above threshold the $\varepsilon(\ell+1)$-partial amplitude dominates photoionization from the $n\ell$ shell, i.e. d $>>$ s. This yields

$$\mathcal{A}_{10}(2p_{1/2}) = -0.5 . \tag{15}$$

As pointed out by Lohmann et al[2], this value of orientation usually occurs in a region of photon energy where the cross section is high. Applying this result, predictions for the spin polarization vector can be given.

The components P_x and P_z of the spin polarization vector have been extensively discussed for the Ar L_2MM Auger transitions by Lohmann and Larkins[18]. The P_z-component for the considered Ar $L_2M_{2,3}M_{2,3}$ multiplet has been plotted against the angle of Auger emission in figure 3. The data show that an experiment should observe a non-zero spin polarization even for isotropic Auger lines and a degree of 50% spin polarization can be found for the single channel transitions to $J_f = 0$ final states.

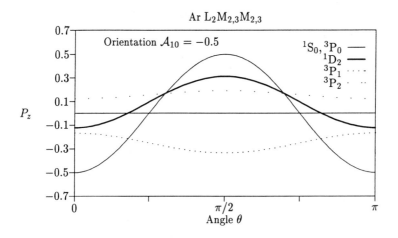

Figure 3: P_z dependent on the Auger electron emission angle.

However, spin polarizations up to 35% can be predicted for most of the other Auger lines.

Since the approximation for the orientation parameter holds for any $np_{1/2}$ initial hole state, the same method can be applied for the $L_2M_{2,3}M_{4,5}$ Auger transitions on krypton. Our results for the final singlet state transitions are plotted in figure 4. A

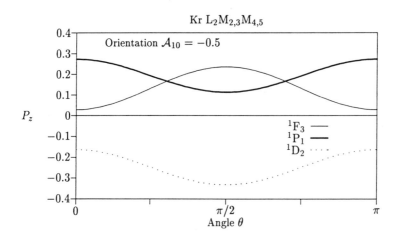

Figure 4: P_z dependent on the Auger electron emission angle.

complete discussion of the results will be published elsewhere[20]. However, as can be seen from the figure, a degree of up to 35% spin polarization can be expected even for the Kr L_2MM Auger transitions. This clearly shows, that the suggested experiment is not restricted to the argon case only. Thus, it can yield a measurable intrinsic spin polarization, even in an area of photon energy where a sufficiently high partial cross section can be expected.

THE EXPERIMENTAL SET-UP

With the results of equations (9–11) the suggested experiment can be carried out as follows:

1. The orientation transferred to the atom by the circularly polarized photon beam has to be determined. This can be done, for instance, by measuring the P_z-component of the spin polarization vector under the 'magic' angle $\theta_m = 54.74^o$, for a single channel transition[‡], e.g. to a final state with $J_f = 0$.

2. Using the obtained value of orientation for the whole multiplet, the intrinsic spin polarization parameters β_1 and γ_1 can be determined in two different ways:

    (a) By measuring P_z only but under two different emission angles θ, e.g. under the magic angle and $\theta = 90^o$. The first measurement directly yields information on β_1, with the second, the value of γ_1 can be obtained.

    (b) Alternatively, measuring both components P_x and P_z under one angle, e.g. the magic angle.

With the first method, only one component of the spin polarization vector has to be determined but under, at least, two different angles. With the latter, both components need to be measured, however, their determination under one certain angle is sufficient.

CONCLUSION

We have shown that generally the measurement and interpretation of the intrinsic spin polarization vector components is not without problems. This is due to the fact that for most Auger transitions the spin polarization vector is a complex function of several independent parameters one has to determine, for properly interpreting the experimental results. However, for certain Auger transitions the number of independent parameters is drastically reduced which makes an experimental interpretation more simpler. We have demonstrated this behaviour for some selected Auger transitions of the argon and krypton L_2MM spectra. It has been shown that for such type of experiments a measurable spin polarization of up to 50% can be expected. A possible experimental set-up has been outlined.

Acknowledgments

A Habilitation grant by the Deutsche Forschungsgemeinschaft is acknowledged. I am thankful to C Lehmann for her ongoing support.

References

1. Mehlhorn W 1990 *X-ray and Inner Shell Processes* (AIP Conf. Proc. 215) ed T A Carlson, M O Krause and S T Manson (New York: AIP) pp 465

2. Lohmann B, Hergenhahn U and Kabachnik N M 1993 *J. Phys. B: At. Mol. Opt. Phys.* **26** 3327

3. Hahn U, Semke J, Merz H and Kessler J 1985 *J. Phys. B: At. Mol. Opt. Phys.* **18** L417

[‡]i.e. independent of the Auger transition matrix elements

4. Merz H, Semke J 1990 *X-ray and Inner Shell Processes* (AIP Conf. Proc. 215) ed T A Carlson, M O Krause and S T Manson (New York: AIP) pp 719

5. Becker U 1992, private communication

6. Heinzmann U 1992, private communication

7. Kuntze R, Salzmann M, Böwering N and Heinzmann U 1993 *Phys. Rev. Lett.* **70** 3716

8. Kabachnik N M and Lee O V 1989 *J. Phys. B: At. Mol. Opt. Phys.* **22** 2705

9. Kessler J 1985 *Polarized Electrons*, 2nd edition (Berlin: Springer)

10. Kabachnik N M and Sazhina I P 1984 *J. Phys. B: At. Mol. Opt. Phys.* **17** 1335

11. Lohmann B 1990 *J. Phys. B: At. Mol. Opt. Phys.* **23** 3147

12. Grant I P, McKenzie B, Norrington P, Mayers D and Pyper N 1980 *Comp. Phys. Comm.* **21** 207

13. Aksela H, Aksela S and Pulkkinen H 1984 unpublished data, quoted in Tulkki *et al* 1993

14. Tulkki J, Kabachnik N M and Aksela H 1993 *Phys. Rev. A* **48** 1277 and private communication

15. Berezhko E G, Ivanov V K and Kabachnik N M 1978 *Phys. Lett.* **66A** 474

16. Becker U, Szostak D, Kerkhoff H G, Kupsch M, Langer B, Wehlitz R, Yagishita A and Hayaishi T 1989 *Phys. Rev. A***39** 3902

17. Lindle D W, Ferrett T A, Heimann P A and Shirley D A 1988 *Phys. Rev. A* **37** 3808

18. Lohmann B and Larkins F P 1994 *J. Phys. B: At. Mol. Opt. Phys.* **27** L143

19. Mehlhorn W and Stalherm D 1968 *Z. f. Physik* **217** 294

20. Lohmann B 1995, in progress

21. Bussert W and Klar H 1983 *Z. f. Physik A* **312** 315

22. Berezhko E G, Kabachnik N M and Rostovsky V S 1978 *J. Phys. B: At. Mol. Opt. Phys.* **11** 1749

PAULI BLOCKING IN RELATIVISTIC (E,2E) COLLISIONS

H Ast[1,2], Colm T Whelan[1], S Keller[2], J Rasch[1], HRJ Walters[3] and RM Dreizler[2]

[1]Department of Applied Mathematics and Theoretical Physics, University of Cambridge, Silver Street, Cambridge, CB3 9EW, UK

[2]Institut für Theoretische Physik der Universität, Robert Mayer Straße 8–10, D–60054 Frankfurt am Main, Germany

[3]Department of Applied Mathematics and Theoretical Physics, The Queen's University of Belfast, Belfast BT7 1NN, Northern Ireland

Introduction

One cannot have a relativistic description of angular momentum without spin. Indeed even in a non–quantum theory orbital angular momentum is an operationally unsound concept - because of the nature of Lorentz transformation one cannot set about defining such a quantity which would admit a measurement[1]. In quantum mechanics a full consistent theory is only possible if spin is included[2]; but spin on its own is only properly defined in the rest frame of the particle. When we say that we are going to discuss spin dependent effects in relativistic (e,2e) processes we have to be very careful about what we really mean since it is the total angular momentum which is the only sensible physical quantity that we can properly talk about.

In this paper we will be concerned with relativistic (e,2e) processes. In such a process a beam of electrons is fired at a target, ionizes it and two electrons are detected in coincidence i.e. it is a fully resolved experiment in angle and energy[3]. At relativistic energies it is a delicate matter to discuss experiments where the spin is also resolved[4-6]. In this contribution we will consider (e,2e) experiments with unpolarized beams on unpolarized targets and demonstrate the role of spin dependent effects i.e. we are going to take care to elucidate the implicit effects of the different spin channels on the cross section.

Inner Shell Ionization

Nakel and his colaborators in Tübingen have measured the triple differential cross section (TDCS) for the K-shell ionization of copper, silver, tantalum and gold at relativistic impact energies[7-10]. As these data are on an absolute scale with a systematic error of less than 15% they present a serious challenge to theory. For some time, despite a large number of calculations, agreement between experimental data and theoretical results was very poor[10-16], due either to the use of semirelativistic wave functions or the neglection of the strong field of the nucleus and the spectator electrons in which the ionization of a K–shell electron takes place. In Ast et al[17] and subsequent papers[18-20] it was shown that the ionizing process can be described to first order in the electron electron interaction but the strong field of the nucleus and the atomic electrons has to be taken into account in the incident and final channels. A computer code to calculate TDCS for inner shell ionization in the relativistic energy regime, including the full photon propagator and exact eigenstates of the Dirac equation with an effective atomic potential was produced. The agreement with the results of the Tübingen experiments[9,10] notably improved.

Quite generally for an unpolarized beam incident on an unpolarized target the spin averaged TDCS is given by

$$\frac{d^3\sigma(e,2e)}{d\Omega_1 d\Omega_2 dE_2} = \frac{(2\pi)^2}{c^6} \frac{k_1 k_2}{k_0} E_0 E_1 E_2 \frac{1}{2} \sum_{s_1,s_2,s_0,s_b=-\frac{1}{2}}^{\frac{1}{2}} |< k_1 s_1 \ k_2 s_2 \ | \ \hat{S} \ | \ k_0 s_0 \ \kappa s_b >|^2$$

(1)

where the indices 0,1,2,b refer to incoming, the two outgoing and the bound electron respectively. E_0, E_1, E_2 and k_0, k_1, k_2 are on shell total energies and momenta of the unbound particles $E^2 = k^2 c^2 + c^4$. κ denotes the quantum numbers of the atomic bound state and s are the spin projections with respect to the quantization axis, which we define to be the beam axis. In the relativistic distorted wave Born approximation (RDWBA)

$$< k_1 s_1 \ k_2 s_2 \ | \ \hat{S} \ | \ k_0 s_0 \ \kappa s_b >^{DWBA} = S^{Dir} - S^{Ex}$$

(2)

where the direct and exchange matrix elements are defined in Keller et al[18]. In eq. (1) we average over the initial and sum over the final spin states. Among 16 possible terms a number would of course be not allowed in a non relativistic theory, namely those associated with changes of spin, e.g.

$$s_0 = s_b = -s_1 = -s_2 \qquad (3)$$

However, as stressed above, the spin projection is not a conserved quantum number in relativistic physics and in a collision process a change of the spin projection quantum number of one or both particles involved is always possible. We will refer to these channels which do not occur in a non relativistic approximation as "spin flip" channels.

Let us first consider the importance of these channels in kinematical arrangements which have already been studied experimentally. In the relativistic regime experiments have been performed under either highly asymmetric or fully symmetric conditions. In the former case it is usual to work in "Ehrhardt" geometry where the incident , scattered and ejected electrons lie in one plane. One electron, normally the faster, is detected at a fixed angel θ_1 while the angle of the second one is varied. In fully symmetric or "Pochat" geometry both electrons share energy and are detected with the same angle to the left and the right of the beam direction.

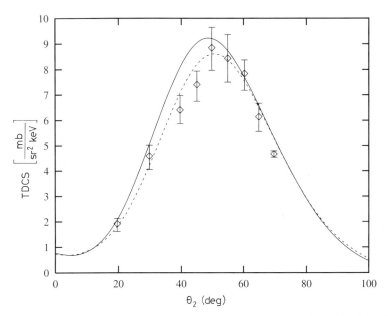

Figure 1. TDCS for the K shell ionization of gold by electron impact plotted against the angle of the slow outgoing electron. Impact energy 500 keV, energy of the slow outgoing electron 100 keV, detection angle of the fast outgoing electron $\theta_1 = -15^0$, $\diamond$ Bonfert et al[9],—- complete RDWBA[19], - - - - RDWBA but non spinflip channels only

In Figure 1 an asymmetric collision is considered. The TDCS obtained in RDWBA[17,18] are plotted against the detection angle of the slow outgoing electron for a K shell ionization of gold at an impact energy of 500 keV. The detection angle of the fast outgoing electron ($E_1 = 319.3\ keV$) is fixed at $\theta_1 = -15^0$, the slow electron carries an energy of 100 keV. The solid curve represents the complete TDCS for all channels and the dashed one the partial TDCS for the non spin flip contributions. Here spin flip processes contribute only a little to the TDCS.

However the picture changes completely for a fully symmetric collision. Here the outgoing electrons share energy and they are detected under the same angles left and right of the beam axis. If the spins are equal and unchanged during the collisions, direct and exchange term have to cancel each other due to the indistinguishable nature of the electron. Hence the non spin flip channels with equal spins of both electrons are blocked. Therefore the relative importance of the spin flip channels is enhanced compared to the "Ehrhardt" kinematics. This was first noted by Walters et al[10] while investigating the semirelativistic first Born approximation.

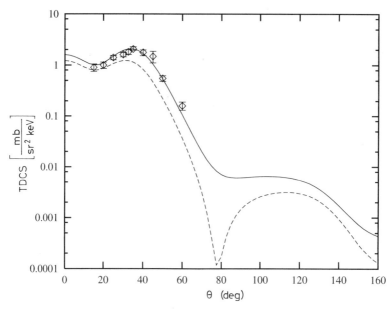

Figure 2. TDCS for (e,2e) on gold 500 keV impact energy in coplanar symmetric geometry plotted against the detection angle of one outgoing electron. ◇ Bonfert et al[9], — RDWBA[19](all channels), - - - - RDWBA but non spinflip channels only

In Figure 2 the TDCS for gold at an impact energy of 500 keV is plotted against the detection angle θ of one of the outgoing electrons who share energy ($E_1 = E_2 = 209.65\ keV$). The solid line represents the TDCS obtained in RDWBA[14] including all spin flip contributions and the dashed curve represents the partial TDCS for the non spin flip channels only. The spin flip channels contribute up to 40% to the TDCS in the binary region. Both calculated cross sections are compared with the absolute experimental results from Nakel and coworkers in Tübingen[9]. Whereas the calculated data of the TDCS (solid line) agree very well with the experimental results, the partial non spin flip TDCS underestimates the experimental results significantly. It is particularly interesting to compare the situation near to the minimum between the binary and large angle peaks. A deep minimum in "Pochat" geometry was first predicted by Zhang et al[21] in their study of helium at an impact energy of 200 eV and was observed by Frost et al[22]. We note that in the full RDWBA the minimum is much less deep than in the non-relativistic case and that this is almost entirely due to the contribution of the spin flip channels whose contribution is two orders of magnitude larger than the non spin flip ones at this point.

To extract further information about specific spin channels, complete spin sensitive experiments would be highly desirable. However we will show in this contribution, that it is possible to make significant statements about specific spin channels by a different choice of the collision geometry although none of the spins will be analyzed.

Instead of performing the experiments in "Ehrhardt" geometry or fully symmetric in "Pochat" geometry we keep the relative angle between the outgoing electrons constant and rotate both detectors in the plane. The detection angles of both electrons are then related by the simple expression

$$\theta_2 = \theta_1 + \Theta_{1,2}$$

This coplanar constant $\Theta_{1,2}$ geometry was proposed by Whelan et al[23] as an ideal one to study polarization effects in low energy (e,2e) collisions. Experiments in this geometry have been recently performed at Kaiserslautern[24] and confirmed its utility in studying incident channel effects while keeping PCI to some extend constant. In the relativistic case PCI is neglagable over a wide angular range[25] but we have previously argued[20] that such a geometry would still be of value for the high energy case because it would allow one to emphazise the role of distortion effects in the incident and final channels.

In Figure 3 the TDCS for the K–shell ionization of gold obtained in the relativistic distorted wave Born approximation (RDWBA) is plotted against the detection angle of the second electron. The relative angle $\Theta_{1,2}$ between the two outgoing electrons is fixed at 60 degrees, the impact energy is 300 keV and the two outgoing electrons share energy ($E_1 = E_2 = 109.65\ keV$). One recognizes a double peaked primary structure and secondary maxima to the left and the right of the primary. Each point accessed in the coplanar constant $\Theta_{1,2}$ geometry has one point in common with a specific "Ehrhardt" geometry. By comparison with the TDCS obtained in "Ehrhardt" geometry one recognizes that the primary structure samples over binary and the secondary maxima over recoil peaks in diffrerent "Ehrhardt" geometries. We further note that the saddle point in the primary structure has the coordinates $-\theta_1 = \theta_2 = \frac{\Theta_{1,2}}{2}$ and is therefore also the point in common with the "Pochat" geometry. Because of the symmetry of the problem the point $-\theta_1 = \theta_2 = \frac{\Theta_{1,2}}{2}$ must be either an extremum or the TDCS must be a constant in the neighbourhood of this point. It is very instructive to see why a minimum is realized here. To calculate the TDCS we have summed over asymptotic spin projection channels for the outgoing

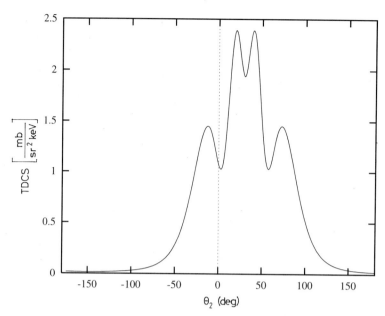

Figure 3. TDCS obtained in RDWBA plotted against the detection angle of electron 2 for gold at 300 keV impact energy, the outgoing electrons share energy, the relative angle $\Theta_{1,2}$ between the detectors is fixed at 60^0

electrons and averaged over the initial ones (eq (1)). Among the 16 channels , 6 are dominant, namely

$$(i) \qquad s_1 = s_0 = s_2 = s_b$$

$$(ii) \qquad s_1 = s_0 = -s_2 = -s_b$$

$$(iii) \qquad s_1 = -s_0 = s_2 = -s_b$$

where s_1 is in atomic units $\frac{1}{2}$ or $-\frac{1}{2}$. The remaining channels, which contain only spin flip amplitudes, make a much smaller contribution. For channel (i) above, both the direct and exchange S-matrices are non spin flip while for (ii) S^D and for (iii) only S^{Ex} is non spin flip. For channel (i) where all spins are equal it is clear that at the symmetric point $-\theta_1 = \theta_2 = \frac{\Theta_{1,2}}{2}$ the S matrix is exactly zero, as the direct and exchange terms have to be equal there due to the symmetry of the problem. However for the two other channels the S matrix elements are finite at this point. Indeed for (ii) $\quad |\ S^D\ |^2 >> |\ S^{Ex}\ |^2$ and for (iii) $\quad |\ S^D\ |^2 << |\ S^{Ex}\ |^2$ over the entire angular range, because spin flip amplitudes are smaller compared to non spin flip amplitudes at the impact energies under consideration here. In Figure 4 we plot the partial TDCS in RDWBA corresponding to the spin channels (i), (ii) and (iii) with $s_1 = \frac{1}{2}$ and $\Theta_{1,2} = 60^0$ for gold at 300 keV impact energy. We see that the dip observed in the TDCS arises because of the relative importance of the individual channel contributions as we move away from the point of symmetry.

It is now instructive to vary the relative angle between the two outgoing electrons.

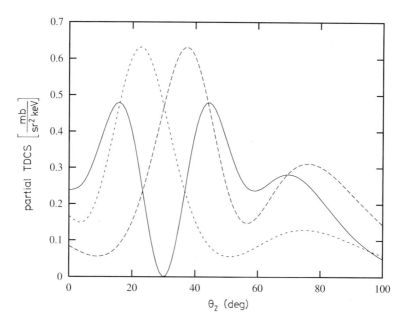

Figure 4. Partial TDCS obtained in RDWBA plotted against the detection angle of electron 2 for gold at 300 kev impact energy, the outgoing electrons share energy, the relative angle $\Theta_{1,2}$ between the detectors is fixed at 60^0 — partial cross section channel (i), - - - partial cs channel (ii), – – – partial cs channel (iii)

In Figure 5 the TDCS is plotted as a function of the detection angle θ_2 for gold at 300 keV impact energy for a)$\Theta_{1,2} = 10^0$, b) 30^0 , c) 90^0 , d) 115^0 , e) 140^0 and f) 180^0 . Significant changes in the shape of the curve are observed if the relative angle $\Theta_{1,2}$ is enlarged or decreased with respect to our earlier example $\Theta_{1,2} = 60^0$. If $\Theta_{1,2}$ is enlarged the primary structure decreases with respect to the secondary maximum which remain more or less unchanged and for $\Theta_{1,2} = 140^0$ the primary structure completely vanishes. With increasing relative angle $\Theta_{1,2}$ the geometrical conditions allow us only to sample over the outer edges of binary maxima in the respective "Ehrhardt" geometries and therefore the primary structure decreases. The case f) $\Theta_{1,2} = 180^0$ is somewhat special since for an energy sharing collision the direction of the momentum transfer $\vec{\Delta}$ with respect to one electron coincides with minus the direction of momentum transfer $(-\vec{\Delta})$ with respect to the other electron. Therefore relating a structure to the binary or recoil peak in an "Ehrhardt" geometry becomes meaningless.

However , physically it is more interesting to study the decrease of the fixed relative angle $\Theta_{1,2}$. Here the primary structure remains strong but the dip vanishes and the point of symmetry converts from a local minimum to a maximum. To investigate this in greater detail we analyse once more (see Figure 4) the contribution of the different spin channels to the TDCS, but now for the case 5a) $\Theta_{1,2} = 10^0$. From Figure 6 we see that the contribution of spin channel (i) is dramatically decreased

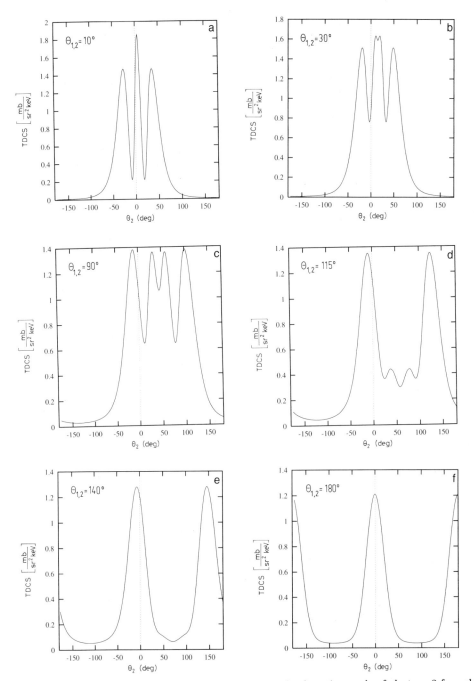

Figure 5. TDCS obtained in RDWBA plotted against the detection angle of electron 2 for gold at 300 kev impact energy, the outgoing electrons share energy, the relative angle $\Theta_{1,2}$ is fixed at (a) 15^0 , (b) 30^0 , (c) 90^0 , (d) 115^0 , (e) 140^0 , (f) 180^0

in absolute size and with respect to the channels (ii) and (iii), not only in the neighbourhood of the point of symmetry $-\theta_1 = \theta_2 = \frac{\Theta_{1,2}}{2}$ but over the entire angular range. This is a clear example of Pauli blocking. In channel (i) the outgoing electrons have the same spin projection and for small relative angles $\Theta_{1,2}$ the spatial quantum numbers are nearly equal. Therefore the probability of finding the two electrons in this channel is supressed.

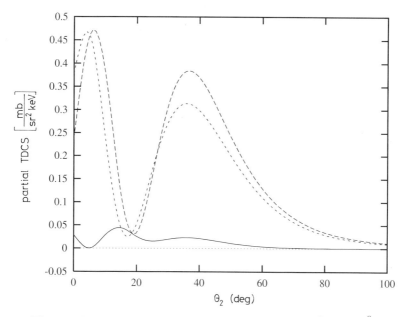

Figure 6. same as figure 4, but for a fixed relative angel $\Theta_{1,2} = 10^0$

In principal this information could also be deduced from a sufficient complete set of measurements in different Ehrhardt geometries by comparing data from measurements with different fixed angles θ_1 . This would always require accurately internormalized data or absolute TDCSs. In the coplanar constant $\Theta_{1,2}$ geometry the shape of the curves changes significantly. This provides a much more straightforward way of studying the Pauli blocking of specific channels.

Finally we remark that since Pauli blocking is not a relativistic phenomenon it should also be observable at low energies[26]. In this case, while the spin flip amplitudes no longer contribute we still have the same effect, i.e. the channels with all spins equal contribute zero at the symmetric point and this relative contribution to the TDCS is much smaller for small angles of $\Theta_{1,2}$ than for large. In Figure 7 a-d we show data for the K shell ionization of Neon at an impact energy of 3.2 keV. The outgoing electrons share energy ($E_1 = E_2 = 1.178 \ keV$). The ratio between the impact energy and the K shell binding energy for neon is the same as the ratio of impact to K shell binding energy for gold at 300 keV. In Figure 7(a) $\Theta_{1,2} = 15^0$, (b) $\Theta_{1,2} = 30^0$, (c)

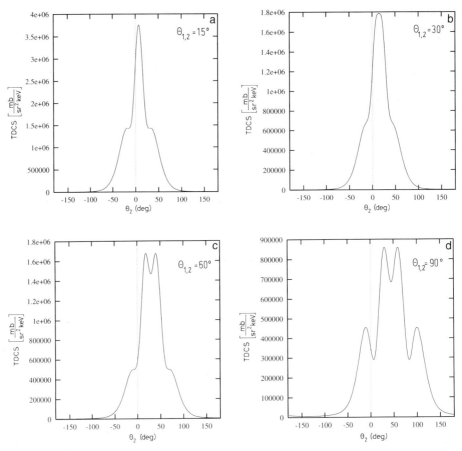

Figure 7. TDCS plotted against the detection angle of electron 2 for neon at 3.2 keV impact energy, the outgoing electrons share energy ($E_1 = E_2 = 1.178$ keV), the relative angle $\Theta_{1,2}$ is fixed at (a) 10^0 , (b) 30^0 , (c) 60^0 , (d) 90^0

$\Theta_{1,2} = 60^0$ and in (d) $\Theta_{1,2} = 90^0$. Similar observations compared to those obtained in the collisions on gold at relativistic could be made here. At the symmetric point ($\theta_1 = -\theta_2$) the TDCS as a function of the detection angle of one electron converts from a local minimum to a maximum if the relative angle $\Theta_{1,2}$ between the detectors decreases. The effect of Pauli blocking should be clearly observable.

Conclusions

We have considered the ionization of gold at relativistic energies and neon at nonrelativistic energies. We have shown that many of the same structures should be observable and we have predicted that one should observe a clear example of Pauli blocking in both cases if one were to work in the coplanar constant $\Theta_{1,2}$ geometry proposed by Whelan et al[20,23] and vary the angle $\Theta_{1,2}$. In the high energy case we have demonstrated the importance of carrying out a fully relativistic calculation and the importance of not neglecting any of the spin channels. We have shown that in coplanar symmetric geometry the spin flip channels contribute 40% to the cross section at binary maximum and two orders of magnitude at the minimum of the cross section.

Acknowledgements

We would like to thank Prof Nakel and his colleagues for many fruitful discussion. We gratefully acknowledge financial support for this work by the Deutsche Forschungs Gemeinschaft and Royal Society, the European Science Foundation (REHE), the Nato (crg 950407) and the EC (HCM CHRX-CT93-0350). The numerical calculations of this work have been performed on the workstation clusters of the Gesellschaft für Schwerionenforschung (GSI) Darmstadt, the Hochschulrechenzentrum Universität Frankfurt, the Institut für Theoretische Physik, Universität Frankfurt and the Department of Applied Mathematics and Theoretical Physics , University of Cambridge.

References

[1] A Einstein , *The Meaning of Relativity* 5th ed. Princeton Unviversity Press: Princeton NJ (1955).

[2] PAM Dirac , *The Principles of Quantum Mechanics* Clarendon: Oxford, (4^{th}ed. 1957).

[3] Colm T Whelan, HRJ Walters and X Zhang , (e,2e), effective charges, distorted waves and all that *in (e,2e) and related processes* eds. CT Whelan, HRJ Walters, A Lahmam-Bennani and H Ehrhardt p.1 Kluwer: Dordrecht (1993).

[4] HT Prinz, KH Besch and W Nakel, Spin orbit interaction of the continuum electrons in relativistic (e,2e) measurements *Phys. Rev. Lett.* **74** , 243(1995).

[5] Colm T Whelan, HRJ Walters, S Keller, H Ast, J Rasch and RM Dreizler, *Can. J. Phys.* , in press

[6] S Keller, RM Dreizler, H Ast, CT Whelan and HRJ Walters, Theory of (e,2e) processes with spin polarized relativistic electrons *Phys. Rev.A* , submitted

[7] E Schüle and W Nakel, Triply differential cross section for K shell ionization of silver by relativistic electron impact *J. Phys. B: At. Mol. Phys.* **15** , L639 (1982).

[8] H Ruoff and W Nakel, Absolute triply differential cross section for K-shell ionisation by relativistic electron impact for high atomic number *J. Phys. B: At. Mol. Phys.* **20** , 2290 (1987).

[9] J Bonfert, H Graf and W Nakel, Relativistic (e,2e) processes on atomic inner shells *J. Phys. B: At. Mol. Opt. Phys.* **24** , 1423 (1991).

[10] HRJ Walters, H Ast, CT Whelan, RM Dreizler, H Graf, CD Schröter, J Bonfert

and W Nakel, Relativistiv (e,2e) collisions on atomic inner shells in symmetric geometry *Z. Phys.D* **23** , 253, (1992).

[11] JN Das and AN Konar, Inner shell ionization cross sections when Sommerfeld-Maue wavefunction is used *J. Phys. B: At. Mol. Phys.* **7** , 2417, (1974).

[12] F Bell, Double and triple differential cross sections for K shell ionization by relativistic electron impact *J. Phys. B: At. Mol. Opt. Phys.* **22** , 287, (1989).

[13] D Jakubassa-Amundsen, Relativistic theory of K shell ionization by fast electrons *Z. Phys.D* **11** , 305, (1889).

[14] D Jakubassa-Amundsen, A systematic study of relativistic (e,2e) collsions in comparison with experiment *J. Phys. B: At. Mol. Opt. Phys.* **25** , 1297, (1992).

[15] YV Popov and NM Kuzmina, EBWA calculations for the relativistic (e,2e) experiments *J. Phys. B: At. Mol. Opt. Phys.* **26** , 1215, (1993).

[16] A Cavalli and L Avaldi, Symmetric (e,2e) experiments at 300 and 500 keV on Cu, Ag,and Au K shell and impulsive models *Nuovo Cim.* **16 D** , 1, (1994).

[17] H Ast, S Keller, CT Whelan, HRJ Walters and RM Dreizler, Electron impact ionization of the K shell of silber and gold in coplanar asymmetric geometry *Phys. Rev.A* **50** , R1, (1994).

[18] S Keller, CT Whelan , H Ast, HRJ Walters and RM Dreizler, Relativistic distorted wave Born calculations for (e,2e) processes on inner shells of heavy atoms *Phys. Rev.A* **50** , 3865, (1994).

[19] Colm T Whelan, H Ast, S Keller , HRJ Walters and RM Dreizler, Triple differential cross sections in energy sharing symmetric geometry for gold and uranium at relativistic impact energies *J. Phys. B: At. Mol. Opt. Phys.* **28** , L33, (1995).

[20] Colm T Whelan, H Ast, HRJ Walters, S Keller and RM Dreizler, Relativistic energy sharing (e,2e) processes in a coplanar constant $\Theta_{1,2}$ geometry *Phys. Rev.A* , in press

[21] X Zhang, CT Whelan and HRJ Walters, (e,2e) cross sections for the ionization of He in coplanar symmetric geometry *J. Phys. B: At. Mol. Opt. Phys.* **23** , L509, (1990).

[22] L Frost, P Freienstein and M Wagner, 200 eV coplanar symmetric (e,2e) on helium: a sensitive test of reaction models *J. Phys. B: At. Mol. Opt. Phys.* **23** , L715, (1990).

[23] Colm T Whelan, RJ Allan, HRJ Walters, PCI, polarization and exchange effects in (e,2e) collisions *Joural de Physique IV, Colloq.* **3** , C6-39, (1993).

[24] J Röder, J Rasch, K Jung, CT Whelan, H Ehrhardt, RJ Allan and HRJ Walters, *Phys. Rev.A* , in press

[25] B Joulakain, S Keller, J Hanssen and H Ast, PCI effects in inner shell (e,2e) collisions , in preparation, (1995).

[26] J Rasch, CT Whelan, RJ Allan, HRJ Walters and H Ast, *J. Phys. B: At. Mol. Opt. Phys.* , in preparation, (1995).

A DENSITY MATRIX TREATMENT
OF ATOMIC LINE RADIATION

Klaus Bartschat

Department of Physics and Astronomy
Drake University
Des Moines
IA 50311, U.S.A.

INTRODUCTION

Since the early days of quantum mechanics, the theory of atomic line radiation has been of central interest in atomic physics. Percival and Seaton[1] presented a general treatment of the subject that has been widely used by both experimentalists and theoreticians. They briefly review the "Oppenheimer-Penney (O.-P.) theory of polarization" and later criticize it for not satisfying the "principle of spectroscopic stability". This principle requires, for example, that results obtained in a $|J, M_J\rangle$ coupling scheme involving the total electronic angular momentum J and its component M_J with regard to a specified quantization axis, must be identical to those obtained in an $|L, M_L; S, M_S\rangle$ coupling scheme (where L and S are the orbital and spin angular momenta with components M_L and M_S, respectively) in the limit of vanishing fine-structure splitting in the target states. A similar argument can be made for hyperfine-structure effects due to a non-vanishing nuclear spin of the target.

Recently,[2] we have commented in some detail on the Percival and Seaton criticism to the O.-P. theory. Here we outline a modern version of the theory of line radiation in terms of "integrated state multipoles".[3,4] These multipoles represent the irreducible components of the reduced density matrix that describes the ensemble of excited atoms. In particular, we show that some non-vanishing off-diagonal elements of the reduced density matrix ensure the conservation of the principle of spectroscopic stability.

For simplicity and clarity of presentation, we restrict our formulation to fine-structure transitions, particularly the excitation $(ns)^2 S_{1/2} \longrightarrow (n'p)^2 P^o_{1/2,3/2}$ with subsequent radiation back to the ground state. An extension to account for hyperfine-structure effects is straightforward and can be obtained by supplementing the present approach with formulas given, for example, by Blum.[5]

FORMULATION OF THE PROBLEM

We consider atomic line radiation after impact excitation by unpolarized electrons. In the standard experimental arrangement (see, for example, the early work by Skinner and Appleyard[6]), one observes radiation emitted at right angle with regard to the incident beam axis while the scattered electrons are not detected. In this case, one can measure a non-vanishing linear polarization fraction defined by

$$\eta_3 = P_1 \equiv \frac{I(0°) - I(90°)}{I(0°) + I(90°)}, \tag{1}$$

where $I(\beta)$ denotes the intensity transmitted by a Nicol prism oriented at an angle β with respect to the incident beam axis. In Eq. (1), η_3 corresponds to the notation of Blum[5] while P_1 is used, for example, by Born and Wolf.[7]

We begin with the excitation process and write the wavefunction of the combined projectile + target system after the collision process as

$$|\Psi(L_1, S_1; \Theta)\rangle = \sum_{M_{L_1}, M_{S_1}, m_1} f(M_{L_1} M_{S_1} m_1, M_{L_0} M_{S_0} m_0; \Theta) |M_{L_1}, M_{S_1}; \mathbf{k}_1, m_1\rangle, \tag{2}$$

where $f(M_{L_1} M_{S_1} m_1, M_{L_0} M_{S_0} m_0; \Theta)$ is the scattering amplitude for a transition from an initial state $|L_0 M_{L_0}; S_0 M_{S_0}; \mathbf{k}_0 m_0\rangle$ to a final state $|L_1, M_{L_1}; S_1, M_{S_1}; \mathbf{k}_1 m_1\rangle$. The scattering angle Θ is the angle between the initial and final projectile momenta $\mathbf{k}_0$ and $\mathbf{k}_1$, respectively, and we have also included the projectile spin components m_0 and m_1 with regard to the quantization axis.

Equation (2) describes the *coherent* excitation of the target atom by the incident electron. It is only valid if all explicitly spin dependent forces, such as the spin-orbit interaction between the continuum electron and the target nucleus, can be neglected during the excitation process, and if one looks at excited target states with fixed total orbital angular momentum L_1 and total spin S_1, respectively. Only the *combined system* of target atom plus projectile electron can be represented by a fully coherent quantum state, and such a representation also requires well-defined values of M_{L_0}, M_{S_0} and m_0 in the initial state. This "principle of non-separability" is in agreement with the findings of Percival and Seaton[1] and has been discussed in detail in Chapter 3 of Blum.[5]

Since we will eventually deal with transitions between fine- (and hyperfine-) structure levels, it is important to discuss the necessary modifications to the scattering amplitudes. In the above LS-approximation, excitation of the various fine-structure levels is described by a purely algebraic re-coupling procedure for the scattering amplitudes. For a transition between states with total electronic angular momenta J_0 and J_1 with magnetic quantum numbers M_0 and M_1, respectively, this transformation is given by[8]

$$f(J_1 M_1 m_1; J_0 M_0 m_0) = \sum_{L_1, M_{L_1}, L_0, M_{L_0}, S, M_S} f^S(M_{L_1}, M_{L_0})$$
$$\times (L_0, M_{L_0}; S_0, M_{S_0} | J_0, M_0)(S_0, M_{S_0}; \tfrac{1}{2}, m_0 | S, M_S)$$
$$\times (L_1, M_{L_1}; S_1, M_{S_1} | J_1, M_1)(S_1, M_{S_1}; \tfrac{1}{2}, m_1 | S, M_S). \tag{3}$$

Equation (3) expresses the conservation of the total spin S and its component

$$M_S = M_{S_1} + m_1 = M_{S_0} + m_0 \tag{4}$$

through the Clebsch-Gordan coefficients $(j_1, m_1; j_2, m_2 | j_3, m_3)$.

The transformation (3) *may* keep some coherence between excitation of fine-structure levels with *different* J values; in fact, the coherence is (and remains) complete if these levels are exactly degenerate, i.e., in the case of vanishing fine-structure splitting. However, coherence in the *observed signal* can be destroyed because of various effects. The two most important possibilities are:

1) If the level separation is already resolved in the electronic excitation process, by an incident beam with narrow energy width whose energy loss is determined with an accuracy higher than the fine-structure splitting, a definite J value of the excited state can be selected. In this case, the excitation of the individual multiplet levels is already incoherent, and thus no coherence between radiation from different J levels can be expected.

2) Integration over different excitation and decay times in a steady-state experiment may lead to a loss of coherence in the observed photon signal — even if such a coherence were present in an individual collision event. The extent of this loss of coherence between contributions from different J values, however, is determined by the relationship between the fine-structure (or hyperfine-structure) splitting ΔE_{fs} of the target states and the width γ of the individual lines. In fact, the loss is essentially complete if the separation is much larger than the width(s) of the lines (see below).

It should be noted that the "rate equation" formalism of Penney[9] implicitly contains this argument since he only accounts for radiation originating from well-defined fine-structure or hyperfine-structure levels. In essence, his ansatz corresponds to the special case of an infinitely narrow energy width of the incident beam and $\gamma / \Delta E_{fs} = 0$. It can, therefore, never be applied to the case of $\Delta E_{fs} / \gamma = 0$ and is unable to describe the smooth transition between these two extremes, as required by the "principle of spectroscopic stability".

REDUCED DENSITY MATRIX THEORY

We now present a density matrix formulation for the theory of radiation. Such a formulation is not only elegant, but the various steps and approximations also become very transparent.

We begin by assuming the validity of LS-coupling for the excitation process and write down the full density matrix of the system immediately after excitation of states labeled by $|L_1, M_{L_1}; S_1, M_{S_1}; \mathbf{k}_1, m_1\rangle$. Generalizing Eq. (2.4) of Bartschat[10] to our case of interest, we find

$$
\rho_{out}(\Theta', \Theta)_{m_1' m_1}^{M_{L_1}' M_{L_1}; M_{S_1}' M_{S_1}} = \frac{1}{2 \cdot (2S_0 + 1) \cdot (2L_0 + 1)}
$$
$$
\times \sum_{M_{L_0} M_{S_0} m_0} f(M_{L_1}' M_{S_1}' m_1', M_{L_0} M_{S_0} m_0; \Theta')\, f^*(M_{L_1} M_{S_1} m_1, M_{L_0} M_{S_0} m_0; \Theta),
$$

$$(5)$$

where a star denotes the complex conjugate quantity. Note that we have specialized Eq. (5) to the case of unpolarized initial spin and orbital angular momenta in both the target and the projectile beams. We have temporarily kept coherences between different scattering angles $\Theta' \neq \Theta$ which would, however, disappear in an angle-differential measurement that selects a fixed scattering angle.

In order to calculate the outcome that is measured in the experimental arrangement described above, we have to construct the "reduced density matrix" by summing *incoherently* over all non-observed quantum numbers in the system and by integrating over all possible scattering angles Θ as well as over all possible orientations of the scattering plane (i.e., an azimuthal integration over Φ). This *incoherent* summation leads, for example, to $\sum_{m_1'=m_1}$ since the spin component of the scattered projectile is not observed, and to setting $\Theta' = \Theta$ in the angular integration.

If there are no explicitly spin dependent forces to be taken into account, the spin (S_1) and the orbital (L_1) angular momentum systems in the target can be treated separately (see the discussion above). This leads to the spin selection rule

$$
M_{S_1}' = M_{S_1}. \tag{6}
$$

Furthermore, the integral over all azimuthal angles yields

$$
M_{L_1}' = M_{L_1}. \tag{7}
$$

Equations (6) and (7) express the cylindrical symmetry of the problem around the incident beam axis. In fact, if the spin system is completely decoupled from the orbital angular momentum system, this system has spherical symmetry since it does not know anything about the axis defined by the incident beam direction. The average over the initial spins and the summation over all final spin components can then be performed with the trivial result of unity, leaving us with reduced density matrix elements given by

$$
\rho^{M_{L_1}} \equiv Q(M_{L_1}) = \frac{1}{2L_0 + 1} \sum_{M_{L_0}} \int d\Omega_{\mathbf{k}_1} f(M_{L_1}, M_{L_0}; \Theta) f^*(M_{L_1}, M_{L_0}; \Theta), \tag{8}
$$

where $Q(M_{L_1})$ is the angle-integrated (total) magnetic sublevel cross section for an excited state with orbital angular momentum projection M_{L_1}. (For simplicity, we have assumed that normalization factors such as k_1/k_0 are included in the definition of the scattering amplitudes.)

To simplify the notation further, we now treat the special case of $L_0 = M_{L_0} = 0$ and use L and M_L instead of L_1 and M_{L_1}, respectively. Furthermore, the transformation from the $|L, M_L; S, M_S\rangle$ to the $|J, M_J\rangle$ coupling scheme becomes most transparent if we temporarily keep the spin of the excited target state in the equations. Simplifying again from (S_1, M_{S_1}) to (S, M_S), we find

$$\rho^{M_L M_S} = Q(M_L, M_S) = \frac{1}{2S+1} Q(M_L). \tag{9}$$

(See also Eq. (3.4) of Percival and Seaton.[1])

Consequently, the reduced density matrix for the excited atoms in our case of interest can be written in the $|L, M_L; S, M_S\rangle$ coupling scheme as

$$\rho = \sum_{M_L, M_S} |LM_L, SM_S\rangle \, \rho^{M_L M_S} \, \langle LM_L, SM_S|. \tag{10}$$

Note that this density matrix is diagonal in both M_L and M_S, and recall that L and S are assumed to be well-defined fixed quantum numbers.

According to the "principle of spectroscopic stability", one must now be able to perform a purely algebraic transformation into the $|J, M_J\rangle$ coupling scheme and obtain the same results, as long as Nature cannot distinguish between the two coupling schemes. This is the case when the line width is much larger than the splitting of the fine-structure levels in the excited target states. (Very similar arguments can be made to account for hyperfine-structure effects.) The above transformation is achieved by inserting the unit operator $\mathbf{1} \equiv \sum_{J, M_J} |J, M_J\rangle\langle J, M_J|$ twice into Eq. (10). This yields

$$\rho = \sum_{J, M_J; J', M_J'} |J, M_J\rangle \, \langle J, M_J | LM_L, SM_S\rangle \, \rho^{M_L M_S}$$
$$\times \langle LM_L, SM_S | J', M_J'\rangle \, \langle J', M_J'|$$

$$= \sum_{J, J', M_J} |J, M_J\rangle \, (L, M_L; S, M_S | J, M_J) \, \rho^{M_L M_S}$$
$$\times (L, M_L; S, M_S | J', M_J) \, \langle J', M_J|$$

$$= \sum_{J, J', M_J} |J, M_J\rangle \, \rho^{M_J}_{JJ'} \, \langle J', M_J|, \tag{11}$$

where we have used the fact that the transformation is performed through the Clebsch-Gordan coefficients $(L, M_L; S, M_S | J, M_J)$, and that the selection rules for these coefficients lead to $M_J' = M_J$. This is not surprising since the density matrix must reflect the axial symmetry of the problem.

The most important point to be seen from Eq. (11), however, is the fact that the density matrix after the transformation *is not diagonal in the $|J, M_J\rangle$ coupling scheme* since $J' \neq J$ is possible ! As long as these off-diagonal elements are kept in the subsequent description of the optical decay, the "principle of spectroscopic stability" will be fulfilled. The diagonal elements alone, namely the fine-structure resolved cross sections $Q(SLJM_J)$, *do not represent the complete reduced density matrix of the excited state* ! If only those diagonal density matrix elements are taken into account, the spectroscopic stability is lost and can never be regained by subsequently

145

taking the limit of vanishing fine-structure splitting. The latter approximation is essentially the approach of Penney.[9] It *sometimes* yields the correct result for the *observed radiation*, namely in cases where the line separations are large compared to the widths.

We finish this section with some general remarks about exploring the symmetry properties of the problem. The reduced density matrix formalism reflects *directly* the characteristics of the *ensemble* of excited atoms from which radiation is observed. The symmetry properties of this ensemble, therefore, determine the maximum amount of coherence that can be detected. Coherence can only be destroyed but not be generated through the radiative transition.

FROM EXCITATION TO DECAY: TIME-INTEGRATED PERTURBATION COEFFICIENTS

To describe the time evolution of the excited atomic ensemble, we use the method of "perturbation coefficients" (for an introduction, see Blum,[5] Sections 4.7 and 5.5). We will illustrate our results for electron impact induced transitions of the kind $(ns)^2S_{1/2} \longrightarrow (n'p)^2P^o_{1/2,3/2}$ with subsequent optical decay back to the $(ns)^2S_{1/2}$ state. The generalization of this formulation to more complicated situations, particularly the inclusion of effects due to hyperfine-structure in the target, is straightforward. Again we refer to the book by Blum,[5] as well as to the pioneering papers by Macek and Jaecks[11] and by Fano and Macek.[12]

We begin with Eq. (8), specialize again to $L_0 = M_{L_0} = 0$, and replace L_1, M_{L_1} by L, M_L, respectively. From this equation we construct the "integrated state multipoles"[3,4]

$$\langle T(L)^+_{K0} \rangle \equiv \sum_M (-1)^{L-M_L} (LM_L, L-M_L|K0)\, Q(M_L). \tag{12}$$

Equation (12) is based on the fact that the problem of interest exhibits cylindrical symmetry around the incident beam axis which we choose to be the quantization axis. Consequently, the density matrix must be diagonal in all magnetic quantum numbers and thus only state multipoles $\langle T(L)^+_{KQ} \rangle$ with $Q = 0$ may be non-zero. Since neither the incident projectile beam nor the target atoms in the initial state are spin-polarized, the excitation part of the problem is also mirror-symmetric with regard to any plane through the incident beam axis. Non-vanishing state multipoles, therefore, require an even rank $K = 0, 2, 4, \ldots$ Finally, if only electric dipole radiation is observed and no external fields are used to mix multipoles with different ranks, $\langle T(L)^+_{00} \rangle$ and $\langle T(L)^+_{20} \rangle$ completely determine the linear polarization of the line radiation.[5]

Before one can calculate the polarization characteristics of this radiation, however, it is necessary to couple the state multipoles, which describe the system immediately after the excitation process, to the radiation field. For projectile energies of up to a few 100 eV, the bremsstrahlung effects treated in detail by Percival and Seaton[1] can be neglected and, in the simplest approximation, a semi-empirical first-order perturbative approach may be applied by adding a term

$$H_{rad} = e^{-\frac{\gamma}{2}t} \tag{13}$$

to the Hamiltonian that describes the time evolution of the system. In Eq. (13), γ is the line width and $1/\gamma$ is the lifetime of the excited state.

The time evolution of the density matrix is given by[5]

$$\rho(t) = U(t)\,\rho(t=0)\,U(t)^\dagger, \qquad (14)$$

where

$$U(t) = exp^{-\frac{i}{\hbar}Ht} \qquad (15)$$

is the time evolution operator and the dagger denotes its adjoint. After performing an integration over all excitation and decay times in a steady-state experimental arrangement without further time resolution, one finds[5] that the multipoles $\langle T(L)_{KQ}^+ \rangle$ should be replaced by $\bar{G}_K(L)\langle T(L)_{KQ}^+ \rangle$ in the formulas for the observed radiation. For our case of interest, the radiation from various fine-structure levels without further corrections due to hyperfine-structure effects, the time-integrated perturbation coefficients are given by (see, for example, Blum,[5] Section 5.5)

$$\bar{G}_K(L) = \frac{1}{2S+1}\sum_{J'J}(2J'+1)(2J+1)\begin{Bmatrix} L & J' & S \\ J & L & K \end{Bmatrix}\frac{\gamma}{\gamma^2 + \omega_{J'J}^2}. \qquad (16)$$

Here $\omega_{J'J}$ is the separation between levels with J' and J. For simplicity we have assumed that the line width γ is independent of the individual fine-structure level. A generalization is again straightforward, but this simplification will enhance the points to be made with regard to interference effects between levels with $J' \neq J$.

The most important point is the fact that after the algebraic transformation from the $|L, M_L; S, M_S\rangle$ to the $|J, M_J\rangle$ coupling scheme, some coherence between levels with $J' \neq J$ is retained. Furthermore, the integration over excitation and decay times (in fact, an integration over quantum beats) leads to a partial loss of that coherence. The amount of this loss is determined by the relationship between the line width γ and the level splitting $\omega_{J'J}$.

To illustrate these points, we consider electron impact excitation $(ns)^2S_{1/2} \longrightarrow (n'p)^2P^o_{1/2,3/2}$ and subsequent radiative transition back to the initial state $(ns)^2S_{1/2}$. This is a very important process for electron scattering from alkali atoms. It was used by Percival and Seaton[1] as the counter-example to demonstrate that the "principle of spectroscopic stability" was violated in Penney's[9] approach. Without accounting for hyperfine-structure effects, this case is an ideal example to demonstrate the basic principles.

The formalism outlined above can be applied directly to the radiative P $\longrightarrow$ S radiative transition. Using $Q(M_L) = Q(-M_L)$, we find

$$\langle T(L)_{00}^+ \rangle = \frac{1}{\sqrt{3}}[2Q(1) + Q(0)] \qquad (17a)$$

and

$$\langle T(L)_{20}^+ \rangle = \frac{2}{\sqrt{6}}[Q(1) - Q(0)]. \qquad (17b)$$

Inserting these results into Eqs. (6.1.1.a/b) of Blum,[5] we obtain for photon detector angles $(\theta, \varphi) = (90°, 90°)$ (i.e., observation in a direction perpendicular to the incident beam axis)

$$P_1 = \eta_3 = \frac{(I\eta_3)}{I} = \frac{3\bar{G}_2[Q(0) - Q(1)]}{2\bar{G}_0[2Q(1) + Q(0)] - \bar{G}_2[Q(1) - Q(0)]}. \qquad (18)$$

While the selection rules in the 6j-symbol of Eq. (16) require $J' = J$ for multipoles with rank $K = 0$ and thus lead to $\bar{G}_0 \equiv 1/\gamma$, spectroscopic stability is conserved through the contributions from $J' \neq J$ in $\bar{G}_2$. In the one extreme case of $\omega_{J'J} \ll \gamma$ (i.e., when the fine-structure splitting is much smaller than the line width), we can neglect $\omega_{J'J}^2$ in the denominator of Eq. (16) and then use the sum rule for 6j-symbols to obtain $\bar{G}_2 = 1/\gamma$ as well. Consequently,

$$\eta_3 = \frac{Q(0) - Q(1)}{Q(1) + Q(0)}, \tag{19}$$

in agreement with Eq. (3.54) of Percival and Seaton.[1]

On the other hand, if $\omega_{J'J} \gg \gamma$ (i.e., when the level separation is much larger than the line width), the main contributions in Eq. (16) originate from the terms with $J' = J$. Hence, $\bar{G}_2 \approx 1/3\gamma$ and

$$\eta_3 = \frac{3[Q(0) - Q(1)]}{11Q(1) + 7Q(0)}, \tag{20}$$

which is Penney's result.[9]

Finally, let us look at the case where the fine-structure separation is large compared to the line width, but nevertheless not resolved in the photon channel. In this case, we need to average the results obtained for the individual multiplet members. A straightforward way to calculate coupled multipoles in the LS-approximation for excitation of a fine-structure level with a fixed value of the total electronic angular momentum J was given in Eq. (3) of Bartschat and Blum.[4] As mentioned above, the spin system is spherically symmetric in LS-coupling, so only the multipole $\langle T(S)_{00}^+ \rangle \equiv 1/\sqrt{2S+1}$ of the spin system will contribute while $\langle T(L)_{00}^+ \rangle$ and $\langle T(L)_{20}^+ \rangle$ are the same as before. After calculating the multipoles $\langle T(J)_{00}^+ \rangle$ and $\langle T(J)_{20}^+ \rangle$ for $J = 1/2$ and $J = 3/2$, respectively, one finds

$$\eta_3(J = \tfrac{1}{2}) \equiv 0 \tag{21a}$$

and

$$\eta_3(J = \tfrac{3}{2}) = \frac{3[Q(0) - Q(1)]}{7Q(1) + 5Q(0)}. \tag{21b}$$

Note that $\bar{G}_K(J) \equiv 1/\gamma$ for both $K = 0$ and $K = 2$ in this case, since excitation of a definite J value was assumed from the very beginning.

If the individual fine-structure levels are not resolved in the photon channel, one actually measures

$$\begin{aligned}
\langle P_1 \rangle = \langle \eta_3 \rangle &= \frac{I\eta_3(J = \tfrac{1}{2}) + I\eta_3(J = \tfrac{3}{2})}{I(J = \tfrac{1}{2}) + I(J = \tfrac{3}{2})} \\
&= \frac{0 + 3[Q(0) - Q(1)]}{\tfrac{1}{2}[8Q(1) + 4Q(0)] + [7Q(1) + 5Q(0)]} = \frac{3[Q(0) - Q(1)]}{11Q(1) + 7Q(0)},
\end{aligned} \tag{22}$$

in agreement with (20). Note that a factor of 1/2 had to be introduced into the first part of the denominator of Eq. (22) to account correctly for the statistical branching ratio of the oscillator strengths for the two fine-structure levels. This factor appears in the *absolute* reduced dipole matrix element of the intensity formula but cancels out in the *relative* polarization formulas for individual levels.

CONCLUSIONS

We have summarized the theory of atomic line radiation in terms of "integrated state multipoles" that represent the irreducible components of the reduced density matrix for the ensemble of excited atoms. Our formulation conserves the principle of spectroscopic stability, as shown explicitly for a simple model case. Extension of the formalism to more complicated situations is straightforward.

Acknowledgments

I would like to thank Dr. G. Csanak for bringing this problem to my attention and encouraging me to write it up, and Prof. H. Kleinpoppen for inviting me to publish the derivation in the present proceedings. Financial support from the National Science Foundation under grant # PHY-9318377 is gratefully acknowledged.

References

1. I.C. Percival and M.J. Seaton, Phil. Trans. Roy. Soc. A**251**, 113 (1958).
2. K. Bartschat and G. Csanak, Comments At. Mol. Phys. (1996), in press
3. K. Bartschat, K. Blum, G.F. Hanne and J. Kessler, J. Phys. B **14**, 3761 (1981)
4. K. Bartschat and K. Blum, Z. Phys. A **304**, 85 (1982)
5. K. Blum, *Density Matrix Theory and Applications* (Plenum, New York 1981)
6. H.W.B. Skinner and E.T.S. Appleyard, Proc. Roy. Soc. A **117**, 224 (1927)
7. M. Born and E. Wolf, *Principles of Optics* (Pergamon, New York 1970)
8. See, for example, H.-J. Goerss, R.-P. Nordbeck and K. Bartschat, J. Phys. B **24**, 2833 (1991)
9. W.G. Penney, Proc. Nat. Acad. Sci. **18**, 231 (1932)
10. K. Bartschat, Phys. Rep. **180**, 1 (1989)
11. J. Macek and D.J. Jaecks, Phys. Rev. A **4**, 2288 (1971)
12. U. Fano and J.H. Macek, Rev. Mod. Phys. **43**, 553 (1973)

ELECTRON OPTIC DICHROISM*

Joachim Kessler

Physikalisches Institut
Universität Münster
Wilhelm-Klemm-Str. 10
48149 Münster
Germany

It is well known that screws played an important role in Peter Farago's life. He not only knows how to handle a screw driver (which even for a physicist cannot be taken for granted), but his relations to screws are deeper. He is obviously fascinated by the old question why mirror symmetry is often absent in nature.

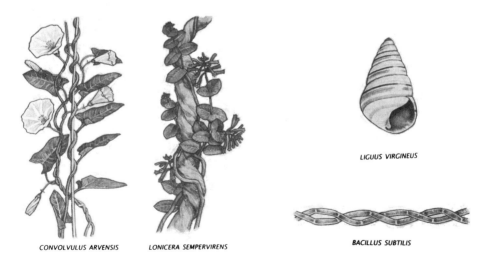

LIGUUS VIRGINEUS

CONVOLVULUS ARVENSIS   LONICERA SEMPERVIRENS   BACILLUS SUBTILIS

Figure 1. Handedness of living things.

---

*After-dinner speech.

I imagine that he wondered why the trumpet honeysuckle (Fig. 1) in his parents' garden in Hungary always winds to the left. On the other hand he may have found that the bindweed always winds as a right-handed helix. Being in the middle of his life, he may have walked along the shores of Scotland and found that certain sea-shells are generally right-handed; left-handed individuals exist only as a result of mutations which appear with a frequency of one in a million. And when one day he had a bad infection he may have wondered whether the bacillus subtilis had inflicted him, which at normal temperatures exists only in the right-handed form (Fig. 1).

All this may or may have not been the reason for an interesting encounter I had with Peter Farago about 15 years ago. The FOM Institute in Amsterdam had invited me for a talk and when they showed me around we suddenly hit upon Peter sitting in one of the labs. He spent a sabbatical year there and was just doing an electron-scattering experiment from camphor ($C_{10}H_{16}O$), a chiral molecule (Fig. 2). We both knew that this was a highly complex problem and that the chances for a positive result were not too good. "But", he said, "I can afford to do such an experiment, because I am an old man and at the end of his career. To a young man at the beginning of his career such a risky experiment would not appeal."

He had started this experiment because there is some hope that electron scattering from chiral molecules can be a cue to the mysteries of non-mirror-symmetric matter. He had focused his attention on the polarization effects that have their origin in the fact that such molecus do not posses space-reflection symmetry. Fig. 3 gives an example of a chiral molecule simpler than camphor. The mirror image of this molecule is not identical to the original molecule. There is no way of turning this picture around so that it coincides with its mirror image. It is just the same situation one has with screws: there is no way to make their mirror image coincide with the original.

This has the consequence that there should be polarization phenomena in electron scattering from such molecules which do not appear with molecules that possess reflection symmetry. In the FOM Institute Peter Farago together with Beerlage and van der Wiel[1] scattered unpolarized electrons from camphor and looked for polarization components of the scattered electrons lying in the scattering plane. Such in-plane polarization components can appear, when the target does not possess reflection symmetry. However, they did not find such polarization above their detection limit of $5 \cdot 10^{-3}$.

But there should be several other structure-related polarization effects, as a recent fundamental analysis has shown[2]. I will discuss here only one of them, which was suggested by Peter Farago[3].

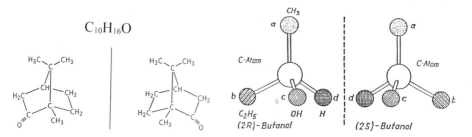

Figure 2. The camphor molecule and its mirror image

Figure 3. Simple molecule and its mirror image demonstrating the chiral structure

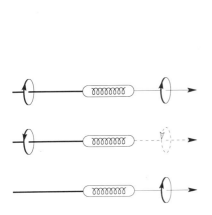

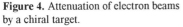

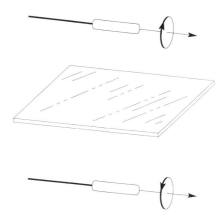

Figure 4. Attenuation of electron beams by a chiral target.

Figure 5. Conceivable violation of parity conservation with non-chiral target.

When a longitudinally polarized electron beam passes through a chiral target, it is attenuated by a certain amount (Fig. 4). If a beam of the opposite helicity passes through the same target, symmetry does not forbid that it becomes attenuated by a different amount. If you add up the two processes, i.e., if you have an unpolarized incident beam, which is a mixture of two beams of opposite helicities and equal intensities, then the outgoing beam should be polarized because the two component beams have different intensities. This is a direct analogue to optical circular dichroism so that it was christened by Peter Farago "electron optic dichroism".

Fig. 5 explains why this cannot happen with a non-chiral target. If it did happen, as indicated there, with a target possessing reflection symmetry we could look at this experiment in a mirror and we would see the same incident beam and the same target as in the lab, but a right-handed beam emerging in the mirror while a left-handed beam emerges in the lab. This violates parity conservation, because one and the same initial state (same beam, same target) must not yield different results in the lab and in the mirror.

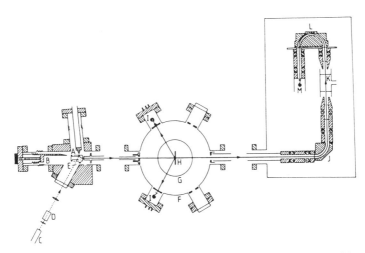

Figure 6. Schematic diagram of the apparatus used by Campbell and Farago for studying electron optic dichroism in camphor.

Fig. 6 shows the apparatus which Campbell and Farago[4] used a few years ago for such a transmission experiment: 5 eV-electrons of 28% polarization pass through a polarization detector which also monitors the stability of the beam current. The polarization is then made longitudinal and the electrons pass through a gas cell which is filled with (unoriented) camphor of one handedness (D- or L-enantiomer). From the transmitted beam intensities I(P) and I(−P) for electron polarization P parallel and antiparallel to the beam axis the transmission asymmetry

$$A = \frac{I(P) - I(-P)}{I(P) + I(-P)} \tag{1}$$

was determined. For the right-handed enantiomer the result was A(D) = (23 ± 11) × 10^{-4} while for the L-enantiomer the authors found A(L) = (−50 ± 17) × 10^{-4}. The quoted uncertainties are single standard deviations. The measured values differ from one another by more than three standard deviations. Accordingly, the authors conclude that the existence of electron optic dichroism has been confirmed with a confidence limit exceeding 95%.

This asymmetry is much larger than could be anticipated from the theoretical estimates made so far, which – for camphor – predicted much smaller values.

In an attempt to throw light on this mystery, Stephen, Burrow[5] and coworkers sent slow electrons through camphor, and they found resonance features in the transmission spectrum. It is conceivable that the temporary-ion formation indicated by such resonances could enhance the polarization effect, because it lengthens the time over which spin-orbit interaction, which produces the effects, takes place. On the other hand, the energies at which these resonanses occur are clearly smaller than the 5 eV of the Campbell-Farago experiment.

Personally, I had never planned to include chiral molecules into the spectrum of experiments we did with polarized electrons. I remembered the numereous experiments where chiral molecules had been bombarded with polarized electrons in order to find out, whether the degradation of the molecules depends on the helicity of the electrons. The results were frustrating because they were highly controversial: positive results were always counterbalanced by negative ones. So I knew: this is a risky field. But when I reached the age of 60, I remembered what Peter Farago told me, when I met him at the FOM Institute and I thought: "Well, now I am also an old man at the end of his career so that I can afford to do such an experiment".

I felt that we were in a good position for doing such a type of experiment, because in one of our projects of the past years we had focused on reliable measurements of small asymmetries and polarizations. A detailed study of the proper position of the monitor detectors which correct for instrumental asymmetries turned out to be an important improvement, just as the increased stability of our polarized electron beam.

This enabled us to perform experiments that were improved in 3 respects:
i)   The uncertainty limits were smaller by a factor of about 50 compared to the Edinburgh experiment.
ii) We made measurements not only at 5 eV, but at energies down to 0.9 eV, so that we measured in the range where the resonances appear, which are supposed to increase the polarization.
iii) We used not only camphor as the target, but aimed at chiral molecules containing atoms of higher atomic number Z so that the effect should be much larger due to the larger spin-orbit interaction.

Fig. 7 shows the basic arrangement of our experiment which is the Ph.D. thesis of S. Mayer. An electron beam of 40% longitudinal polarization from a GaAs source passes through a gas cell with the chiral molecules, where it is attenuated by a factor of 10 or so. The transmitted beam can either be sent into the Mott detector for polarization analysis; or, in another mode of operation, the electrons are caught in a Faraday cup, where the current I_t that is transmitted through the gas cell is measured with an electrometer amplifier. Its output is

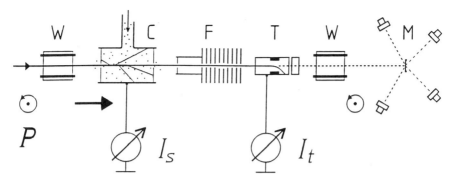

Figure 7. Schematic diagram of our apparatus for the verification of electron optic dichroism. W: Wien filters, C: gas cell, F: filter lens, T: transport optics/Faraday cup, M: Mott analyzer.

digitized and fed into a computer which calculates the asymmetry according to equation (1). We also measure the scattered current I_s which is not transmitted through the gas. These two currents are proportional to the incident current, so that we could eliminate spurious asymmetries caused by fluctuations of the incident current by normalizing the currents to each other. This reduced the spurious asymmetries to 10^{-3} which was tested with non-chiral molecules or the empty gas cell. The remaining asymmetries were much more troublesome. They were caused by minute changes of the position or the angular spread of the incident beam when the electrons were switched between positive and negative helicity. The reduction of these asymmetries to 10^{-4} took us many months. Important steps for this were
- thorough adjustment of the Pockels cell in the GaAs polarized electron source,
- temperature stabilization of the diode laser in the GaAs polarized electron source,
- selection of a non-birefringent vacuum window for entrance of the laser beam.

After these time-consuming preparations the measurements were straightforward. The camphor measurements were made at 5.0 eV for comparison with the data of Campbell and Farago, and at 0.9 and 4.0 eV where the resonances appear. At each energy 30 runs were taken for both right- and left-handed camphor. The handedness of the target was changed every second run. During each run the polarization was reversed with frequencies up to 270 Hz. The more than 10.000 asymmetries measured in a single run show a Gaussian distribution and were averaged for each run.

Fig. 8 shows the asymmetry for each of the 30 runs; filled symbols for right-handed, open symbols for left-handed camphor. The values are randomly distributed around zero. The error bars indicate the statistical uncertainty, but usually they are smaller than the symbol itself. It is obvious that we could not reproduce the results of Campbell and Farago which are shown with their error bars. This agrees with what Tim Gay's group has recently found[6].

While the principle of this experiment is the same as that of the Edinburgh experiment, though the accuracy is much higher, we have made a quite different type of experiment which serves as a cross check. We have sent an unpolarized electron beam through camphor of one or the other handedness and tried to measure the polarization which the transmitted beam should acquire if the Edinburgh data are correct.

The result was again negative: the polarization was zero at each of the three energies. I must, however, say that the detection limit for the polarization was only $\leq 8 \cdot 10^{-4}$ compared to $< 10^{-4}$ for the asymmetry measurement. But still it was somewhat below the uncertainty limits

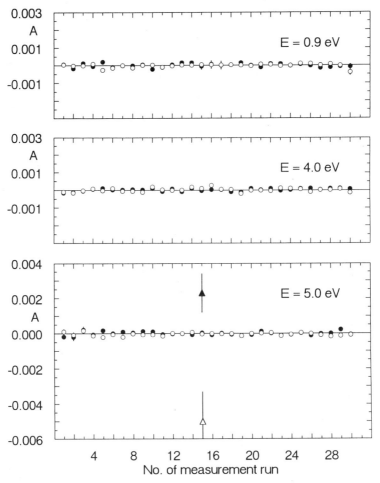

Figure 8. Transmission asymmetry A measured for D-camphor (filled symols) and L-camphor (open symbols) at different values of the electron energy E. The error bars indicate the statistical uncertainty. Where no error bars are given they are smaller than the symbols denoting the measured asymmetries. The triangles are results from Ref. 4.

of the Edinburgh transmission experiment.

After all this had failed we extended our experiment towards chiral molecules containing heavy atoms. We tried $C_5H_{11}I$, and saw no effect. Then we switched to other substances which were more inconvenient because they had to be heated to high temperatures in order to reach a sufficient vapour pressure. For this, the apparatus had to be considerably modified. But it was worth doing it. We found a positive result with Yb(hfc)$_3$ which stands for Tris[3-(heptafluoropropylhydroxy- methylene) camphorato]ytterbium. It contains ytterbium as a high-Z atom surrounded by three camphor-like ligands (Fig. 9). After we had seen an effect we felt

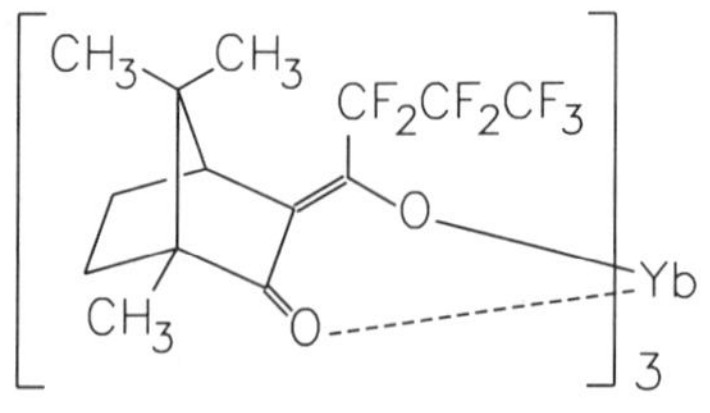

Figure 9. The Yb(hfc)$_3$ molecule.

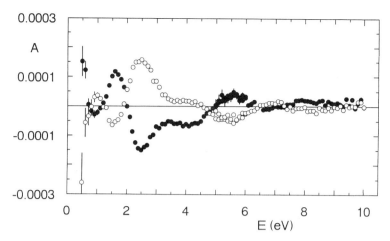

Figure 10. Transmission asymmetry A measured for D-Yb(hfc)$_3$ (filled symbols) and L-Yb(hfc)$_3$ (open symbols) vs. electron energy E. The statistical uncertainty is indicated by error bars, whenever they are larger than the symbols denoting the measured asymmetry.

that it was worth to study the whole energy range between 0.5 and 10.0 eV in steps of 100 meV. Fig. 10 shows the result. The measured asymmetries differ significantly from zero and have opposite signs for the right- and left-handed molecules, as required by symmetry. In a separate study we have searched for the cross-section resonances of our molecules and it turns out that they occur near those energies where we have the pronounced maxima and minima of the asymmetry.

If the measurements had no systematic errors, the sum of the black and white data of Fig. 10 should be zero. This is, however, not the case as Fig. 11 shows which is the sum of the data. The deviation of this curve from zero is caused by instrumental asymmetries which even in the worst case do not exceed $5 \cdot 10^{-5}$ at energies above 0.5 eV.

The asymmetries of order 10^{-4} are in agreement with the theoretical estimates for a model molecule[7], so that we can say now that the severe discrepances between theory and experiment have been resolved.

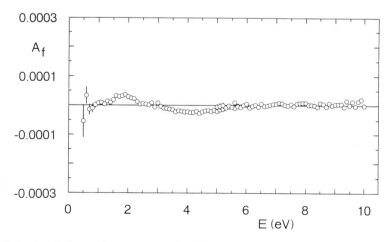

Figure 11. Estimate of the spurious asymmetry A_f which limits the accuracy of the measurement shown in Fig. 10 vs. electron energy E.

Let me conclude by emphasizing the merits of the experiment made in Peter Farago's group: If this experiment had not been done

- there had been no stimulus for the broad theoretical work that has been done on this topic in the meantime
- nobody would have speculated about the connection between the resonances and the structure-related polarization phenomena
- we would never have made this great effort to obtain such accurate information on the fascinating topic of electron optic dichroism.

We are grateful to Peter Farago for the stimulation.

REFERENCES

1. M.J.M.Beerlage, P.S.Farago, and M.J.Van der Wiel, J. Phys. B **14**, 3245 (1981).
2. C.Johnston, K.Blum, and D.Thomson, J. Phys. B **26**, 965 (1993).
3. P.S.Farago, J. Phys. B **13** L567 (1980), B **14**, L 743 (1981).
4. D.M.Campbell and P.S.Farago, J. Phys. B **20**, 5133 (1987).
5. T.M.Stephen, Xueying Shi, and P.D.Burrow, J. Phys. B **21**, L 169 (1988).
6. T. Gay, talk at the present Symposium.
7. R.Fandreyer, D.Thompson, and K. Blum, J. Phys. B **23**, 3031 (1990).

SCATTERING OF CHIRAL ELECTRONS BY

CHIRAL MOLECULES

T.J. Gay, M.E. Johnston,[1] K.W. Trantham, and G.A. Gallup

Behlen Laboratory of Physics
University of Nebraska
Lincoln, Nebraska 68588-0111 U.S.A.

INTRODUCTION

In 1811, Arago demonstrated that when linearly-polarized light passes through a chiral medium, its plane of polarization is rotated. This experimental result, called *optical activity*, was analyzed by Fresnel in 1825, who showed that it was complementary to a second effect, *circular dichroism*, in which, e.g., left-handed circularly polarized light is scattered more strongly in a chiral medium than is light with right-handed circular polarization.   Optically active substances that cause clockwise rotation of the polarization plane with respect to an observer looking at the light source are called *dextrorotatory*, or "right-handed." Substances that cause anti-clockwise rotations are *levorotatory*, or "left-handed." These phenomena have been well-characterized in the intervening years.[2] Since electrons, like photons, can be polarized,[3] it is interesting to speculate that similar phenomena might be observed in electron scattering from chiral targets. Such electronic analogs would be of physical significance in their own right, but would also have interesting implications for biophysics.

It is known that all naturally-occurring biological amino acids are left-handed, while organic sugars are right-handed. Since the primordial precursors of organic materials on earth were presumably synthesized electrochemically, and hence with initially equal numbers of left- and right-handed molecules, the mechanism whereby one chirality ultimately "won out" over the other has been the subject of much debate. One model,[4,5] often referred to as the "Vester-Ulbricht hypothesis of biological homochirality," postulates that beta-radiation, which is longitudinally polarized and hence chiral, preferentially destroyed right-handed amino acids in the initially racemic primordial mixture. This idea led to a number of experiments designed to look for asymmetries in the interactions between longitudinally-polarized electrons or positrons and solid chiral biological targets. While some evidence has been found for such effects,[6] from the standpoint of basic physics it would be more interesting to observe them in electron scattering by single molecules, as opposed to bulk targets, so that there would be some hope of identifying specific chiral dynamical mechanisms responsible for the asymmetries.

In two pioneering papers published in the early 1980's, Farago discussed mathematically the symmetry principles relevant for observing electronic analogs of optical activity and circular dichroism in electron-molecule scattering.[7,8] His development is summarized here. Consider the simplest case of elastic scattering of non-relativistic electrons by a spinless target, in which the initial and final electron momenta, $\bar{k}_i$ and $\bar{k}_f$ respectively, are used to define a collisional coordinate system (figure 1):

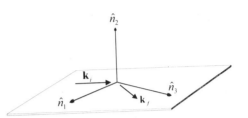

Figure 1. Collision coordinate system defined in equation 1.

$$\hat{n}_1 = (\bar{k}_f - \bar{k}_i)/|(\bar{k}_f - \bar{k}_i)|,$$

$$\hat{n}_2 = (\bar{k}_i \times \bar{k}_f)/|(\bar{k}_f \times \bar{k}_i)|, \text{ and} \qquad (1)$$

$$\hat{n}_3 = (\bar{k}_i + \bar{k}_f)/|(\bar{k}_i + \bar{k}_f)|.$$

In this case, the final-state spin density matrix ρ_f can be defined in terms of a spin scattering matrix M and the initial density matrix ρ_i:

$$\rho_f = M \rho_i M^\dagger. \qquad (2)$$

The scattering matrix can, in turn, be expanded in terms of the Pauli spin matrices, the unit matrix, and a series of amplitudes that are implicitly dependent on $\bar{k}_i$ and $\bar{k}_f$:

$$M = f\sigma_0 + f'\vec{\sigma}\cdot\hat{n}_1 + g\vec{\sigma}\cdot\hat{n}_2 + h\vec{\sigma}\cdot\hat{n}_3, \qquad (3)$$

where

$$\vec{\sigma}\cdot\hat{n}_1 = \begin{pmatrix} 0 & 1 \\ 1 & 0 \end{pmatrix}, \quad \vec{\sigma}\cdot\hat{n}_2 = \begin{pmatrix} 0 & -i \\ i & 0 \end{pmatrix}, \quad \vec{\sigma}\cdot\hat{n}_3 = \begin{pmatrix} 1 & 0 \\ 0 & -1 \end{pmatrix}, \qquad (4)$$

and σ_0 is the 2×2 unit matrix. We now assume that the scattering interaction is invariant under the operation of spatial rotations, time-reversal, and inversion (parity). This means that under the action of a given symmetry operator, each term of M must remain invariant, or be identically zero. This is manifestly true for spatial rotations. Under time reversal, however, $\vec{\sigma}$ flips sign, as do $\hat{n}_2$ and $\hat{n}_3$. Thus while the first, third, and fourth terms of eq.3 remain the same under time reversal, the second flips its sign, meaning that f' must be zero. Finally, we note that $\hat{n}_2$ is an axial vector, whereas $\hat{n}_3$ is polar. The three components of $\vec{\sigma}$ are axial vectors. Thus under inversion $\vec{\sigma} \cdot \hat{n}_2$ will not change sign but $\vec{\sigma} \cdot \hat{n}_3$ will. Therefore h must be identically zero *unless* it also flips sign under inversion. This can happen only if the target is chiral, i.e., lacks any element of symmetry that is an improper rotation. Target chirality, in turn, can be the result of electroweak interactions (which we neglect), spatial orientation (which we do not consider), or of the stereochemical arrangement of the atoms making up the molecule itself. At least four atoms in a non-planar geometry are required for a molecule to be chiral.

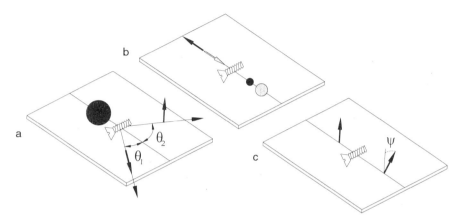

Figure 2. Chiral effects in electron-molecule scattering (see text). a) Production of in-plane polarization of electrons scattered to Θ_1. Transverse polarization production (electrons scattered to Θ_2) does not require chiral targets. b) Preferential transmission of electrons with a given helicity. c) Rotation of initially transverse polarization about the incident momentum.

Farago discussed three classes of "parity-violating," or chiral effects, illustrated schematically in figure 2. Each requires that the amplitude h be non-zero.

(1) Production of polarization of the scattered beam in the plane of scattering. If the incident beam of unit intensity is unpolarized, the scattered flux at (θ, ϕ) is given by

$$I(\theta, \phi) = |f(\theta, \phi)|^2 + |g(\theta, \phi)|^2 + |h(\theta, \phi)|^2, \qquad (5)$$

and its polarization components along the three unit vectors are

$$\vec{P} \cdot \hat{n}_1 = \frac{1}{2I}\left(|g + ih|^2 - |g - ih|^2\right),$$

$$\vec{P} \cdot \hat{n}_2 = \frac{1}{2I}\left(|f + g|^2 - |f - g|^2\right), \text{ and} \qquad (6)$$

$$\vec{P} \cdot \hat{n}_3 = \frac{1}{2I}\left(|f + h|^2 - |f - h|^2\right).$$

The polarization along $\hat{n}_2$ is independent of the chiral amplitude h, and corresponds simply to

Mott scattering, i.e. the production of polarization by spin-orbit coupling between the continuum electron and one (or perhaps more) of the target nuclei. The remaining two polarization components are in the plane of scattering; they will be present only if the chiral scattering amplitude is non-zero. Notice that the magnitude of these components depends on cross terms involving either gh or fh, so that parity-violating effects will be reduced generally by simple ratios of the parity-violating and non-parity-violating amplitudes, as opposed to the ratios of their squares.

(2) Rotation of incident transverse polarization. For incident electrons that are transversely polarized and that are scattered into the forward direction ($\theta=0^0$), the initial polarization vector can be rotated in the plane perpendicular to $\bar{k}_i$ by an angle

$$\psi = \frac{4\pi\lambda}{k}\, dz\, \mathrm{Re}(h), \tag{7}$$

where $k=k_i=k_f$, λ is the target's areal density, and dz is the path length of the electron beam through the target. This effect is thus analogous to optical activity; in that case the rotation angle is also proportional to target thickness and the real part of the optical index of refraction.

(3) Polarization-dependent beam attenuation. Consider now an incident beam with longitudinal polarization either parallel or anti-parallel to its momentum (i.e. with positive or negative helicity). In analogy with photon circular dichroism, a chiral target will preferentially attenuate one helicity over the other. The attenuation asymmetry, A, is given by

$$A = \frac{I^+ - I^-}{I^+ - I^-} = -P_e\, \frac{4\pi\lambda}{k}\, dz\, \mathrm{Im}\,(h), \tag{8}$$

where $I^\pm$ is the intensity of electrons with positive or negative helicity exiting the target undeflected (scattered to 0°), and P_e is the beam's initial longitudinal polarization. The formula for optical circular dichroism is essentially identical to this, involving the imaginary part of the index of refraction.

For chiral targets, the effects listed above are permitted on the basis of symmetry considerations alone. But it is instructive to consider the dynamical mechanisms that cause these effects, i.e. that yield a non-zero value for h. To do this, we consider the specific case of production of longitudinal polarization in the scattered beam. There are essentially three clearly distinguishable mechanisms, which we have illustrated schematically in figure 3. The first was proposed by Kessler[9] in 1982 (fig.3a), and involves a combination of Mott scattering followed by plural electrostatic scattering. For the molecular orientation shown, the incident unpolarized beam is scattered from a first atom. Because of the spin-orbit interaction between the continuum electron and the atomic nucleus, a transverse polarization is generated in the first (Mott) scattering. Neglecting any further spin-orbit coupling, a second, purely electrostatic scattering now deflects the electron to its final path without affecting its spin. Thus the intermediate transverse polarization is partially converted to longitudinal polarization. This scattering sequence could of course happen with a non-chiral target with this specific orientation. But it is only with chiral targets that an average of the longitudinal polarization over all target orientations is not reduced to zero. These

162

considerations imply that chiral effects will be biggest if there is an atom with high Z (to enhance the spin-orbit coupling) located at or near the molecule's chiral center.

A second mechanism, involving induced electric and magnetic moments in the target, is analogous to that responsible for chiral optical effects[2], and has been discussed by Walker[10] and Gallup[11] (fig.3b). We consider again a specific orientation of the chiral target and outgoing electron trajectory. As the incident electron approaches the molecule, it induces a time-varying electric dipole moment on the target. Because of the molecule's chirality, this electric moment causes a net orbital angular momentum of the target electrons, yielding a magnetic dipole. This dipole in turn interacts with the incident electron spin, yielding a "spin-other-orbit" term in the scattering Hamiltonian. The sign of this term, and the resultant differential scattering cross section, depends on whether the incident electron helicity is positive or

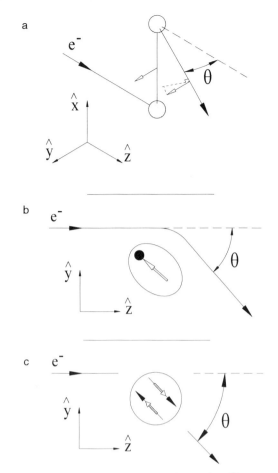

Figure 3. Dynamical causes of chiral scattering effects (see text).

negative; i.e. the probability of an electron being scattered to a specific angle will depend on its initial spin. The scattering process thus acts as a "spin filter," and yields a longitudinal polarization of the scattered beam. In this model we neglect spin-flip processes. The important quantities for determining the magnitude of the expected effects are the electric and magnetic polarizabilities; individual nuclear charges are important only to the extent that there is a correlation between nuclear charge and atomic size, and hence polarizability. This offers the possibility that the judicious choice of a series of targets could allow a determination of the relative importance of mechanisms one and two.

A final mechanism invokes the concept of the target's "helicity density," an idea developed by Hegstrom, Rich, and coworkers.[12,13] In a closed-shell chiral target with a high-Z atom near the chiral center, the singlet and triplet character of the molecular wave functions are partially lost due to the spin-orbit coupling with the high-Z nucleus. This spin mixing results in a non-zero local expectation value of the helicity operator

$$\bar{H} = \hbar^{-1}\bar{\sigma} \cdot \bar{\nabla}; \qquad (9)$$

in these regions there is a correlation between an electron's momentum and its spin. Thus for the target as a whole, there is a net electron helicity, i.e. on average the electrons of the target will be spinning in (for example) the same direction they are moving. While the helicity density can be non-zero locally in achiral molecules (around any local chiral center), the net helicity, integrated over the entire molecular electron density, must be zero. The helicity density of the chiral molecule camphor is shown in figure 4.

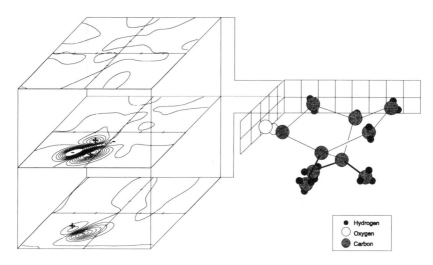

Figure 4. Helicity density in three cross sections of the (right-handed) camphor molecule.

We consider again scattering to a specific angle (fig. 3c), but now that a net helicity of the target has been assumed, we need not posit a given target orientation. The scattered electrons will have originated either in the incident beam ("direct" scattering) or the target ("exchange" scattering). Now it will be generally true that for exchange scattering there will be a dynamical preference for electrons initially on the target whose momenta during the scattering process are parallel (or *vice versa*) to the final momentum vector selected by the scattering experiment. The component of the scattered beam due to direct processes will be unpolarized, but the exchange component will have a longitudinal polarization reflecting this dynamical preference in conjunction with the target's net helicity.

In 1981, Beerlage *et al.*[14] reported a search for longitudinal polarization in initially unpolarized electrons scattered by a camphor target. At an incident energy of 25 eV, at scattering angles between 40° and 70°, their results were consistent with zero to a precision of 5×10^{-3}. But in 1985, Campbell and Farago[15,16] reported a non-zero scattering asymmetry when they measured the current of 28% longitudinally-polarized 5eV electrons transmitted through a target of camphor vapor. Their values of A (eq.8) were $50(17) \times 10^{-4}$ for right-handed camphor and $-23(11) \times 10^{-4}$ for left-handed camphor. We note that the left- and right-handed results have opposite sign, as they must, and that when the absolute values of the two results are combined, they yield a result more than three standard deviations away from zero: $31(9) \times 10^{-4}$. This result was both important and surprising, offering as it did the first evidence of an electronic analog of circular dichroism, and in a molecular target, as opposed to the bulk, where some hope could be held for a good understanding of the origins of the result.

On the other hand, the magnitude of A reported was significantly larger than that which one would expect theoretically. From very basic, general considerations, Kessler[9] and Hayashi[17] used the Mott/plural scattering model to predict chiral production of longitudinal polarization in scattered electrons (the Beerlage *et al.* experiment) to be on the order of 10^{-4}-10^{-5}. Corresponding values of A would be expected to be of the same size. More recent work by Blum, Thompson, and coworkers[18,19] have confirmed this order-of-magnitude result. Rich *et al.*,[14] using the completely different dynamical approach of the helicity density model, also arrived at an estimate of chiral effects on the order of 10^{-5}. With an optical polarizability model, we have estimated effects below the 10^{-4} level.[11]

Because of the importance and controversy of the Campbell and Farago result, we decided to repeat their experiment. A further motivation was the experiment of Stephen *et al.*,[20], in which the camphor molecule was shown to have a strong negative-molecular ion resonance at 0.9 eV, well below 5 eV where Campbell and Farago made their measurements. Hayashi[17] has shown that such resonances can dramatically enhance chiral effects, presumably because the incident electron and the target have a longer time in which to "sample each other's chirality."

EXPERIMENTAL PROCEDURE

The apparatus we used consisted of a GaAs polarized electron source, an in-line optical electron polarimeter, and a camphor target chamber. Care was taken to minimize departures from axial symmetry in the apparatus. The electron beam was undeflected (with the exception of small steering fields) from its point of origin at the GaAs crystal to the electron multiplier where it was detected downstream from the camphor target, and was coincident with the apparatus' symmetry axis. The electron source was thus arranged so that the GaAs crystal plane was perpendicular to this symmetry axis. The diode laser beam (780nm wavelength) that produced the photoemitted electrons entered the source vacuum chamber from behind the crystal, was reflected from a polished electron beam aperture downstream from the crystal, and struck the crystal at near-normal incidence. The electrons were therefore longitudinally polarized at every point along their trajectories through the apparatus.

In order to reduce the effects of stray magnetic fields (which were kept below 10 μT at all points along the beamline), the extracted electrons were transported at 200 eV into the optical polarimeter. This device has been described in detail elsewhere.[21,22] Essentially, it consists of an effusive noble gas target (in this experiment we used neon) which is crossed by the electron beam. An optical polarimeter, whose transmission axis is at a polar angle of 135° to the electron beam, monitors the 640nm fluorescence of the Ne $2p^53p$ $^3D_3$–$2p^53s$ $^3P_2$ transition. By rotating the quarter-wave-plate of the optical train about its transmission axis, the three Stokes parameters of the flourescence radiation, η_1, η_2, and η_3, can be determined. The electron polarization is in turn given by

$$P_e = \frac{3}{\sqrt{2}}\eta_2\left(1 - (17/9)\eta_3\right)^{-1}. \qquad (10)$$

In this experiment, we measured electron polarizations ranging from 26% for bulk GaAs crystals to 47% for epitaxially-grown GaAs.

The electron beam was transported at 200 eV into the camphor target chamber, shown schematically in figure 5. The chamber has a cylindrical glass vacuum housing that is concentric with the electron optical system. This lens system was used to decelerate the beam to its collision energy between 0.9 and 10 eV. The camphor collision cell was fed by a heated vapor line, and the vapor pressure in the cell was monitored by a capacitance manometer. Immediately downstream from the target is a retarding field analyzer (RFA), followed by an electron multiplier. In this experiment, we used devices with both discrete and continuous dynodes. The RFA and the aperture immediately succeeding it served to make cuts in energy and angle on the transmitted electrons. Campbell and Farago's experiment detected electrons that had not suffered angular deviations greater than 2° or energy losses greater than 200 meV. Using the program SIMION[23] to model the axially symmetric RFA and its exit aperture, we calculated the appropriate aperture diameter and electrode voltages necessary to ensure similar conditions with our apparatus, assuming the worst case of isotropic scattering. The resulting electron transmission histogram for one set of parameters is shown in figure 6. Subsequent RFA spectra taken with the apparatus agreed well with our model calculations.

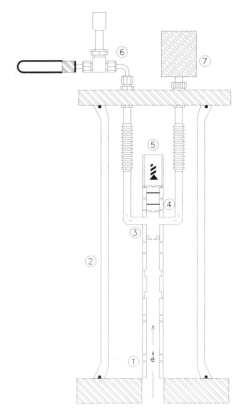

Figure 5. Camphor target chamber showing 1) input electron lenses; 2) glass vacuum chamber; 3) camphor target cell; 4) post-collision retarding field analyzer; 5) electron multiplier; 6) camphor reservoir system; and 7) capacitance manometer.

Camphor can cause problems when used as a target in low-energy electron scattering experiments. The most pernicious of these is changing electron-beam tuning characteristics and energies as camphor coats various electron-optical elements. Two other concerns in this experiment were poisoning of the GaAs crystal and a reduction of the detection efficiency of the electron multipliers used to monitor the transmitted electrons. In order to minimize the effusion of camphor vapor into the rest of the apparatus, the entrance and exit apertures of the target cell were of small diameter; 1.6 and 2.0 mm respectively. Another precaution we took was to heat the internal components of the target chamber to about 50°C through the glass vacuum wall with infrared incandescent lamps. These measures were helpful but did not eliminate all problems. The GaAs photocathode activation lifetimes, which were effectively infinite with no camphor in the target cell, were at most 12h during data runs. This was true in spite of the fact that the GaAs was "two vacuum chambers away" from the target. A

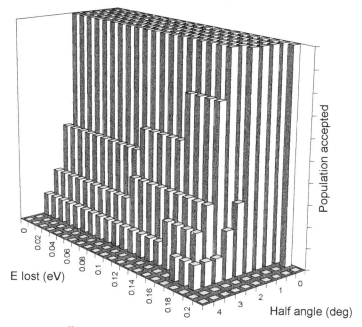

Figure 6. SIMION analysis[23] of electrons scattered in the target cell and subsequently detected (see text).

second, more serious problem was the contamination of the electron multipliers we used. At typical camphor operating pressures, we found the channel electron multipliers to have useful detection lifetimes between 36 and 100h. This led us to try "active film" discrete-dynode electron multipliers, but these devices exhibited even shorter lifetimes. On the other hand, they could be operated at much higher count rates. A possible remedy for the contamination problem would have been to bend the transmitted beam away from the target axis into the detectors, but we were reluctant to do anything to reduce our apparatus' axial symmetry.

DATA ACQUISITION PROTOCOL

Data taking proceeded in the following fashion. The electron beam from a freshly activated crystal was guided into the polarimeter where P_e was measured. With no camphor in the target cell, the beam was then directed onto the electron detector, after being defocussed so that the maximum permissible count rate of the electron multiplier was not exceeded. (This was about 50 kHz for the channeltrons and 1.5 MHz for the discrete dynode multipliers.[24]) At this point, camphor was admitted to the target cell until the incident beam had been attenuated by a factor of two. This required a target pressure of approximately 1 mTorr. After camphor had been in the target cell for a long enough time that the electron-beam tuning potentials had equilibrated, the collision energy was set by adjusting the potential on the scattering cell, and comparing it with that required to completely block the incident beam. Then the RFA potential required to pass only electrons that had lost less than 0.2 eV was determined by taking a complete RFA spectrum.

A typical data-taking sequence for a given target chirality involved ~4 sets of 25 +/– electron helicity pairs. Data was taken for 2s with each helicity; the first helicity was selected randomly by computer. After a 4-set run, the target handedness was changed automatically. This was accomplished with a target handling system that had heated reservoirs of both antipodes of camphor. These reservoirs were connected to the target cell and to a dedicated vacuum line by means of heated stainless steel pipes and solenoid-controlled valves. After four more sets of data were collected with the other enantiomorph, the target helicity was toggled back and another group of data was acquired. This process was repeated until ~400 helicity pairs were accumulated for each handedness of camphor. A histogram of the individual asymmetries for each target chirality was then constructed. We found that while these histograms were Gaussian distributions, the standard deviations of the mean that characterized them were usually larger than would be expected on the basis of counting statistics alone. In all cases, we have therefore used the larger standard deviations of the mean for our quoted uncertainty limits.

The average of the two data sets of opposite handedness at a given incident energy was usually not zero to within our statistical uncertainty. This offset, which was reproducible and slightly energy-dependent, was taken to be a measure of our instrumental asymmetry. The offset values were consistent with instrumental asymmetries we measured using argon as a target.

Table 1. Camphor Asymmetries.

	Electron Energy (eV)	A (left) × 10^{-4}	A (right) × 10^{-4}	[A(right) – A(left)]/2 × 10^{-4}
This work	1.0	–2(5)	+2(5)	2(4)
	3.0	0(5)	0(5)	0(4)
	5.0	–1(4)	+1(5)	+1(3)
	7.0	+3(6)	–3(6)	–3(4)
	10.0	0(2)	+1(7)	+1(4)
Ref. 16	5.0	–178(61)	82(39)	260(36)

RESULTS

The values of A we measured (eq.1) as a function of incident electron energy are given in Table 1. These results have been normalized to the electron polarization, and are compared with the results of Campbell and Farago, normalized to their quoted value of P_e: 28%. It is apparent that we have failed to reproduce Campbell and Farago's results. Our values of A are more consistent with the theoretical upper limits on chiral effects discussed previously. It is disappointing that we found no evidence for non-zero asymmetries, even at the lowest energy, 1eV, where a strong negative ion resonance has been shown to exist.[20]

Another article in this volume discusses evidence for non-zero chiral effects in a ytterbium-based compound[25], and confirms our result in camphor. The next experimental challenge in this area will be to choose targets that elucidate which of the dynamical mechanisms discussed above is responsible for these effects.

ACKNOWLEDGEMENTS

The authors would like to thank P.D. Burrow for numerous useful discussions. This work was supported by the University of Missouri, the University of Nebraska, and by the National Science Foundation (Grant # PHY-9504350).

REFERENCES

1. Permanent address: Physics Department, University of St. Thomas, St. Paul, MN 55105-1096.
2. L.D. Barron. "Molecular Light Scattering and Optical Activity," Cambridge University Press, Cambridge (1982).
3. J. Kessler. "Polarized Electrons," 2nd ed., Springer, Berlin (1985).
4. S.L. Miller and L.E. Orgel. "The Origins of Life on the Earth," Prentice-Hall, Englewood Cliffs (1974).
5. D.C. Walker (editor). "Origins of Optical Activity in Nature," Elsevier, Amsterdam (1979).
6. A.S. Garay and J.A. Ahlgren-Beckendorf, Differential interaction of chiral β-particles with enantiomers, *Nature* 346:451(1990), and references therein.
7. P.S. Farago, Spin-dependent features of electron scattering from optically active molecules, *J.Phys.B* 13:L567(1980).
8. P.S. Farago, Electron optic dichroism and electron optical activity, *J.Phys.B* 14:L743(1981).
9. J. Kessler, Polarisation components violating reflection symmetry in electron scattering from optically active molecules, *J.Phys.B* 15:L101(1982).
10. D.W. Walker, Electron scattering from optically active molecules, *J.Phys.B* 15:L289(1982).
11. G.A. Gallup, The scattering of longitudinally polarized electrons from chiral molecules and optical rotatory power, *in*: "Electron Collisions with Molecules, Clusters, and Surfaces," H. Ehrhardt and L.A. Morgan, eds., Plenum, New York (1994).
12. A. Rich, J. Van House, and R.A. Hegstrom, Calculation of a mirror asymmetric effect in electron scattering from chiral targets, *Phys.Rev.Lett.* 48:1341(1982).
13. R.A. Hegstrom, β decay and the origins chirality: theoretical results, *Nature* 297:643(1982).
14. M.J.M. Beerlage, P.S. Farago, and M.J. Van der Wiel, A search for spin effects in low-energy electron scattering from optically active camphor, *J.Phys.B* 14:3245(1981).
15. D.M. Campbell and P.S. Farago, Spin-dependent electron scattering from optically active molecules, *Nature*, 258:419(1985).
16. D.M. Campbell and P.S. Farago, Electron optic dichroism in camphor, *J.Phys.B* 20:5133(1987).

17. S. Hayashi, Asymmetry in elastic scattering of polarized electrons by optically active molecules, *J.Phys.B* 21:1037(1988).
18. R. Fandreyer, D. Thompson, and K. Blum, Attenuation of longitudinally polarized electron beams by chiral molecules, *J.Phys.B* 23:3031(1990).
19. K. Blum, Chiral effects in elastic electron-molecule collisions, these proceedings.
20. T.M. Stephen, X. Shi, and P.D. Burrow, Temporary negative-ion states of chiral molecules: camphor and 3-methylcyclopentanone, *J.Phys.B* 21:L169(1988).
21. J.E. Furst, W.M.K.P. Wijayaratna, D.H. Madison, and T.J. Gay, Investigation of spin-orbit effects in the excitation of noble gases by spin-polarized electrons, *Phys.Rev.A* 47:3775(1993).
22. T.J. Gay, J.E. Furst, K.W. Trantham, and W.M.K.P. Wijayaratna, Optical electron polarimetry with heavy noble gases, submitted to *Phys. Rev. A.*
23. SIMION is available from D.A. Dahl, Idaho National Engineering Laboratory, Idaho Falls, ID 83415.
24. K.W. Trantham, A count-rate safety circuit for high-voltage single-particle detectors, *Rev.Sci.Instrum.,* in press, to appear in November 1995.
25. S. Mayer and J. Kessler, Experimental study of spin-dependent electron scattering from chiral molecules, these proceedings.

CHIRAL EFFECTS IN ELASTIC ELECTRON-MOLECULE COLLISIONS

K. Blum

Westfälische Wilhelms-Universität
Institut für Theoretische Physik I
Wilhelm-Klemm-Str. 9
48149 Münster

1. Introduction

The concept of chirality, established about 100 years ago, plays an important role in many domains of recent science. It was first discovered as a biological phenomenon. Most of the macromolecules in living organisms possess two stereo isomeric forms (L and D), which are space-reflected images of each other. When these molecules are synthesised in the laboratory under circumstances simulating those of the earth in the early stages of biological evolution the two isomers appear with equal probability. Nevertheless, as discovered by Pasteur, macromolecules occuring in living organisms belong exclusively to one of the possible isomeric forms usually the L-form in amino-acids and the D-form for sugars.

Ever since the early days of biochemistry the question of the origin of the biomolecular homochirality has been posed and is more topical than ever before. Until 1956 scientists had been convinced that the laws of nature do not distinguish between left and right. Hence, there seemed to be no a-priori reason why a chiral molecule should be superior to its mirror-image. A non-probabilistic explanation became possible in principle after the discovery in 1956 that parity is not conserved for the weak interactions governing β-decay. Electrons emitted in β-decay processes possess an intrinsic left-chirality (that is, momentum and spin vectors are antiparallel), and positions a right-handed chirality. Therefore the hypothesis was put forward by Garay[1] and Vestor and Ulbright[2] that the electrons of beta-decay interact differently with the two optical isomers. The molecular dissymmetry may have originated through preferential destruction of one enantiomer of a racemic mixture by spin-polarised electrons arising from β-decay, or circularly polarised bremsstrahlung produced during their deceleration. However, experiments to test this hypothesis have been largely inconclusive.

A new aspect occured with the discovery of parity-violating weak neutral currents (PNC) in 1971. PNC interactions can produce an energy difference between a chiral molecule and its optical isomer (see e.g. Mason[3], for a review). The energy splitting depends on spin-dependent interaction and its extremly small ($\sim 10^{-19}$eV) but can give rise to small differences in their Arrhenius reaction rates. However, autocatalytic chemical systems can effectively amplify these small differences in reaction rates[4] and provide a mechanism for producing the observed optical purity of macromolecules over the long periods of prebiotic evolution, at least in principle (see Kesztheli[19] for a critical review).

In all these investigations the spin-dependence of the interactions between polarised electrons and chiral molecules plays an essential part. It seems to be worthwhile to consider the purely physical problem of the spin-dependence of electron-molecule collisions under the simplest possible conditions. These studies have been initiated by Peter Farago[5] and the subsequent developments in this field can be traced back to his stimulating papers. Collisions between right- (or left-) handed electrons and chiral molecules can be considered as a simple prototype of an interaction between two chiral systems. Farago proposed that in these collisions new spin-dependent phenomena should occur in analogy with optical activity and circular dicroism. Experiments to find one of these effects have been designed and Campbell and Farago[6] reported a positive result. They studied the attenuation of a longitudinally polarised electron beam traversing through optically active camphor vapor and found a non-vanishing asymmetry. This result stimulated further experiments with increased accuracy and we refer to the papers by Kessler[16], and by Gay[24] in this volume for information on the present experimental results and also to the papers by Gay et al.[15] and Mayer and Kessler[16].

Typical chiral effects can also be obtained in collisions from optically inactive, but oriented molecules. Similar phenomena are well known from the theory of optical activity (see e.g. Barron[7]) and first experimental proofs have been obtained for crystals of AgGaS[8], and certain planar nematic liquid crystals[9].

It is the purpose of the present paper to give a review of recent theoretical work on chiral effects in electron molecule scattering. The general framework for the discussion of spin-dependent electron-molecule collisions will be derived in sections 2 and 3. Scattering from chiral molecules, and scattering from non-chiral but oriented molecules will be treated on the same footing. Results of numerical calculations have been obtained in order to obtain estimate for some of the predicted novel chiral effects.

In performing numerical calculations a hypothesis most be put forward on the dynamics responsible for chiral effects. Several mechanisms have been suggested in the literature We refer in particular to the „helicity density model" discussed by Garey and Hrasko[20], Hegstrom et al.[21], and Rich et al.[22].

Our numerical calculations have so far concentrated on closed-shell molecules and are based on the assumption that the usual spin-orbit coupling (in relation with certain geometrical factors) is the dominant mechanism for producing chiral effects. The obtained numerical results for a model molecule show that this mechanism can indeed create significant chiral phenomena for oriented[11],[18] and also for randomly oriented[17] molecules. The same conclusion can be drawn from recent experimental results[16].

We note that the model used by Hayashi[23] is also based on the assumption that spin-orbit coupling is the dominant mechanism. However, these calculations have been based on the „independent atom model". Furthermore, Hayashi has considered only light atoms so that the predicted chiral effects are very small. In our calculations the collision has been treated as one single molecular event, and the influence of heavy atoms within the molecule is discussed in detail.

A brief discussion on our recent numerical results will be given in section 4. Finally,

2. General theory. Definition of „chiral effects"

2.1 Chiral Objects

Let us start with the definition of the basic concepts. A system is called chiral, or handed, if it is distinguishable from its image under the operation of spatial inversion, that is, if it is not possible to superimpose the mirror image on the original by a rotation.

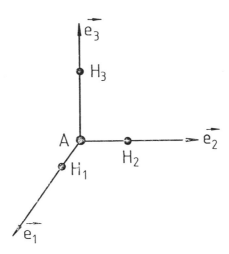

Figure 1. Chiral model molecule.

A simple example is a model molecule of type AH_3 shown in fig. 1 when the three bond lengths AH_1, AH_2, AH_3 are different from each other. The three unit vectors $\vec{e}_1, \vec{e}_2, \vec{e}_3$ are mutually orthogonal and define a molecule-fixed coordinate system.

The chirality of the molecule 1 is exhibited by the left-handed (anticlockwise) or right-handed (clockwise) sequence of the groups AH_1 and AH_2 when viewed in the direction of AH_3. The given sequence defines an axial vector $\vec{e}_1 \times \vec{e}_2$ (or a „sense of rotation") and when multiplied with the polar vector $\vec{e}_3$ gives the pseudoscalar $[\vec{e}_1 \times \vec{e}_2] \cdot \vec{e}_3$.

Pseudoscalar quantities are of central importance in the theory of chiral phenomena. We recall that the reflection of a right-handed srew in a mirror appears to be a left-handed srew but there is no way of rotating or reorienting a srew which changes its handedness. This behaviour is reflected by the properties of the pseudoscalar $[\vec{e}_1 \times \vec{e}_2] \cdot \vec{e}_3$: it changes its sign under reflections (and spatial inversions) but remains invariant under rotations. All molecules of a given isomer have therefore the same value of $[e_1 \times e_2] \cdot \vec{e}_3$, independent of the molecular orientation in space. The pseudoscalar defines a „srew sense" which is positive for the molecule 1, and negative for its optical antipode.

As stated in the introduction, non-chiral but oriented molecules can give rise to chiral effects in scattering with light or electrons. In the rest frame of the projectile electrons the molecule moves around the electron with axis $\vec{n}$ fixed in space (assuming that the molecular orientation does not change during the collision). In this case a „srew sense" is defined not by the structure of the molecule but by the geometry of the experiment, namely by the initial wave vector $\vec{k}_0$, the final wave vector $\vec{k}_1$, and the molecular axis $\vec{n}$. If $\vec{n}$ lies outside the scattering plane ($\vec{k}_0 - \vec{k}_1$-plane) then a reflection in this plane transform $\vec{n}$ into its mirror position $\vec{n}'$, and the three vectors $\vec{k}_0, \vec{k}_1$, and $\vec{n}'$ can not be brought back into the initial position $\vec{k}_0, \vec{k}_1, \vec{n}$ by a rotation. Hence, $\vec{k}_0, \vec{k}_1$, and $\vec{n}$ define a chiral system. (In fact, a srew sense is defined by the pseudoscalar $[\vec{k}_1 \times \vec{k}_0] \cdot \vec{n}$.) The spin of the electron „sees" this handed structure and interacts with it, and chiral effects can be produced. We will return to this case in more detail in subsection 2.3.

The same principle applies to other cases. It is clear from our discussion that chiral effects can also be produced in electron collisions with excited atoms with an anisotropic charge distribution. Initial and final momentum and the alignment axis of the atom can define a screw sense for certain orientations of the alignment axis. Other examples are photoionisation processes of oriented diatomic molecules or Auger processes. In general, if a srew sense is defined by the geometry of the experiments, then chiral effects can be expected.

In order to discuss these and related phenomena we first have to develop a general theoretical framework. Besides longitudinally polarised electrons also transverse polarisation has to be taken into account. Furthermore, the theory has to be developed in such a way that chiral molecules, or non-chiral but oriented molecules are treated on the same footing. That is, the theory must apply to both cases, whether a srew-sense is defined by the structure of the molecule, or whether a srew sense is given by the geometry of the experiments. We will describe the basic results in the following subsection.

2.2 General theory

We will first consider elastic electron collisions from spinless oriented (chiral or non-chiral) molecules in a Σ^+ groundstate. Following Farago[5] we introduce a space-fixed coordinate system by the three orthogonal unit vectors

$$\vec{n} = \frac{\vec{k}_1 + \vec{k}_0}{|\vec{k}_1 + \vec{k}_0|}, \vec{n}_2 = \frac{\vec{k}_1 - \vec{k}_0}{|\vec{k}_1 - \vec{k}_0|}, \vec{n}_3 = \frac{\vec{k}_1 \times \vec{k}_0}{|\vec{k}_1 \times \vec{k}_0|} \tag{1}$$

where $\vec{k}_0$ and $\vec{k}_1$ are the wave vectors of initial and scattered electrons respectively. A set of molecule-fixed axes is specified by three orthogonal unit vectors $\vec{e}_1, \vec{e}_2, \vec{e}_3$ (see e.g. fig. 1).

Let us denote by m_0 and m_1 spin components of initial and scattered electrons respectively relative to a suitably choosen quantisation axis. Let us further denote by $f(\vec{k}_1 m_1, \vec{k}_0 m_0; \vec{e}_i)$ the corresponding scattering amplitude for a collision with an oriented molecule, assuming that the orientation does not significantly change during the scattering. The orientation of the molecule is fixed by giving the relations between the molecule-fixed axes $\vec{e}_i = \vec{e}_1, \vec{e}_2, \vec{e}_3$, and the axes (1).

We define an operator $M = M(\vec{k}_1, \vec{k}_0, \vec{e}_i)$ in spin space by the condition

$$< m_1|M|m_0 >= f(\vec{k}_1 m_1, \vec{k}_0 m_0, \vec{e}_i) \tag{2}$$

The spin scattering matrix M contains all information on the elastic collision and has been used extensively in atomic collision physics (see e.g. Kessler[10]).

The scattering of electrons by atoms and molecules is dominated by electromagnetic interactions. Contributions from weak interactions which are parity-violating can be neglected. It follows that the electron-molecule collision system is invariant under rotations, spatial inversion, and time reversal, that is, M must transform as a proper scalar. Taking this symmetry condition into account it has been shown by Johnston et al.[11] that the most general form of M is given by the expression

$$M = g_0 1 + g_1 \vec{n}_1 \cdot \vec{\sigma} + g_2 \vec{n}_2 \cdot \vec{\sigma} + g_3 \vec{n}_3 \cdot \vec{\sigma} \tag{3}$$

where the four (complex) functions $g_i (i = 0, \ldots 3)$ are functions of $\vec{k}_1, \vec{k}_0, \vec{e}_i$:

$$g_i = g_i(\vec{k}_1, \vec{k}_0, \vec{e}_i) \tag{4}$$

and $\vec{\sigma} = (\sigma_1, \sigma_2, \sigma_3)$ denote the Pauli matrices.

Table 1. Transformation properties of vectors.

	Spatial inversion	Time reversal
n_1	-	-
n_2	-	-
n_3	+	-
e_i	-	+
σ	+	-

From the fundamental condition that M must transform as a proper scalar, and from the transformation properties of the basic vectors (see table 1) we obtain the following symmetry requirements of the g_i-functions[11]. Under spatial inversion we have

$$g_j(-\vec{k}_1, -\vec{k}_0, -\vec{e}_i) = g_j(\vec{k}_1, \vec{k}_0, \vec{e}_i), j = 0, 3 \tag{5}$$

$$g_j(-\vec{k}_1, -\vec{k}_0, -\vec{e}_i) = -g_j(+\vec{k}_1, +\vec{k}_0, +\vec{e}_i), j = 1, 2 \tag{6}$$

and for time reversal $(\vec{k}_1 \leftrightarrow -\vec{k}_0)$:

$$g_j(-\vec{k}_0, -\vec{k}_1, \vec{e}_i) = g_j(\vec{k}_1, \vec{k}_0, \vec{e}_i), j = 0, 1, 3 \tag{7}$$

$$g_2(-\vec{k}_0, -\vec{k}_1, \vec{e}_i) = -g_2(\vec{k}_1, \vec{k}_0, \vec{e}_i) \tag{8}$$

Under rotations of the total system $\vec{k}_1, \vec{k}_0, \vec{e}_i$ all g_i must remain invariant.

It follows from eqs. (5) – (8) that g_0 and g_3 transform as proper scalar functions (invariant under rotations, inversion, and time reversal), while g_1 and g_2 transform as time-even and time-odd pseudoscalars respectively.

g_1 and g_2 will therefore be central for the study of chirality. Let us consider this property from a different point of view. Consider a collision between an electron and a non-chiral molecule and assume that no srew sense is defined by the geometry of the collision (for example, a diatomic molecule with its axis in the scattering plane). This means that the mirror image $(-\vec{k}_1, -\vec{k}_0, -\vec{e}_i)$ of the vectors $\vec{k}_1, \vec{k}_0, \vec{e}_i$ can be brought back into its original positions $(\vec{k}_1, \vec{k}_0, \vec{e}_i)$ by a rotation. Under these two operations g_1 and g_2 transform in the following way (see eqs. (5),(6)):

$$g_i(+\vec{k}_1, +\vec{k}_0, \vec{e}_i) = -g_i(-\vec{k}_1, -\vec{k}_0; -\vec{e}_i) \tag{9}$$
$$= -g_i(\vec{k}_1, \vec{k}_0, \vec{e}_i)$$

for $i = 1, 2$, where the first equality describes the effect of the spatial inversion, and the second equality follows from our assumption that $(-\vec{k}_1, -\vec{k}, -\vec{e}_i)$ can be brought back into the original positions by a rotation (which leaves the value of g_1 and g_2 unchanged). Hence, g_1 and g_2 vanish for non-chiral systems.

In conclusion, a necessary condition for g_1 and g_2 to be different from zero is that a srew sense is defined, either by the structure of the molecule, or by the geometry of the experiment, or both. This result allows to give a definiton of a chiral observable with regard to electron scattering: An observable will be called chiral if (and only if) it depends linearly on g_1, or g_2, or both.

A further classification of chiral observables can be obtained by considering the transformation properties under time reversal. Barron[13] was the first to point out the significance of time reversal for any discussions of chiral effects, and he has used these concepts to clarity many aspects of chirality in optics, in particular in the presence of external fields[14].

Developing these concepts, and adapting them to our present case of interest, we obtain the following hierarchy of spin-dependent effects[11],[12]:

i) A spatially inverted system can be transformed back into its original position by a rotation (that is, the system is non-chiral). In this case, we have in general

$$g_0 \neq 0, \, g_3 \neq 0$$

$$g_1 = 0, \, g_2 = 0$$

ii) A spatially inverted system can be transformed back into its original position by a rotation followed by time reversal. In this case we have in general the relations:

$$g_0 \neq 0, \, g_3 \neq 0, \, g_2 \neq 0$$

$$g_1 = 0$$

That g_1 is necessarity zero follows from eqs. (5) – (8). Applying successively spatial inversion (eq. (6)), time reversal (eq. (7)), and then such a rotation which transforms all vectors back into their original position (applying here the stated assumption). This gives

$$g_1(\vec{k}_1, \vec{k}_0, \vec{e}_i) \quad = -g_1(-\vec{k}_1, -\vec{k}_0, -\vec{e}_i) \tag{10}$$

$$= -g_1(\vec{k}_0, \vec{k}_1, -\vec{e}_i) \tag{11}$$

$$= -g_1(\vec{k}_1, \vec{k}_0, \vec{e}_i) \tag{12}$$

Hence, g_1 vanishes necessarily.

iii) Neither (i) nor (ii) is possible. In this case, all four functions g_i are different from zero.

We will say that a system exhibits „time-odd" chirality if it depends linearly on g_2 but not on g_1, a system exhibits „time-even chirality" if it depends linearly on g_1 but not on g_2. Time-even and time-odd chirality correspond to „true" and „false" chirality in Barrons notation respectively. We note that for randomly oriented molecules all terms linear in g_2 vanish, so that only „time-even" chirality can be observed.[11]

2.3 Examples

In order to illustrate eq. (3) we will give some examples. Since the set g_j must transform as scalars under rotations of the total collision system they can only depend on scalar products constructed from $\vec{k}_0, \vec{k}_1$ and the three vectors $\vec{e}_i$. From equations (6) and (7) it follows that g_0 and g_3 can depend only on proper scalar functions, g_1 can depend only on time-even pseudoscalars such as $(\vec{k}_1 + \vec{k}_0) \cdot \vec{e}_1)((\vec{k}_0 \times \vec{k}_1) \cdot \vec{e}_1)$ and g_2 only on time-odd pseudoscalars like $(\vec{k}_0 \times \vec{k}_1) \cdot \vec{e}_1$.

For elastic scattering from spinless atoms the only scalar conbinations available are $k_0^2 = k_1^2 = k^2$ and $\vec{k}_1 \cdot \vec{k}_0$ which are invariant under spatial inversion and time reversal. No pseudoscalars can be contructed from $\vec{k}_0$ and $\vec{k}_1$. From equation (6) follows that

$$g_j(k^2, \vec{k}_1 \cdot \vec{k}_0) = -g_j(k^2, \vec{k}_1 \cdot \vec{k}_0) \qquad j = 1, 2.$$

Hence $g_1 = g_2 = 0$ because of the invariance of the interaction under spatial inversion. Although this is a sufficient condition, it can further be shown that invariance under time reversal requires that $g_2 = 0$ and invariance under a combination of spatial inversion and time reversal requires that $g_1 = 0$. Equation (3) then reduces to the well known expression found in electron-atom scattering:

$$M = g_0 \mathbf{1} + g_3 \vec{n}_3 \cdot \vec{\sigma} \,. \tag{13}$$

As a second example let us briefly consider interactions violating parity invariance. In this case M will not remain invariant under spatial inversion and condition (i) and equations (5) and (6) will not hold. However, assuming that time-reversal symmetry applies, equations (7) and (8) will still be valid. The spherical symmetry of the target dictates that the g_j functions can depend again only on the proper scalars k^2 and $\vec{k}_1 \cdot \vec{k}_0$ and from equation (8) it follows that

$$g_2(k^2, k_1 \cdot k_0) = -g_2(k^2, k_1 \cdot k_0) \,.$$

Hence time-reversal invariance requires g_2 to vanish. The most general form of M which allows for parity violation but is invariant under time reversal is given by

$$M = g_0 \mathbf{1} + g_1 \vec{n}_1 \cdot \vec{\sigma} + g_3 \vec{n}_3 \cdot \vec{\sigma} \tag{14}$$

if the target possesses spherical symmetry. This case has been discussed in detail by Kessler (1985).

As a final example consider electron collisions with an oriented diatomic hetero-nuclear molecule (for example, a molecule adsorbed at a surface). As discussed in subsection 2.1 a srew sense is defined by the geometry of the experiment (that is, by the vectors $\vec{k}_1, \vec{k}_0$ and the molecular axis $\vec{n}$) for certain orientation of $\vec{n}$. In this simple case it is easy to give a more explicit form for the scattering matrix M, and the pseudo scalar functions g_1 and g_2. We write

$$g_1 = (\vec{n}_3 \cdot \vec{n})(\vec{n}_1 \cdot \vec{n}) g_1' \tag{15}$$

$$g_2 = (\vec{n}_3 \cdot \vec{n}) g_2' \tag{16}$$

where now g_1' and g_2' are proper scalars and remain invariant under spatial and time inversion and rotations of the total system. In fact, $\vec{n}_3 \cdot \vec{n} \sim [\vec{k}_1 \times \vec{k}_0] \cdot \vec{n}$ transforms as time-odd pseudo scalar, and $\vec{n}_1 \cdot \vec{n}$ transforms as time-odd scalar. Hence, eqs. (15) and

(16) give explicitly the symmetry properties (6) - (8) of g_1 and g_2. The matrix M can then be written in the form

$$M = g_0 1 + g_1'(\vec{n}_3 \cdot \vec{n})(\vec{n}_1 \cdot \vec{n})\vec{n}_1 \cdot \vec{\sigma} + g_2'(\vec{n}_3 \cdot \vec{n})\vec{n}_2 \cdot \vec{\sigma} + g_3(\vec{n}_3 \cdot \vec{\sigma}) \tag{17}$$

We can read off from eqs. (17) the essential geometrical conditions which must be satisfied in order to observe chiral effects. If $\vec{n}$ lies in the scattering plane, then $\vec{n}_3 \cdot \vec{n} = 0$ and g_1 and g_2 vanish identically. No srew sense is defined by the geometry of the collision, and the system exhibits no chirality.

Furthermore, g_1 vanishes also if $n_1 \cdot n = 0$ that is, if $\vec{n}$ lies in the $\vec{n}_2 - \vec{n}_3$-plane (in particular, if $\vec{n}$ is perpendicular to the scattering plane). This case is an example of an oriented system which exhibits time-odd but no time-even chirality[12].

In the present case chirality is defined by the relationship between the vectors $\vec{k}_1, \vec{k}_0$ and $\vec{n}$. The „isomer" is obtained by spatial inversion, or simpler by a reflection for example in the scattering plane which transforms $\vec{n}$ into its mirror image $\vec{n}'$. The functions g_0, g_1', g_2', g_3 remain invariant under this operation because of their scalar character. Hence, the spin scattering matrix M, describing the „antipodal system", is simply obtained by substituting $\vec{n}'$ for $\vec{n}$ in eq. (17). This result illustrate the well-known fact that the dynamics (expressed by g_0, g_1', g_2', g_3) remains unchanged when a chiral system is transformed into its mirror image.

3. Expressions for polarisation and asymmetry

General expressions for spin-dependent observables for oriented and randously oriented molecular ensembles have been derived by Johnston et al.[11] and we refer to this paper for the main results. Here, we will only give a few examples. Consider an initially unpolarised electron beam scattering from an ensemble of oriented molecules. The longitudinal polarisation of the scattered electrons $P_{\parallel}$ is then given by the expression

$$I P_{\parallel} = 2 \left[Re(g_0 g_1^*) \cos\frac{1}{2}\Theta - Im(g_3 g_1^*) \sin\frac{1}{2}\Theta + Re(g_0 g_2^*) \sin\frac{1}{2}\Theta - Im(g_2 g_3^*) \cos\frac{1}{2}\Theta \right] \tag{18}$$

where Θ is the scattering angle, and

$$I = \sum_i |g_i|^2 \tag{19}$$

is the differential cross section. Now consider an incident beam which is completely longitudinally polarised parallel or antiparallel to the incident beam direction. The asymmetry A is defined by

$$A = \frac{I_+ - I_-}{I} \tag{20}$$

where I_+ and I_- are the differential cross sections for initially parallel or antiparallel longitudinal polarisation. One obtains[11]

$$I A = 2 \left[Re(g_0 g_1^*) \cos\frac{1}{2}\Theta - Im(g_3 g_1^*) \sin\frac{1}{2}\Theta - Re(g_0 g_2^*) \sin\frac{1}{2}\Theta + Im(g_2 g_3^*) \cos\frac{1}{2}\Theta \right] \tag{21}$$

Comparing with equation (18) we see that $P_{\parallel}$ and A differ from each other in the sign of the g_2-dependent terms. $P_{\parallel}$ and A vanish identically if the system exhibits no chirality, that is, if g_1 and g_2 vanish simultaneously.

We note that $P_{||}$ is the sum of a term describing time-even chirality and a term describing time-odd chirality, while A is constructed from the difference of these terms. Hence, the contributions of time-even and time-odd chirality can be obtained by summing and subtracting experimental results of $P_{||}$ and A.

It is interesting to express $P_{||}$ and A in the helicity representation. Let us denote by $\sigma(+-)$ the differential cross section where the initial electrons have „spin down" with respect to the initial momentum $\vec{k}_0$, and where the scattered electrons have „spin up" with respect to the final momentum $\vec{k}_1$, and similarly for the other combinations. We obtain[12]

$$IP_{||} = [\sigma(++) - \sigma(--)] + [\sigma(+-) - \sigma(-+)] \qquad (22)$$

and the asymmetry A is obtained by reversing the sign in front of the second bracket. It can be shown that the first team of eq. (22) vanishes if $g_1 = 0$, and the second term vanishes if $g_2 = 0$. Time-even chirality is therefore responsible for the difference between the non-flip cross sections, and if the system exhibits time-odd chirality than the helicity-flip-terms will differ.

Suppose now that one wishes to perform the same experiment for a target ensemble of randomly oriented chiral molecules of one mirror image form. The required expressions follow from eqs. (18) and (21) by averaging over all molecular orientations. Note that all terms linearly in g_2 vanish when the average is taken which follows from symmetry requirements[11]. Hence, in randomly oriented ensembles of chiral molecules, only time-even chirality can be observed. One obtains[11]:

$$< P_{||} > = < A > = \frac{2\left[< Reg_0 g_1^* > \cos\frac{1}{2}\Theta - < Img_3 g_1^* > \sin\frac{1}{2}\Theta\right]}{\sum_i < |g_i|^2 >} \qquad (23)$$

where the brackets $< \ldots >$ denote the average over molecular orientations. Note that $< P_n >$ and the asymmetry $< A >$ are equal in this case. It is evident from eq. (23) that $< P_k >$ and $< A >$ are time-even pseudoscalars. Furthermore, $< P_{||} >$ and $< A >$ change their sign if the handedness of the molecules is inversed, and both parameters vanish for non-chiral molecules, or if the target system is a racemic mixture[11]. Both parameters can therefore be considered as a measure of the amount by which longitudinally polarised electrons can distinguish between the two optical isomers.

Finally, we note that chiral effects can also be observed by measuring cross sections, or transverse polarisations. The relevant terms depend linearly on g_1 and for g_2 and can be identified with the help of the corresponding general equations given in ref. 11.

4. Numerical results

In order to obtain some insight into the magnitudes of the predicted chiral effects we have performed numerical calculations for the chiral model molecule shown in fig. 1 where A is a heavy atom (Bi). The bound lengths have been choosen as $AH_1 = 2a.u.$, $AH_2 = 3a.u.$, $AH_3 = 4a.u.$. It has been assumed that the usual spin-orbit coupling between projectile and target, derived from the nuclear potential of the heavy atom only, is the dominant interaction for producing chiral effects. The scattering is treated as one molecular event. Note that the projectile must also be influenced by the H-atoms. If the electrons would be scattered on the heavy atom only, without „seeing" the H-atoms, then g_1 and g_2 would both automatically vanish according to our general analysis, and no chiral effects would be obtained. For all information on the numerical procedure, we refer to Johnston et al.[11], and Fandreyer et al.[17].

Results for the polarisation $P_\parallel$ and the asymmetry A, defined in eqs. (18) and (21) respectively, have been obtained by Johnston et al.[11] as functions of energy, scattering angle, and the orientation of the molecule. Here we will only present results for the asymmetry A for randomly oriented molecules, obtained by Fandreyer et al.[17].

The theory of the experiment has been developed by Farago[5], Fandreyer et al.[17], and within an extended context by Thompson and Kinnin[18]. The asymmetry is given in good approximation by the equation

$$A = -P < Q > \rho d < x >\tag{24}$$

where P denotes the longitudinal polarisation of the initial electrons, $< Q >$ is the total cross section, ρ the pressure, and d the length of the sample, and

$$< x > = \frac{< Img_1 >}{< Img_0 >}\tag{25}$$

where the brackets denote the average over all molecular orientations, and „Im" the imaginary part.

Table 2. Numerical results.

Energy (eV)	$< Img_1 > \,|\, < Img_0 >$
0.01	-7.27×10^{-3}
0.1	-7.63×10^{-3}
0.2	-6.40×10^{-3}
0.3	
0.4	
0.5	-4.43×10^{-3}
1.0	-1.03×10^{-3}
1.5	-2.72×10^{-5}
2	2.13×10^{-4}
3	1.64×10^{-4}
4	1.26×10^{-5}
6	-5.76×10^{-5}
8	-4.42×10^{-5}
10	-4.72×10^{-5}

Numerical results are shown in table 2. Our model molecule has a resonance around $0.5eV$[11]. The data indicate that, for randomly oriented molecules, the effects are small but measurable in resonance regions, and if the molecules contain at least one heavy atom. We note that, for oriented molecules, the effects can be significantly larger[11].

5. Conclusions

A review has been given on recent theoretical results on chiral effects in collisions between electrons and closed-shell molecules. It has been shown that, compared to the atomic case, the spin-scattering matrix contains two new functions, g_1 and g_2, which transform as time-even and time-odd pseudo scalars respectively. Chirality can be defined either by the structure of the molecule, or the geometry of the experiment, or both. Using fundamental symmetry properties a classification of chiral effects in electron-molecule collisions has been given.

Numerical calculations have been performed for a chiral model molecule, where the collision has been treated as one single molecular event, and the spin-orbit interaction is derived by the nuclear potential of the heavy atom only. The dynamics responsible for producing chiral effects in our model can therefore be characterised as follows: i) the orbital motion of the electron is influenced by the full chiral structure of the molecule (in a simple trajectory picture the electron would move on a helical orbit through the molecule), ii) the spin of the electron changes because of the usual spin-orbit interaction with the heavy atom. Our analytical and numerical results are a clear proof that this combination of „chiral" orbital motion and spin-orbit coupling is sufficient to create chiral effects with a reasonable order of magnitude.

The dynamical mechanism discussed here is of course only one of several as mentioned in the introduction. A comparison between these mechanisms is in progress[24]. It should further be noted that our theoretical results, presented here, apply only to spinless molecules. The dynamics might be different for collisions with open shell molecules (where exchange processes must be taken into account), or in inelastic collisions. A considerable amount of further theoretical and experimental studies is required in order to clarify all details of chiral interactions in electron-molecule collisions.

References

1. H.S. Garay: Nature 219, 338 (1968)

2. F. Vester, T. Ulbricht, H. Kraus: Naturwissenschaften 46, 68 (1959)

3. S.F. Mason: Molecular optical activity and the chiral discrimination, Cambridge University Press (1982)

4. S.F. Mason, G. Tranter: Mol. Phys. 53, 1091 (1984)

5. P.S. Farago: J. Phys. B13, L567 (1980), J. Phys. B14, L743 (1981)

6. D.M. Campbell, P.S. Farago: Nature 318, 52 (1985)

7. L.D. Barron: Molecular light scattering and optical activity, Cambridge University Press (1982).

8. M.V. Hobden: Nature 216, 678 (1967)

9. R. Williams: Phys. Rev. Lett. 21, 342 (1968)

10. J. Kessler: Polarised Electrons, second edition, Springer

11. C. Johnston, K. Blum, D. Thompson: J. Phys. B26, 965 (1993)

12. K. Blum: J. Phys. B23, L253 (1990)

13. L.D. Barron: Chem. Phys. Lett. 123, 423 (1986)

14. L.D. Barron: New Developments in Molecular Chirality, (ed. P.G. Mezey), (Dordrecht: Reidel) (1990)

15. T.J. Gay, K. Trentheim, M.L. Johnston: to be published

16. S. Mayer, J. Kessler: Phys. Rev. Lett., to be published

17. R. Fandreyer, D. Thompson, K. Blum: J. Phys. B23, 3031 (1990)

18. D. Thompson, M. Kinnin: J. Phys. B, in press

19. L. Keszthelyi: in: „Selected Topics in Electron Physics (P. Farago symposium), Plenum Press, to be published

20. A.S. Garay, P. Hrasko: J. Mol. Evolution 6, 77 (1975)

21. R.A. Hegstrom, D.W. Rein, P.G. Sandars: J. Chem. Phys. 73, 2329 (1980)

22. A. Rich, J. v. House, R.A. Hegstrom: Phys. Rev. Lett. 48, 1341 (1983)

23. S. Hayashi: J. Phys. B18, 1229 (1985), J. Phys. B21, 1037 (1988)

24. D. Thompson: private communication

ELECTRON SCATTERING FROM EXCITED VIBRATIONAL LEVELS IN CO_2 AND N_2O

Dr W R Newell

Department of Physics and Astronomy
University College London
Gower Street
LONDON WC1E 6BT

INTRODUCTION

The study of molecules in vibrationally excited states is important simply because the molecules contain internal energy which can be released in reactions. This will influence chemical and scattering processes in low temperature plasmas in atmospheric reactions and in surface processes. The study of hot molecules is important not only for their role in determining energy path ways but also as a means of understanding the dynamics of molecular transitions. Transitions between excited states will have different Frank-Condon factors than those between ground and excited states and different vibrational modes will have different transition moments. Molecular symmetry changes will also influence differential scattering and resonance processes. It is known that electron attachment to electronically excited molecules can be larger than to the ground state configuration (Christophorou 1991) and that total electron scattering cross sections from vibrationally excited CO_2 increase by 30% in the region of the $^2\Pi_u$ resonance (Fersch et al (1989), Buckman et al (1987)). Consequently, a knowledge of cross section values for electron scattering from specific vibrational states is necessary. This paper discusses high resolution differential electron scattering from vibrationally excited CO_2 and N_2O.

PRODUCTION OF EXCITED VIBRATIONAL STATES

Since this work is concerned with electron scattering from excited vibrational states of molecules, a brief outline of the different methods which have been used to produce molecules in excited vibrational states will be given. The methods used can be broadly classified as either selective, in which a well defined vibrational state is populated, and non-selective in which a distribution of vibrational levels is populated. The majority of methods used to date fall into the non-selective class.

Thermal excitation of a molecular gas is simple and direct. The molecular gas can be in a gas cell, for use in total cross section measurements, or produced as a beam for differential cross section measurements. Direct ohmic heating of a gas cell will produce temperatures of 1600°K if iridium is used to construct the cell. Such methods have been used by Spence and Schulz (1969), Fersch et al (1989) and Allan and Wong (1978) in total cross sections measurements. Typical vibrational populations produced in H_2 are given in Table 1 for an oven temperature of 1400°K. In the case of a gas beam ohmic or RF heating a separate oven source can be used. In addition thermal excitation of a gas can be accomplished by indirectly heating a low mass hypodermic needle gas beam source (Johnstone and Newell 1995) using a quartz lamp radiative source. Temperatures of 800°K are easily achieved by this method and typical vibrational populations produced in CO_2 are given in Table 1. It is clear, Table 1, that the thermal excitation method is more efficient for lower energy vibrational levels which occur in triatomic molecules. The Boltzmann distributions of molecules produced can of course be easily varied by changing the source temperature.

Table 1. Population P_v of vibrational level v

Vibrational Level (v)	H_2[1] at 1400°K		CO_2[2] at 673°K	
0	~ 9.85	$10^{-1}(0)$[3]	0.55	(0)[3]
1	1.4	$10^{-2}(0.5)$	0.26	(0.082)
2	2.4	$10^{-4}(1)$	0.04	(0.159)
3	5.5	$10^{-6}(1.4)$	0.06	(0.165)
4	1.6	10^7	0.03	(0.172)
5			0.02	(0.239)
6			0.01	(0.248)

(1) Allan and Wong, 1978
(2) Johnstone et al 1993
(3) Energies of vibrational levels in eV

Discharge sources, both microwave and RF, are a convenient, but non selective, source of 'hot' molecules. Using a 2 GHz source with a gas pressure of 0.1 to 0.2 torr typically 40% of the molecular flow is excited with a power input of 100 W. The higher effective temperature ~ 5000°K of a microwave discharge makes it more suitable for exciting the vibrational levels of diatomic molecules. Microwave sources have been employed by White and Ross (1976) and Anketell and Brocklehurst (1974) for the production of excited N_2. In these cases, however, the vibrational population distribution produced was a non-Boltzmann distribution. It would appear that these sources produce non-relaxed distributions which could be dependent on the gas flow conditions in the source. The efficiency of microwave discharge sources have recently been improved by redistributing the microwave power around the source and by the introduction of magnetic fields to confine the discharge (Geddes et al 1994). Excited species number densities of 10^{12}-10^{13}/cc are achievable with microwave discharges.

In order to have some degree of state selection in vibrational excitation it is necessary to use optical pumping. Selection rules forbid the optical direct excitation of ground state vibrational of homonuclear molecules so consequently, a decay scheme must be employed in this case. Using a N_2 pumped dye laser producing 10 μJ pulses of 5 ns duration Geoch and Schlier (1986) pumped either the $A^1\Sigma_u^+$ (v = 4) or the $B^1\Pi_u$ (v = 6) levels of Li_2 which then decayed to the long lived vibrational levels around v = 11 of the ground state $X^1\Sigma_g^+$. The final vibrational population depends on the Franck-Condon overlap between the selected upper level and the ground state and typically 2% of the molecules were indirectly pumped to excited vibrational states of $X^1\Sigma_g^+$. A similar pumping scheme in

Li_2 has also been reported by Fuchs and Toennies (1987) with 7% excitation.

Vibrationally excited H_2 and D_2 have been prepared by surface recombinations of H and D on the walls of the gas cell (Hall et al 1988). The gas is leaked into the cell which contains a tungsten filament. Dissociation of the $H_2(D_2)$ occurs on the hot filament and this is followed by recombinative desorption on the surface of the gas cell.

$$H_2 \rightarrow H + H$$
$$H + SURFACE \rightarrow H_2^* \ (v > 0)$$

The excited molecules effuse out through an aperture and provide a usable flux for electron scattering experiments. The population of the $v = 1$ level in H_2 and D_2 correspond to vibrational temperatures of roughly $1850°K$ giving a Boltzmann population, P_1, of $\sim 2 \ 10^{-2}$. In another application of surface chemistry Foner and Hudson (1984) have used the catalytic decomposition of NH_3 on a heated platinum surface to produce N_2 in vibrational states up to $v = 9$. The full mechanisms of the process is not fully understood but believed to proceed by recombination of chemisorbed NH radicals

$$NH_{ad} + NH_{ad} \rightarrow N_2^* \ (v = 1 \text{ to } 9) + H_2$$

with the possibility that the hydrogen remains as absorbed H. Operating the platinum surface at temperatures of $300°K$ to $1400°K$ with NH_3 gas pressures of 0.1 to 1.4 torr produces substantial quantities of N_2^*. There has however been no detailed analysis of the intensity distribution of the excited molecular flux.

In general, most chemical reaction product molecules are produced in vibrational states, eg HI* from fluorine and hydrogen but such procedures do not constitute effective sources suitable for scattering experiments.

A method for the direct electron beam excitation of vibrational states in a molecular beam has been given by Srivastava and Orient (1983) using a high current (mA) magnetically confined electron beam to excite the molecules. These vibrational states are then probed by scanning the energy of a low current (μA) electron beam.

Gas expansion and compression methods have also been employed to promote molecules to their excited states. Shock tube experiments produce molecules in excited vibrational states but these are part of the reaction under study where as supersonic pulsed gas expansion sources will cool the beam to lower rotational and vibrational states.

HIGH RESOLUTION ELECTRON ENERGY LOSS SCATTERING APPARATUS

The technique of high resolution electron energy loss spectroscopy (HREELS) of molecules has been used successfully for many years in an effort to understand the scattering dynamics of molecular excitation into rotational (J), vibrational (V) and electronic states (Λ) (Brown 1979). The method also provides a means of determining absolute cross sections for elastic scattering and the excitation of allowed and forbidden states which are coupled to the ground state configuration (Johnstone and Newell 1991). However, information on electron scattering from excited molecular states is limited (Johnstone et al 1993).

Molecules in vibrationally excited states (v) of the ground electronic state (Λ) contain energy which can influence molecular reactions in plasmas, atmospheres and on surfaces. Consequently a study of such 'hot' molecules by HREELS is important in applied applications such as diamond film production as well as in the fundamental understanding of scattering processes. It is known that electron attachment to electronically excited molecules can be larger than to the ground state configuration and that the total electron scattering cross section from vibrationally excited CO_2 increase by

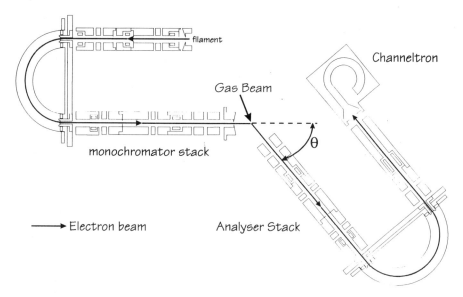

Figure 1. The electron spectrometer used in these experiments.

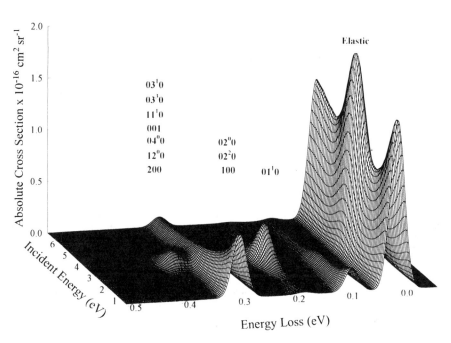

Figure 2. Absolute cross section for the excitation of CO_2 as a function of incident energy and energy loss at a scattering angle of $20°$.

30% in the region of the $^2\Pi_u$ resonance at 4 eV (Buckman et al 1987, Fersch et al 1989). Consequently a knowledge of cross section values for electron scattering from specific vibrational states is necessary.

The high resolution electron spectrometer, Figure 1, used in the present (Mapstone and Newell 1992) work produces a primary electron beam current of 2 nA which scatters from a gas beam produced using a titanium tube (ID = 0.5mm and length 10mm); the typical gas beam density is 10^{13} molecules/cc. The overall energy resolution of the spectrometer is 25 meV, FWHM, with an angular revolution of 1°. Contact potentials were determined by scanning the negative ion resonance at 2.3 eV in N_2 ($^2\Pi g$). A typical absolute cross section for excitation of the modes in CO_2 is given in Figure 2.

In the present research CO_2 and N_2O molecules were excited to vibrational modes N_2O (v = 1 to 7) CO_2 (v = 1 to 11) by thermally heating the titanium tube gas source using a quartz-halogen lamp. Temperatures (T) of 200°C were readily attained and these were monitored using a Cu-Co therocouple located 5mm from the exit of the titanium tube. A typical population P_v distribution between the modes produced by the non selective pumping process, is given in Table 1. Higher temperatures (~ 400°C) are attained by heating the tube with a focused laser beam.

The populations were determined using the formula (Herzberg (1945))

$$P_v = g_v \, e^{-\Delta E_{(v)}/kT} \left[1 + \sum_{v=2} g_v \, e^{-\Delta E_{(v)}/kT} \right]$$

where ΔE is the energy of the vibrational mode (v) above the ground state (v = o) and g is the statistical weight of the vibrational mode. The fundamental vibrational modes, symmetric stretch (noo), asymmetric stretch (oon) and bending (ono) exist in CO_2 and N_2O in addition to mixed modes. Typical energy loss/gain spectra for CO_2 and N_2O are given in Figures 3 and 4 which clearly show the presence of super elastic peaks with intensities dependent on the temperature ie the excited state populations.

The first peak on the energy loss side of the elastic peak in CO_2 arises from excitation of the bending mode (010) while the second peak is a composite peak of the second bending (020) and first symmetric stretch (100) modes. The wide composite feature at 0.3 eV arises from the excitation of seven other modes. The positions of these energy loss features are in good agreement with those previously measured by Register et al (1980) and Johnstone et al (1995); the energies are given in Table 1. On the energy gain side of the elastic peak the first feature (010)* arises from superelastic scattering from the excited mode (010) to the ground state (000). However, due to the harmonic nature of the modes contributions to the feature (010)* could also arise from super elastic scattering from the excited (020)* (100)* modes to the (010) mode. With the energy resolution of the present spectrometer (25 meV) the transitions (020) (100) $\rightarrow$ (010) and (010) $\rightarrow$ (000) would be indistinguishable. A similar pattern exists in N_2O with the isolated bend mode (010) and the unresolved modes (020) and (100) occurring on the energy loss side followed by a well structured composite feature at 0.3 eV composed of at least four modes (110), (120), (200), (011). The energies and positions of these modes are in good agreement with the work of Asira et al (1975).

Again superelastic features are observed at 73 meV and 160 meV which will arise from superelastic scattering from the excited (010)* and (020)* (100)* modes producing transitions to the ground vibrational mode (000). The same problem with the harmonic nature of the modes also applies in N_2O. However, it should be remembered that CO_2 has no dipole moment in the symmetrics stretch ground state whereas N_2O has a dipole moment.

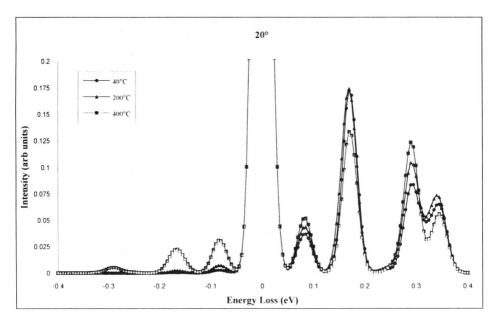

Figure 3. Energy loss spectra for electron scattering from CO_2 at 3.8eV and 20° at three different temperatures.

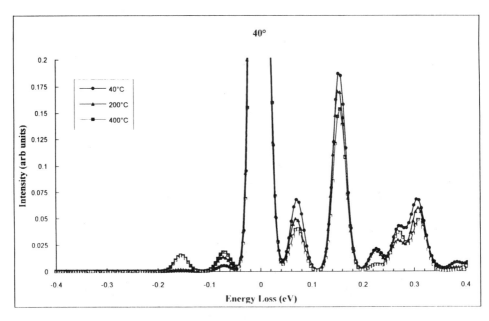

Figure 4. Energy loss spectra for electron scattering from N_2O at 2.2eV and 40° at three different temperatures.

Currently superelastic differential electron scattering measurements have been made on CO_2 for 3.8 eV incident electron energy and at scattering angles of 20, 30, 40, 60 and 80°. In N_2O the incident electron scattering energy is 2.3 eV with measurements at a scattering angle of 30°. In both cases the analysis of the experimental data has been confined to scattering processes involving the first excited bending mode (010) only.

DISCUSSION AND ANALYSIS OF RESULTS

The spectra for CO_2 and N_2O were analysed by comparing the ratios of the intensities of the measured spectral features to that of the elastic feature.

In each case the relative intensities of the bending mode (010), the elastic peak (000), and the first superelastic feature (010)* were determined by curve fitting the spectra (Johnstone and Newell 1995). The intensity ratios R = (010)/(000) and R* = (010)*/(000) for scattering at 30° for N_2O are plotted in Figures 5 and 6, for the temperature range 25°C to 220°C (Akther and Newell 1995); similar plots were determined for the scattering angles in CO_2. It is clear that R* increases with temperature as the initial population of the first excited mode (010) increases whereas R decreases. Superelastic, inelastic and elastic differential scattering processes give rise to the measured ratios R and R* which can be written as:

$$R = \frac{(010)}{(000)} = \frac{P_o \, d\sigma_{01} + P_1 \, d\sigma_{12} + P_x \, d\sigma_{xy}}{P_o \, d\sigma_{00} + P_1 \, d\sigma_{11} + P_x \, d\sigma_{xx}} \qquad (1)$$

$$R* = \frac{(010)*}{(000)} = \frac{P_1 \, d\sigma_{10} + P_x \, d\sigma_{yx}}{P_o \, P\sigma_{00} + P_1 \, d\sigma_{11} + P_x \, d\sigma_{zx}} \qquad (2)$$

The subscripts indicate the initial and final mode of the molecule before and after scattering. For example, $d\sigma_{10}$ is the superelastic differential scattering cross sections from the (010)* mode to the ground state (000) mode whereas $d\sigma_{11}$ is the differential elastic scattering cross section from the (010)* mode. P_o and P_1 are the populations of the (000) and (010) modes respectively and P_x is the number of molecules not in the ground nor the first vibrational state. It is clear that when there is no excited state population ($P_1 = P_x$ = O, P_o = 1) R* is zero. The terms $d\sigma_{xy}$, $d\sigma_{yx}$ and $d\sigma_{xy}$ represent inelastic, superelastic and elastic collision processes from the higher vibrational modes and are technically averages over all possible higher mode transitions.

In order to determine the cross section values $d\sigma_{10}$, $d\sigma_{11}$, $d\sigma_{12}$ for scattering from the excited mode (010) the equation 1 and 2 were fitted to the measured values of R and R* respectively for each of the scattering angles) θ = 20°, 30°, 40°, 60°, 80°. The data was fitted in two stages with initial values for $d\sigma_{10}$ and $d\sigma_{11}$ obtained, using equation 2, for ratios of R* at temperatures less than 60°C where $P_x \approx 0$. The complete curve was then fitted and the final values of $d\sigma_{10}$ and $d\sigma_{11}$ obtained. Equation 1 was fitted second using this final value of $d\sigma_{11}$ in the denominator.

The results of the curve fitting are given in Tables 2 and 3 for CO_2 and N_2O respectively. In all cases the cross section values determined have been normalized to the elastic scattering differential cross section (DCS) $d\sigma_{00}$ at that scattering angle.

In the case of CO_2, $d\sigma_{01}$, which involves both direct and resonance excitation, follows the trend of the data of Andrick et al (1969) and Andrick and Read (1971). The determined excitation DCS $d\sigma_{12}$ for inelastic scattering from (010)* shows an angular trend similar to $d\sigma_{01}$. The superelastic scattering from (010)* → (000) produces a DCS, $d\sigma_{10}$, which is approximately half of that of $d\sigma_{10}$. This is generally what would be expected when the effects of detailed balance are included. It also appears quite

189

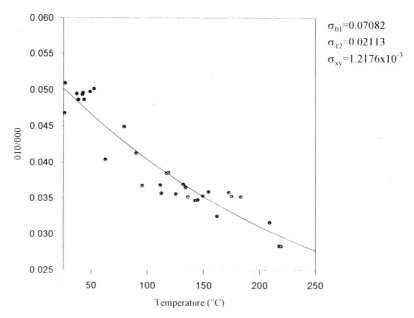

$\sigma_{01}=0.07082$

$\sigma_{12}=0.02113$

$\sigma_{xy}=1.2176 \times 10^{-3}$

Figure 5. Ratio of inelastic intensity to elastic intensity R at 30° in N_2O.

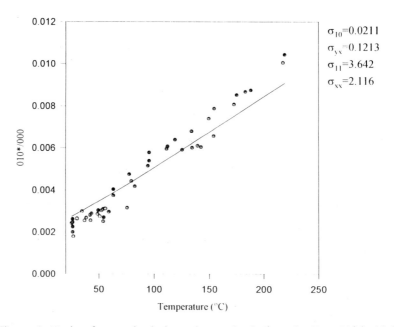

$\sigma_{10}=0.0211$

$\sigma_{yx}=0.1213$

$\sigma_{11}=3.642$

$\sigma_{xx}=2.116$

Figure 6. Ratio of superelastic intensity to elastic intensity R at 30° in N_2O.

Table 2. Determined values for DCS in CO_2 at 3.8 eV

Scattering Angle	$d\sigma_{01}$	$d\sigma_{12}$	$d\sigma_{11}$	$d\sigma_{10}$
80	0.349	0.549	2.18	0.197
60	0.273	0.310	2.24	0.138
40	0.156	0.107	2.52	0.077
30	0.078	0.089	2.40	0.034
20	0.054	0.063	2.24	0.026

Table 3. Determined values for DCS in N_2O at 2.3 eV

Scattering Angle	$d\sigma_{01}$	$d\sigma_{12}$	$d\sigma_{11}$	$d\sigma_{10}$
30	0.071	0.021	3.642	0.021

remarkable that elastic scattering σ_{11} from the excited mode (010)* gives DCS which are 2.1 to 2.5 greater than $d\sigma_{00}$.

This may seem surprising but when the populations are included it is in good agreement with the work of Fersch et al who reported a 1.3 factor increased in the total scattering in CO_2 at 240°C. A similar pattern of derived differential cross sections is seen in Table 3 for N_2O. The $d\sigma_{10}$ DCS is again approximately 30% of $d\sigma_{11}$ consistent with the degeneracies of the modes in the detailed balance. Elastic scattering $d\sigma_{11}$ from (010)* is 3.6 greater than $d\sigma_{00}$. Fersch (1994) have also made total cross section measurements in N_2O and reported an increase of 1.12, at 247°C. If we assure that all the increase in the total cross section is due to elastic scattering then when the populations are included the present data gives an increase of 1.5. This assumes that the factor of 30% for $d\sigma_{11}$ persists over the full angular range.

CONCLUSION

These preliminary measurements on electron scattering from excited states demonstrate the importance of excited or hot molecules in collision processes. At present there is no theory available for comparison.

ACKNOWLEDGEMENTS

WRN wishes to acknowledge the help of his colleagues Dr W M Johnstone and Miss P Akther in the preparation of this article.

REFERENCES

Akther, A. and Newell, W.R., 1995, *J.Phys.B. At.Mol.Phys.* (to be published).

Allan, M. and Wong, S.F., 1978, *Phys. Rev. Lett* **41** *1791-4.*

Andrick, A., Danner, D. and Ehrhardt, H., 1969, *Physics Letters,* **29a** *346-347.*

Andrick, D. and Read, F.H., 1971, *J.Phys.B: At. Mol. Phys.* **4** *389-395.*

Anketell, J. and Brocklehurst, B., 1974, *J.Phys.* **7** *1937-47.*

Asira, R., Wung, S.F. and Schulz, G.J., 1975, *Phys.Rev.A.* **11** *1309-13.*

Brown, S.C., ed. Electron Molecule Scattering, Wiley 1979.

Buckman, S.J., Elford, M.T. and Newman, D.S., 1987, *J.Phys.B: At. Mol. Opt. Phys.* **20** *5157*

Cadez, I., Hall, R.I., Landau, M. Pichou and F. Schermann, C., 1988, *J.Phys.B:At.Mol.Opt.Phys.* **21** *3271-84.*

Christophorou, L.G., 1991, 20th Int. Conf. on Phenomena in Ionised Gases-Pisa - P3, ed. Polleski, V.,
    Singh, D.P. and Vaselli, M. *Instuto di Fisica Atomica et Molecdare.*

Ferch, J., Masche, C., Raith, W. and Wiemann, L., 1989, *Phys. Rev. A.* **40** *5407.*

Fersch, J., 1994, *Private Communications.*

Foner, S.N. and Hudson, R.L., 1984, *J.Chem.Phys.* **80** *518-23.*

Fuchs, M. and Toennies, J.P., 1987, *J.Phys.B: At. Mol. Phys.* **20** *605-16.*

Geddes, J., McCollough, R.W., Higgins, D.P., Woolsey, J.M. and Gilbody, H.B., 1994, *Plasma Sources Sci.
    Technol.* **3** *58-60.*

Herzberg, G., 1945, Molecular spectra and molecular structure Vol 2, *New York: Van Nostrand.*

Johnstone, W.M., Mason, N.J. and Newell, W.R., 1993, *J.Phys.B: At. Mol. Opt. Phys.* **26** *L147-152.*

Johnstone, W.M. and Newell, W.R., 1991, *J.Phys.B: At.Mol. Opt. Phys.* **24** *3633-43.*

Johnstone, W.M., Akther, P. and Newell, W.R., 1995, *J.Phys.B: At.Mol.Opt.Phys* **28**
    743-7 53.

Johnstone, W.M. and Newell, W.R., 1995, *J.Phys.B: At.Mol. Opt.Phys* **28** *743-53*

Kochem, K.H., Sohn, W., Hebel, N. and Ehrhardt, H., 1985, *J.Phys.B. At.Mol.Opt Phys* **18** *4455-67.*

Mapstone, I.M. and Newell, W.R., 1992, *J.Phys.B: At.Mol. Opt. Phys.* **25** *491.*

McGroch, M.W. and Schlier, R.E., 1986, *Phys Rev. A.* **33** *1708-17.*

Register, D.F., Nishimura, H., and Trajmar, S., 1980, *J.Phy.B. At.Mol.Phys.* **13** *1651.*

Spence, D. and Schulz, G.J., 1969, *Phys.Rev.* **188** *280-7*

Srivastava, S.K. and Orient, O.J., 1983, *Phys.Rev.A.* **27** *1209-12.*

White, M.D. and Ross, K.J., 1976, *J.Phys.B.* **9** *2147-51.*

GAS-PHASE ELECTRON SCATTERING FROM FREE SPATIALLY ORIENTED MOLECULES *

Norbert Böwering

Fakultät für Physik der Universität Bielefeld
D-33501 Bielefeld, Germany

INTRODUCTION

Diffraction methods are often employed for structure analysis. While x-ray diffraction is widely used to analyze condensed matter, electron scattering [1-3] is the preferred method of investigating gaseous compounds. In the vapor phase it is superior to x-ray scattering since electrons interact much more strongly with the target atoms than photons. This was already demonstrated in the early pioneering work of Mark and Wierl [4]. A comparison with x-ray scattering can serve to illustrate the role of orientation in diffraction experiments. In the powder diffraction (Debye-Scherrer) technique a sample consisting of randomly oriented crystallites is exposed to a monochromatic x-ray beam. This leads to a characteristic pattern of interference rings which is cylindrically symmetric due to the random orientation of the target components. In this respect the result of electron diffraction from a beam of unoriented molecules with a statistical distribution of their molecular axes is quite similar, although the resulting circular scattering pattern is usually more diffuse. Since there is no preferred orientation, the structure information is only one-dimensional in these cases.

On the other hand, in x-ray scattering from a single crystal a Laue spot pattern arises. It is the result of the interference of scattered rays from the highly oriented crystal planes when the Bragg condition is satisfied. An analysis of the diffraction from single crystals provides detailed three-dimensional information on the crystal structure. Likewise, diffraction of an electron beam from molecules which are oriented with their axis in space contains multi-

* This article is dedicated to Peter Farago in admiration of his seminal contributions to electron physics.

dimensional information about the molecular structure. In such a case not only the absolute values of the internuclear separations but also the components parallel and perpendicular to the electron beam direction may be extracted thus leading directly to the molecular bond angles.

Generally, for sufficiently high electron energy the scattering from unoriented gas molecules can be expressed as a superposition of complex spherical amplitudes originating from independent atomic scattering centers. In this so-called independent atom model (IAM) [1-3] the diffraction pattern which depends on the momentum transfer s consists of a dominating atomic part, I_{atom} (s), the sum of single-scattering contributions of all atoms within the molecules, and a molecular interference part, $M(s)$, resulting predominantly from pairwise interference between the various atomic scattering amplitudes. Only this second term contains the molecular structure information. By taking its Fourier transform a radial distribution curve is obtained yielding the interatomic distances $|\vec{r}_{ij}|$ and also the vibrational

amplitudes $|\vec{l}_{ij}|$ between atoms i and j [1-3]. However, it can occur that several distances have nearly the same length; therefore, the molecular bond lengths may not always be determined unambiguously.

If a molecule is oriented with respect to some external direction (e.g. an electric field direction $\hat{E}$) the differential cross section for elastic electron scattering, $d\sigma/d\Omega$, depends on α, the angle between $\hat{E}$ and $\hat{s}$. Then a new additional interference term arises originating from the orientation of the molecule in a specific rotational state $|JKM\rangle$:

$$\frac{d\sigma}{d\Omega}(s,\alpha) = I_{atom}(s) + M(s) + M_{JKM}(s,\alpha)$$

$$= \sum_i |f_i(s)|^2 + 2\sum_{i<j} |f_i(s)| \, |f_j(s)| \, j_0(sr_{ij}) \cos(\eta_i(s) - \eta_j(s)) \tag{1}$$

$$+ \sum_{i<j} |f_i(s)| \, |f_j(s)| \, (2J+1) \sum_{n=1}^{2J+1} C_{JKMn} \, Re \left\{ F_{nij}(s,\alpha) \, exp(i(\eta_i(s) - \eta_j(s))) \right\}$$

Here, f_i and η_i are the direct scattering amplitudes and phases of atom i, and C_{JKMn} are expansion coefficients characterizing the orientational probability distribution of the internuclear axis [5]. When vibrational damping is neglected the functions F_{nij} (s,α) are defined in terms of Legendre polynomials P_n and spherical Bessel functions j_n as:

$$F_{nij}(s,\alpha) = 2(2\pi)^2 i^n P_n(\cos\alpha) P_n(\cos\beta_{ij}) j_n(sr_{ij}) \tag{2}$$

where $\cos\beta_{ij} = z_{ij}/r_{ij}$ describes the projections of the bond distances on the molecular symmetry axis.

Following first calculations by Fink and coworkers [6] for molecules held rigidly fixed in space, the IAM-based theoretical description for electron diffraction from oriented molecules outlined above was developed by Kohl and Shipsey [7]. As an example, Fig. 1 shows the predicted normalized diffraction pattern for the $|212\rangle$ state of CH_3I calculated according to this model for a scattering geometry where the electric field is perpendicular to the direction of the incident electron beam. The ordinate gives the ratio of the cross section for this state to the one for the $|000\rangle$ state. The resulting scattering distribution is not cylindrically symmetrical; it depends on both polar and azimuthal scattering angles. Both the periodicity and the amplitude of the oscillations are characteristic of each oriented rotational state.

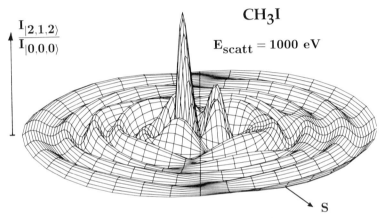

Figure 1. Calculated two-dimensional scattering distribution for the $|212\rangle$ state of CH_3I (normalized to the distribution for the $|000\rangle$ state) in perpendicular scattering geometry. The momentum transfer s ranges from 0 Å^{-1} at the center of the figure to 16 Å^{-1} at the rim.

EXPERIMENTAL

Experimentally, beams of oriented molecules can be produced by state selection of polar molecules which have the properties of symmetric tops via the linear Stark effect in a strong electrostatic hexapole field, as reviewed, e.g. in Ref. 8. The hexapole acts as a lens and, depending on its voltage U_o, different states exhibiting a positive Stark effect are focussed. By adiabatic transfer to a weak homogeneous electric orientation field a substantial average degree of orientation can be produced in the scattering region.

By combining such a molecular beam machine with an electron diffraction unit, the first measurements of the differential cross section for oriented molecules have been obtained in 1992 [9]. Fig. 2 shows a scheme of this apparatus described in greater detail elsewhere [10]. A continuous supersonic jet of CH_3I or CH_3Cl molecules is generated with rotational temperatures of ca. 70K or ca. 25K by a nozzle-skimmer arrangement. After passage through the hexapole and the guiding field, the molecules are oriented preferentially parallel or antiparallel to the electron beam. For a positive polarity of the voltage applied to the orientation field plate nearest the electron gun the CH_3 group points out predominantly towards the incident electrons whereas for reversed field polarities it is the halogen atom.

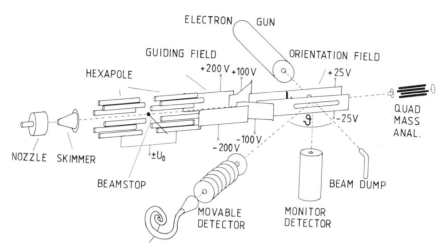

Figure 2. Experimental apparatus for elastic electron scattering from oriented molecules in a partially state-selected molecular beam.

The detailed shape of the orientational probability distribution of the molecular axes depends on the particular state ensemble selected by the hexapole. At a given voltage U_o and a rotational temperature T_{rot} of the molecules it can be calculated by taking into account the geometric dimensions of the experiment. The electron beam (typically 1000 eV electron energy and 10 μA current) passes perpendicularly through the molecular beam focus in the orientation field. The angular dependence of the electronically elastic scattering intensity is recorded by a rotatable detector consisting of a retarding field analyzer and a channeltron.

EXPERIMENTAL RESULTS

Parallel/antiparallel scattering geometry

By switching only the guiding field off (and not the orientation field), the molecular orientation can be turned off without influencing the electron beam. Thus, the additional molecular interference term $\overline{M}(s)$ for the focussed state ensemble can be determined from two intensity measurements as a fraction of the differential cross section $\sigma_{unor} = (d\sigma/d\Omega)_{unor}$ according to

$$\frac{\overline{M}(s)}{\sigma_{unor}} = \frac{I_{or}(s) - I_{unor}(s)}{I_{unor}(s)} \tag{3}$$

In this quantity most apparatus related effects cancel; in particular, all contributions from unoriented background molecules are removed. The mean interference terms obtained for both polarities of the orientation field, $\overline{M}^+$ and $\overline{M}^-$, can be divided into even and odd parts (representing pure alignment and pure orientation contributions):

$$\overline{M}^{\pm}(s) \;=\; \overline{M}_A(s) \pm \overline{M}_O(s) \tag{4}$$

These interference terms were measured for oriented CH_3I molecules and compared with model calculations [9,10]. The results obtained at an electron energy of 1 keV, a hexapole voltage of $U_o = 7kV$ and under expansion conditions such that $T_{rot} \approx 70K$ are shown in Fig. 3. The full curves in the figures represent the numerical results obtained by applying the IAM of Ref. 7 to the state ensemble determined by hexapole transmission calculations. As a function of momentum transfer a distinct oscillatory pattern is observed. This is also qualitatively predicted by theory. The numerical results indicate that the pure orientational contributions are modeled quite accurately, whereas a deviation occurs in the alignment part at small momentum transfer.

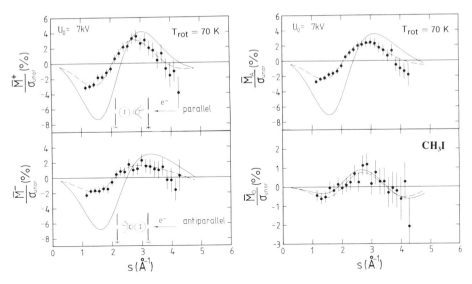

Figure 3. Molecular orientational interference contributions for CH_3I molecules as a function of momentum transfer for parallel and antiparallel orientation, as well as pure alignment and orientation parts, normalized to the cross section for unoriented molecules. The data were obtained at $U_o = 7kV$ for a molecular beam with $T_{rot} \approx 70K$. The full drawn lines represent the model calculations, the dashed lines indicate the result of the Legendre polynomial fit. Insets: Illustration of the scattering geometry.

It is quite conceivable that the model is sufficiently accurate and the deviations can be attributed to the limited knowledge of the population fractions of the large number of rotational states focussed into the scattering region under the conditions of the experiment. The experimental results can then be fitted (see dashed curves in Fig. 3) using a Legendre polynomial expansion for the description of the particular mean orientational probability distribution of the molecules present in the scattering region. The first 6 Legendre moments were extracted [10] demonstrating the quite general use of this scattering technique to analyze an oriented molecular state ensemble. The only other method which has been developed for this purpose is the laser-induced photodissociation technique first carried out by Gandhi et al. [11].

The experiments were also performed for CH_3Cl molecules under the same conditions as described above ($U_o = 7kV$, $T_{rot} \approx 70K$) for parallel and antiparallel orientation with respect to the electron beam [12]. The experimental data for $\overline{M}^+/\sigma_{unor}$ and $\overline{M}^-/\sigma_{unor}$ are shown in Fig. 4a in comparison with the corresponding model calculations. They resemble those obtained for CH_3I molecules (Fig. 3) which have a similar molecular structure. However, in the case of CH_3Cl the results are almost identical for parallel and antiparallel orientation, i.e. the experimental data and the model calculations both indicate that there is essentially only an alignment part and the pure orientation contribution is nearly zero. Nevertheless, the shape of the diffraction pattern is very sensitive to the rotational state distribution and the orientational properties of the target molecules.

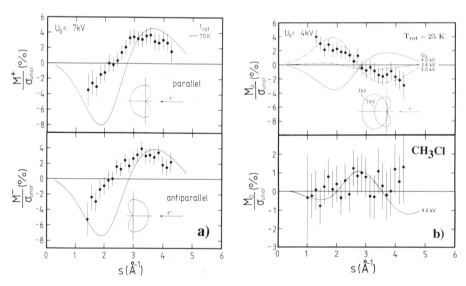

Figure 4. Normalized molecular orientational interference contributions for CH_3Cl molecules as a function of momentum transfer. In Fig. 4a the data points are for parallel and antiparallel orientation at $U_o = 7kV$ and $T_{rot} \approx 70K$; in Fig. 4b pure alignment and orientation parts are shown at $U_o = 4kV$ and $T_{rot} \approx 25K$. The lines represent the corresponding model calculations at hexapole voltages U_o as indicated. Insets: Polar plots of the symmetry axis orientational distribution.

When using different expansion conditions to create a colder beam of CH_3Cl ($T_{rot} \approx 25K$), such that the number of contributing states is substantially reduced, and a hexapole voltage of $U_o = 4kV$, which leads to an increase of the calculated average degree of orientation from 0.22 to 0.40, the shape of the interference pattern changes strongly and a "turnover" of the phase of the oscillations is observed (see $\overline{M}_A / \sigma_{unor}$ of Fig. 4b). The calculations indicate that as U_o is decreased the contributions from the $|111\rangle$ and $|222\rangle$ states, which have opposite trends compared to the majority of the other states, increase, thus leading to the change of the oscillations. However, the relative weights of these states seem to be underestimated since the numerical results for $U_o = 3kV$ and not $U_o = 4kV$ agree with the data. The fact that the contributions $\overline{M}_O / \sigma_{unor}$ are small can be explained by stronger scattering asymmetries for the sideways geometry compared to the parallel configuration. As seen from the polar plots in the insets of Fig. 4, sideways collisions are quite probable.

Perpendicular scattering geometry

The distribution for the scattering from an anisotropic target ensemble should also depend on the azimuthal angle if the electric field is not parallel or antiparallel to the electron beam direction. Therefore, using orientation field plates rotated by 90 degrees and scanning a detector at fixed polar angle on a circle around the electron beam direction, the azimuthal dependence of the molecular interference term was examined for preferentially perpendicular incidence of the electrons with respect to the molecular symmetry axis of CH_3Cl [13]. As shown in Fig. 5, pronounced angular variations have been obtained in both

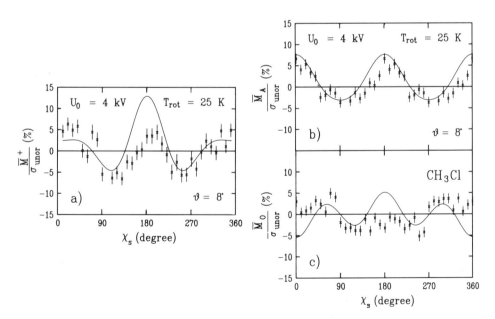

Figure 5. Azimuthal dependence of the normalized molecular orientational interference contributions for CH_3Cl molecules at $U_o = 4kV$, $T_{rot} \approx 25K$, and at a fixed polar scattering angle of 8°. a): Data in comparison with model calculations. b): Pure alignment contribution. c): Pure orientation contribution.

experiment and corresponding model calculations at a fixed polar scattering angle of 8° for $U_o = 4kV$ and $T_{rot} \approx 25K$. The distribution, $\overline{M}^+$, can also be divided into even and odd parts, $\overline{M}_A$ and $\overline{M}_O$, respectively. In this particular case the alignment contributions show better agreement with the calculated dependence.

FOURIER TRANSFORM CALCULATIONS

The perpendicular scattering geometry is most suitable for future structure analysis since then the diffraction pattern depends on both polar and azimuthal scattering angles. Kohl and Shipsey [14] have first shown that two-dimensional scattering data of oriented molecules may be directly inverted using two-dimensional Fourier transforms (2D-FT) to obtain real-space images, independent of a model. Such an inversion, based on diffraction data, is illustrated here for the case of CH_3I [15] where the molecular structure [16] is of course well-known. The IAM is used only to calculate the two-dimensional dependence of the orientational molecular interference pattern for the experimental conditions of the data of Fig. 3, except that the perpendicular scattering geometry is now assumed. These calculated distributions are shown in Fig. 6 for a square of momentum transfer up to values of $s = \pm 5 \text{ Å}^{-1}$, using either the state population fractions obtained from the hexapole transmission calculations (Fig. 6a) or the fitted Legendre moments (Fig. 6b) as input data. (The two cases differ mainly in the alignment coefficients.) Experimentally, such distributions could be obtained by a two-dimensional detection system and a measurement procedure according to Eq. (4). Starting from the x-axis which coincides with the electric field direction, the distributions are anisotropic for any given value of s on a semi-circle around the origin with the two halves on either side of $\hat{E}$ being mirror images.

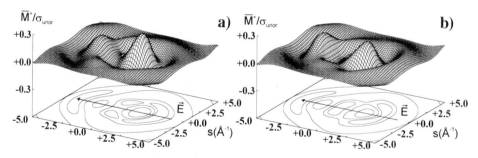

Figure 6. Two-dimensional plot of the normalized orientational interference pattern for CH_3I molecules in momentum space. The results were determined numerically for perpendicular orientation of a) the calculated and b) the fitted state distribution.

By applying 2D-FT methods to data like those of Fig. 6 the corresponding patterns in real space can be obtained. Ideally, the complete scattering space would have to be probed so that the 2D-FT could be taken with respect to s and α, the angle between $\hat{E}$ and $\hat{s}$ [14]. However, in the experiment the direction of $\hat{E}$ is kept fixed and the additional interference term is measured at definite polar and azimuthal scattering angles. A range of small values of α cannot be sampled in this way [14, 17]; therefore, the scattering space with respect to s and α is not completely covered. Nevertheless, the missing regions can be included to a good approximation by linear interpolation, making use of the fact that the integrand of the 2D-FT is zero for α= 0°. The corresponding 2D-FT for the diffraction patterns of Fig. 6a and 6b are plotted in Fig. 7a and 7b. They give the spatial probability distribution of the internuclear separations for atom pairs within the molecule in real space. Different distances from the origin correspond to different atom pair separations. The two-dimensional images of Fig. 7 thus represent a convolution of the spatial probability distribution of the molecular axis due to the rotational motion of the oriented molecules of the scattering state ensemble with the geometric molecular structure.

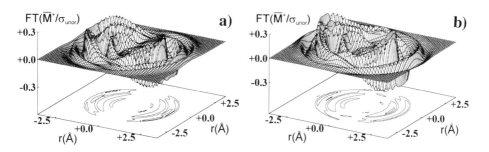

Figure 7. Two-dimensional Fourier transforms of the data of Fig. 6: Spatial distributions of atom-pair separations in real space.

In contrast to diffraction patterns discussed and predicted for laser-aligned molecules by Williamson and Zewail [18] the left and right halves of Fig. 7a,b are not mirror images since there are also orientational contributions. The slightly different axis distributions used as input for the two examples discussed here give rise to the subtle differences between Fig. 7a and 7b. In particular, the distribution of the C-I bond distance on a radius of 2.12 Å around the origin has a different shape: in Fig. 7b it is peaked along the x-direction; in Fig. 7a maxima occur at about ± 35° with respect to the x-axis, in agreement with the polar distributions of the molecular symmetry axis for the oriented state ensembles used as input. Similar numerical calculations were reported for oriented CH_3Cl molecules [17].

In general, independent of the IAM description, results of the type shown in Fig. 7 are useful in two respects: If the molecular structure is known, the spatial axis distribution of the oriented state ensemble may be extracted; vice versa, if the orientational properties of the scattering target molecules are well-defined and known independently, the molecular structure information on bond distances, bond angles and, in principle also vibrational amplitudes may be obtained. If, in addition, the scattering pattern is accurately described by a model, both pieces of information may be determined from an analysis of the data. On the other hand, if only a single rotational state is selected, it is always possible to reconstruct the molecular structure [14]. In this context, the scattering phase shifts can be exploited to highlight a particular atom pair separation [14].

CONCLUSIONS AND OUTLOOK

Provided the scattering target molecules are (at least partially) oriented or aligned, additional molecular interference terms contribute to the electron scattering cross sections. They are very sensitive to the spatial distribution of the molecular axis orientations. For CH_3I and CH_3Cl, normalized differential cross sections have been determined experimentally in recent years. They agree qualitatively and sometimes even quantitatively with numerical calculations based on the IAM. For perpendicular orientation of the molecules with respect to the electron beam the diffraction pattern depends also on the azimuthal angle. Information about the molecular structure and the orientational probability distribution is inherent in these two-dimensional images which can be analyzed by Fourier transform techniques.

In the future, sensitive tests of scattering models may be possible by performing experiments with pulsed seeded beams where the molecules may be focussed in predominantly single rotational states. By developing two-dimensional detection schemes [19,20], a more efficient and simultaneous collection of the scattering pattern should be feasible. This would also enable the use of this method for analysis of larger and more complex polar molecules for which the structure is not known very accurately.

ACKNOWLEDGMENTS

I am greatly indebted to M. Volkmer, Ch. Meier, J. Lieschke, A. Mihill, R. Dreier and M. Fink for their essential contributions to the work presented here. I would also like to thank U. Heinzmann and D.A. Kohl for helpful suggestions, and A.K. Lofthouse for help in preparation of this manuscript. Financial support by the Deutsche Forschungsgemeinschaft through SFB 216 is gratefully acknowledged.

REFERENCES

1. R.A. Bonham and M. Fink, "High Energy Electron Scattering", Chap. 1, Van Nostrand Reinhold, New York (1974).
2. E.A. Ebsworth, D.W. Rankin, S. Cradock, "Structural Methods in Inorganic Chemistry", Chap. 8, Blackwell Scientific, Oxford (1987).
3. I. Hargittai and M. Hargittai, eds., "Stereochemical Applications of Gas-Phase Electron Diffraction", Part A, VCH Publishers, Weinheim (1988).
4. H. Mark and R. Wierl, *Naturwissenschaften* 18: 205, 778 (1930).
5. S.E. Choi and R.B. Bernstein, *J. Chem Phys.* 85: 150 (1986).

6.   M. Fink, A.W. Ross and R.J. Fink, Z. *Phys. D* 11: 231 (1989).
      A. Mihill and M. Fink, Z. *Phys. D* 14: 77 (1989).
7.   D.A. Kohl and E.J. Shipsey, Z. *Phys. D* 24: 33 (1992).
8.   D.H. Parker and R.B. Bernstein, *Annu. Rev. Phys. Chem.* 40: 561 (1989).
9.   M. Volkmer, Ch. Meier, A. Mihill, M. Fink and N. Böwering, *Phys. Rev. Lett.* 68: 2289 (1992).
10.   M. Volkmer, Ch. Meier, J. Lieschke, A. Mihill, M. Fink and N. Böwering, *submitted to Phys. Rev. A.*
11.   S.R. Gandhi, T.J. Curtiss and R.B. Bernstein, *Phys. Rev. Lett.* 59: 2951 (1987).
12.   N. Böwering, M. Volkmer, C. Meier, J. Lieschke, M. Fink, Z. *Phys. D* 30: 177 (1994).
13.   M. Volkmer, Ch. Meier, J. Lieschke, R. Dreier, M. Fink and N. Böwering, *to be published.*
14.   D.A. Kohl and E.J. Shipsey, Z. *Phys. D* 24: 39 (1992).
15.   M. Volkmer, Ch. Meier, J. Lieschke, R. Dreier, N. Böwering (*unpublished*).
16.   The equilibrium bond lengths are: r_e(C-H) = 1.06 Å, r_e (C-I) = 2.12 Å, r_e (H-H) = 1.76 Å , r_e (H-I) = 2.63 Å; the bond angle (H-C-H) is 112.0°. M. Fink and A. Mihill, *private communication* (1990).
17.   N. Böwering, M. Volkmer, Ch. Meier, J. Lieschke, R. Dreier, *J. Mol. Struct.* 348: 49 (1995).
18.   J.C. Williamson and A.H. Zewail, *J. Phys. Chem.* 98: 2766 (1994).
19.   J.C. Williamson, M. Dantus, S.B. Kim and A.H. Zewail, *Chem. Phys. Lett.* 196: 529 (1992).
20.   R. Dreier, Ch. Meier, M. Volkmer, N. Böwering in: "Abstracts of Contributed Papers, 19th. Int. Conf. on the Physics of Electronic and Atomic Collisions", Whistler, Canada, J.B.A. Mitchell, J.W. McConkey, C.E. Brion, eds., Vol. 2, p.799 (1995).

A POLARIZED ELECTRON SOURCE USING GALLIUM ARSENIDE STRAINED BY DIFFERENTIAL THERMAL CONTRACTION

D.M.Campbell* and G.Lampel[+]

*Department of Physics and Astronomy,
University of Edinburgh,
U.K.

[+]Laboratoire de Physique de la Matière Condensée,
URA CNRS 1254D,
Ecole Polytechnique,
France.

INTRODUCTION

Since 1975, when efficient spin-polarized photoemission from a semiconductor was first achieved (Pierce et al., 1975a, 1975b; Pierce and Meier, 1976), there have been many semiconductor spin-polarized electron sources constructed in laboratories studying atomic, solid-state and high energy physics. The polarization of these sources, based on cubic III-V semiconductors (usually highly p-doped GaAs), has been intrinsically limited to a maximum of 50% by the band structure of the cubic crystals. In recent years, measured polarizations considerably exceeding 50% have been reported for a new generation of sources; these make use of III-V semiconductor based crystals in which the cubic symmetry has been lowered either by straining or by growing anisotropic heterostructures. In general, the yield of these high polarization sources (measured as the ratio of the emitted electron flux to the incident photon flux in the semiconductor) has been substantially below that obtained with conventional GaAs cathodes.

The aim of the present paper is twofold. Firstly, we discuss briefly some specific issues arising in the physics of the semiconductor spin-polarized electron source. Although there are no new results in this part, we feel that the presentation could be useful to those users of polarized electron sources without an extensive background in semiconductor physics. Secondly, we present some preliminary results from a study of a novel type of photocathode, in which GaAs is strained on a substrate of CaF_2. The method by which strain is achieved, which is based on the difference in the thermal expansion coefficients of GaAs and CaF$_2$, allows strained layers of GaAs several microns thick to be constructed, offering the promise of high efficiency, high polarization electron sources.

THE PHYSICS OF THE GaAs SPIN-POLARIZED ELECTRON SOURCE

The basic physical process which governs the operation of semiconductor electron sources is photoemission from the semiconductor. This process is well described by the Three Step Model, introduced about thirty years ago; a comprehensive description of this model is given in Spicer and Herrera-Gomez (1994). The photoemission of electrons into vacuum is considered to take place in three stages: (i) absorption of a photon of energy hν in the solid, promoting an electron into the conduction band with an energy above the vacuum level; (ii) energy relaxation and transport of this electron towards the surface; (iii) emission into vacuum with an escape probability P_{esc}.

The semiconductor spin-polarized source is very attractive due to the fact that, in process (i), absorption of circularly polarized light can create spin-polarized electrons in the conduction band by optical pumping (Meier and Zakharchenya, 1984). A discussion of process (ii) for such electrons must take into account spin relaxation during transport to the surface. Process (iii) can be very efficient in semiconductors, and particularly in GaAs, because it is possible to lower the vacuum level to an energy below the bottom of the conduction band in the bulk. This is the situation described as negative electron affinity (NEA) (Bell, 1973; Escher, 1981), in which all the photoexcited conduction electrons have sufficient energy to escape into the vacuum. Activation to NEA involves coadsorption of caesium and oxygen on a clean surface of GaAs in ultrahigh vacuum.

There have been many papers dealing with the design and operation of semiconductor polarized electron sources. Among these, the early paper by Pierce et al. (1980) is still relevant, and the recent review by Pierce (1995) includes discussion of most of the new sources in which polarizations substantially higher than 50% have been obtained using suitably designed cathode structures. The physics of spin-polarized photoemission from NEA GaAs, on the other hand, is dealt with in great detail by Drouhin(1985a; 1985b), and a review article concerned with photoelectronic processes in semiconductors has recently appeared (Hermann et al., 1992).

The purpose of this part of the present paper is not to repeat what can be found in more detail in these review articles, but rather to concentrate on a few points which appear to us to be important for the understanding and improvement of existing sources. In Section (a) we discuss in some detail the absorption of circularly polarized light in a III-V direct cubic semiconductor such as GaAs; the discussion is extended to include situations where the symmetry is lowered by an axial perturbation. Section (b) is devoted to diffusive transport of polarized electrons towards the surface, and presents a simple physical picture of this process. In Section (c), we briefly raise some issues concerned with the escape probability of electrons, and the maximum polarizations typically measured in the new generation of high polarization electron sources.

(a) Excitation of spin polarized electrons by absorption of circularly polarized light

The band structure of GaAs has been described in a number of papers (Chelikowsky and Cohen, 1976; Pierce, 1995). The fundamental optical transitions take place near

206

the centre of the Brillouin zone at $\vec{k} \simeq 0$ between the three valence bands and the conduction band. For $h\nu$ close to the direct gap energy E_g ($E_g = 1.42\,\mathrm{eV}$ in undoped GaAs at 300 K), only transitions from the heavy hole (hh) and light hole (lh) bands may occur. Spin-orbit coupling splits the valence bands by an amount Δ; for GaAs at 300K, $\Delta = 0.34\,\mathrm{eV}$. When $h\nu \geq E_g + \Delta$, additional transitions from the spin-orbit split-off band (so) also take place.

The wave functions of the conduction band (Kane, 1957) are s-type near $\vec{k} = 0$. When the spin-orbit splitting is taken into account, the valence states at $\vec{k} = 0$ split between the fourfold Γ_8 representation (hh and lh) and the twofold Γ_7 representation (so). It is usual to associate the Γ_6, Γ_8 and Γ_7 states with the $^2S_{1/2}$, $^2P_{3/2}$ and $^2P_{1/2}$ atomic levels respectively. Then the optical transitions allowed for $\sigma_\pm$ circularly polarized light correspond to $\Delta M_J = \pm 1$, and the well known optical transition diagram shown in Figure 1 may be drawn. The $|\frac{3}{2} \pm \frac{3}{2}\rangle$ states are associated with the heavy holes, while the $|\frac{3}{2} \pm \frac{1}{2}\rangle$ are associated with the light holes. The theoretical maximum polarization is then

$$P_i = \left(\frac{n_+ - n_-}{n_+ + n_-}\right)_i$$

where $n_\pm$ are the electronic populations of the states $|\frac{1}{2} \pm \frac{1}{2}\rangle$ in the conduction band. The suffix i refers to the initial electron population, before relaxation processes have come into play.

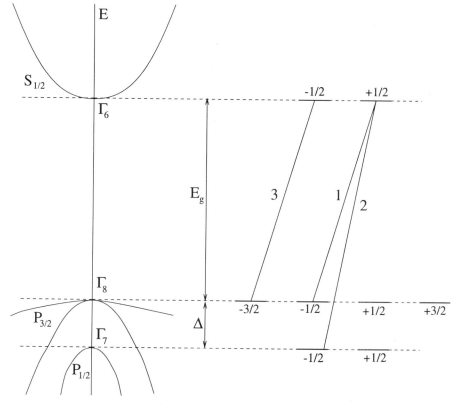

Figure 1. Optical transitions induced in GaAs by photons with angular momentum $+\hbar$. The magnetic sublevels are labelled by m_J, and the relative probability of each transition is shown.

This picture is somewhat oversimplified, but has become very popular because it yields the correct results:

$$|P_i| = 50\% \text{ if } E_g \leq h\nu \ll E_g + \Delta;$$
$$|P_i| \simeq 0 \text{ if } h\nu \gg E_g + \Delta.$$

Without going into too much detail, we shall stress below why this picture is oversimplified, and how the question can be correctly treated.

First of all, let us recall that within the isotropic effective mass approximation the constant energy surfaces in $\vec{k}$-space are spheres. Defining m_c^*, m_{hh}^*, m_{lh}^* and m_{so}^* as the effective masses of the conduction electrons, the the heavy, light and spin-orbit-split holes, and taking the origin of energy at the top of the Γ_8 valence band, we can write the various energies as functions of k.

For the conduction electrons:

$$\epsilon_c = E_g + \frac{\hbar^2 k^2}{2m_c^*}.$$

For the heavy (light) holes:

$$\epsilon_h = -\frac{\hbar^2 k^2}{2m_h^*},$$

where h stands for hh (lh).

For the spin-orbit split holes:

$$\epsilon_{so} = -\Delta - \frac{\hbar^2 k^2}{2m_{so}^*}.$$

It should also be recalled that, due to the small magnitude of the photon $\vec{q}$ vector, conservation of momentum in the process of absorption or emission leads to so-called "vertical" transitions. This means that the allowed optical transitions between valence and conduction bands take place between any states which satisfy the condition $\epsilon_c - \epsilon_v = h\nu$ and which have the same $\vec{k}$ vector, as shown in Figure 2.

The electronic polarization in the conduction band is the polarization calculated by averaging over all the $\vec{k}$-vector directions. This would be trivial if the states $|J\,M_J\rangle$ involved in the optical transitions had a common quantisation axis fixed with respect to the direction of propagation of the light, as is implicitly assumed in the simple atomic picture usually given. This picture is not, however, valid, since when crystal states are described in the $|J\,M_J\rangle$ representation it is implicitly understood that the quantisation axis lies along the $\vec{k}$-vector. This can been taken into account by averaging over all the optical transitions which take place at a given photon energy (Dymnikov et al., 1976; Zakharchenya et al., 1982; Meier and Zakharchenya, 1984).

In the following discussion, we take a given direction $\vec{z}$ in the laboratory frame as the reference axis for the electronic polarizations $P_i(v \to c)$ obtained for the transitions from the three valence bands ($v = hh, lh, so$) to the conduction band c. Then for absorption of a photon carrying an angular momentum $+\hbar$ along $\vec{z}$, the results depend on the magnitude of $\vec{k}$.

For small k the Γ_8 and Γ_7 states are well separated, so

$$P_i(hh \to c) = P_i(lh \to c) = -50\%;$$

$$P_i(so \to c) = +100\%.$$

When k increases the heavy hole band remains unperturbed, so

$$P_i(hh \to c) = -50\%.$$

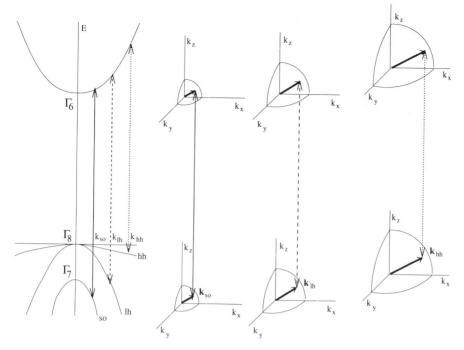

Figure 2. Transitions from the three valence bands to the conduction band which can be induced by a photon with $h\nu > E_g + \Delta$.

However, the light hole and spin-orbit split bands tend to hybridise; at large k, $P_i(lh \to c)$ has changed from -50% to +50% and $P_i(so \to c)$ from 100% to 0%, as shown in Figure 2 of Drouhin et al. (1985b)

Thus in the limiting cases we recover the well known results:

$$P_i = -50\% \text{ for } h\nu \simeq E_g;$$
$$P_i \sim 0 \text{ for } h\nu \gg E_g + \Delta.$$

The situation is much simpler when a uniaxial distortion perpendicular to the surface is applied to the crystal, as is the case for strained layers or superlattice structures. In a strained layer, the distortion is usually due to a compressive (tensile) stress in the plane of the surface, which is equivalent to an isotropic compression (expansion) plus an axial tensile (compressive) stress perpendicular to the surface (Pollak and Cardona, 1968).

The main effect of the uniaxial distortion on the band structure is to lift the degeneracy between the heavy hole and light hole bands near $\vec{k} = 0$ by an amount proportional to the stress (Figure 3). There are then two different gaps, $E_g^{\pm 3/2}$ and $E_g^{\pm 1/2}$, corresponding to the onset of optical transitions from the heavy (light) hole bands to the conduction band at $h\nu = E_g^{\pm 3/2}$ ($E_g^{\pm 1/2}$). The sign of $\Delta E = E_g^{\pm 1/2} - E_g^{\pm 3/2}$ is determined by the sign of the distortion. This is positive for compressive stress in the plane of the surface, as in the GaAs/CaF$_2$ and GaAs/GaAsP systems, and negative

209

for tensile planar stress, found for example in GaAs/Si.

The quantisation axis for the $|J\,M_J\rangle$ states is now given by the axis of distortion, and no longer by the $\vec{k}$-vector, provided that the kinetic energy of the holes is smaller than the splitting of the valence bands:

$$\frac{\hbar^2 k^2}{2m_h^*} \ll |\Delta E|.$$

Then, to zeroth order in perturbation, the eigenfunctions describing the valence states are quantised perpendicular to the surface, that is, along the light propagation direction. This case is essentially equivalent to the atomic picture in the presence of a cylindrical distortion, yielding the results

$$P_i(hh \to c) = -100\%; \; P_i(lh \to c) = +100\%.$$

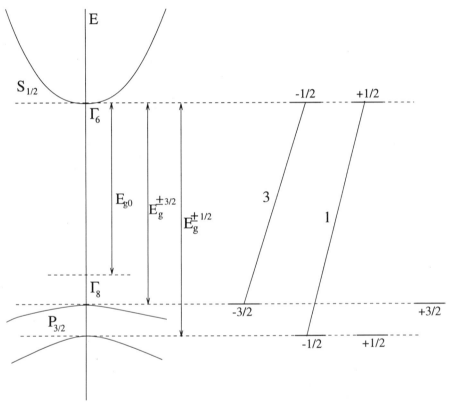

Figure 3. Optical transitions between the Γ_6 conduction band and the Γ_8 valence bands of GaAs, split by a uniaxial distortion caused by a planar compressive stress.

A crude model emerges, which states that the absorption coefficient $\alpha(h\nu)$ near threshold decomposes into two components corresponding to transitions from heavy hole states ($M_{3/2} = \pm3/2$) and from light hole states ($M_{3/2} = \pm1/2$). The first (second) component corresponds to creation of electrons with spin -(+):

$$\alpha(h\nu) \simeq \alpha_0(h\nu - E_g^{\pm3/2}) + \alpha_0(h\nu - E_g^{\pm1/2})/3,$$

where the function α_0 describes the energy dependence of the absorption coefficient, and the factor $1/3$ arises from the fact that the transition probabilities from the $M_{3/2} = \pm3/2$

states are three times larger than those from the $M_{3/2} = \pm 1/2$ states. Of course, one should also take into account the joint density of states between valence and conduction bands, which are different for heavy and light hole bands. However, since the joint density of states is mainly determined by the effective mass of the conduction band electrons, this difference should be small.

The following results are then obtained for $E_g^{\pm 3/2} < E_g^{\pm 1/2}$:

$$P_i = -100\% \text{ for } E_g^{\pm 3/2} \leq h\nu < E_g^{\pm 1/2};$$
$$P_i = -50\% \text{ for } E_g^{\pm 1/2} < h\nu \ll (E_g^{\pm 1/2} + E_g^{\pm 3/2})/2 + \Delta;$$
$$P_i \sim 0 \text{ for } (E_g^{\pm 1/2} + E_g^{\pm 3/2})/2 + \Delta \ll h\nu.$$

The fact that the polarization is reduced by a factor of approximately 2 when transitions from the light hole band become possible has been experimentally verified in a number of cases (Maruyama et al., 1991, 1992; Nakanishi et al., 1991; Aoyagi et al., 1992; Meier et al., 1993); the measurements presented in this paper confirm this behaviour in GaAs/CaF$_2$.

When the ordering of $E_g^{\pm 3/2}$ and $E_g^{\pm 1/2}$ is reversed, for example in GaAs/Si, P_i changes sign when transitions from the heavy hole band become possible. This has also been experimentally verified (Lassailly et al., 1990).

(b) Transport and Relaxation

We have seen in the preceding paragraph that it is mandatory to use photons with energies close to the bandgap in order to obtain a large initial polarization P_i. We are then concerned with transport and relaxation of conduction band electrons with very small kinetic energies, and the present discussion is limited to this case.

The transport of electrons towards the surface is discussed in a number of textbooks (see, for example, Bell (1973)). The number of electrons per unit area arriving at the surface, $J(0)$, can be calculated in terms of the total number of carriers $n_+ + n_-$ photoexcited by a given photon flux J_{ph} in the semiconductor. For NEA photocathodes, in which any electron reaching the surface has a probability P_{esc} of being emitted into vacuum, the flux of photoemitted electrons is just $J_e = P_{esc}J(0)$. From the solution of the transport equation, it is easily shown that

$$J(0) = J_{ph}\frac{\alpha}{\alpha + 1/L} = J_{ph}\frac{L}{L + 1/\alpha} \tag{1}$$

where the diffusion length $L = \sqrt{D\tau}$, D being the diffusion coefficient and τ the lifetime of the excited electron. In the highly p-doped GaAs used for efficient photocathodes, typically $\tau \lesssim 10^{-9}$s and $L \lesssim 2\,\mu$m. Strictly speaking, Equation 1 is true only if the surface recombination velocity is large in comparison with the diffusion velocity L/τ, but this is valid for the usual NEA photocathodes.

Equation 1 tells us that, of all the excited electrons distributed over a region of depth $1/\alpha + L$ by photon absorption and diffusion, only those within a distance L of the surface can reach it. In other words, the escape depth for NEA photocathodes is just the diffusion length. For non-NEA photocathodes, the escape depth is normally several orders of magnitude smaller; it is given essentially by the mean free path for inelastic collision, since a single inelastic collision is usually sufficient to reduce the energy of the photoexcited electron below the vacuum level (Spicer and Herrera-Gómez, 1994)

For a direct gap semiconductor, the absorption coefficient is

$$\alpha = \beta\sqrt{h\nu - E_g};$$

the value of β for GaAs is approximately $2\,\mu\mathrm{m}^{-1}\,\mathrm{eV}^{-1/2}$. For NEA photocathodes, the quantum yield is

$$Y = J_e/J_{ph} = P_{esc}\frac{\alpha}{\alpha + 1/L}.$$

$Y \simeq P_{esc}$ when $\alpha \gg 1/L$; for a GaAs sample with $L = 2\,\mu\mathrm{m}$, this condition will be satisfied for $h\nu - E_g \gg 60\,\mathrm{meV}$. Thus for photon energies a few hundred meV above the excitation threshold, the yield is limited mainly by the escape probability at the surface; it is this feature which gives NEA GaAs its value as a high yield photocathode.

The electronic polarization in the bulk decays with a spin relaxation time τ_s. Various relaxation mechanisms have been studied, both theoretically and experimentally (Meier and Zakharchenya, 1984; Fishman and Lampel, 1977). In heavily doped materials, the main relaxation mechanism for conduction electrons arises from exchange collisions with the unpolarized holes. For thermalised electrons in heavily doped GaAs at 300 K, τ_s is of the order of $10^{-9} - 10^{-10}\,\mathrm{s}$ (Pikus and Titkov, 1984).

The rate equations describing the time dependence of $(n_+ \pm n_-)$ show that $(n_+ + n_-)$ will survive for a time τ after creation, while $(n_+ - n_-)$ will survive for a shorter time

$$\tau_s^* = (1/\tau + 1/\tau_s)^{-1},$$

as a result of the combined effects of spin relaxation and recombination. Thus the steady state total population $(n_+ + n_-)_{ss}$ is proportional to $J_{ph}\tau$, while the steady state population difference $(n_+ - n_-)_{ss}$ is proportional to $P_i J_{ph}\tau_s^*$.

The steady state polarization in the bulk is then

$$P_B = \left(\frac{n_+ - n_-}{n_+ + n_-}\right)_{ss} = P_i\frac{\tau_s^*}{\tau} = P_i\frac{\tau_s}{\tau + \tau_s},$$

which shows that high P_B is obtained in materials for which τ_s/τ is large, and little spin relaxation occurs during the electron lifetime.

To obtain the steady state polarization of the electrons emitted into vacuum, it is necessary to calculate the polarization flux $\Delta J(0)$ (that is, the flux of $(n_+ - n_-)$ at the surface) (Pierce et al., 1980). The appropriate depth is the spin lifetime diffusion length $L_s^* = \sqrt{D\tau_s^*}$ (in contrast to the depth $L = \sqrt{D\tau}$ appropriate to $J(0)$). In a semi-infinite slab,

$$\Delta J(0) = P_i J_{ph}\frac{L_s^*}{L_s^* + 1/\alpha} = P_i J_{ph}\frac{\alpha}{\alpha + 1/L_s^*},$$

and the steady state polarization of the emitted electron beam is

$$P = \frac{\Delta J(0)}{J(0)} = P_i\frac{\alpha(h\nu) + 1/L}{\alpha(h\nu) + 1/L_s^*}.$$

Near the threshold, α^{-1} is usually larger than L and L_s^*, so that

$$P = P_i\sqrt{\frac{\tau_s}{\tau_s + \tau}};$$

this means that the steady state polarization of the emitted electron beam is larger than the bulk polarization P_B measured by photoluminescence experiments (Lampel and Eminyan, 1980). At higher photon energies, α^{-1} may be much smaller than L and L_s^*, so that $P \simeq P_i$; however, P_i decreases strongly with increasing energy. It is therefore always advantageous to use photon energies close to the gap.

For layers whose thickness d is much less than α, L and L_s^*, the fluxes $J(0)$ and $\Delta J(0)$ are simply equal to $J_{ph}\alpha d$ and $P_i J_{ph}\alpha d$ respectively, so that the polarization of the emitted electron beam should be $P = P_i$. This has been experimentally verified for thin unstrained GaAs layers, of thickness $d = 0.2\,\mu m$, epitaxially grown on GaAlAs; with $h\nu \sim E_g$, electron polarizations very close to 50% have been measured (Maruyama et al., 1989).

(c) Transmission through the surface

The last step occurring in the photoemission process is the transmission of electrons through the band bending region and emission into vacuum. Although a great deal of work has been done in surface physics under highly controlled conditions, very little has been published concerning the quantitative understanding of the NEA activation. On the other hand, there is a substantial literature of tricks and recipes designed to improve the yield and stability of NEA photocathodes. A most interesting survey of this subject is given in Pierce (1995).

For bulk GaAs , an escape probability $P_{esc} \simeq 0.1$ is considered adequate, whereas $P_{esc} \geq 0.5$ is currently achieved in commercial photomultiplier tubes. For thin layers, with reported quantum yields in the $10^{-3} - 10^{-4}$ range and $\alpha d \sim 0.1$, it may be deduced that $P_{esc} < 10^{-2}$; there should be considerable scope for future improvement in this figure.

Concerning the polarization obtained from strained layers, it is not at all clear why it seems to be difficult to reach the 100% value predicted for $h\nu$ close to the absorption threshold (Maruyama et al., 1991, 1992; Nakanishi et al., 1991; Aoyagi et al., 1992; Meier et al., 1993). The problem may be additional relaxation near the interface with the substrate; it is perhaps more probable that the discrepancy arises from the crudity of the theoretical prediction of 100% polarization, as discussed in Section (a). In fact, a detailed study of $P_i(h\nu, \Delta E)$ has not yet been carried out, except for a heuristic discussion by Zorabedian (1982). He concluded that, when the heavy and light hole bands are too close (small ΔE), $|P_i|$ is no longer 100%, because of the possibility of contribution from the other neighbouring valence band with opposite polarization. It does indeed seem to be experimentally established that the larger the value of ΔE, the larger is the maximum polarization achieved for a given type of structure (Aoyagi et al., 1992).

A STRAINED PHOTOCATHODE USING GaAs/CaF$_2$

It was pointed out in the introduction to this paper that a large increase in the electron polarization available from semiconductor photocathodes has been achieved in recent years by using thin strained layers grown on a substrate with a lattice mismatch. GaAs on GaAsP (Nakanishi et al., 1991; Maruyama et al., 1992) and GaInAs on GaAs (Maruyama, 1991; Meier et al., 1993) are examples of such systems. Unfortunately, the strain induced by lattice mismatch cannot be maintained for layer thicknesses greater than a critical value of the order of 100 nm. On the other hand, it was shown in the previous section that the optimum yield can only be obtained from a photocathode if the depth of the active layer is at least as great as the absorption length $1/\alpha$ for the appropriate photon. Since in GaAs, $1/\alpha > 1\,\mu m$ for photon energies close to the band gap, it is impossible to satisfy this criterion for strained photocathodes which rely on lattice mismatch.

An alternative method of inducing strain is to grow the active layer on a substrate which has a substantially different coefficient of thermal expansion. GaAs on CaF_2 is such a system, and samples of $GaAs/CaF_2$ have recently been prepared as part of a research programme in optoelectronics (Campbell et al., 1994). A relatively thick layer of GaAs epitaxially grown on a CaF_2 substrate is essentially strain-free at the growing temperature of 600°C, since dislocations at the interface relax the stress induced by lattice mismatch (Figure 4a). As the crystal cools down, the substrate contracts more rapidly than the upper layer, creating a biaxial compressive stress in the GaAs (Figure 4b).

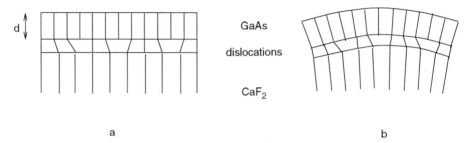

Figure 4. A GaAs layer grown on a CaF_2 substrate (a) at the growing temperature, the lattice mismatch strain relaxed by interface dislocations; (b) strained by differential contraction during cooling.

As explained earlier, the effect of the stress is to lift the degeneracy at the top of the valence band; the $M_{3/2} = \pm3/2$ heavy hole states lie higher in energy than the $M_{3/2} = \pm1/2$ light hole states, as illustrated in Figure 3.

In contrast to the case in which stress is induced by lattice mismatch, the stress created by differential thermal contraction can be maintained over layers several microns thick. This has been verified experimentally for $GaAs/CaF_2$ by Tessler et al. (1994), who studied the photoluminescence obtained from a $2\,\mu m$ layer of GaAs epitaxially grown on a CaF_2 substrate. From the photoluminescence excitation spectra of a sample cooled to 77 K, it was possible to deduce that $E_g^{\pm3/2} = 1.515\,\text{eV}$ and $E_g^{\pm1/2} = 1.575\,\text{eV}$, giving a valence band splitting of 60 meV. The corresponding stress was estimated to be -12 kbar. For photon energies between 1.52 eV and 1.57 eV the steady state photoluminescence polarization was greater than 70%; as expected, it fell by a factor of approximately 2 for photon energies much greater than $E_g^{\pm1/2}$.

In the present study, several similar samples of $GaAs/CaF_2$ were activated to negative electron affinity, and measurements were carried out to determine the dependence of the yield and the polarization of the photoemitted electrons on photon energy.

Measurements of photoelectron yield

To establish the optimum conditions for activation of $GaAs/CaF_2$ photocathodes, the yield was measured as a function of photon wavelength for several different samples and states of activation. The samples were mounted in a rotating carrousel in an ion-pumped vacuum chamber with a base pressure of 2×10^{-10} mb. Two tunable light sources were used: a tungsten lamp with a grating monochromator, and an argon-ion-pumped titanium sapphire laser. In each case, the light could be focused either on to the photocathode surface or on to a photodiode power meter. The photocathode was biased at -35 V relative to the chamber wall, and the emitted current was measured by an electrometer. Scanning of the photon wavelength and data acquisition were controlled by an Apple Macintosh computer.

As expected, the yield curves decrease monotonically with decreasing photon energy; an example is shown in Figure 5. The magnitude of the yield obtained at a given photon energy depended strongly on the activation procedure, which consisted of a heating and cooling cycle followed by deposition of Cs and O_2. The crystal was mounted on a heater consisting of a 500 μm thick strip of silicon mounted between molybdenum electrodes; the heating temperature could be accurately controlled by varying the direct current through the silicon strip. The highest yield was obtained for a sample heated to a maximum temperature of 638°C, as measured by an optical pyrometer. In this case, the yield after caesiation and oxygenation was 2×10^{-3} for $h\nu = 1.46$ eV, rising to 4.5×10^{-2} at the HeNe photon energy of 1.96 eV.

Early in the yield measurements, it became clear that there was a problem in maintaining the adhesion of the GaAs layer to the substrate during the heating and cooling cycle. While different samples behaved differently in this respect, it was frequently found that after several activation cycles a large fraction of the GaAs layer had peeled away from the CaF_2. Despite careful efforts to ensure that the rate of change of temperature did not exceed 2°C per minute, this problem persisted, seriously limiting the scope of the measurements.

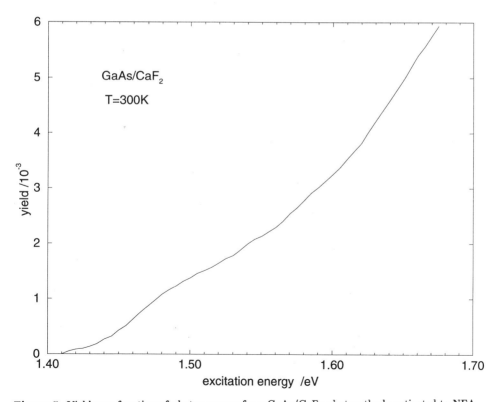

Figure 5. Yield as a function of photon energy for a GaAs/CaF$_2$ photocathode activated to NEA.

Measurements of electron polarization

To measure the polarization of the photoemitted electrons, samples of GaAs on CaF$_2$ were mounted in an apparatus previously used for studies of polarized photoemission from GaAs photocathodes (Drouhin et al., 1985b). A schematic diagram of the experimental arrangement is shown in Figure 6. The cathode was retracted by several centimetres for activation, which was carried out by the process described in the previous section. It was then moved into the photoemission position in front of the apertured anode. Light from the titanium sapphire laser, attenuated by neutral filters, was circularly polarized using a Babinet-Soleil retarder and focused on to the photocathode through small apertures in the electrode structure. The polarization vector of the light could be reversed in direction by rotation of the retarder; this rotation was controlled by an Apple Macintosh microcomputer, which also controlled the scanning of the laser wavelength.

An electron transport system, consisting of two orthogonal 90° electrostatic rotators and a series of coaxial cylinder electrostatic lenses, was used to accelerate the electron beam and deliver it to a cylindrical Mott polarimeter. The polarimeter was normally used at an electron incidence energy of 30 kV, but was calibrated before the start of the measurements using an extrapolation to zero energy loss with an electron incidence energy of 100 kV and a 60 nm gold foil. In the extrapolation procedure, a correction was made to account for finite foil thickness, using data from Gay et al.(1992).

The dependence of electron polarization on photon wavelength for a GaAs/CaF$_2$ photocathode at room temperature is shown in Figure 7. The error bars include an estimated ±3% uncertainty in the Sherman function of the polarimeter. As expected, the electron polarization exceeds 50% for photon energies near the excitation threshold, although the maximum polarization of 58% is substantially lower than that predicted

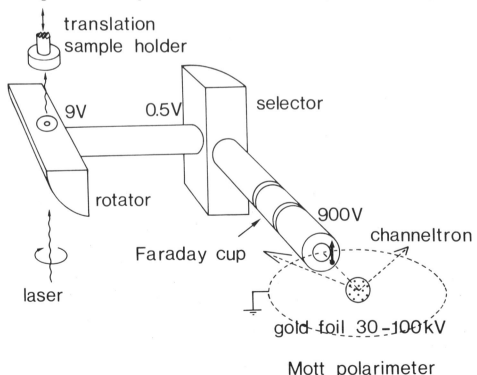

Figure 6. Schematic diagram of apparatus used for photoelectron polarization measurement.

from the photoluminescence measurements of Tessler et al. (1994). On the other hand, the overall shape of the polarization spectrum is in accord with the theoretical prediction, with the polarization falling to half its maximum value around 100 meV above the threshold. The measurements carried out after the first activation of this photocathode (circles in Figure 7) showed a reduced polarization very close to the threshold; after a second activation, the polarization remained essentially constant at its maximum value as the threshold was approached, in accord with the theory of the previous section (squares in Figure 7).

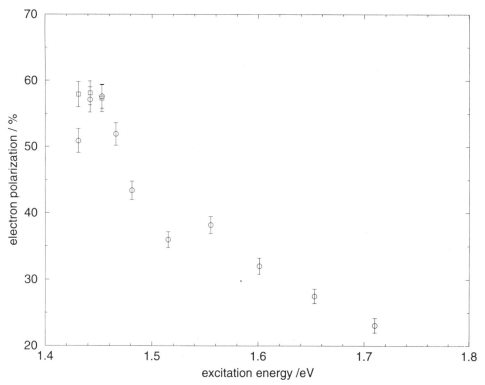

Figure 7. Electron polarization as a function of photon energy for a GaAs/CaF$_2$ photocathode activated to NEA.

Estimation of stress in the GaAs layer

The threshold of the yield curve (figure 5) provides an estimate of the heavy hole bandgap in the strained GaAs layer:

$$E_g^{\pm 3/2} = 1.415 \pm 0.005 \, \text{eV}.$$

From the sharp decrease in the polarization curve (Figure 7), it can be deduced that the light hole bandgap is

$$E_g^{\pm 1/2} = 1.46 \pm 0.005 \, \text{eV}.$$

From these measurements, we can obtain numerical values for two quantities which are related directly to the stress in the GaAs layer: the splitting between light and heavy hole bands

$$\Delta E = E_g^{\pm 1/2} - E_g^{\pm 3/2} = 0.045 \pm 0.008 \, \text{eV},$$

and the average band gap

$$E_{g0} + \Delta E_g = (E_g^{\pm 1/2} + E_g^{\pm 3/2})/2 = 1.437 \pm 0.008 \, \text{eV}.$$

Following the analysis given in Bernier et al. (1989), and using their values for the stress tensor components and deformation potentials, we find

$$\Delta E = -5.2 \times 10^{-3} X \, \text{eV} \qquad (2)$$

and

$$\Delta E_g = -7.1 \times 10^{-3} X \, \text{eV} \qquad (3)$$

where the stress X (in kbar) is positive for a biaxial tensile stress, and therefore negative for the biaxial compression in GaAs/CaF$_2$.

Substituting the measured value of ΔE in 2 gives

$$X = -9 \pm 2 \, \text{kbar}.$$

Using this value of X in equation 3, the size of the unstrained band gap in the heavily doped GaAs ($p = 5 \times 10^{18} \, \text{cm}^{-3}$) can also be calculated:

$$E_{g0} = 1.38 \pm 0.01 \, \text{eV}.$$

This value is consistent with previous measurements on highly p-doped GaAs (Olego and Cardona, 1980).

CONCLUSION

The measurements reported and discussed in the previous section, together with the photoluminescence studies of Tessler et al. (1994), show that differential thermal contraction in GaAs/CaF$_2$ can be exploited to create a high degree of stress in a GaAs layer of thickness $2 \, \mu$m. The results of the photoluminescence and photoemission studies, carried out at temperatures of 77 K and 300 K respectively, are consistent with the assumption that the induced stress is linearly proportional to the difference between the growing temperature (873 K) and the temperature of study:

$$X = \xi \Delta T,$$

with $\xi = -1.5 \times 10^{-2} \, \text{kbar} \, \text{K}^{-1}$.

The experimental work described here is still at an early stage, and many questions remain to be resolved. In particular, there is at present no satisfactory explanation for the fact that the polarization of the photoemitted electrons is substantially lower than the mean polarization of the conduction band electrons, as measured by photoluminescence. The theory presented above predicts that the photoelectron polarization should exceed the photoluminescence polarization by a factor $\sqrt{\tau_s/(\tau_s + \tau)}$. A similar discrepancy has been observed in studies of GaAs/Si photocathodes (Lassailly et al., 1990). It is worth noting, however, that the maximum photoelectron polarization reported here is a factor of two greater than the highest polarizations routinely observed from unstrained GaAs photocathodes in the same apparatus. Several other groups have reported electron polarizations substantially below the theoretically predicted values for both strained and unstrained GaAs photocathodes (Meier et al., 1993; Plutzer et al., 1994; Pierce, 1995). Further work on this question is clearly required.

The GaAs/CaF$_2$ photocathode remains a promising candidate as a high yield, high polarization electron source. The optimisation of its yield and polarization will depend on a solution being found to the peeling problem described earlier; a further generation of samples is now in process of investigation. The fact that the CaF$_2$ substrate is transparent to photons capable of exciting GaAs with high polarization provides an additional incentive for this work, since it opens the possibility of a practically useful cathode design in which the laser beam excites the active GaAs layer from the rear.

ACKNOWLEDGEMENTS

The experimental study of GaAs/CaF$_2$ photocathodes has been made possible only through the skilful and pioneering fabrication techniques of the group at the Laboratoire d'Automatique et d'Analyse des Systèmes, L.P.8001 du CNRS, Toulouse, France: Chantal Fontaine, Emmanuelle Daran and Antonio Muñoz-Yagüe. We are also grateful for the collaboration of our colleagues in the Laboratoire de Physique de la Matiére Condensée at Ecole Polytechnique: Henri-Jean Drouhin, Heinrich Gentner, Claudine Hermann, Yves Lassailly, Daniel Paget and Leandro Tessler. One of the authors (DMC) acknowledges the support of the CNRS during a visit to the Ecole Polytechnique to participate in the experimental work described here.

REFERENCES

Aoyagi, H., Horinaka, H., Kamiya, Y., Kato, T., Kosugoh, T., Nakamura, S., Nakanishi, T., Okumi, S., Saka, T., Tawada, M. and Tsubata, M, 1992, *Phys.Lett.A* 167:415.
Bell, R.L., 1973, "Negative Electron Affinity Devices", Clarendon, Oxford.
Bernier, G., Beerens, J., De Boeck, J., Deneffe, K., van Hoof, C. and Borghs, G., 1989, *Solid State Comm.* 69:727.
Campbell, D.M., Tessler, L.R., Lassailly, Y., De Laharpe, P., Hermann, C., Lampel, G., Puech, P., Pisani, P., Landa, G., Carles, R., Daran, E., Munoz-Yague, A. and Fontaine, C., 1994, Ecole Thematique CNRS-PIRMAT, Marseille, September 12-14 1994.
Chelikowsky, J.R. and Cohen, M.L., 1976, *Phys.Rev.B* 14:556.
Ciccacci, F., De Rossi, S. and Campbell, D.M., 1995, *Rev.Sci.Instrum.* 66:4161.
Drouhin, H.-J., 1984, Thése de Doctorat d'Etat, Université de Paris-Sud, Centre d'Orsay.
Drouhin, H.-J., Hermann, C. and Lampel, G., 1985a, *Phys.Rev.B* 31:3859.
Drouhin, H.-J., Hermann, C. and Lampel, G., 1985b, *Phys.Rev.B* 31:3872.
Dymnikov, V.D., D'yakonov, M.I. and Perel', V.I., 1976, *Zh.Eksp.Teor.Fiz.* 71:2373 (1976, *Sov.Phys.JETP* 44:1252).
Escher, J.S., 1981, NEA Semiconductor Photoemitters, *in* "Semiconductors and Semimetals 15", R.K.Willardson and A.C.Beer, eds., Academic Press, New York.
Fishman, G., and Lampel, G., 1977, *Phys.Rev.B* 16:820.
Gay, T.J., Khakoo, M.A., Brand, J.A., Furst, J.E., Meyer, W.V., Wijayaratna, W.M.K.P. and Dunning, F.B., 1992, *Rev.Sci.Instrum.* 63:114.
Hermann, C., Drouhin, H.-J., Lampel, G., Lassailly, Y., Paget, D., Peretti, J., Houdré, R., Ciccacci, F. and Riechert, H., 1992, Photoelectronic Processes in Semiconductors Activated to Negative Electron Affinity, *in* "Spectroscopy of Non-equilibrium Electrons and Phonons", C.V. Shank and B.P.Zakharchenya, eds., Elsevier, Amsterdam.
Kane, E.O., 1957, *J.Phys.Chem.Solids* 1:249.
Lampel, G. and Eminyan, M., 1980, Proc.15th Int.Conf. Physics of Semiconductors, *J.Phys.Soc.Japan* 49(Suppl.A):627.
Lassailly, Y., Hermann, C., Lampel, G., Fontaine, C. and Muñoz-Yagüe, 1990, Proceedings of the 20th International Conference on the Physics of Semiconductors, Anastassakis, E.M. and Joanopoulos, J.D., eds., World Scientific, Singapore, 1053.
Maruyama, T., Prepost, R., Garwin, E.L., Sinclair, C.K., Dunham, B. and Kalem, S., 1989, *Appl.Phys.Lett.* 55:1686.

Maruyama, T., Garwin, E.L., Prepost, R., Zapalac, G.H., Smith, J.S. and Walker, J.D., 1991, *Phys.Rev.Lett.* 66:2376.

Maruyama, T., Garwin, E.L., Prepost, R. and Zapalac, G.H., 1992, *Phys.Rev.B* 46:4261.

Meier, F. and Zakharchenya, B.P., eds., 1984, "Optical Orientation in Solid State Physics", North Holland, Amsterdam.

Meier, F., Grobli, J.C., Guarisco, D., Vaterlaus, A., Yashin, Y., Mamaev, Y., Yavich, B. and Kochnev, I., 1993, *Physica Scripta* T49:574.

Nakanishi, T., Aoyagi, H., Horinaka, H., Kamiya, K., Kato, T., Nakamura, S., Saka, T. and Tsubata, M., 1991, *Phys.Lett.A* 158:345.

Olego, D. and Cardona, M., 1980, *Phys.Rev.B* 22:886.

Pierce, D.T., 1995, Spin Polarized Electron Sources, *in* "Experimental Methods in the Physical Sciences, Vol.29A", F.B.Dunning and R.G.Hulet, eds., Academic Press, New York.

Pierce, D.T. and Meier, F., 1976, *Phys.Rev.B* 13:5484.

Pierce, D.T., Meier, F. and Zürcher, P., 1975a, *Phys.Lett.A* 51:465.

Pierce, D.T., Meier, F. and Zürcher, P., 1975b, *Appl.Phys.Lett.* 26:670.

Pierce, D.T., Celotta, R.J., Wang, G.-C., Unertl, W.N., Galejs, A., Kuyatt, C.E. and Mielczarek, S.R., 1980, *Rev.Sci.Instrum.* 51:478.

Pierce, D.T., Garvin, S.M., Unguris, J. and Celotta, R.J., 1981, *Rev.Sci.Instrum.* 52:1437.

Pikus, G.E. and Titkov, A.N., 1984, Chapter 3 *in* "Optical Orientation in Solid State Physics", F.Meier and B.P.Zakharchenya, eds., North Holland, Amsterdam.

Plützer, S., Dockendorf, C. and Reichert, E., 1994, Workshop on Polarized Electron Beams and Targets, Les Houches, France, June 7-10 1994.

Pollak, F.H. and Cardona, M., 1968, *Phys.Rev.* 172:816.

Spicer, W.E. and Herrera-Gómez, A., 1994, SLAC Report 6306.

Tessler, L.R., Hermann, C., Lampel, G., Lassailly, Y., Fontaine, C., Daran, E. and Munoz-Yague, A., 1994, *Appl.Phys.Lett.* 64:895.

Zakharchenya, B.P., Mirlin, D.N., Perel', V.I. and Reshina, I.I., 1982, *Usp.Phys.Nauk.* 136:459 (1982, *Sov.Phys.Usp.* 25:143).

Zorabedian, P., 1982, SLAC Report 248, Stanford University.

ULTRATHIN MAGNETIC STRUCTURES

AND SPIN FILTER

H.C. Siegmann

Swiss Federal Institute of Technology
Laboratory for Solid State Physics
CH-8093 Zürich, Switzerland

INTRODUCTION

Strained GaAs-photocathodes make it possible to produce almost completely spin-polarized electron beams of high brillance and monochromasy[1]. But one problem still to be solved in the field of spin polarized electron physics is efficient spin filtering; that is one needs some kind of device which transmits one spin state but rejects the other. Ultrathin ferromagnetic films have a potential to provide the basic mechanism to construct spin filters with an efficiency approaching the one of filters for polarized light. This will be shown in the present paper.

P.S. Farago's review paper of 1965[2] marks the beginning of the era in which solids began to play the leading rôle in the development of the field of spin polarized electron beams. The following formal description of a spin filter is taken from his paper. Provided we restrict the considerations to nonrelativistic electrons, all definitions and formulae are valid for electrons bound in atoms or within solids as well as for free electrons in vacuo.

SPIN FILTER

Partly polarized electron states may be looked upon as the superposition of two completely polarized electron states, respectively parallel and antiparallel to a given direction. The density matrix characterizing the ensemble of electrons is given by

$$\rho = \frac{1}{2} \cdot \begin{pmatrix} 1 + P_z & P_x - iP_y \\ P_x + iP_y & 1 - P_z \end{pmatrix} \tag{1}$$

where $\mathbf{P} = (P_x, P_y, P_z)$ is the vector of electron spin polarization. Four parameters describe the properties of a spin filter:

(1) A direction in space, given by a unit vector $\mathbf{e}$, along which the electrons must be polarized in order best to be transmitted.
(2) The transmission w_{max} of electrons with $\mathbf{P} \parallel \mathbf{e}$ and with $|\mathbf{P}| = 1$.
(3) The transmission w_{min} of electrons with $-\mathbf{P} \parallel \mathbf{e}$ and $|\mathbf{P}| = 1$.

When dealing with ferromagnets, **e** is antiparallel to the magnetization **M**. We choose the z axis of the coordinate system parallel to **e**, that is parallel to the majority spins within the ferromagnet. Such spin filters are described by the matrix:

$$F = w \cdot \begin{pmatrix} 1 + \frac{1}{2}(\Delta w/w) & 0 \\ 0 & 1 - \frac{1}{2}(\Delta w/w) \end{pmatrix} \tag{2}$$

where $w = \frac{1}{2}(w_{max} + w_{min})$ and $\Delta w = (w_{max} - w_{min})$. The transmission probability for a beam of unit intensity and polarization **P** is given by the trace of the product $(\rho \cdot F)$. If $\mathbf{P} = 0$ we have $(\rho \cdot F) = \frac{1}{2} F$, $\mathrm{Tr}(\rho \cdot F) = w$. It follows that an initially unpolarized beam of unit intensity, after passing through a spin filter, has the intensity w and polarization $\mathbf{P} = (0, 0, \frac{1}{2} \Delta w/w)$.

Based on the formal analogy between polarizing and analyzing spin filter, one might expect that GaAs is not only a source, but also a detector of spin polarization. In the source operation, circularly polarized light is used to induce the emission of spin polarized electrons. Consequently, spin polarized electrons should produce circularly polarized luminescence. S.F. Alvarado and Ph. Renaud have shown[3] that this is indeed the case. However, the luminescence yield $w \lesssim 10^{-6}$ achieved so far is too low for general application of GaAs as an analyzing spin filter.

FERROMAGNETIC LAYERS AS SPIN FILTERS

When a beam of electrons strikes a solid, there is inelastic and elastic reflection, absorption, and, if the solid is sufficiently thin, also transmission of electrons. In the case of a ferromagnetic material, it is well known that reflection and absorption depend on the spin polarization of the incident electron beam, and that this phenomenon may be used for effective and simple detection of spin polarization[4]. Kisker and coworkers have indeed constructed a detector for electron spin polarization based on the reflectivity of a beam of low energy electrons incident along the surface normal of Fe(100) that is superior by one order of magnitude to previous devices based on spin orbit scattering of electrons[5]. With this spin detector, new applications of electron spectroscopy in, e.g., probing spin polarization of core levels have become more practicable[6]. Further improvement in detection of electron spin polarization by one order of magnitude can be achieved if transmission of low energy electrons through ferromagnetic material is observed instead of reflection and/or absorption.

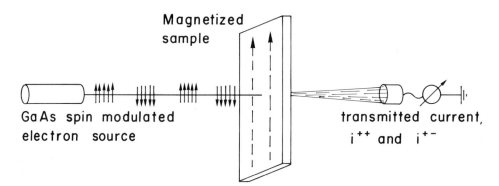

Figure 1. Prinziple of the experiment with a ferromagnetic spin filter acting as analyzer for the degree of polarization P of the incident electron beam along the direction of the sample magnetization **M**. For P parallel to M, the transmitted current is $i^{++} = w(1 + P(\frac{1}{2} \Delta w/w))$, for P antiparallel to M, $i^{+-} = w(1 - P(\frac{1}{2} \Delta w/w))$

Fig. 1 shows the principle on an experiment in which the electron current transmitted by a very thin ferromagnetic sample is measured. To observe the current, the sample can only be a few atomic layers thick. This means that Fig. 1 shows a Gedankenexperiment that cannot be realized in this form with a free standing ferromagnetic film. It will be used merely to evaluate the basic ideas. The spin dependence of the transmission of low energy electrons through ferromagnetic material has been studied previously[7]. The key idea of electron transport in transition metals has been recognized by N.F. Mott[8] who proposed that the mean free path λ is dominated by scattering into the holes of the d-bands. The total inelastic scattering cross section $\sigma = 1/\lambda$ of electrons in transition metals is given by[7]

$$\sigma = \sigma_0 + \sigma_d(5 - n) \tag{3}$$

The dominant term $\sigma_d(5 - n)$ accounts for scattering into the partly occupied d-states. With n = number of occupied d-states, $(5 - n)$ is the number of d-holes/atom. σ_0 describes the remaining scattering into states other than the d-states; it accounts for the electron transport in the noble metals Cu, Ag and Au where the d-bands are fully occupied (n = 5). Eq.(3) has been experimentally veryfied in overlayer experiments[7].

When a transition metal becomes ferromagnetic, the occupancy of the spin up part of the d-shell increases by Δn, while the occupancy of the spin down subshell decreases by Δn. It follows from Eq.(3) that the total inelastic cross section becomes spin dependent according to

$$\sigma^{(\pm)} = \sigma_0 + \sigma_d \, [5 - (n(\pm) \, \Delta n)] \tag{4}$$

where $\sigma^{(\pm)}$ is the cross section for majority (minority spins). For the bulk of the material, Δn can be calculated from the Bohr magneton number n_B obtained in conventional magnetometry according to $n_B = n^+ - n^- = 2\Delta n$, because the contribution of s-p-electrons to the magnetization is small. It follows that $\sigma^- - \sigma^+ = n_B \sigma_d$.

Eq.(4) postulates that a ferromagnetic film of thickness x is a spin filter with the following characteristics: $w_{max} = I_0 \exp(-\sigma^+ x)$, $w_{min} = I_0 \exp(-\sigma^-)$, and $w \cong I_0 \exp(-\sigma x)$, where $\sigma = 1/2 \, (\sigma^+ + \sigma^-)$ is the spin averaged total scattering cross section of eq.(3) and I_0 the current of electrons entering the film.

The spin polarization a(x) acquired by initially unpolarized electrons after passing through the filter is $a(x) = (w_{max} - w_{min})/(w_{max} + w_{min})$, hence

$$a(x) = \{[\exp(\sigma^- - \sigma^+)x]-1\}/\{[\exp(\sigma^- - \sigma^+)x]+1\} \tag{5}$$

The optimum thickness x of the spin filter can be found by maximizing the figure of merit[9]

$$\varepsilon = a^2(x) \cdot w(x) \tag{6}$$

The condition for the maximum is $d\varepsilon/dx = 0$. It occurs at

$$x_m = \frac{1}{\Delta\sigma} \ln \left(\frac{2\Delta\sigma}{\sigma} + \sqrt{1+\left(\frac{2\Delta\sigma}{\sigma}\right)^2} \right) \leq 2\lambda \tag{7}$$

where $\Delta\sigma = (\sigma^- - \sigma^+) = \sigma_d \cdot \mu_B$. It follows that $\varepsilon_{max} \cong a^2(x_m) \cdot e^{-2}$ is the best possible figure of merit; for fcc Fe it occurs at a thickness of ~ 1 nm or ~ 6 atomic layers depending on the packing of the crystal planes.

EXPERIMENTS WITH FERROMAGNETIC SPIN FILTERS

The characteristics of ferromagnetic spin filters are the transmission probability w and the transport polarization $a(x) = \frac{1}{2} \Delta w/w$. These constants can be determined by either using the ferromagnetic film as a polarizing or as an analyzing spin filter. In the latter case, the transmission for a beam of degree of polarization P and intensity $I_0 = 1$ is given by

$$Tr(\rho F) = w(x)(1 + Pa(x)) \tag{8}$$

This application is depicted in Fig. 1.

The case of the polarizing spin filter has been realized by depositing a ferromagnetic overlayer of thickness x onto a non-magnetic substrate. Unpolarized electrons are emitted from the substrate by illuminating the surface with photons. Of course, the overlayer emits also photoelectrons. But the substrate electrons are identified by the energy distribution curves specific for the substrate material. In this way, the transport polarization of the substrate electrons acquired in passing through the ferromagnetic overlayer can be determined. Such experiments have been performed with Fe on Cu(100)[10] and Fe and Co an W(110)[11]. For Fe/W(110), a(x) = 0.45 was obtained at a thickness of 6 monolayers of Fe.

It turns out that it is difficult to know whether or not these very thin overlayers are fully and homogeneously magnetized at the temperature of the experiment, usually room temperature. Also, it is unlikely that such ultrathin films exhibit the saturation magnetization of the bulk. Fe on Cu(100) is extremely complicated[12]. It crystallizes initially in the fcc-structure, while bulk Fe is bcc; estimates of the magnetization of fcc-Fe range from nonmagnetic to a state where only the surface layer is magnetic. After a few layers, Fe assumes the bcc-structure again with a magnetization of 2.2 Bohr magnetons at T = 0. Fe and Co on W(110) are easy to produce because they tend to not form islands on W which has a large surface energy. But the lattice mismatch between Fe and Co and the substrate is as large as 10% and 20% respectively. It is not clear after how many layers the crystal structure of the ferromagnetic overlayer relaxes to the bulk values[13], estimates range from 3-6 layers. Therefore, it is not straightforward to know what the thickness of the ferromagnetic films is quite apart from the size of the magnetic moment.

The case of the analyzing spin filter has also been realized. In these experiments, an independent source of spin polarized electrons is needed, and the change in the transmitted intensity is measured when the relative direction between the magnetization of the filter and the spin polarization of the incident electrons is inverted. J.C. Gröbli et al.[14] excited polarized electrons in GaAs by absorption of circularly polarized photons. These electrons are photoemitted into vacuum after having passed through 5 monolayers of Fe. The work function of the Fe-film is reduced below 1.5 eV so that the highly polarized electrons excited at the band gap of 1.5 eV in GaAs can escape. The Fe-film can not be deposited directly onto the surface of GaAs. 5 monolayers of Ag are needed to act as a diffusion barrier between the Fe film and the GaAs inhibiting chemical reactions between Fe and GaAs that would otherwise occur and make it impossible to obtain ferromagnetism in only 5 monolayers of Fe on GaAs. In this experiment it is important that the functioning of the two essential elements of the multilayer structure can be tested independently. These elements are the source and the analyzer of the spin polarization. The source is tested by switching off the external magnetic field. The magnetization direction lies then in the plan of the Fe-film, and no spin filtering occurs for the electrons from GaAs polarized perpendicular to the surface. The spin polarization of the electrons emitted through the Ag/Fe/Cs-layers into vacuum is now measured vs the energy of the circularly polarized photons. One observes the typical signature of the GaAs source of electrons: the highest polarization for hv = 1.5 eV, and a sign reversal at hv = 3 eV. From this special dependence one can conclude that some of the electrons excited in GaAs have traversed the surface barrier. The reduced degree of spin polarization indicates which percentage of the emitted electrons have originated in the Ag/Fe/Cs overlayers, as these latter ones are unpolarized in the direction perpendicular to the surface. The analyzing ferromagnetic spin filter is tested by measuring the spin polarization of the photoelectrons emitted with unpolarized light as it depends on the strength of the external magnetic field applied perpendicular to the film surface. The electrons originating in the Fe as well as the ones having been transported through it will exhibit spin polarization under these conditions. Magnetic saturation perpendicular to the surface is reached at an external field of 1.5 Tesla. If this field is applied and the photons are circularly polarized, one observes circular dichroism in photoemission: The photoelectron intensity changes when the handedness of the photons is inverted. The same change occurs when the direction of the magnetic field is inverted. The experiment yields the analyzing power a(x) = 26% of 5 monolayers of Fe at electron energies $1.5 \leq E \leq 2$ eV above the Fermi energy. Particularly interesting and novel in this experiment is the fact that it can be done with very low energy electrons.

A conceptually quite similar experiment testing the analyzing power of a 5-6 ML Co-film was reported by Y. Lasailly et al.[15]. The spin polarized electron beam is again prepared with a GaAs photocathode, and the intensity transmitted through a spacially separated, free standing metallic target is measured in a Faraday cup. The 3 mm wide metallic target consists of a 1 nm thick Co film sandwiched between 21 and 2 nm thick gold layers. At low electron energy, close to the clean surface vacuum level, the transmission of majority spin electrons is found to be 1.43 times higher than the one of minority spin electrons. This yields an analyzing power of a(x) = 18% for the electron energy $E \cong 5$ eV above E_F. The intensity of the electrons was attenuated by 5 orders of magnitude mainly during passage through the first Au-film of 21 nm, in reasonable agreement with eq.(3) from which an attenuation of 2.10^{-6} is expected. This result is really unique in that it points the way to new and much simpler detectors of spin polarization[16]. The task is to reduce the thickness of the supporting substrate. Cs deposition on both sides of the target was shown to increase the transmission ratio up to 3×10^{-4} and also the analyzing power to a maximum of 40%. This structure is already a competitive spin detector[17].

Both of the experiments with the GaAs source suffer again from the uncertainty inherent in the crystalline structure and the magnetization of ultrathin magnetic films on nonmagnetic substrates.

A quite elegant method to determine the characteristics of ferromagnetic spin filters relies on the enhancement of the spin polarization of low energy cascade electrons emitted from a thick ferromagnetic sample. The low energy secondary or cascade electrons are emitted in copious amounts when the surface of a metal is irradiated with a primary beam of electrons or photons. The spin polarization P_C of the cascade electrons has been measured in a number of laboratories. It is of great value for magnetometry on surfaces[7,12], and it is the basis for high resolution imaging of magnetic structures in secondary electron microscopy with polarization analysis (SEMPA)[18].

The polarization P_C of the cascade electrons shows enhancement over the average spin polarization P_0 of the valence electrons producing the magnetization. The N electrons in the valence bands of 3d-transition metals occupy orbitals of s-, p-, and d-parentage. With the assumption that all these electrons are excited to an energy level from which they can escape into vacuum with equal probability in the process of the formation of the cascade, one obtains $P_0 = 2\Delta n/N$. While N includes the s- and p-electrons, $2\Delta n$ is made up almost entirely by the 3d-electrons alone, and can be determined with conventional magnetometry. It has been proposed right after the first observation of the enhancement of P_C over P_0 that the scattering into the unoccupied d-orbitals could be responsible for the enhancement. A summary of these first arguments is given in Ref. 19.

The polarization $P_C(x)$ of electrons originating at a depth x from the surface at $x = 0$ with polarization $P_0 = (N^+ - N^-)/(N^+ + N^-)$ is given by:

$$P_C(x) = \frac{N^+ w_{max} - N^- w_{min}}{N^+ w_{max} + N^- w_{min}}$$

In all practical cases, P_0 is sufficiently small to approximate this by $P_C(x) = P_0 + a(x)$. The integral polarization P_C of the cascade electrons is then given by $P_C = P_0 + (\int_\infty a(x)i(x)dx)/\int_\infty i(x)dx$ yielding[7]:

$$P_C = P_0 + A \quad \text{with } A = \tfrac{1}{2}\Delta\sigma/\sigma \qquad (9)$$

The enhancement A of the cascade polarization over the polarization $P_0 = 2\Delta n/N$ of the electrons in the valence bands can then be determined from the measurement of P_C and from the magnetization $2\Delta n = \mu_B$. From the enhancement A, an experimental value of $\Delta\sigma$ is obtained. If σ is known from other sources, one can calculate w(x) and a(x) from eq.(5).

The question of what the magnetization of the surface of Fe or Co on its own proper substrate really amounts to is of course not eliminated, but it is much safer to assume that the crystal structure and magnetization exhibit the values of the bulk except maybe for corrections at the very last layer; surface distortions are known to heal very fast in metals. It is however important to measure P_C at very clean surfaces and to extrapolate to $T \to 0$ as severe reductions of P_C can occur already in the spin wave regime, that is at temperatures far below the Curie point[12].

An independent experimental verification results from the hot electron spin-valve effect in coupled magnetic layers recently described by Celotta, Unguris, and Pierce[20]. The phenomenon occurs if two ferromagnetic layers are coupled over a nonmagnetic spacer layer such as Ag or Au. Depending on the thickness of the spacer layer, the coupling can be such that the magnetizations in the two films are parallel to each other or antiparallel. If now secondary electrons are excited with a primary electron beam, one observes that the intensity of the low energy cascade electrons depends on the relative orientation of the magnetizations. It is higher when the magnetizations are parallel. A related phenomenon, known as giant magnetoresistance occurs when an electric current flows through such trilayer or multilayer structures: The electrical resistance is lower when the magnetizations are parallel. The underlying physics of both observations is spin dependent scattering into the d-holes of the ferromagnet. In the case of the electrons remaining within the material, there are promising applications as magnetic field sensors[21] and spin valve transistors[22].

Celotta, Unguris and Pierce[20] use a Fe single crystal wisker as a starting substrate. It exhibits a high degree of crystalline perfection and surface flatness, and it is magnetized along the axis of the wisker. An Ag- or Au-film is grown epitaxially on the surface of the wisker in the shape of a wedge with its thickness varying from 0 to 25 monolayers. Finally, an Fe-overlayer is grown with a uniform thickness acting as the spin filter. Due to the periodic change of sign of the exchange coupling as a function of the Ag or Au interlayer thickness, the domains in the Fe overlayer will be oriented either parallel or antiparallel to the substrate magnetization. The resulting domain pattern directly reflects the periodicity of the coupling. The magnetization pattern is detected with scanning electron microscopy with polarization analysis (SEMPA). Fig. 2 shows the nanostructure employed in this experiment. The intensity

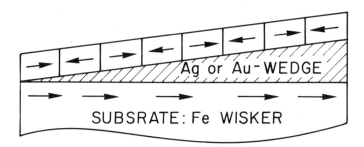

Figure 2. Fe/Spacer/Fe sandwich structure in the experiment of Celotta, Unguris, and Pierce[20]. The wedge angle is much exaggerated in the drawing

of the low energy cascade is reduced in those locations where the cascade polarization P_C is inverted with respect to the substrate. This is due to the spin filtering of the spin polarized cascade electrons originating in the substrate when passing through the overlayer as expected from eq.(8). However, the nonmagnetic Au- or Ag-spacer layer contributes unpolarized electrons and also attenuates the intensity of the spin polarized electrons from the substrate. Furthermore, spin polarized quantum well states may be set up in the spacer layer[23]. It is therefore necessary to eliminate the contribution of the spacer layer by extrapolation of the relative intensity variations $\frac{1}{2}\Delta c/c$ of the cascade electrons to cero spacer layer thickness. In this limit, the Fe-layer on top acts as a spinfilter for the polarized electrons emitted from the substrate Fe-layer. Even if top and bottom layer are magnetized antiparallel, there is no transition region or domain wall that would normally occur. This transition region has been eliminated by the extrapolation to cero spacer layer thickness. As the spacer layer contributes unpolarized electrons with an exponential dependence on its thickness, the extrapolation must be done using an exponential. Fig. 3 shows this extrapolation for the case of 6 layers of Fe on

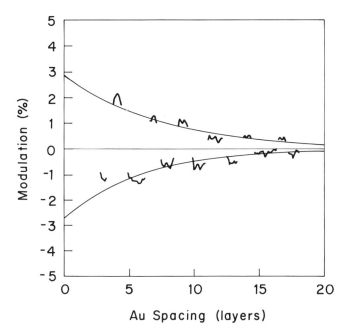

Figure 3. The modulation of the intensity c of cascade electrons emitted from the sandwich structure Fig. 2 vs the thickness of the Au-spacer. Data from Ref. 20

an Au-wedge; it is apparent that the extrapolation is reasonable and leads to $\frac{1}{2}\Delta c/c = 2.81\%$ in this case. The extrapolated relative asymmetry should be independent of the nature of the spacer. In agreement with this expectation, one obtains $\frac{1}{2}\Delta c/c = 2.58\%$ for 6 layers of Fe on an arbitrarily thin Ag-spacer[20].

Before applying eq.(8), one has to take account of the fact that the Fe-overlayer also emits cascade electrons of intensity $i_o = \int_{x_o}^o I_o e^{-\sigma x} dx$ where x_o is the thickness of the overlayer. i_o does not change when the magnetization is inverted with respect to the substrate. Spin filtering occurs only for the intensity $i_s = \int_\infty^{x_o} I_o e^{-\sigma x} dx$ emitted from the substrate. Hence the fraction of the total current contributing to $\frac{1}{2}\Delta c/c$ is $i_s/(i_o + i_s) = e^{-\sigma x_o}$. Therefore one obtains from eq.(8)

$$\tfrac{1}{2}\Delta c/c = \exp(-\sigma x_o) \cdot P_C \cdot a(x_o)$$

If the total scattering cross section σ and the degree P_C of the low energy cascade electrons are known, one obtains the transport polarization $a(x_o)$. The greatest uncertainty arises from the fact that different values are quoted in the litterature for the spin averaged mean free path $\lambda = 1/\sigma$. It is $0.5 \leq \lambda \leq 0.7$ nm[7]. If the Fe-overlayer is too thin, the magnetization is reduced at room temperature; if it is too thick, the uncertainty in the term $e^{-\sigma x}$ becomes overwhelmingly large. Therefore, from the data with 3, 6 and 12 ML Fe, only the ones with the 6 ML Fe-overlayer were employed. Corrections were made for the fact that 6 ML showed only 75% of the polarization P_C observed with 12 ML which one can assume to be bulklike.

Fig. 4 shows a comparison of the experimental results for the transport polarization $a(x)$ or the analyzing power $a(x) = \frac{1}{2}\Delta w/w$ of Fe-films obtained with various techniques. The experiments employing thin films appear to have a tendency to yield lower $a(x)$-values compared to the results obtained from the polarization P_C of the low energy cascade on bulk

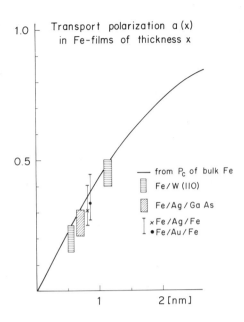

Figure 4: The transport polarization or the analyzing power $a(x) = \frac{1}{2} \Delta w/w$ of Fe-ferromagnetic films of thickness x from various experiments. The full line is calculated from the spin polarization $P_C = 53\%$ of the low energy cascade electrons at $T = 0$ as observed on bulk samples. The total inelastic spin averaged scattering cross section was assumed to be $\sigma = 1.67$ nm^{-1}. In the other experiments, Fe films are used as polarizing or analyzing spin filters as described in the text

materials. This may certainly be explained by the difficulty of producing fully magnetized ultrathin films of Fe on nonmagnetic substrates. Nevertheless, the agreement between the various techniques is astonishingly good. There is no reasonable doubt on the reality of spinfiltering in ferromagnets.

DISCUSSION

The most detailed information on magnetism and the electronic structure of magnetic materials comes from photoemission (PES) and inverse photoemission (IPS). These techniques can determine the dispersion relation E(k) of electronic states below and above the Fermi level. The energy states of majority and minority spin electrons are separated in the ferromagnetic state by the energy- and momentum-dependent exchange splitting. Spectral features result from optical transitions into bulk as well as into surface- or adsorbate-induced states. By analyzing the spin polarization of the emitted electrons (in PES) or by observing the bremsstrahlung emitted when a spin polarized electron strikes the surface (in IPS), the majority and minority bands in ferromagnets can be investigated separately. With the advent of synchrotron radiation sources, core levels are also accessible to spin polarized PES.

However, with very low energy electrons close to the Fermi-energy E_F, the situation is still quite nebulous. Meservey and Tedrow[24] point out that the spin polarization of electrons extracted from states close to E_F does not seem to fit into a simple pattern. Fig. 5 shows as an example results obtained with Ni using various techniques. The filled points give the polarization of threshold photoelectrons obtained with the Ni(111) surface[25]. In the limit $h\nu \rightarrow \Phi$ one obtains emission from the states near E_F only, and the polarization tends to increasingly negative values in agreement with the predictions of Hunds rule. But if states from E_F tunnel into superconducting Al, a positive polarization is observed[24]. Positive is also the polarization P_C of the very low energy cascade electrons and of course the average

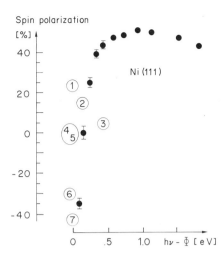

Figure 5. Spin polarization of electrons extracted from states close to the Fermi energy ($E_F = 0$) in Ni. Filled points: Photoemission from Ni(111) + Cs with photons of energy hν; the work function is $\Phi = (1.56 \pm 0.05)$ eV[25]. 1: Electrons tunneling into superconducting Al[24]; 2: P_C; 3: P_0; 4: thermionic emission [27]; 5: field emission[26]; 6: electrons tunneling into GaAs[3]; 7: electron capture by swift H-atoms[28]

polarization P_0 of all the valence electrons (see Table). In field emission[26] and thermionic emission[27], the polarization is close to cero, while tunneling into GaAs[3] and capture of electrons by swift hydrogen atoms[28] delivers negative polarizations, both compatible with the photoemission result.

A number of authors have discussed the enhancement of the polarization P_C of the low energy cascade electrons over the polarization P_0 of the electrons in the valence bands and there is a variety of different explanations for the enhancement. In one model[29] it is assumed that quasi-elastic scattering occurs which transforms a minority spin into a majority spin electron leaving behind a reversed spin (Stoner-excitation). Already in Ref. 19 it was pointed out that these Stoner excitations cannot be important to the enhancement of the secondary electron polarization because they produce appreciable polarization only for energy losses of the order of the exchange splitting. Meanwhile, there is direct experimental proof that the Stoner excitations are indeed negligible for the enhancement of P_C. Namely, the enhancement is still present in Co with electrons excited by photons of energy within 0 - 0.1 eV from photoelectric threshold[25]. With such low energy losses of well below 0.1 eV it is impossible to induce Stoner excitations in Co in which the exchange splitting is of the order of ~ 1 eV. In yet another explanation of the high value of P_C it is proposed that the spin dependence of the *elastic* scattering of the electrons on the magnetic atoms is responsible for it. While the experiments confirm without doubt that the spin dependence of elastic electron scattering can be quite large at certain electron energies and in some crystal directions, it is impossible to explain the uniformly *positive* sign of the polarization enhancement over a wide range of energies with all materials alike irrespective of whether they are crystalline or amorphous.

Getzlaff, Bansmann and Schönhense[11] explain the spin polarization P_C of the cascade in eq.(3) of their paper by spin filtering alone. This is certainly incorrect as the initial state from which the cascade originates is already polarized. This initial state polarization P_0 has to be added to the enhancement $A = \frac{1}{2}\Delta\sigma/\sigma$ which in turn must be distinguished from the thickness dependent transport polarization $a(x) = \frac{1}{2}\Delta w/w$. The agreement between P_C and $a(x')$ at one particular thickness x' is fortuitous.

At the time when Ref. 19 appeared, the extremely large value of the total spin averaged inelastic scattering cross section σ was not yet established. Hence the very general dependence of electron transport on the number of d-holes, eq.(3), was not apparent. The all inclusive spin filtering model can explain a variety of seemingly different observations[7]:

1.	The spin polarization P_C of the cascade is proportional to the magnetization $2\Delta n = \mu_B$ in the limit of the approximation $P_C = P_O + A$.

2.	The enhancement factor $f = P_C/P_O$ increases with the filling of the d-shell, that is f increases from 1.22 in Gd to 1.89 in Fe to 2.21 in Co and to 2.80 in Ni.

3.	The enhancement A of P_C over P_O decreases with increasing electron energy.

4.	The spin filter function a(x) of ferromagnetic layers as observed in direct experiments agrees with the a(x) calculated from the enhancement within experimental uncertainty (Fig. 4). In contrast to the enhancement A, a(x) is not a strong function of the energy.

A number of older experiments with spin polarized electrons of low energy can also be understood with spin filtering. One example among many others[31] is the appearance of negative spin polarization with photoelectrons at photothreshold in the case of Ni as shown in Fig. 5. The puzzle is that Co, which has a fully occupied majority spin band just like Ni, nevertheless exhibits positive polarization of threshold photoelectrons[25]. The explanation is simple: The enhancement in Co is much larger due to the larger number of minority holes compared to Ni. The transport polarization A has to be added to the initial state polarization according to eq.(9). This leads to positive values of the spin polarization of threshold photoelectrons in Co, but not in Ni. The table summarizes the enhancement A and other data relevant to the understanding of the low energy cascade polarization P_C in the elemental ferromagnets listed according to increasing occupancy of the d-shell. The only adjustable parameter needed to calculate A in agreement with the observations is the ratio σ_o/σ_d of the scattering σ_o into all states other than the d-states to the scattering σ_d into the d-states[7].

	P_O	n	Δn	λ(nm)	A	P_c	P_t
Gd	0.18	0.5	0.27	0.24	0.04	0.22	0.14
Fe	0.28	3.50	1.10	0.6	0.28	0.53	0.40
Co	0.19	4.10	0.85	0.8	0.25	0.42	0.35
Ni	0.05	4.75	0.25	1.1	0.09	0.14	0.23

Table. Total spin polarization P_O of s-, p-, and d-electrons in the bulk, estimated number n of occupied d-orbitals in the paramagnetic state and their increase (decrease) Δn due to the spontaneous magnetization at T = 0 in Gd, Fe, Co, and Ni. λ is the spin averaged mean free path, A the enhancement and P_C the total polarization of the low energy cascade electrons[7]. P_t is the spin polarization of the electrons tunneling into superconducting Al[24].

With electrons of energy below the top of the d-bands, the situation is not so simple. This applies to tunnel-junctions, electron transport across different ferromagnets, and multilayers. In this case, the scattering is possible only to the portion of the d-band that is lower in energy. Therefore, one looses the simple relation $\Delta\sigma = \sigma^- - \sigma^+ = n_B \cdot \sigma_d$, that is the spin dependence of the scattering is no more proportional to the (spin part) of the magnetization. Instead, one has to write $\Delta\sigma = \sigma_d \cdot (D^-(E) - D^+(E))$, where $D^\pm(E)$ is the accessible density of states for ($\pm$) spins. Depending on electron energy and crystal direction, $\Delta\sigma$ can now have various magnitudes and even change sign. Therefore, spin filtering with low energy electrons is less predictable.

The only really disturbing discrepancy in Fig. 5 is the positive value of the polarization observed in tunneling from Ni into superconducting Al. Meservey[32] already pointed out that the high positive polarization is common to tunneling and to the low energy cascade as evident from the table. It is tempting to explain this by an identical physical process: scattering into d-states. In the case of the tunneling experiments performed at very low temperatures, the scattering will introduce lifetime broadening. In the case of Ni, broadening will occur only with the minority states. It is conceivable that this leads to a reduced tunneling probability of the minority states into the extremely sharp states of the superconductor. One does not expect rigid scaling between P_c and P_t in this picture, but only similar trends because of the complications introduced by the partial density of states.

ACKNOWLEDGEMENT

R.J. Celotta, J. Unguris, and D.T. Pierce kindly put at my disposal the original data of the hot electron spin valve effect. I am grateful to Peter Farago for the fruitful time at the University of Munich and Edinburgh at the dawn of spin polarized electron physics.

REFERENCES

1.   E.L. Garwin, D.T. Pierce, and H.C. Siegmann, Helv. Phys. Acta 47, 29 (1974); T. Maruyama et al., Phys. Rev. Lett. 66, 2376 (1991); T. Nakanishi et al., Phys. Lett. A 158, 345 (1991).
2.   P.S. Farago, Advan. in Electronics & Electron Physics 21, 1-66 (1965).
3.   S. F. Alvarado and P. Renaud, Phys. Rev. Lett. 68, 1387 (1992).
4.   H.C. Siegmann, D.T. Pierce, and R.J. Celotta, Phys. Rev. Lett. 46, 452 (1981).
5.   D. Tillmann, R. Thiel, E. Kisker, Z. Phys. B 77, 1 (1989).
6.   R. Jungbluth et al., Surf. Sci. 269/270, 615 (1992.)
7.   H.C. Siegmann, Surf. Sci. 307-309, 1076 (1994).
8.   N. Mott, Adv. Phys. 13, 325 (1964).
9.   Di-Jing Huang et al., Rev. Sci. Instrum. 64, 3474 (1993).
10. D.P. Pappas et al., Phys. Rev. Lett. 66, 504 (1991).
11. M. Getzlaff, J. Bansmann, and G. Schönhense, Solid State Comm. 87, 467 (1993).
12. H.C. Siegmann, J. Phys.: Condensed Matter 4, 8375 (1992).
13. H.J. Elmers and U. Gradmann, Appl. Phys. A 51, 255 (1990), K.P. Kamper, Dissertation RWTH Aachen, 1989.
14. J. Gröbli et al., Phys. Rev. B 51, 2945 (1995).
15. Y. Lassailly et al., Phys. Rev. B 50, 17 (1994).
16. G. Schönhense and H.C. Siegmann, Ann. Phys. 2, 498 (1993).
17. Annemarie van der Sluijs et al., C.R. Acad. Sci. Paris, t. 319, Série II, 753-759 (1994).
18. K. Koike and K. Hayakawa, Appl. Phys. Lett. 45, 585 (1984); J. Unguris et al., J. Microscopy 139 , RP 1(1985); R. Allenspach, J. Magn. Magn. Mat. 129, 160 (1994).
19. D.R. Penn, S.P. Apell, and S.M. Girvin, Phys. Rev. B 32, 7753 (1985).
20. R.J. Celotta, J. Unguris, and D.T. Pierce, J. Appl. Phys. 75, 6452 (1994).
21. B.A. Gurney et al., Phys. Rev. Lett. 71, 4023 (1993).
22. D.J. Monsma et al., Phys. Rev. Lett. 74, 5260 (1995).
23. K. Koike et al., Phys. Rev. B 50, 4816 (1994).
24. R. Meservey and P.M. Tedrow, Physics Rep. 238, 173 (1994).
25. J.C. Gröbli, et al., Physica B 204, 359 (1995).
26. M. Landolt and M. Campagna, Phys. Rev. Lett. 38, 663 (1977).
27. A. Vaterlaus, F. Milani, and F. Meier, Phys. Rev. Lett. 65, 3041 (1990); J.S. Helman and W. Baltensperger, Mod. Phys. Lett. B 5, 1769 (1991).
28. C. Rau, J. Magn. Magn. Mater. 31, 141 (1982).
29. J. Glazer and E. Tosatti, Solid State Comm. 52, 11507 (1984).
30. M.P. Gokhale and D.L. Mills, Phys. Rev. Lett. 66, 2251 (1991).
31. H.C. Siegmann in: Core Level Spectroscopy for Magnetic Phenomena, P. Bagus, F. Parmigiani, and G. Pacchione ed., Plenum 1995.
32. R. Meservey, J. Appl. Phys. 61, 3709 (1987).

INTERACTION OF SPIN-POLARIZED HELIUM 2³S-ATOMS WITH FERROMAGNETIC SURFACES

H. Steidl and G. Baum

Fakultät für Physik
Universität Bielefeld
D-33501 Bielefeld

INTRODUCTION

In recent years numerous investigations of clean and adsorbate covered substrates have been carried out by different methods. As most investigations of electronic properties use methods which give information averaged over a depth of a few atomic layers, below the surface, there is, in comparison not so much knowledge about the electronic properties at the surface. A distinct surface sensitivity, however, can be achieved by electron emission caused by impact of metastable He(2^3S)-atoms of thermal energy, a method called metastable deexcitation spectroscopy (MDS) (e.g. [1]). This technique probes predominantly the outermost atomic layer. The spin selective version of MDS uses an electron spin polarized He(2^3S) atomic beam (SPMDS) and was experimentally pioneered at Rice University [2]. Using the spin selectivity in the deexcitation process (at the surface), one has an excellent tool for obtaining information on the magnetic properties of the outermost region of the surface. Here we report on studies of electron emission from clean and oxygen covered iron (110) and cobalt (0001) films. Due to the high intensity of our atomic beam we are able to carry out angle and energy resolving SPMDS measurements. With our spectrometer the energy distribution of the emitted electrons is determined directly.

DEEXCITATION MECHANISMS.

Depending on the work function ϕ of the surface the deexcitation of metastable helium atoms occurs either by resonance ionization (RI) with a subsequent Auger neutralization (AN), or by Auger deexcitation (AD), see Figure 1. If the wave function of the 2s-electron of the He(2^3S) atom overlaps sufficiently with an empty level of the surface, then a tunneling into this state will occur (RI). The resulting positive ion continues towards the surface and subsequently AN takes place with an electron from the solid tunneling into the 1s hole of the helium ion. The energy released is transferred to another electron which may be ejected from the solid. The kinetic energy of the escaping electron depends on the binding energy of the 1s hole. This binding energy decreases due to the image potential as the ion approaches the surface. Thus, the closer to the surface this process occurs, the smaller the maximum energy of the emitted electron. Without RI taking place the metastable atoms come close to the surface and AD will then occur as dominant deexcitation process. RI is suppressed if the excited 2s-Helium level lies below the Fermi level or if there is an insufficient overlap with empty states due to an adsorbate layer. In this case the 1s hole is filled by an electron from the solid or the adsorbate layer with an ejection of the excited 2s atomic electron. In the vacuum the excess-energy of the ejected electrons can be measured as kinetic energy. States close to the Fermi-edge contribute more because the tails of their wavefunctions, due to the smaller potential barrier at the Fermi edge, overlap much more with the 1s-hole than the deeper lying states with higher binding energy. For the interpretation of results, it must be taken into account that in the case of RI + AN a <u>pair</u> of electrons is determining the excess energy of the ejected electron. If we confine our studies to the electrons with maximum kinetic energy this pair of electrons involved in the deexcitation originates from the Fermi edge. Regarding electrons with lower kinetic energy, a deconvolution of the measured electron energy distribution of the emitted electrons would be necessary to gain information about the density of states below the Fermi edge. In comparison to this, the AD process is more like photoelectron spectroscopy, apart from the distance dependent effective ionization potential.

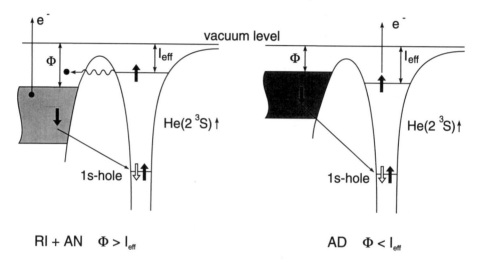

Figure 1: Deexcitation processes, for spin polarized metastable He(2^3S)-atoms at a metal surface, (for details see text).

METHOD OF MEASUREMENT

The method of measurement is shown schematically in Figure 2. Shown in the middle of the figure are the spin resolved density of states (D(E)) which are different for majority and minority electrons (D(E)^{maj} resp D(E)^{min}), the direction of magnetization M is indicated, E_F is the Fermi energy. (Note that the arrows with respect to the electrons indicate the spin direction not the direction of the magnetic moment.) The ejected electron yield observed with the incident H(2^3S) spin-polarization parallel (antiparallel) to the marjority spin direction are N↓↓ (N↑↓). In the case of the parallel (antiparallel) configuration only a minority (majority) electron can tunnel into the 2s-hole, and give rise to the ejection of another electron.

We define the asymmetry A of ejected electrons by

$$A = \frac{1}{P_A} \frac{N\downarrow\downarrow - N\uparrow\downarrow}{N\downarrow\downarrow + N\uparrow\downarrow},$$

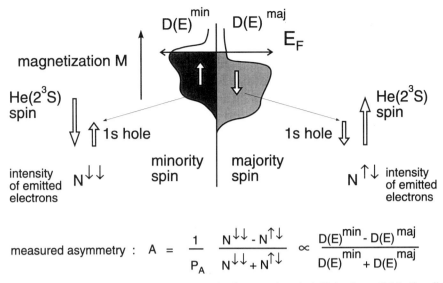

measured asymmetry : $A = \dfrac{1}{P_A} \dfrac{N^{\downarrow\downarrow} - N^{\uparrow\downarrow}}{N^{\downarrow\downarrow} + N^{\uparrow\downarrow}} \propto \dfrac{D(E)^{min} - D(E)^{maj}}{D(E)^{min} + D(E)^{maj}}$

Figure 2: Method of measurements (schematic) (for details see text). $\downarrow\downarrow(\uparrow\downarrow)$: He* spin parallel (antiparallel) to the majority spin.

assuming that full single-domain magnetization of the target is preserved. P_A is the polarization of the atomic beam. On the assumption that the probability for tunneling is the same for both spin orientations the detected asymmetry A is therefore seen to be a measure for the difference of the density of states of minority and majority electrons in the ferromagnetic target

$$A \propto \frac{D(E)^{min} - D(E)^{maj}}{D(E)^{min} + D(E)^{maj}} .$$

A measured <u>positive asymmetry</u> means a dominance of <u>minority electrons</u>.

SPINPOLARIZED He(2³S) BEAM AND EXPERIMENTAL ARRANGEMENT

A schematic view of the experimental setup is shown in Figure 3. The He(2³S) beam line runs from left to right.

A helium *discharge source* produces a beam which contains ground-state He(1¹S) atoms, metastable He(2¹S) and He(2³S) atoms as well as long lived Rydberg atoms. The radiative lifetime of the He(2³S) atoms of about 10^4 s [3] is "infinitely" long compared with the millisecond flight time between source and detector.

A *sextupole magnet* acts as a *polarizer* by preferentially transmitting the atoms with a magnetic quantum number $m_s = +1$ and strongly suppressing the ones with $m_s = -1$. For atoms with a field-strength-independent magnetic moment, such as He(2³S), the inhomogenous magnetic field in the pole gap of a sextupole magnet provides a force which is proportional to the atoms' radial distance from the axis and is directed towards it if $m_s > 0$, away from the axis if $m_s < 0$, and no force is acting if $m_s = 0$. For atoms with $m_s > 0$, this magnet acts like a positive lens. This lens is free of aberrations except for the chromatic one which restricts the focussing for thermal atomic beams to a portion of their velocity distribution.

Without a central stop at the magnetic exit, the atoms with $m_s = 0$, including all the ground-state and the metastable singlet atoms, are transmitted if their straight-line trajectories can clear the magnet. With a central stop in place, ideally only the $He(2^3S)$ atoms in Zeeman state $m_s = +1$ are transmitted. Between the components along the beamline a magnetic guiding field is provided by a longitudinal field of about 50 µT. In the target region this field is lowered to about 15 µT and applied vertically to the beam axis.

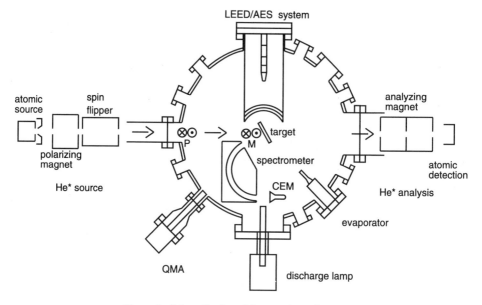

Figure 3: Schematic view of the experimental setup.

We take measurements with parallel and antiparallel spin orientations of the atomic beam electrons and of the majority electrons of the ferromagnetic surface (see Figure 2). The use of a *spin flipper* [4] provides a method for reversing the polarization of the incoming atomic beam while keeping the magnetic guiding field constant and systematic errors small. Such a reversal corresponds to a Zeeman transition from $m_s = +1$ to $m_s = -1$ (diabatic case) [5]. The efficiency of the spin flipper is determined to be very close to 100%, that is, one obtains equal beam polarization for the reversed and the non-reversed case.

Figure 4 shows the profiles of the polarized beam, taken by a Stern-Gerlach dipole magnet, for the adiabatic (a and c) and the diabatic (b and d) setting of the spin flipper, each taken without (a and b) and with the central stop (c and d) at the polarizing sextupole exit. Evaluation of the profiles of Figure 4 c and d with respect to their composition of atoms with $m_s = +1$, $m_s = 0$ and $m_s = -1$ components yields the $He(2^3S)$ beam polarization $P_A = 0.90 \pm 0.02$. The flux density on the target amounts to about $\sim 10^{12}$ s^{-1} cm^{-2}. The kinetic energy of the $He(2^3S)$ atoms is about 70 meV [5].

The ferromagnetic target can be pulled out vertically in order to measure the beam polarization by means of the analyzing magnet. The atomic beam direction has an angle of $\Theta = 30°$ with respect to the surface normal of the target. The discharge lamp (HeI: $h\nu = 21.22$ eV) is used for photoelectron spectroscopy (UPS) in order to compare with the MDS measurements at identically prepared targets. All spectra are taken in normal emission. The electron-optical entrance system provides an angular resolution of the emitted electrons of ±5°. A 150° sphercial spectrometer with a radius of 100 mm serves as the energy dispersing element. The resolution is set to about 180 meV. Detection of the transmitted electrons is achieved by a channel-electron multiplier (CEM).

The experiment is performed in a UHV chamber with a base pressure in the upper 10^{-11} mbar range. The chamber is equipped with a LEED-Auger system for surface characterization and a quadrupole mass analyzer (QMA). A W(110) crystal serves as the target substrate and is cleaned by heating in oxygen and flashing up to 2600 K. The iron and cobalt films are evaporated without using a crucible by means of a small electron-beam evaporator with a thin rod (typically 2 mm diameter) of the evaporating material. An integrated ionization-gauge-like flux monitor facilitates a reproducable growth rate; typically the rate is set to a third of a layer per minute. The quantity of the evaporated iron or cobalt can be controlled to better than 5%. The films are grown epitaxially on the W(110) crystal in the low 10^{-10} mbar pressure range. A subsequent annealing of these films to about 400 K leads to patterns corresponding to well-defined bcc iron (110) and hcp cobalt (110) surfaces, as is checked by means of LEED.

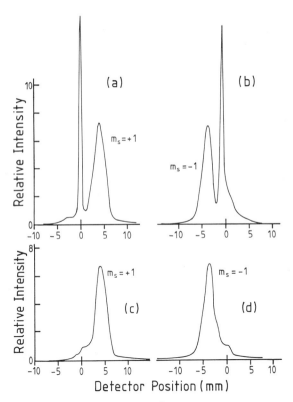

Figure 4a-d: Stern-Gerlach profiles of the metastable He(2^3S) beam (From [5]):
a) without central stop, adiabatic mode of spin flipper; b) without central stop, diabatic mode; c) with central stop, adiabatic mode; d) with central stop, diabatic mode.

The thickness of the films is determined with an oscillating quartz crystal and can additionally be estimated from satellite LEED spots. The thickness of the films was typically about 20 Å. In these films the surface anisotropy causes the easy magnetization axis to lie in-plane along the [1$\bar{1}$0] direction of the substrate. The films were uniformly magnetized in this direction, collinear to the atomic beam polarization, by a current pulse through a coil situated close to the sample. Oxygen exposures were done after preparation of the surface.

RESULTS AND DISCUSSIONS

Figure 5 shows the measurement of the electron intensity and asymmetry as a function of the vacuum kinetic energy of the ejected electrons for a Fe(110) film with a thickness of about 20 Å. Indicated is the statististical error. The behavior at high kinetic energies near the Fermi level is attributed to 3d-electrons. For these electrons a positive asymmetry is measured, which corresponds to a dominance of minority electrons in the surface vacuum region. Theoretical calculations [6] predict for bulk-like Fe atoms a dominance of <u>majority</u> spin states at the Fermi edge. These calculations show that the surface layer DOS curves are narrowed due to the lower coordination number, with the result that the <u>minority</u> spin contribution overwhelms the majority spin at E_F for the surface layer. In the vacuum, minority spin states become dominant especially in the region close to E_F [6]. This calculation is in agreement with our experimental results here and also with the measurements of the Rice group from Fe(110) on a GaAs substrate [2].

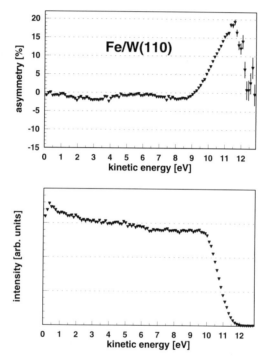

Figure 5: Electron spectrum (lower panel) and spin asymmetry curve (upper panel) of a Fe(110) surface.

We measured the asymmetry of the Fe/W(110) film as function of oxygen exposure up to 24 L (1 L equals 10^{-6} Torr s), the results are shown in Figure 6. With oxygen exposure significant differences are observed. At an exposure of about 3 L the positive asymmetry for the high kinetic electrons drops to zero and a transient structure occurs with a positive asymmetry which peaks at about 5.5 eV kinetic energy. This may be related to the exchange split oxygen derived bands, found by spin resolving photoelectron spectroscopy [7]. With further exposure this structure decreases to zero, which may be interpreted that the chemisorbed oxygen has been transformed to non ferromagnetic iron oxide. From about 3 L through 4 L oxygen exposure the asymmetry for electrons near the Fermi edge changes sign. This change of sign is in agreement with measurements performed on O/Fe/GaAs(110) [2] and with a theoretical calculation [8] which predicts a dominance of majority electrons in the surface vacuum region of O/Fe(110). It should be noted that this calculation is based on a p(1x1) structure of the oxygen, whereas experimentally c(2x2) and c(3x1) structures are

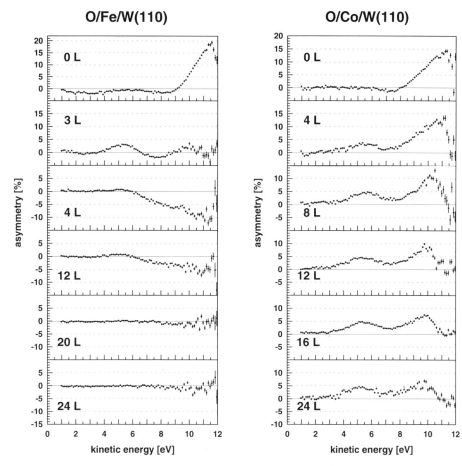

O/Fe/W(110)

O/Co/W(110)

Figure 6: Spin asymmetry curve of Fe/W(110) as a function of oxygen exposure. At the Fermi edge a change in the sign of the asymmetry occurs after an exposure of about 3 L (1 L ≙ 10⁻⁶ Torr s).

Figure 7: Spin asymmetry curve of Co/W(100) as a function of oxygen exposure. A change in the sign of the asymmetry at the Fermi edge does not occur, in contrast to oxygen exposure on an iron surface, see Figure 6.

present. With further increasing coverage the asymmetry at the Fermi edge decrease and is finally reduced to zero, pointing to an complete transformation of the surface to iron oxide.

Measured asymmetries as a function of oxygen exposure for Co(0001)/W(110) are shown in Figure 7. There is only one theoretical calculation of the surface layer projected spin density of states for uncovered cobalt available [9]. This calculation predicts a dominance of minority spin electrons near the Fermi energy. The measured positive yield asymemtry agrees with that. For oxygen covered Co(0001) there is, as far as we know, no surface layer projected spin density calculation in the literature. The oxygen induced structure at about 5.5 eV is also present here. The asymmetry increases at this point from zero up to 5% and remains constant up to an exposure of 24 L. This reflects the lower chemical reactivity of oxygen with the cobalt film than with iron. In the structure near the Fermi energy the asymmetry decreases slowly with increasing exposure <u>without changing sign</u>. This behavior is markedly different from that for iron.

Investigations carried out by spin resolved photoelectron spectroscopy (SPUPS) [10] showed that for Fe(110) films on W(110), in contrast to our SPMDS measurement here, the density of majority electrons is higher than for minority electrons at the Fermi energy and within an interval of about 1 eV below, independent of whether the excitation is done with HeI (hv = 21.22 eV) or with NeI (hv = 16.85 eV) radiation. Also SPUPS studies of clean and oxygen covered Co(0001) film showed [10] that more majority electrons are emitted in an energy interval of about 1 eV at the Fermi edge after excitation with NeI and ArI (hv = 11.83 eV) light, but more minority electrons are emitted using HeI light. This behavior and its comparison to our SPMDS findings show that SPUPS studies are inherently more related to several layers below the surface than with the outermost surface vacuum region, as SPMDS.

SUMMARY AND CONCLUSION

The energy resolved measurement of the asymmetry in this experiment gives information about the spin density at the Fermi level in the outermost region of ferromagnetic surfaces and overlayers. Concerning the dominance of majority or minority spin our measurements are in agreement with surface layer projected and Fermi energy sliced theoretical calculations of ferromagnetic Fe(110), O/Fe(110) and Co(0001) films [6,8,9]. There are only few Fermi energy sliced surface specific studies. Our measurements will be very helpful in the development of reliable theoretical calculations of energy-sliced and spin resolved densities for the surface vacuum region.

This experiment has been supported by the Deutsche Forschungsgemeinschaft through Sonderforschungsbereich 216 Bielefeld/Münster Polarization and Correlation in Atomic Collision Complexes. We acknowledge the work of D. Egert, P. Rappolt, M. Wilhelm, Dr. M. Getzlaff and Prof. Dr. W. Raith who participate with us in this research project.

REFERENCES

[1]   H. Conrad, G. Ertl, J. Küppers, W. Sesselmann, H. Haberland, *Surf.Sci.* **100**, L461 (1980)
       B. Woratschek, W. Sesselman, J. Küppers, G. Ertl, H. Haberland, *Phys.Rev.Lett.* **55**, 611 (1985)
       W. Sesselmann, B. Woratschek, J. Küppers, G. Ertl, H. Haberland, *Phys.Rev.* **B35**, 1547 (1987).

[2]   M. Onellion, M.W. Hart, F.B. Dunning, G.K. Walters, *Phys.Rev.Lett.* **52**, 380 (1984)
       M.S. Hammond, F.B. Dunning, G.K. Walters, G.A. Prinz, *Phys.Rev.* **B45**, 3674 (1992).

[3]   C.W.F. Drake, *in:* "Atomic physics III", S.J. Smith, G.K. Walters, ed., p. 269, New York, Plenum Press 1977.

[4]   W. Schröder, G. Baum, *J.Phys.E* **16**, 52 (1983).

[5]   G. Baum, W. Raith, H. Steidl, *Z.Phys.D* **10**, 171 (1988).

[6]   R. Wu, A.J. Freeman, *Phys.Rev.Lett.* **69**, 2867 (1992).

[7]   M. Getzlaff, J. Bansmann, C. Westphal, G. Schönhense, *J.Magn.Magn.Mater.* **104-107**, 1781 (1992).

[8]   R. Wu, A.J. Freeman, *J.Appl.Phys.* **73**, 6739 (1993).

[9]   R. Wu, D.S. Wang, A.J. Freeman, *J.Magn.Magn.Mater.* **132**, 103 (1994).

[10]   M. Getzlaff, J. Bansmann, G. Schönhense, *Solid State Comm.* **87**, 467 (1993);
       M. Getzlaff, J. Bansmann, G. Schönhense, *Phys.Rev.Lett.* **71**, 793 (1993);
       M. Getzlaff, J. Bansmann, G. Schönhense, *J.Magn.Magn.Mater.* **140-144**, 729 (1995).

SPIN ASYMMETRIES IN RELATIVISTIC (e,2e) PROCESSES AND BREMSSTRAHLUNG

Werner Nakel

Physikalisches Institut der Universität Tübingen
D-72076 Tübingen, Germany

INTRODUCTION

The aim of our experiments with transversely polarized electron beams of 300 keV is to investigate the role of the spin-orbit interaction of the continuum electrons in the processes of electron-impact ionization and bremsstrahlung.

The spin-orbit interaction, well-known from the elastic (Mott)scattering, arises from the interaction of the magnetic moment of the electrons with the magnetic field felt in the rest frame of the electrons because of their motion in the Coulomb field of the target nucleus. As a result, spin up-down asymmetries are to be expected in the processes of electron-impact ionization and bremsstrahlung. To get detailed information on the *elementary* collision process we performed coincidence measurements between the outgoing particles. This means in the case of electron-impact ionization to perform coincidence measurements between the outgoing electrons, so-called (e,2e) experiments,[1,2] in the case of bremsstrahlung electron-photon coincidence measurements.[3] The experiments and the observed spin up-down asymmetries of the triply differential cross sections are discussed in comparison with pertinent calculations in the following two sections.

SPIN ASYMMETRIES IN RELATIVISTIC (e,2e) PROCESSES

In nonrelativistic (e,2e) experiments spin-dependent asymmetries have been studied which are due to exchange interaction[4] and to the so-called fine-structure effect.[5,6] In this section we discuss a relativistic (e,2e) experiment[7] with a transversely polarized electron beam designed to look for a spin asymmetry in the triply differential cross section of electron impact ionization caused by spin-orbit interaction of the *continuum* electrons in the Coulomb field of the atomic nucleus. In particular we were interested to investigate the angular distribution in view of the question whether the asymmetry is larger within the region of the so-called recoil peak than the asymmetry within the binary peak. Our assumption was based on the following intuitive argument. As the binary peak has a large contribution from a direct binary collision between the incoming electron and the atomic electron with the nucleus in the role of a 'spectator', the

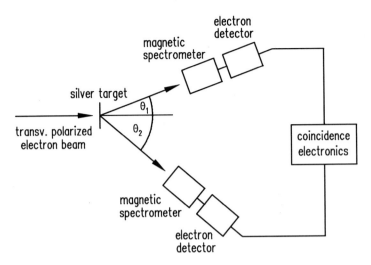

Figure 1. Sketch of the coplanar electron-electron coincidence experiment. The spin direction of the primary electron beam (300 keV) is perpendicular to the scattering plane (from Prinz et al.[7]).

spin-orbit interaction will be weak. The recoil peak, however, cannot be explained unless an electron-nucleus interaction is taken into account. Consequently, here a spin-orbit interaction must contribute and a spin asymmetry is to be expected.

A sketch of the experimental arrangement is shown in Fig. 1. The source for the polarized electron beam (described in detail elsewhere[8]) used the photoemission of electrons from a GaAsP crystal irradiated by circularly polarized light of a helium-neon laser. After being deflected by a 90° cylindrical deflector, the extracted electrons are transversely polarized. The spin-flip of the electron beam can be easily realized by reversing the helicity of the laser light. The source is installed in a high voltage terminal of a 300-kV accelerator tube and produces a continuous transversely polarized beam with a polarization degree in the range of 35% to 40%. The degree of spin polarization was measured by a Mott analyzer put into the beam line in front of the entrance of the scattering chamber. In the Mott analyzer the electrons scattered through 120° by a gold foil were detected by a pair of ion-implanted silicon detectors. The beam was focused to a 1-mm-diam spot on the target foil placed at the center of a vacuum chamber. As the target we used silver foil with a thickness of 50 μg/cm^2, for which plural scattering was found to be small in previous cross section measurements. Each of the two electron detector systems consisted of a magnetic spectrometer for the energy analysis combined with a plastic scintillation detector. Each magnet is a doubly focusing homogenous sector field shaped by an iron core. The fast signals from the detectors were fed into a time-to-amplitude converter via constant fraction discriminators. The quantity measured directly is the counting rate of the true coincidences alternately for spin-up and spin-down electrons of the primary beam. The spin asymmetry is defined as the relative cross section difference

$$A = \frac{d^3\sigma^\uparrow - d^3\sigma^\downarrow}{d^3\sigma^\uparrow + d^3\sigma^\downarrow} \tag{1}$$

where $d^3\sigma^\uparrow$ and $d^3\sigma^\downarrow$ are the triply differential ionization cross sections for impinging electrons with spin up and spin down perpendicular to the scattering plane. We got the asymmetry A as the ratio

$$A = \frac{N}{P} \qquad (2)$$

where P is the polarization of the beam and, for spin-up and spin down counting rates $N^\uparrow$ and $N^\downarrow$, respectively

$$N = \frac{N^\uparrow - N^\downarrow}{N^\uparrow + N^\downarrow} \; . \qquad (3)$$

Because theoretical predictions[9,10] of A were not available until the end of the present measurements we had used the following arguments to choose suitable atomic number of the target nucleus and kinematical conditions. Since the spin-orbit interaction the continuum electrons experience will increase with the strength of the electric field in which they are moving the resulting spin up-down asymmetry is expected to increase with the atomic number Z of the target nucleus. The triply differential cross section, however, decreases with increasing Z. Balancing asymmetry to be expected against cross section we chose the K shell of silver (Z=47).

We used coplanar asymmetric kinematics (fast outgoing electron is detected under a small scattering angle with regard to the primary beam) where the angular distribution of the coincident slow outgoing electrons consists of a binary peak and a distinct recoil peak. As already shown in a previous measurement,[11] for our relativistic primary energies the recoil peak is relativistically transformed into the forward direction, whereas for nonrelativistic energies it appears in the backward direction.[1,2]

We chose the following parameters for our measurement: A primary electron beam of $E_0 = 300$ keV transversely polarized with the spin perpendicular to the reaction plane impinges onto a silver target. The outgoing fast electrons of $E_1 = 200$ keV are observed at a fixed scattering angle of -10° with respect to the primary beam direction. The detector for the coincident slow electrons was adjusted to an energy of 74.5 keV in order to select (e,2e) processes from the K shell ($E_{bind} = 25.5$ keV). The results of the measurement of the spin asymmetry A as a function of the scattering angle of the outgoing slow electrons is shown in Fig. 2(a). To visualize the angular positions of the recoil and the binary peak Fig. 2(b) shows the measured relative triply differential cross section (averaged over the spin directions of the primary beam).

Comparing the angular dependence of the asymmetry with the pertinent cross section, the intuitive physical picture described above is confirmed. Whereas in the recoil peak the spin-orbit interaction of the continuum electrons generate a distinct asymmetry up to 16%, the asymmetry in the binary peak is close to zero. This is, to our knowledge, the first direct evidence for the influence of spin-orbit coupling of the continuum electrons on the process of electron impact ionization.

The dashed and solid lines in Fig. 2(a) are theoretical predictions of Jakubassa-Amundsen[9] and of Tenzer and Grün,[10] respectively. In both calculations the process is treated in lowest order perturbation theory. For the continuum electrons nonrelativistic Coulomb waves multiplied by a free Dirac spinor are used. For the bound state of the atomic electron Jakubassa-Amundsen uses a semirelativistic Darwin function, whereas Tenzer and Grün use a hydrogenic 1s Dirac wavefunction. Exchange and spin-flip processes are accounted for in both calculations. The differences between the results of the two calculations are presumed to be partly due to numerical inaccuracies.

Qualitatively, both calculated curves of Fig. 2(a) follow the tendency of the measured asymmetry, quantitatively, however, strong discrepancies occur. The latter

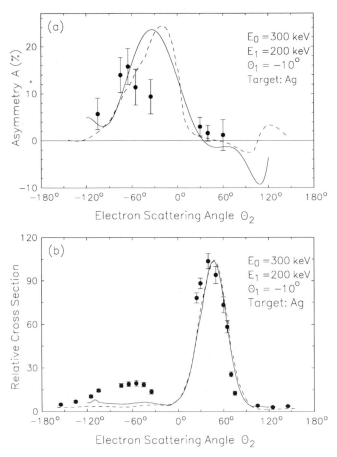

Figure 2. (a) Spin up-down asymmetry A of the triply differential cross section for electron-impact ionization of the K shell of silver as a function of the scattering angle Θ_2 of the outgoing slow electrons of energy $E_2 = 74.5$ keV. The primary electron energy amounted to $E_0 = 300$ keV. The outgoing fast electrons of $E_1 = 200$ keV were observed at an angle of $\Theta_1 = $ -10°. The error bars represent the standard deviations only; the systematic error of the asymmetry scale was estimated to be $\pm$ 2 %. The dashed and solid lines are calculations of Jakubassa-Amundsen[9] and of Tenzer and Grün,[10] respectively.

(b) Measured (full circles) relative triply diffential cross section averaged over the spin directions of the primary beam. The dashed and the solid lines are calculations of Jakubassa-Amundsen[9] and of Tenzer and Grün,[10] respectively, normalized to the measurement in the maximum of the binary peak (from Prinz et al.[7]).

is not surprising if one compares (Fig. 2(b)) the measured relative triply differential cross sections with the calculations of Jakubassa-Amundsen[9] and of Tenzer and Grün[10] (normalized to the measurement in the maximum of the binary peak).

The predicted ratio of the recoil to the binary peak intensities is too low. This is in accordance with the result of a former absolute measurement,[11] where the recoil peak is strongly underestimated by the Coulomb Born approximation.

The theory, so far predicting the absolute cross section of asymmetric relativistic (e,2e) measurements best, is a relativistic distorted wave Born approximation.[12,13] Calculations of the asymmetry according to this approximation are in progress. This approximation allows for elastic (Mott) scattering of the incident and outgoing electrons in the field of the atom. It would be interesting to analyse theoretically the contribution of the incoming, the fast and the slow outgoing electrons to the spin asymmetry.

SPIN ASYMMETRIES IN THE ELEMENTARY PROCESS OF BREMSSTRAHLUNG

The emission of a photon in the scattering of an electron from a (screened) nuclear Coulomb field is called bremsstrahlung. An important observable is the angular asymmetry of the radiation emitted by a transversely polarized electron beam where the spin direction is perpendicular to the photon emission plane, defined by the momenta of the incoming electron and the emitted photon. Disregarding the decelerated outgoing electrons such left-right photon emission asymmetries have been measured by several groups[14,15] and recently in our laboratory[16] in order to clear up discrepancies to calculations which appeared in previous work.[15] We found good agreement with the theoretical predictions of a partial wave calculation of Tseng and Pratt.[17] However, since in all these experiments only the emitted photons have been observed, the results are necessarily averaged over all electron scattering angles. Therefore the test of theoretical predictions is not as strong as it could be. To get detailed information on the *elementary* collision process[3] we performed a coincidence measurement between outgoing electrons and photons emitted by a transversely polarized beam.[18]

Using unpolarized electrons such an electron-photon coincidence experiment was first performed in our laboratory in 1966.[19] Its continuation and the work of other groups were reviewed by Nakel.[3] In all these (coplanar) coincidence measurements it proved that there is a strong angular correlation with the photons being predominantly emitted on the same side relative to the primary beam as the decelerated electrons. This feature can be used to illustrate in a very simple way the photon emission asymmetry from transversely polarized electrons observed in non-coincidence experiments by additionally taking into account the asymmetric scattering of the radiating electrons due to spin-orbit coupling (analogous to Mott scattering). However, to interpret the results of our present electron-photon coincidence experiment revealing the structure of the elementary bremsstrahlung process, the simple picture does not suffice. This becomes particularly clear in the special case where the electron detector is put into the 0° position, i.e. in the direction of the primary beam. Here the outgoing decelerated electrons are not deflected, nevertheless we did find a left-right bremsstrahlung emission asymmetry.

A sketch of the experimental arrangement is shown in Figure 3. The source for the polarized electron beam (described in detail elsewhere[8]) is the same as used in the (e,2e) experiment discussed above. The beam was focused to a 1 mm diameter spot on the target foil placed at the center of a vacuum chamber. The targets consisted of evaporated films of gold on carbon backing with thicknesses typically 50 μg/cm^2, for which plural scattering had been found to be small in previous cross section measurements. Since the target foil consisted of gold the electron polarization could be monitored simultaneously with the bremsstrahlung asymmetry measurements using a second pair of electron detectors, located inside the scattering chamber at 120°.

The bremsstrahlung photons from the target leave the chamber through a thin plastic window before entering a high purity germanium detector. The electron detector system consisted of a magnetic spectrometer for the energy analysis combined with a plastic scintillation detector. The magnet is located inside the chamber and mounted on the lid, the photomultiplier is mounted outside on top of the lid. The angular position can be changed by rotating the lid (under atmospheric-pressure con-

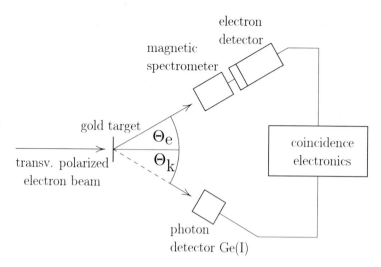

Figure 3. Sketch of the coplanar electron-photon coincidence experiment. The spin direction of the primary electron beam (300 keV) is perpendicular to the emission plane (from Mergl et al.[18]).

ditions). The magnet is a doubly focusing homogeneous sector field shaped by an iron core with a deflection angle of 137°. Its medium plane is perpendicular to the scattering plane. The fast signals from the electron and photon detectors were fed into a time-to-amplitude converter (TAC) via constant fraction discrimimators. The TAC was gated by slow pulses of the photon detector (corresponding to the energy split chosen). There are two experimental ways of obtaining spin dependent bremsstrahlung emission asymmetry: (1) by measuring the relative, triply differential cross sections for fixed photon and electron detector positions but for two opposite orientations of the primary electron spin; (2) by measuring them for fixed spin orientation but changing the angular position of both the photon and the electron detector symmetric to the direction of the primary beam. We used the first method as the arrangement allows us to easily change the orientation of the electron spin by changing the helicity of the circularly polarized light. In this way many possible instrumental asymmetries do not enter.

The quantity measured directly is the counting rate of the true coincidences alternately for spin-up and spin-down electrons of the primary beam. We got the asymmetry coefficient C_{200} as the ratio

$$C_{200} = \frac{A}{P} \tag{4}$$

where P is the polarization of the beam and, for spin-up and spin-down counting rates $n \uparrow$ and $n \downarrow$ respectively,

$$A = \frac{n \uparrow - n \downarrow}{n \uparrow + n \downarrow} . \tag{5}$$

The indices of C_{200} were chosen corresponding to a classification scheme of bremsstrahlung-polarization correlations introduced by Tseng and Pratt.[17] The meaning of the three indices is (from left to right): the primary electron beam is transversely polarized with the spin perpendicular to the emission plane, the polarization of the photon is not measured, the polarization of the outgoing electron is not measured.

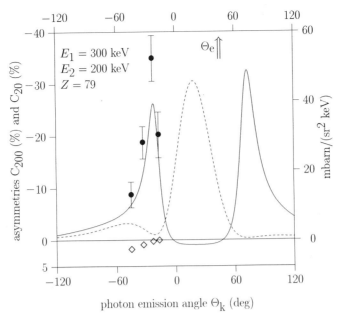

Figure 4. Photon emission asymmetry C_{200} (full circles) as a function of the photon emission angle Θ_k for outgoing electrons of scattering angle $\Theta_e = 45°$ and an energy of $E_2 = 200$ keV. The solid line is a calculation of the emission asymmetry from a theory of Haug,[20] the broken line the pertinent triply differential cross section for unpolarized primary electrons.[21] The error bars represent the standard deviations only; the systematic error of the asymmetry scale was estimated to be $\pm 2\%$. The open diamonds give the non-coincident photon asymmetry coefficients C_{20} measured simultaneously (from Mergl et al.[18]).

The results of the measurements of C_{200} as a function of the photon emission angle Θ_k for fixed outgoing electron angle Θ_e of 45° are shown in Figure 4. The full curve is a calculation of Haug,[20] the dashed curve the pertinent triply differential cross section for unpolarized primary electrons from a theory of Elwert and Haug.[21]

Theoretical predictions of the process of bremsstrahlung production (reviewed by Pratt and Feng[22]) require calculations of the probability that the incident electron will make a transition to a different electronic continuum state with a photon emitted while in the Coulomb field of an atom. The perturbation causing this transition is the interaction of the electron with the Coulomb field and the radiation field. Whereas presently the interaction of electrons with the radiation field can be treated only by perturbation theory, their interaction with the Coulomb field of the atom can, in principle, be handled exactly. In the latter case one has to use exact wave functions, which describe an electron in a screened nuclear Coulomb field. This is particularly important when the field is strong, i.e. for high atomic numbers Z as in the present case. It is, however, not possible to solve the Dirac wave equation in closed form for a free electron in a Coulomb field, let alone in a screened nuclear field. Therefore only approximations with various assumptions had been available before 1971, when Tseng and Pratt[17] obtained an exact calculation of the bremsstrahlung process. They solved the Dirac equation numerically by means of partial wave expansion using a relativistic local self-consistent screened central potential. This theory, the best currently available for bremsstrahlung, yields *doubly* differential cross sections we used for comparison with our *non*-coincident asymmetry measurements.[16] However, this theory does not yet yield numerical results for the *triply* differential cross section needed for electron-photon coincidence experiments. Therefore we can compare our measurements only with approximate calculations. Calculations in first order Born approximation (classical Bethe-Heitler formula[23]) give zero emission asymmetry. Measurements of the asymmetry, therefore, are a proper test for theories going beyond the first Born approximation. In a theory of Elwert and Haug,[21] Sommerfeld-Maue wave functions were used to solve the Dirac equation. On the basis of this theory Haug[20] calculated the bremsstrahlung emission asymmetry for transversely polarized electrons. The region of validity can be expressed $Z \cdot \alpha \ll 1 (\alpha \approx 1/137)$. Although this condition is not fulfilled for gold ($Z = 79$), the measurement is in qualitative agreement with the calculation. Comparison of the calculated triply differential cross sections with the pertinent photon emission asymmetries shows that high asymmetries occur in regions of small cross sections, whereas in the region of the large lobe the asymmetry is very small. Notice in this context the open diamonds in Figure 4, giving non-coincident photon asymmetries C_{20} measured simultaneously. The values of C_{20} are very small compared to C_{200} (and have the opposite sign), as this particular measurement integrates over all angles of the outgoing electrons. Thus small asymmetries associated with large cross sections mask the large asymmetries with small cross sections. In comparison with it, the results of the coincidence experiment yield very detailed information on the elementary collision process.

We have also studied the interesting special case of the electron detector being put into the direction of the primary beam to detect outgoing decelerated electrons with deflection angle $\Theta_e = 0°$. Then, for an unpolarized primary beam the corresponding photon angular distribution has to be symmetric about the beam (see e.g.[24]). For a transversely polarized beam, however, we did find an emission asymmetry[18] of up to 10%. After the analysis of Sobolak and Stehle[25] the sign of the asymmetry is dependent on whether or not the electron spin flips in the bremsstrahlung emission process. If for primary electrons with spin up, the radiation is more right than left — which was the case in our experiment — the spin-flip processes should be dominant. (These authors also give a classical argument for the origin of the asymmetry considering the force on a magnetic dipole moving in the Coulomb field of an atomic nucleus). One might conceive of a future "complete" scattering experiment with a polarized primary beam and detectors sensitive to polarization of the photons and of the outgoing electrons.

Acknowledgements

The support of the Deutsche Forschungsgemeinschaft and of the Deutscher Akademischer Austauschdienst is gratefully acknowledged.

REFERENCES

1. A. Lahmam-Bennani, J. Phys. B **24**, 2401 (1991)
2. C. T. Whelan, H. R. J. Walters, A. Lahmam-Bennani, and H. Erhardt eds., *"(e,2e) and related processes"* Kluwer, Dordrecht (1993)
3. W. Nakel, Phys. Rep. **243**, 317 (1994)
4. G. Baum, W. Blask, P. Freienstein, L. Frost, S. Hesse, W. Raith, P. Rappolt, and M. Streun, Phys. Rev. Lett. **69**, 3037 (1992)
5. S. Jones, D. H. Madison, and G. F. Hanne, Phys. Rev. Lett. **72**, 2554 (1994)
   T. Simon, M. Laumeyer, G. F. Hanne, and J. Kessler, Contribution to 7th International Symposium on Polarization and Correlation in Electronic and Atomic

Collisions, Universität Bielefeld, Bielefeld, Germany, 29-31 July 1993 (unpublished).

6. B. Granitza, X. Guo, J. Hurn, Y. Shen, and E. Weigold, in *Proceedings of the XVIII International Conference on the Physics of Electronic and Atomic Collisions, Abstracts of Contributed Papers*, p. 201.
7. H.-Th. Prinz, K.-H. Besch, and W. Nakel, Phys. Rev. Lett. **74**, 243 (1995)
8. E. Mergl, E. Geisenhofer, and W. Nakel, Rev. Sci. Instrum. **62**, 2318 (1991)
9. D. H. Jakubassa-Amundsen, J. Phys. B **27**, 1 (1994)
10. R. Tenzer and N. Grün, Phys. Lett. **A 194**, 300 (1994)
11. J. Bonfert, H. Graf, and W. Nakel, J. Phys. B **24**, 1423 (1991)
12. H. Ast, S. Keller, C. T. Whelan, H. R. J. Walters, and R. M. Dreizler, Phys. Rev. **A50**, R1 (1994)
13. S. Keller, C. T. Whelan, H. Ast, H. R. J. Walters, and R. M. Dreizler, Phys. Rev. **A50**, 3865 (1994)
14. K. Güthner, Z. Phys. **182**, 278 (1965);
    P.E. Pencynski, H.L. Wehner, Z. Phys. **237**, 75 (1970);
    A. Aehlig, Z. Phys. A - Atom and Nuclei **294**, 291 (1980).
15. H.R. Schaefer, W. von Drachenfels, and W. Paul, Z. Phys. A - Atoms and Nuclei **305**, 213 (1982);
16. E. Mergl and W. Nakel, Z. Phys. **D 17**, 271 (1990);
17. H.K. Tseng and R.H. Pratt, Phys. Rev. **A 7**, 1502 (1973);
18. E. Mergl, H.-Th. Prinz, C.D. Schröter, and W. Nakel, Phys. Rev. Lett **69**, 901 (1992)
19. W. Nakel, Phys. Lett. **22**, 614 (1966);
20. E. Haug (private communication cf. ref. 21.).
21. G. Elwert and E. Haug, Phys. Rev. **183**, 90 (1969).
22. R.H. Pratt and I.J. Feng, in: "Atomic inner-shell physics", B. Crasemann, ed., Plenum, New York (1985) p. 533.
23. H. Bethe and W. Heitler, Proc. R. Soc. (London) Ser. A **146**, 83 (1934).
24. W. Nakel, Phys. Lett. **25A**, 569 (1967).
25. E.S. Sobolak and P. Stehle, Phys. Rev. **129**, 403 (1963).

SPIN POLARIZED AUGER ELECTRONS FROM ATOMS AND NON-MAGNETIC CONDENSED MATTER

Ulrich Heinzmann

University of Bielefeld
Faculty of Physics
D-33501 Bielefeld, Germany

ABSTRACT

The paper reviews the recent progress in the spin resolved experiments of Auger electron emission and their angular distribution behaviours. The Auger decay takes place at spin oriented hole states (ions) produced by means of photoemission using circularly polarized radiation. The cross comparison of photoelectron and Auger electron spin polarization allows the study of the spin-spin coupling of the two holes final state and the validity of the two step process for photoemission and Auger decay. Detailed quantitative examples are given for Auger processes in Barium atoms and condensed Alkali layers.

INTRODUCTION

Auger electrons emitted in the decay of an inner-shell vacancy may be spin polarized if the hole states created have a nonstatistical distribution of their magnetic sublevels. An alignment of the vacancies can arise even when unpolarized light or particles are used for ionization of an unpolarized target (for a review, see Mehlhorn [1,2]). For this case, a so-called dynamical Auger-electron polarization occurring only in the direction perpendicular to the plane of reaction was predicted [3,4] and subsequently observed [5,6], although it was found to be small. The alignment, however, may influence the electron angular distribution considerably, as was observed for the rare gases Ar, Kr and Xe [7]. If the incident light is circularly polarized, an efficient polarization transfer can take place leading to inner-shell vacancies with not only aligned but, moreover, oriented angular momenta. The polarization of the ionic states may then be partially passed to the outgoing Auger electrons generally giving rise to non-zero values for all three components of the spin-polarization vector of the Auger-electrons [3]. This case was discussed in detail theoretically [8] and recently observed experimentally for free barium atoms [9] and in the Auger-electron emission from rubidium adlayers [10] using circularly polarized synchrotron radiaton. The angular distributions of the Auger-electron polarization were also discussed.

The data of spin-and angle resolved Auger spectroscopy can be used to determine alignment and orientation of the intermediate photoion state from the experiment [9]. Cross-comparison of these results with the corresponding dynamcial parameters characterizing the primary photoelectron emission process, angular distribution and spin polarization parameters which form a "complete experiment", should answer whether they have to be seen as a dynamical combination. A mutual influence of Auger- and photoelectrons would give rise to the so-called post-collision interaction, indicating a break-down of the two-step model. A reference case is the 4d ionzation of xenon for which both the occurrence [12] and the vanishing [13] of post-collision interaction was discussed in detail.

In condensed matter Auger spectroscopy provides a way of studying the coupling and correlation of electronic states. However, in experiments on non-magnetic materials, only the energetic position and the shape of Auger lines have been analysed so far [14]. The spin polarization of Auger electrons has been measured with ferromagnetic materials [15-18] mainly. Recently first experiments with spin polarized Auger-electrons from non-magnetic alkali layers [10, 19, 20] and Cu(100) [20] after photoemission by means of circularly polarized synchrotron radiation have been reported.

EXPERIMENT

For the ionization of the Ba atoms and alkali layers circularly polarized out-of-plane synchrotron radiation from the 6.5m normal incidence monochromator (degree of circular polarization: 0.92 ± 0.03) at the Berlin electron storage ring BESSY was used [21]. For spin polarization as well as for intensity measurements an osmium-coated grating with 3600 lines/mm was employed. It has a maximum transmission at $h\nu \approx 21$ eV and a resolution of $\Delta\lambda = 0.18$ mm with a 2mm wide exit slit and of $\Delta\lambda = 0.36$ mm with a 4 mm wide exit slit. The effusive barium atom beam was generated with a boron nitride oven, resistively heated to about 850°C. The photoelectrons and Auger electrons emitted in the magnetic-field-free ionization region were detected energy- and angle-resolved by means of a rotatable simulated hemispherical spectrometer [22, 23]. The electron spin polarization was measured by detecting $A(\theta)$, parallel to the photon beam and $P_\perp(\theta)$, normal to the reaction plane, which is defined by the k vectors of the incident light and the outgoing electron [24].

$$A(\theta) = \gamma \, \frac{A - \alpha \, P_2(\cos\theta)}{1 - \frac{\beta}{2} P_2(\cos\theta)} \tag{1}$$

$$P_\perp(\theta) = \frac{2 \, \xi \sin\theta \cos\theta}{1 - \frac{\beta}{2} P_2(\cos\theta)} \tag{2}$$

where θ is the angle between the incoming light and the outgoing electron. A, ξ, α and β are dynamical parameters describing the electron emission process, γ is the helicity = ± 1, $P_2(\cos\theta)$ the second Legendre polynomial.

The new crossed double undulator U2-FSGM beamline [25] at BESSY now allows experiments at higher photon energies. It is designed to produce intense, circularly polarized light in the photon energy range of 20- 200 eV with a degree of polarization up to 50% [26]. The monochromator resolution $\lambda/\Delta\lambda$ extends up to about 10^4 [27]. This beamline has been used for spin-resolved Auger spectroscopy in Xe atoms ($N_{45}O_{23}O_{23}$) and Cu(100)(M_3VV) [20]. Very recently circularly polarized soft x-rays in the energy range of 500eV up to 950eV have been analyzed and used at the new helical undulator HELIOS of the ESRF for spin resolved Auger electron spectroscopy of free Xe atoms [28,29].

In the latter cases we used two new experimental setups developed and used for spin-resolved spectroscopy at this beamline: a time-of-flight (TOF) electron energy analyzer combined with a retarding spherical Mott polarimeter [30] to detect the electron spin polarization, schematically shown in Fig. 1 [20].

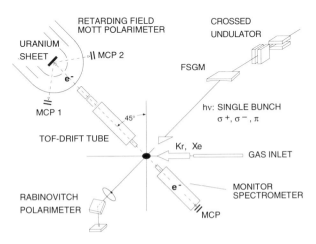

Figure 1: Experimental setup for spin resolved studies at free atoms with circularly polarized soft x-rays [20].

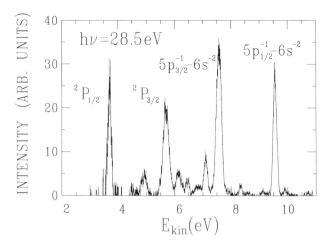

Figure 2: Electron kinetic energy spectrum obtained at hv = 28.5 eV, showing the two $5p^{-1}_{1/2,\ 3/2} - 6s^{-2}$ Auger lines and the corresponding photopeaks of Barium atoms [31].

255

RESULTS

In Fig. 2 an electron kinetic energy spectrum obtained for free Ba atoms [31] at $h\nu = 28.5$ eV und θ_{mag} is shown. This spectrum is background corrected and normalized to the electron current in the storage ring. The spectrometer transmission was also taken into account by measurements of the Xe 5p photoionization using the known Xe 5p cross-section [32].

Four main peaks occur: at $E_{kin} = 9.5$ eV the $5p_{1/2}^{-1} - 6s^{-2}$ Auger line with the corresponding photoline at $E_{kin} = 3.8$ eV, and at $E_{kin} = 7.5$ eV the $5p_{3/2}^{-1} - 6s^{-2}$ Auger line with the corresponding photoline at $E_{kin} = 5.8$ eV.

In Fig. 3 we show the measured angular distributions of the spin-polarization component $A(\theta)$ for the dominant Auger line at $h\nu = 23.5$ eV and and for the $5p_{1/2}^{-1} - 6s^{-2}$ Auger line at $h\nu = 25.8$ eV of Ba [9]. The count rate was typically only 1 Hz or less. The polarization component $P_{\perp}(\theta)$ is not displayed, since the measured values were always lower than 2%, which is compatible with the theoretical value of $\xi = 0$. (Dynamical polarization is a result of the interference of at least two outgoing electron waves and therefore has to vanish for single-channel Auger transitions.) By means of a non-linear least-squares-fit procedure according to Eq. (1), the values of A, α and β are determined from these distributions as given as insets in Fig. 3(a) and 3(b).

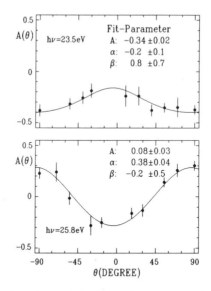

Figure 3: Measured angular dependence of the Auger electron spin polarization at $h\nu = 23.5$ eV for the Auger line with 7.5 eV kinetic energy (a) and at $h\nu = 28.5$ eV for the line with 9.5 eV kinetic energy (b). The solid lines represent a least squares fit to the data according to Eq.(1) [9].

The measured values for A_{Auger} obtained at different photon energies for $O_{2,3}$-P_1P_1 Auger electrons of Ba [31] are shown in Fig. 4. The absolute value of the integral spin polarization A_{Auger} resulting from the $5p_{1/2}^{-1}-6s^{-2}$ Auger decay is lower than for $5p_{3/2}^{-1}-6s^{-2}$, in accordance with the nonrelativistic approximation. In the case of a $^1S_0$ final ionic state the parameter A_{Auger} is a direct measure for the degree of orientation of the photoion as was discussed in detail [9]. The orientation coefficients $A_{10}(1/2)$ und $A_{10}(3/2)$ may be obtained by multiplication of the values of $A_{Auger}(1/2)$ and $A_{Auger}(3/2)$ in Fig. 4 by a factor of -3 and $3/\sqrt{5}$, respectively, resulting in orientation coefficients of the order of -0.3 for both spin-orbit states. The experimental values are compared with results of theoretical calculations using the Hartree-Fock and the RPAE approaches [31]. For the $5p_{1/2}^{-1}-6s^{-2}$ Auger parameter A_{Auger} both theories are in agreement with the experimental data, with the exception of the most precisely determined data point at hv = 25.8 eV where only the RPAE approach coincides with the experimental result.

The results for free Ba atoms represent the first example of how the spin-polarization transfer to Auger electrons can be used to advantage to analyze the dynamics of a two-step double-ionization process. The angular distributions of the polarization component $A(\theta)$ reflect a substantial orientation transfer from the photoion to the ejected Auger electron. The results confirm the theoretical framework and predictions discussed by Kabachnik and Lee [8]. The cross-comparison of the results for photoelectrons and Auger-electrons verifies quantitatively the assumed two-step model where both electron-emission processes occur in sequence [33].

This latter behaviour is described in detail here at the surprising case of Auger electron emission of condensed Rb on Pt. The layer is very thick and non-ordered.

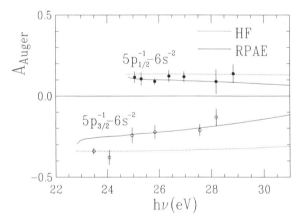

Figure 4: Energy dependence of the calculated spin-polarization parameter A_{Auger} for the $5p_{1/2,3/2}^{-1}-6s^{-2}$ Auger lines in comparison with experimental data [31]. Below and above hv = 25 eV the spectral resolution was $\Delta\lambda$ = 0.18 nm and $\Delta\lambda$ = 0.36 nm, respectively.

Fig. 5 shows in the upper and in the lower part the angular dependence of the spin polarization component $A(\theta)$ (in the direction of the light helicity) of the photoelectrons and the corresponding Auger electrons, respectively, after having made the background correction [19]. While the Auger electrons do not show a pronounced angular distribution of the spin polarization, the photoelectron polarizations switch their sign at around 30°, as usual [34]. It is worth noting that the photoelectrons corresponding to the $p_{3/2}$ and $p_{1/2}$ hole states always have opposite signs. It is worth noting that as in the case of xenon adsorbates [35] the angular dependence of the spin polarization of photoelectrons can be fitted by the angular distribution valid for free atoms and molecules [34] given by Eq. 1.

For hv = 23 eV we obtain

$A_{\frac{3}{2}} = 0.09 \pm 0.024$

$\alpha_{\frac{3}{2}} = 0.012 \pm 0.02$

$\beta_{\frac{3}{2}} = 0.13 \pm 0.3$

$A_{\frac{1}{2}} = -0.12 \pm 0.04$

$\alpha_{\frac{1}{2}} = -0.17 \pm 0.03$

$\beta_{\frac{1}{2}} = -0.6 \pm 0.4$

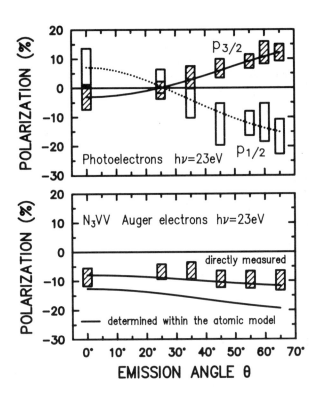

Figure 5: Measured angular dependences of the spin-polarization component in the direction of the helicity of circularly polarized 23eV radiation for photoelectrons and Auger electrons: Upper part: photoelectrons leaving a $p_{3/2}$ or $p_{1/2}$ hole state with a fit using the atomic model (full and dotted curve, respectively). Lower part: Auger electrons in cross comparison with the prediction of the atomic model using the data of the upper part (two curves as upper and lower limit). The rectangles describe the experimental uncertainties, including uncertainties in the background correction performed [19].

From the dynamical parameters for photoemission given above, the Auger electron spin polarization can be calculated within an atomic model using the non-relativistic LS-approximation (neglecting any influence of the spin-orbit interaction in the continuum states) [33]. Thereby the two-step model [8] is applied, treating the emission processs of photoelectrons and Auger electrons as independent and subsequent processes. They are connected only by the aligned and oriented hole states resulting from spin-dependent photoexcitation [36]. These calculated values for the Auger spin polarization as a function of the emission angle is given in the lower part of Fig. 5 including the range of uncertainty determined by the error bars of the corresponding photoelectron polarization values. They show agreement with the direct experimental values of the Auger electron polarization measured, also given in Fig. 5 lower part, within the experimental uncertainty [19].

Thus the spin oriented Auger electron emission subsequently following a photoemission process by means of circularly polarized light, also yielding polarized electrons, can be quantitatively explained by use of an atomic model not only with respect to the the angle integrated results but also to the complete angular dependence of the electron spin polarization. The circularly polarized photons create oriented holes characterized by the photoelectron spin polarizations observed.

It is consistent with the atomic model that - as opposed to the case of Cu(100) [20] - a change of the Auger electron polarization sign across the peak in the electron spectrum has not been observed in all other studies of the M_3VV, N_3VV and O_3VV Auger decay in K, Rb and Cs, respectively. Figure 6 shows spin separated intensity spectra measured at these alkali metals in a normal incidence, normal emission setup [10,19,20,37]. The spin polarization $= (I_+-I_-)/(I_++I_-)$ always appears to be constant across the Auger peak. The change of the preferential spin direction when the photon energy is raised is due to the primary excitation changing from predominantly p $\rightarrow$ s at the threshold to predominantly p $\rightarrow$ d at higher photon energies [37].

CONCLUSION

Auger electrons are in general spin polarized if the primary photoionization or photoexcitation process is performed by means of circularly polarized radiation. In connection with the emission of polarized photoelectrons the remaining hole state or ion is spin-oriented too. In condensed alkali metals the CVV-decay of the hole states results in spin-polarized Auger electrons as a subsequent process; the experimental results are compatible with a coupling of the valence electrons as a singlet state configuration. In agreement with the corresponding photoionization and Auger process of free Ba atoms (also with two outermost s electrons in singlet state configuration) the Auger emission from the Rb adlayer atoms can be quantitatively described within the non-relativistic LS approxmation.

Thus spin-resolved Auger electron spectroscopy provides a tool for studying the spin-coupling of the two-hole final state after the Auger decay and for cross-comparing the angular dependences of photoelectron and Auger-electron emission between isolated atoms and condensed matter under application of an atomic model.

ACKNOWLEDGEMENTS

The author thanks his co-workers: N. Böwering, R. David, M. Drescher, R. Kuntze, N. Müller, M. Salzmann, B. Schmiedeskamp, G. Snell, P. Stoppmanns, and S.-W. Yu for their strong encouragement, N. Brookes, N.A. Cherepkov, U. Becker, H. Griebe, U. Hergenhahn, V.K. Ivanov, N.M. Kabachnik, J. Noffke, J. Viefhaus for the successful cooperation and the institutions BESSY, Hasylab and ESRF and their staff for exellent working conditions with synchrotron radiation and the BMBF and the DFG for financial support.

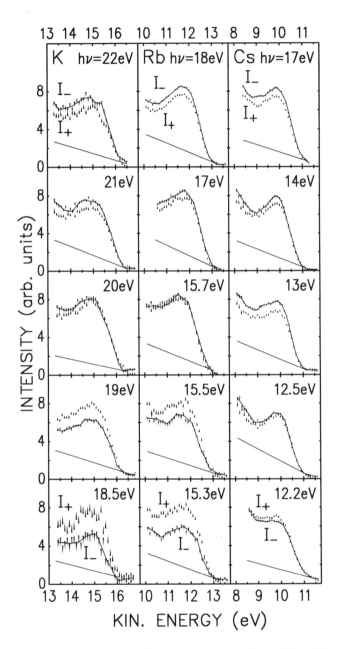

Figure 6: Spin separated partial intensites I_+ and I_- measured across the M_3VV, N_3VV and O_3VV Auger peaks of K, Rb and Cs, respectively, after primary excitation by circularly polarized radiation of varied energies. To guide the eye, the I_--data points are connnected by a line. The straight lines indicate the (unpolarized) background [20].

REFERENCES:

1   W. Mehlhorn, in X-ray and Atomic Inner-Shell Physics, Proceedings of the International Conference on X-ray and Atomic Inner-Shell Physics, edited by B. Crasemann, AIP Conf. Proc. No. 94 (AIP, New York, 1982), p. 55

2   W. Mehlhorn, in Atomic Inner-Shell Physics, edited by B. Crasemann (Plenum, New York, 1985), p. 199

3   H. Klar, J. Phys. B13, 4741 (1980)

4   N.M. Kabachnik, J. Phys. B14, L377 (1981)

5   U. Hahn, J. Semke, H. Merz, and J. Kessler, J. Phys. B18, 417 (1985)

6   H. Merz, and J. Semke, in X-ray and Inner-Shell Processes, Knoxville, 1990, Procceedings of the Fifteenth International Conference on X-ray and Inner-Shell Processes, edited by A. Carlson, M.O. Krause and S.T. Manson, AIP Conf. Proc. No. 215 (AIP, New York 1990), p. 719

7   T.A. Carlson, D.R. Mullins, C.E. Beall, B.W. Yates, J.W. Taylor, D.W. Lindle and F.A. Grimm, Phys. Rev A39, 1170 (1989)

8   N.M. Kabachnik, and O.V. Lee, J. Phys. B22, 2705 (1989)

9   R. Kuntze, M. Salzmann, N. Böwering and U. Heinzmann, Phys. Rev. Lett. 70, 3716 (1993)

10 P. Stoppmanns, B. Schmiedeskamp, B. Vogt, N. Müller and U. Heinzmann, Phys. Scr. T41, 190 (1992)

11 V. Schmidt, Z. Phys. D2, 275 (1986), J. Phys. (Paris) Colloq. 48, C9-401 (1987)

12 V. Schmidt, S. Krummacher, F.J. Wuilleumier and P. Dhez, Phys. Rev. A24, 1803 (1981)

13 M. Borst and V. Schmidt, Phys. Rev. A33, 4456 (1986)

14 J.C. Fuggle, High Resolution Auger Spectroscopy of Solids and Surfaces (New York: Academic) (1981)

15 M. Landolt and D. Mauri, Phys. Rev. Lett. 49, 1783 (1982)

16 M. Taborelli, R. Allenspach, G. Boffa and M. Landolt, Phys. Rev. Lett. 56, 2869 (1986)

17 R. Allenspach, D. Mauri, M. Taborelli and M. Landolt, Phys. Rev. B35, 4801 (1987)

18 K. Schröder, E. Kisker, E and A. Bringer, Solid State Commun. 55, 377 (1985)

19 P. Stoppmanns, R. David, N. Müller and U. Heinzmann, Z. Phys. D30, 251 (1994)

20 N. Müller, R. David, G. Snell, R. Kuntze, M. Drescher, N. Böwering, P. Stoppmanns, S.-W. Yu, U. Heinzmann, J. Viefhaus, U. Hergenhahn and U. Becker, J. Electr. Spectr. 72, 187 (1995)

21 F. Schäfers, W. Peatman, A. Eyers, Ch. Heckenkamp, G. Schönhense and U. Heinzmann, Rev. Sci. Instrum, 57, 1032 (1986)

22 K. Jost, J. Phys. E12, 1006 (1979)

23 Ch. Heckenkamp, A. Eyers, F. Schäfers, G. Schönhense and U. Heinzmann, Nucl. Instrum. Methods A246, 500 (1986)

24 Ch. Heckenkamp, F. Schäfers, G. Schönhense and U. Heinzmann, Phys. Rev. Lett. 52, 241 (1984)

25 J. Bahrdt, A. Gaupp, W. Gudat, M. Mast, K. Molter, W.B. Peatman, M. Scheer, Th. Schröter and Ch. Wang, Rev. Sci. Instr. 63, 339 (1992)

26 R. David, P. Stoppmanns, S.-W. Yu, R. Kuntze, N. Müller and U. Heinzmann, Nucl. Instrum. and Meth. A 343, 650 (1994)

27   W.B. Peatman, J. Bahrdt, F. Eggenstein and F. Senf, BESSY Annual Report, p. 471 (1991)

28   N. Brookes, G. Snell, M. Drescher, N. Müller, U. Heinzmann, F. Heiser, U. Hergenhahn, J. Viefhaus, R. Hentges, O. Geßner and U. Becker, ESRF Newsletter **24** (1995)

29   G. Snell, M. Drescher, N. Müller, U. Heinzmann, U. Hergenhahn, J. Viefhaus, F. Heiser, U. Becker and N. Brookes, to be published

30   L.G. Gray, M.W. Hart, F.B. Dunning and G.K. Walters, Rev. Sci. Instr. **55**, 88 (1984)

31   R. Kuntze, M. Salzmann, N. Böwering, U. Heinzmann, V.K. Ivanov and N.M. Kabachnik, Phys. Rev. A**50**, 489 (1994)

32   J.B. West and J. Morton, At. Nucl. Data Tables **22**, 103 (1975)

33   R. Kuntze, M. Salzmann, N. Böwering and U. Heinzmann, Z. Phys. D**30**, 235 (1994)

34   N.A. Cherepkov,, Sov. Phys. JETP **38**, 463 (1974)

35   B. Kessler, N. Müller, B. Schmiedeskamp, B.Vogt and U. Heinzmann, Z. Phys. D**17**,11 (1990)

36   N.M. Kabachnik and I.P. Sazhina, J. Phys. B**23**, L353 (1990)

37   P. Stoppmanns, R. David, N. Müller, U. Heinzmann, H. Griebe and J. Noffke, J. Phys. Cond. Matter **6**, 4225 (1994)

BREMSSTRAHLUNG EMISSION IN COLLISIONS
OF ELECTRONS WITH ATOMS AND CLUSTERS

A V Korol[1], A G Lyalin[2], and A V Solovy'ov[3]

[1]Department of Physics, Russian Maritime Technical University
[2]Institute of Physics, St Petersburg State University
[3]A.F.Ioffe Physical-Technical Institute, RAS
St Petersburg, Russia

INTRODUCTION

In this paper we present results of recent calculations of the bremsstrahlung (BrS) cross sections formed in the collision of electrons with atoms and clusters.

Two mechanisms of photon radiation during the collision are considered. The first one, illustrated in the left hand side of figure 1, is the 'ordinary' BrS, i.e. the photon emission by a charged projectile decelerated in a static field of a target (see, for example, Berestetskii *et al* 1980). The second mechanism, shown in the right hand side of figure 1, is known as the polarizational, or so called 'atomic', BrS (see e.g. Tsitovich 1993). It considers the photon emission by the target electrons, virtually excited by the projectile.

In this paper we consider the manifestation of the polarizational BrS mechanism in the region of frequencies typical for the giant resonance excitations in atoms and clusters. This region of frequencies is of particular interest, because the polarizability of atoms and clusters in this region is very large and almost all the BrS is formed via the polarizational mechanism.

The giant resonances are characterized by a collective motion of electrons in the target. In atoms like La, Ba, Xe, Eu they appear due to oscillations of electrons of the 4d-subshell. The characteristic frequency of these oscillations is about 100 eV. There are two possible kinds of giant resonances, which may be excited in clusters (see e.g. Bréchigniac and Connerade 1994). Firstly, one may have giant resonances of the electrons localized on individual atoms and, secondly, there are collective oscillations of delocalized electrons belonging to the whole metal cluster. These electrons can oscillate collectively giving rise to giant resonances of an entirely new kind (Bréchigniac et al 1989, Selby 1991). The energy of the giant resonance of the first type in a cluster is comparable with the corresponding giant resonance energy in a single atom, while the energy of the resonance associated with oscillations of delocalized valence electrons is much lower, typically in the range 2-5 eV. This collective oscillation turns out to be

Selected Topics on Electron Physics
Edited by Campbell and Kleinpoppen, Plenum Press, New York, 1996

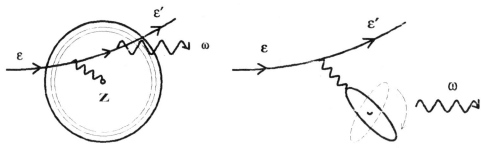

Figure 1: The ordinary bremsstrahlung is the photon emission of a charged projectile decelerated in the static field of the target. Polarizational bremsstrahlung mechanism considers the photon emission of the target electrons, virtually excited by projectile. Virtual excitation of the electrons is equivalent to the polarization of the target.

the precursor of the plasmon in the corresponding metal, thereby providing a new link between physics on the atomic scale and properties of condensed matter.

The energy of the projectile electron ε_1 has been considered sufficient for the excitation of the giant resonance in the target. The choice of the initial energies and the photon frequencies was determined by the fact that a number of experimental results has been obtained for various atomic targets for exactly such ε_1 and ω: for Xe (Verkhovtseva et al 1990) and for La and the lanthanum group (Zimkina et al 1981, 1984). These data proved the result of the earlier theoretical speculations (Amusia et al 1977) showing that a wide maximum can appear in the emission spectrum in the region of ω above the ionization thresholds of atomic many-electron subshells.

Calculations of the BrS spectra of electrons on atoms in the above-mentioned ω regions were performed earlier for Ar, Xe and La within the framework of the non-relativistic Born approximation (Amusia et al 1982, Avdonina et al 1986, Verkhovtseva et al 1990) and for Xe (Amusia et al 1990) using the distorted partial- wave formalism. The polarizational BrS effect is even more prominent in the ion-atom or atom-atom collisions, because in these cases the ordinary BrS is strongly suppressed due to the large mass of the particles involved in the collision. Calculation of the polarizational BrS spectra in the non-relativistic heavy particle collisions was done in Amusia et al (1985a, 1985b), Ishii and Morita (1985) and Solov'yov (1992). The experimental verification of the theoretical results was performed by Ishii and Morita (1985).

The polarizational mechanism of radiation in collisions of electrons with metal clusters was investigated by Connerade and Solov'yov (1995). It was also demonstrated that this mechanism plays the essential role in the BrS and the radiative electron capture processes. In this paper we consider the case of atomic metal clusters, but, in fact, other types of clusters should provide qualitatively similar features in the emission spectra. For example, the giant resonance structure manifests itself also in the spectrum of radiation emitted in collision of electrons with fullerenes as it was discussed by Amusia and Korol (1994).

The difference between various kind of clusters arises from their internal structure. We consider atomic metal clusters, because, on the one hand, the models and methods used for the description of their structure make this sort of clusters similar to a single many-electron atom. On the other hand, the specific metal cluster features which manifest themselves in the emission spectra turn out to be in common with the similar effects arising for the other type of clusters, like semiconducter, van der Waals or molecular clusters, including fullerenes.

BREMSSTRAHLUNG IN ELECTRON–ATOM COLLISIONS

The differential BrS cross section, which characterizes the spectral distribution of radiation, is given by (atomic system of units is used)

$$\omega \frac{d\sigma}{d\omega} = \frac{1}{(2\pi)^4} \frac{\omega^4}{c^3} \frac{p_2}{p_1} \int d\Omega_{\mathbf{p}_2} \int d\Omega_\gamma \sum_\lambda |f_{\mathrm{BrS}}|^2 \tag{1}$$

Here ω is the energy of the photon, p_1 and p_2 are the initial and the final momenta of the projectile electron, $c \approx 137$ is the velocity of light. The integral is carried out over the solid angle of the vector of final momentum $\mathbf{p}_2$, $d\Omega_{\mathbf{p}_2}$, and over the directions of photon emission, $d\Omega_\gamma$. The sum is taken over the photon polarizations, λ, and f_{BrS} is the total amplitude of the BrS process.

An adequate description of the BrS emission formed in the intermediate energy electron-atom collision one obtains using the first order distorted partial-wave approximation (DPWA) (Amusia et al 1990, Amusia and Korol 1991).

Within the frame of the DPWA we consider the BrS process in the lowest order of the non-relativistic perturbation theory in the electron—dipole-photon interaction and in the Coulomb interaction, $\hat{V} = |\mathbf{r} - \mathbf{r}_{\mathrm{a}}|^{-1}$, between the projectile and the atomic electrons which leads to a virtual atomic excitation.

The total amplitude represents by itself a sum of the ordinary, f_{ord}, and the polarizational, f_{pol}, terms:

$$f_{\mathrm{BrS}} = f_{\mathrm{ord}} + f_{\mathrm{pol}} \tag{2}$$

The amplitudes f_{ord} and f_{pol} are written as follows:

$$f_{\mathrm{ord}} = <\mathbf{p}_2|e\mathbf{r}|\mathbf{p}_1> \tag{3}$$

$$f_{\mathrm{pol}} = \sum_n \left\{ \frac{<0|\mathbf{eD}|n><\mathbf{p}_2\, n|\hat{V}|\mathbf{p}_1\, 0>}{\omega - \omega_{n0} + \mathrm{i}0} - \frac{<\mathbf{p}_2\, 0|\hat{V}|\mathbf{p}_1\, n><n|\mathbf{eD}|0>}{\omega + \omega_{n0}} \right\} \tag{4}$$

Here $|\mathbf{p}_1>$ and $|\mathbf{p}_2>$ are the wave functions of the incident and the scattered electrons with asymptotic momenta $\mathbf{p}_1$ and $\mathbf{p}_2$, respectively. The DPWA expansions of the initial ($j = 1$) and the final ($j = 2$) state wave functions of projectile are as follows

$$|\mathbf{p}_j> = 4\pi \sqrt{\frac{\pi}{p_j}} \sum_{l,m} \mathrm{i}^l \, \exp(\pm \mathrm{i}\delta_l(p_j)) \, \frac{P_{\nu_j}(r)}{r} \, Y_{lm}^*(\mathbf{p}_j) Y_{lm}(\mathbf{r}) \tag{5}$$

Here the signs $\pm$ correspond to outgoing (plus) and ingoing (minus) boundary conditions imposed on the asymptotic behaviour of the wave functions, $\delta_l(p)$ are the phase-shifts and the notation ν stands for a set of quantum numbers (p, l). The radial wave functions, $P_\nu(r)$, satisfy to the Schrödinger equation with the "frozen" core. We adopt that they are normalised in the rydberg's scale and this explains the factor in (5).

Vector $\mathbf{D}$ in (4) is the operator of the dipole interaction of the atomic electrons and the electromagnetic field, $\omega_{n0} = E_n - E_0$ is the energy of the atom's transition from the ground state 0 to the excited state n (including excitations into continuum), ω and $\mathbf{e}$ are the photon's energy and the polarization vector, respectively.

265

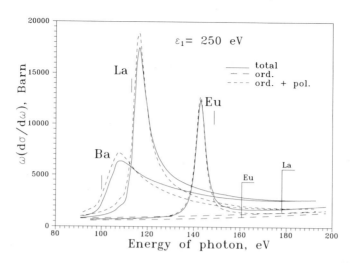

Figure 2: The BrS spectra calculated in DPWA with the RPAE corrections. Incident energy $\varepsilon_1 = 250$ eV.

Within the frame of the DPWA the spectral distribution, $\omega d\sigma/d\omega$, of the BrS can be written as the infinite sum of the partial cross sections which, in turn, contain the ordinary, polarizational and interference parts:

$$\omega\left(\frac{d\sigma}{d\omega}\right)^{tot}_{PW} = \sum_{l=0}^{\infty}\omega\left(\frac{d\sigma}{d\omega}\right)_{l} = \omega\left(\frac{d\sigma}{d\omega}\right)^{ord}_{PW} + \omega\left(\frac{d\sigma}{d\omega}\right)^{pol}_{PW} + \omega\left(\frac{d\sigma}{d\omega}\right)^{int}_{PW} \tag{6}$$

The explicit expressions for the terms from (1) one obtains using, eqs. (2) – (5) in (1). They are too cumbersome to be reproduced here, and may be found in (Amusia and Korol 1991).

In figure 2 we present results of the calculation of the BrS spectra for 250 eV electron on Ba, La and Eu in the vicinity of ionization potentials of the 4d-subshells (marked with vertical lines in the figure). The calculations have been done according to the algorithm described earlier in (Chernysheva *et al* 1989, Amusia *et al* 1990, Amusia and Korol 1991).

The choice of the targets and the regions of photon frequencies were determined by the fact (Amusia *et al* 1977) that a wide maximum in the emission spectrum can appear in the region of ω close to the ionization potentials of many-electron atomic subshells. It was demonstrated (Amusia 1982) that the position of the BrS maximum is closely related to the maximum in the photoionization spectrum, $\sigma_\gamma(\omega)$.

It is known from studying the photoionization of Ba, La and Eu both experimentally (Fomichev *et al* 1967, Becker *et al* 1986, Richter *et al* 1989, Kutzner *et al* 1990) and theoretically (Kelly 1987, Amusia *et al* 1989), that in vicinities of the 4d-thresholds there are powerful maxima in $\sigma_\gamma(\omega)$.

These maxima manifest themselves as well in the spectrum of the total BrS (solid curves in figure 2). In the maximal points the magnitudes of $(\omega d\sigma/d\omega)^{tot}$ greatly exceed the magnitudes of the ordinary BrS, - long-dashed curves in figure 2 (we did not plot the dependence $(\omega d\sigma/d\omega)^{ord}$ for Ba since it almost coincides with that for La).

The solid curves in figure 2 were calculated using various methods: (i) for Ba and La we used the RPAE with relaxation scheme (GRPAE) (Amusia 1990), (ii) for Eu the spin-polarized RPAE (SP RPAE) method (Amusia *et al* 1983) was applied.

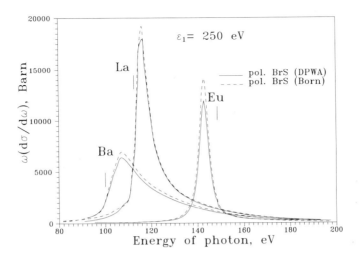

Figure 3: Spectra of the polarizational BrS calculated in the DPWA and the Born approximation as indicated. Incident energy $\varepsilon_1 = 250$ eV.

As it was noted in the cited papers these approximations take into account the major part of the correlation effects and, being applied to the description of the photoionization process, provide a satisfactory agreement with the experimental data. However, the calculation of the BrS spectrum undertaken within the frame of the DPWA and with the simultaneous accounting for the correlation effects in the RPAE (or/and its variations - GRPAE, SP RPAE) are rather laborious and computer time consuming.

It is thus desireable to construct an approximate simplified method which will allow to compute the BrS spectrum more efficiently but without significal loss of accuracy.

To start with we note, firstly, that in the vicinity of the maximum almost all the BrS radiation is formed via the polarizational mechanism. Secondly, the polarizational part of the BrS spectrum is more sensitive to the model of the atomic electron-photon interaction rather than to the choice of the projectile's wave function (see e.g. Amusia et al 1990, Amusia and Korol 1991). To illustrate this in figure 3 the dependences $(\omega d\sigma/d\omega)^{pol}$, calculated within the frame of the DPWA (solid curves) and the plane-wave Born approximation (dashed lines), are compared.

It is seen that for the incident energies as low as 250 eV the Born approximation gives almost the same result as the DPWA. The reason for this coincidence lies in the fact that, in contra to the process of ordinary BrS, the polarizational radiation is formed mainly at large distances, $r \sim p_1/\omega$, between a projectile and an atom (Zon 1979, Amusia *et al* 1982), where the distortive influence of the atomic potential on the projectile's movement is comparatively small. Hence, to calculate the polarizational component of the BrS specrum one may use the Born approximation which results in the formula (Amusia it et al 1977)

$$\left(\omega \frac{d\sigma}{d\omega}\right)^{pol}_B = \frac{16}{3} \frac{\omega^4}{c^3 p_1^2} \int_{p_1-p_2}^{p_1+p_2} \frac{dq}{q} |\alpha(\omega,q)|^2 \qquad (7)$$

Here $\mathbf{q} = \mathbf{p}_1 - \mathbf{p}_2$ is the transferred momentum and $\alpha(\omega,q)$ is the generalized atomic polarizability which reduces to the dynamic dipole polarizability in the limit $q \longrightarrow 0$.

The only characteristics in (7), dependent on the internal dynamics of the target, is the generalized polarizability. Recently (Korol *et al* 1995a) we introduced a simple

267

approximate method for the calculation of $\alpha(\omega, q)$ and, consequently, of $(\omega d\sigma/d\omega)_B^{pol}$. This method allows to avoid rather complicated direct numerical computations of the many-electron correlation effects.

To obtain the magnitude of the generalized polarizability with accounting for the many-electron correlation effects, we first write $\alpha(\omega, q)$ as:

$$\alpha(\omega, q) = \alpha(\omega) \cdot G(\omega, q) \tag{8}$$

This equality is just a definition of a new function $G(\omega, q)$ equal, as it is seen, to the ratio $\alpha(\omega, q)/\alpha(\omega)$ of the *exact* generalized and dipole polarizabilities.

Now let us assume that all the information about the many-electron correlation effects is contained in the dipole polarizability $\alpha(\omega)$, while the factor $G(\omega, q)$ is not that sensitive to the correlation effects and can be calculated in the *Hartree-Fock* approximation. Hence, instead of (8) we obtain the following approximate formula

$$\alpha(\omega, q) \approx \alpha(\omega) \, \frac{\alpha^{HF}(\omega, q)}{\alpha^{HF}(\omega)} \equiv \alpha(\omega) \cdot G^{HF}(\omega, q) \tag{9}$$

This relation is the essential point of the method. Provided the approximate equality in (9) is true, it allows to avoid the complicated direct calculations of the generalized polarizability by means of the many-body theory and reduces the problem to a simpler one: the calculation of the factor $G^{HF}(\omega, q)$ in the Hartree-Fock approximation.

Substituting (9) into (7) one gets:

$$\left(\omega \frac{d\sigma}{d\omega}\right)_B^{pol} = \frac{16}{3} \frac{\omega^4}{c^3 p_1^2} |\alpha(\omega)|^2 \int_{p_1-p_2}^{p_1+p_2} \frac{dq}{q} \, |G^{HF}(\omega, q)|^2 \tag{10}$$

To check the validity of the proposed method we calculated the polarizational part of the spectrum for 0.25 — 10 keV electron on Ba, La and Eu using the exact Born formula (7) and the approximate one (10). In all considered cases the results are close to that presented in figure 4. The discrepancy between the solid curve, corresponding to (7), and the short-dashed one, representing (10), is almost negligible. For the sake of comparison we present also the polarizational cross section calculated via (7) with $\alpha(\omega, q)$ obtained in the independent-particle model with no correlations included (a long-dashed curve).

In figure 4 the short-dashed curve was obtained by using the relation (9) where the factor $\alpha(\omega, q)$ was calculated directly with the GRPAE corrections taken into account.

Another possibility to get $\alpha(\omega, q)$ is to use the relationship between $\mathrm{Im}\,\alpha(\omega, q)$ and the photoabsorption cross section, which is

$$\mathrm{Im}\,\alpha(\omega, q) = \frac{c}{4\pi\omega} \sigma_\gamma(\omega) \tag{11}$$

Provided the dependence $\sigma_\gamma(\omega)$ is known in a wide ω-region one can restore the real part of $\alpha(\omega, q)$ via the dispersion relation. Such a way of obtaining $\alpha(\omega, q)$ is especially useful when the direct calculations calculations can hardly be performed (for example, when the BrS process is investigated in a dense media rather that in a pure "one electron — one atom" collision).

In figure 5 we compare the calculated dependences with the experimentally measured (Zimkina *et al* 1984) BrS cross section formed in the collision of 500 eV electrons with La. Theoretical curves, 1, 1' and 2, represent the ω-dependence of $\omega(d\sigma/d\omega)^{pol}$ obtained within the frame of the Born approximation.

Curve 1 was calculated using formulae (7) where the GRPAE was used to get the generalized polarizability. To compute the dependence 1' we used the approximation

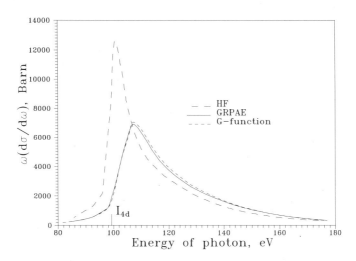

Figure 4: Spectra of the polarizational BrS on Ba calculated in the Born approximation with different methods used for the computation of $\alpha(\omega, q)$. Incident energy $\varepsilon_1 = 250$ eV.

described above by (9) and (10). The factor $G^{HF}(\omega, q)$ was obtained in the Hartree-Fock approximation whereas the dipole polarizability $\alpha(\omega)$ - in the GRPAE. The comparison of 1 and 1' illustrates that the approximation (9) works in this case as well.

We note here that neither 1 nor 1' can be used for the direct comparison with the experimentally measured BrS spectrum (Zimkina et al 1984). The reason is as follows. Curves 1 and 1' were calculated with the correlation effects taken into account within the frame of GRPAE. This approximation, being applied to the computation of the photoionization spectrum of La (see e.g. Amusia 1990), noticably overestimates the magnitude of $\sigma_\gamma(\omega)$, measured experimentally (Zimkina and Gribovskii 1971), where thin La film was used. Since the cross section of the polarizational BrS is (roughly) proportional to $|\alpha(\omega)|^2 \sim Im^2 \alpha(\omega) \sim \sigma_\gamma^2(\omega)$, then one may expect that the discrepancy between experimental and theoretical data for $\omega(d\sigma/d\omega)^{pol}$ will be even greater than in the case of photoionization.

To avoid it we used the experimental data for the photoionization spectrum (Zimkina and Gribovskii 1971) to calculate the dynamic dipole polarizability, which then was substituted into (10). The factor $G^{HF}(\omega, q)$ was obtained in the Hartree-Fock approximation. The results of such a calculation is shown by curve 2 in the figure. In (Zimkina et al 1984) the emission spectrum produced by the 500 eV electrons scattered in the metallic La was registered. Experimental data has no absolute gauge, therefore, we normalized the measured spectrum to the magnitude of theoretical curve 2 in the maximum point. We also substracted the background radiation, i.e. the ordinary BrS, from the measured spectrum. Therefore, the experimental curve 3 in figure 5 represents by itself a sum of the polarizational and the interference terms.

The agreement between curves 2 and 3 is quite good. The main discrepancy is seen on the right wing of the spectrum where the experimental curve lies higher than a calculated one. This discrepancy may be attributed to the contribution of the interference term which was omitted when calculating curve 2.

It looks attractive to modify the formalism in a way that the approximate method (10) of the calculation of the generalized polarizability could be applied to the computation of the total BrS spectrum rather than its polarizational part only. One may do

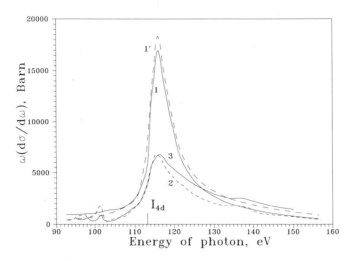

Figure 5: The polarizational BrS spectrum of a 500 eV electron on La. Theoretical curves, - 1, 1' and 2, - are calculated in the Born approximation. The functions $\alpha(\omega)$ and $G(\omega, q)$ were obtained using different methods (see explanations in the text). Curve 3 - experiment (Zimkina et al 1984).

it as follows.

Supposing the incident energy ε_1 is large enough compared with the ionization potential of those atomic subshells which, for a given photon frequency, give the main contribution to the sum over the excited states in (4), it is possible to neglect the exchange between the projectile and the atomic electrons. Then, one may express the amplitude of the polarizational BrS through a generalized atomic polarizability (Korol et al 1995b) as follows:

$$f_{\text{pol}} = -\frac{\text{i}}{2\pi^2} \int d\mathbf{Q} \, \frac{(\mathbf{eQ})}{Q^2} < \mathbf{p}_2 | e^{-\text{i}\mathbf{Qr}} | \mathbf{p}_1 > \cdot \alpha(\omega, Q) \qquad (12)$$

Substituting here the distorted waves $|\mathbf{p}_{1,2}>$ with the wave functions of a free movement, $|\tilde{\mathbf{p}}_{1,2}>= \exp(i\mathbf{p}_{1,2}\mathbf{r})$, and taking into account the relation

$$< \tilde{\mathbf{p}}_2 | \exp(-\text{i}\mathbf{Qr}) | \tilde{\mathbf{p}}_1 >= (2\pi)^3 \delta(\mathbf{q} - \mathbf{Q}),$$

one gets the expression for f_{pol} within the frame of the plane-wave Born approximation (Amusia et al 1977):

$$f_{\text{pol}}^{\text{B}} = -4\pi\text{i} \, \frac{(\mathbf{eq})}{q^2} \cdot \alpha(\omega, q) \qquad (13)$$

To obtain the total BrS cross section within the DPWA frame, one may, in principle, use (12) and (2) in (1). However, because of the computational reasons given below, we found it more convenient to use not the "ordinary + polarizational + interference" representation of the cross sections as in (6), but an alternative one.

It was mentioned above that the polarizational BrS is formed mainly at the large distances between a projectile and a target. Thus, the specific angular momenta in the polarizational BrS process may be estimated as $l_{\text{pol}}^{\text{eff}} \sim p_1^2/\omega$. It follows then that for sufficiently high projectile velocities $l_{\text{pol}}^{\text{eff}} \gg 1$. ¿From the computational viewpoint this inequality means that the sum over l in (6) in the case of the polarizational BrS converges rather slowly, and one has to calculate a large number of partial terms to obtain an accurate result.

To avoid this technical difficulty we take an advantage of the fact that the projectile's radial wave functions of large orbital momenta are close to those of a free movement. Hence, we may re-write expression for f_{pol}, explicitly extracting the Born amplitude (13) from (12). Then, the remainder, $\Delta f_{\text{pol}} = f_{\text{pol}} - f_{\text{pol}}^B$, is represented by a rapidly convergent sum over the orbital momenta.

Thus, instead of (2), let us use the following formula for the total BrS amplitude f_{BrS}:

$$f_{\text{BrS}} = f_{\text{ord}} + f_{\text{pol}}^B + \Delta f_{\text{pol}} \tag{14}$$

For the spectral distribution of radiation one obtains, after some algebra, the expression (Korol $et\ al$ 1995b):

$$\omega \frac{d\sigma}{d\omega} = \frac{16}{3} \frac{\omega^4}{c^3 p_1^2} \int_{p_1-p_2}^{p_1+p_2} \frac{dq}{q} |\alpha(\omega,q)|^2 + \frac{32\pi^2}{3} \frac{\omega^4}{c^3 p_1^2} \sum_{l_1,l_2} l_> |R_{l_2 l_1}^{\text{ord}} - \Delta R_{l_2 l_1}^{\text{pol}}|^2$$

$$+ \frac{32\pi}{3\sqrt{p_1 p_2}} \frac{\omega^4}{c^3 p_1^2} \text{Re} \int_{p_1-p_2}^{p_1+p_2} \frac{dq}{q} \alpha^*(\omega,q) \sum_l (l+1)$$

$$\times \left\{ \exp[i(\delta_l(p_1) + \delta_{l+1}(p_2))] \cdot (R_{l+1 l}^{\text{ord}} - \Delta R_{l+1 l}^{\text{pol}}) \cdot g_l(p_1,p_2,q) \right.$$

$$\left. + \exp[i(\delta_{l+1}(p_1) + \delta_l(p_2))] \cdot (R_{l l+1}^{\text{ord}} - \Delta R_{l l+1}^{\text{pol}}) \cdot g_l(p_2,p_1,q) \right\} \tag{15}$$

Here the partial amplitudes of the ordinary, $R_{l_2 l_1}^{\text{ord}}$, and the polarizational, $R_{l_2 l_1}^{\text{pol}}$, BrS are expressed via the following integrals:

$$R_{l_2 l_1}^{\text{ord}} = \ < \nu_2 || r || \nu_1 > \tag{16}$$

$$R_{l_2 l_1}^{\text{pol}} = \frac{2}{\pi} \int_0^\infty dQ \cdot Q \cdot < \nu_2 || j_1(Qr) || \nu_1 > \cdot \alpha(\omega,Q) \tag{17}$$

where $j_1(Qr)$ is the spherical Bessel function.

The notation, $\Delta R_{l_2 l_1}^{\text{pol}}$, stands for the difference between the polarizational partial amplitudes calculated in the DPWA and in the Born approximation. It reads as:

$$\Delta R_{l_2 l_1}^{\text{pol}} = R_{l_2 l_1}^{\text{pol}} - e^{-i(\delta_{l_1}(p_1)-\delta_{l_2}(p_2))} \cdot \frac{2}{\pi} \int_0^\infty Q dQ \ < \tilde{\nu}_2 || j_1(Qr) || \tilde{\nu}_1 > \cdot \alpha(\omega,Q) \tag{18}$$

The matrix element in the integrand in (18) is calculated between the radial wave functions of a free particle: $||\tilde{\nu}> = (p/\pi)\, r \cdot j_l(pr)$.

Other notations used in (15) are:

$$g_l(p_1,p_2,q) = p_1 P_{l+1}(y) - p_2 P_l(y)$$

with $P_l(y)$ being the Legendre polynomials dependent on the variable y which is expressed via the transferred momentum q as follows:

$$y \equiv \frac{\mathbf{p_1 p_2}}{p_1 p_2} = \frac{p_1^2 + p_2^2 - q^2}{2 p_1 p_2}.$$

It is seen that the first term in (15) is exactly the cross section of the polarizational BrS calculated in the Born approximation (7). The ordinary BrS cross section is "hidden" in the second term in (15). Other terms in this formula represent the interference part of the cross section and the DPWA corrections to the polarizational component.

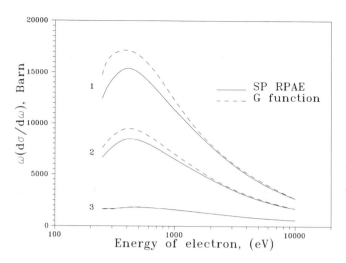

Figure 6: The total BrS spectrum on Eu as a function of ε_1. Curves 1, 2 and 3 correspond to the photon frequencies $\omega_1 = 142.9$ eV, $\omega_2 = 145$ eV and $\omega_3 = 163$ eV, respectively (see explanations in the text).

Although cumbersome, expression (15) allows to calculate the BrS cross section very efficiently and has two advantages as compared with the standart DPWA expansion (6).

Firstly, since the partial amplitudes of the ordinary BrS, (16), and the quantities $\Delta R_{l_2 l_1}^{\text{pol}}$ defined in (18), decrease rapidly with l_1, then the number of partial terms needed for the accurate computation of both sums in (15) is not large. For example, when calculating the BrS cross section formed in electron-Eu collision (figure 6), the account for the first 16 terms, was enough to compute these sums with the relative accuracy of 10^{-4} for the incident energy $\varepsilon_1 = 10$ keV .

The second advantage is that the only characteristics in (15), dependent on the internal dynamics of the target, is the generalized polarizability. It allows to apply the approximate method, discussed above, to the calculation of the total BrS spectrum.

To conclude this section, we present results of the BrS calculations for the electron-Eu collision. The curves in figure 6 show the dependences of the total BrS cross section on the incident electron energy ε_1 calculated for three photon energies, which are (i) $\omega_1 = 142.9$ eV, - the maximum point of the spectrum (see figure 2), (ii) $\omega_2 = 145$ eV, - the ω-point on the right slope of the peak, (iii) $\omega_3 = 163$ eV, - i.e. the ω-point far beyond the maximum.

All the curves were calculated using eq. (15). The generalized polarizability of Eu was obtained in the SP RPAE (solid lines) and using the approximation (9), - dashed lines.

Comparing the sets of the solid and the dashed curves we may state that the approximation (9) combined with the DPWA formula (15) could be effectively used for the calculations of the total BrS spectrum in the whole ω-interval.

PHOTON EMISSION IN ELECTRON-CLUSTER COLLISIONS

In this section we discuss the manifestation of the polarizational BrS mechanism in collisions of an electron with metallic clusters. The particular attention is paid to the photon emission associated with the collective motion of the valence electrons in clusters. This problem has been considered by Connerade and Solov'yov (1995).

The valence electrons play the essential role in metallic clusters defining their global properties. Indeed, with the discovery of electronic shell structure in free alkali clusters by Knight et al (1984, 1985) the main emphasis in the description of this type of clusters has been placed on the quantised motion of the delocalised valence electrons in the mean field created by the ions. This behavior suggests a jellium model (Ekardt 1984, 1985a,1985b, Ivanov et al 1994), which is defined by a Hamiltonian that treats the electrons as usual but the ionic cores as a uniform positively charged background, since detailed ionic structure often does not seem to affect greatly much the properties of alkali and other simple metal clusters. This naturally leads to a description of the electron density in terms of single particle wave functions that extend over the entire cluster. In this way, using relatively simple-minded approaches to the many-body problem, a wealth of experimental data can be classified and often theoretically reproduced at least semiquantitatively (see Brack 1993, de Heer 1993, Bréchigniac and Connerade 1994 for review).

The jellium model makes the description of metal clusters quite similar to the description of atoms or nuclei. Indeed, the equations of the jellium model are analogous to the Hartree-Fock equations for atoms in which the Coulomb potential of the nucleus is replaced by the mean field potential of the ionic background. In our case it means that we can apply, with minor modifications, the approximations and methods developed for atom to describe the polarizational BrS process in collisions with clusters involved.

To prove these statements let us consider the amplitude of the photon emission in collision of an electron with a cluster. In this consideration we follow the treatment performed by Connerade and Solov'yov (1995).

The amplitude f, containing the contributions of the ordinary and the polarizational radiative mechanisms, is equal to

$$f = < \Psi_f |\mathbf{er}| \Psi_i > + \int \frac{d^3q}{(2\pi)^3} \frac{4\pi}{q^2} < \Psi_f | \exp(i\mathbf{qr}) | \Psi_i > \mathbf{eq} \cdot \alpha(\omega; \mathbf{q}) \qquad (19)$$

Here $\alpha(\omega; q)$ is the generalized dynamical polarizability of the cluster; $\omega = \varepsilon_i - \varepsilon_f$ and $\mathbf{e}$ is the frequency and the polarization vector of the emitted photon, $|\Psi_i >$; ε_i and $|\Psi_f >$; ε_f are the initial and the final wave function and energy of the electron. Integration in (19) is performed over the transferred momenta q.

Comparison of (19) with (2-4) and (12) shows that the general analytical form of the amplitude in case of the atomic and the cluster target is the same. The difference is that now, when considering $\alpha(\omega; q)$, we use the wave functions of electrons, involved in a collective motion, calculated within the framework of the spherical jellium model (Ekardt 1985, Ivanov et al 1993, 1994). The wave function of the projectile electron can be described either by a plane wave (Born approximation) or expressed in a form of the partial wave expansion (DPWA), where the partial waves are calculated in the frozen field of the cluster.

The final state of the electron can belong either to the continuous or the discrete spectrum. These two options correspond to the bremsstrahlung and the radiative electron capture processes. In the latter case the wave function of the electron is a solution of the jelluim model for the cluster with $N + 1$ valence electrons, describing either a neutral cluster or a positive or negative cluster ion depending on the initial total charge of the sytem.

Having in mind the analogy between the amplitudes of the polarizational bremsstrahlung on atomic and cluster targets we deduce that the analytical form of the cross sections of the BrS process must be the same in the both cases. Therefore the formalism discussed in the previous section can be directly addressed to the case of the cluster target, if one asumes the wave functions of the valence electrons in a cluster instead of the wave functions of atomic electrons.

We shall not reproduce the formulae of the bremsstrahlung theory presented in the previous section for cluster. But we want to discuss the radiative electron capture process by a cluster in more detail. In this case the final state of the electron belongs to the discrete spectrum and is characterized by the principal quantum number-n and the angular momentum-L. In the dipole approximation this state can only be coupled with $L \overset{+}{-} 1$-partial waves of the projectile electron, when calculating the amplitude and the cross section of the process. For the S final state ($L = 0$) the non-zero contribution to the cross section arises only from the P wave. In this case the cross section of the radiative electron capture has the most simple form (Connerade and Solov'yov 1995):

$$\sigma = \frac{4\omega^3}{3p^3c^3}| < n_f; 0|r|p; 1 > + \frac{2}{\pi} \int_0^\infty dq \cdot q \cdot \alpha(\omega; q) < n_f; 0|j_1(qr)|p; 1 > |^2 \qquad (20)$$

Here $j_1(qr)$ is the radial Bessel function, c is the velocity of light, p is the initial momentum of the electron. The first term in (20) is the contribution of the ordinary mechanism and the second one arises due to accounting for the polarization of the cluster.

Let us compare the contributions of two mechanisms to the cross section of the process, considering the ratio R of the total cross section (20) to the contribution of the ordinary mechanism. The ratio R practically does not depend on the choice of the final state of the electron. Indeed, the main contribution to the integral in (20) arises from the dipole term, being the linear term in the expansion of the Bessel function $j_1(qr) \approx (qr)/3$. In this case the matrix element $< n_f; 0|r|p; 1 >$ disappears from the ratio R, which becomes equal to

$$R = |1 + \frac{2}{3\pi} \int_0^\infty dq \cdot q^2 \cdot \alpha(\omega, q)|^2 \qquad (21)$$

Figure 7 represents R for the clusters Na^{20+} and Ag^{11+}. This figure demonstrates that the polarizational mechanism strongly dominates in the region of the giant resonance. In the higher frequency region R has a minimum, arising due to the interference of the ordinary and the polarizational mechanisms. The interference makes the profile of the giant resonance asymmetric compared with the profile of the resonance arising in the photoabsorption cross section. Such an assymetry is typical for the Fano profiles well known in atomic physics. The ordinary mechanism becomes dominant with the increase of energy of the projectile electron as it is seen from figure 7.

The order of magnitude of R is practically independent on the cluster size. Indeed, the size dependence of the polarizability can be estimated as R_0^3, where R_0 is the size of the cluster. The integration region over q giving the main contribution to the integral in (21) is determined by the equation $q \leq R_0^{-1}$. Therefore the result of the integration in (21) is almost all independent on the cluster size.

The ratio R can be easily estimated in the region of the giant resonance, where the imaginary part of the polarizability dominates over the real part. Using the relationship (11), one derives from (21)

$$R \sim \frac{\sigma_\gamma^2(\omega)c^2}{36\pi^4\omega^2 R_f^6} \qquad (22)$$

Substituting typical resonance values for the photoabsorption cross sections of sodium and silver cluster ions (see e.g. Bréchignac and Connerade 1994), we come to the

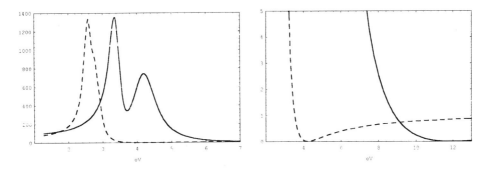

Figure 7: Ratio R for the clusters Na^{20+} (dashed line) and Ag^{11+} (solid line) in various frequency ranges. On the left side of the figure the ratio R in the Na^{20+} case is multiplied by 10.

conclusion that the ratio R reaches the values $10^2 - 10^3$. It means that the polarizational mechanism dominates in the cross section at least by two or three orders of magnitude in the region of giant resonance.

The cross section (20) as well as the cross section of the polarizational bremsstrahlung is expressed via the generalized polarizability of the cluster. The dynamical polarizabilty of cluster in the region of giant resonance can be calculated with the reasonable accuracy only when accounting for the many-body correlation effects. This can be done, for example, within the framework of the random phase approximation developed for clusters by many authors (see e.g. Brack 1993, Ivanov et al 1994).

In the present work we, however, apply the alternative method discussed in the previous section. This method was applied (Korol et al 1995) for the treatment of the bremsstrahlung process in the vicinity of giant atomic resonances. The basic idea of the method is very simple and is explained by the relationship (9). We assume again that $\alpha(\omega)$ carries all the information about the many-electron correlations in the cluster. The function $G(\omega, q)$ describes the dependence of the polarizability on the transferred momentum. This function practicaly does not depend on many-electron correlations and frequency ω. It can be calculated within the framework of the jellium model. The factor $G(\omega, q)$ is approximately equal to $G(\omega, q) \approx W(q)/N$ in the regions $\omega \gg \omega_0$ and $q \ll 1$, where $W(q)$ and N are the form-factor of the cluster and the number of valence electrons respectively; ω_0 is the characteristic frequency of the giant resonance. The numerical calculations performed demonstrate that the form-factor approximation for the function $G(\omega; q)$ turns out to be quite reliable in the whole reqion of ω and q.

We have used the form-factor approximation for the function $G(\omega; q)$ when calculation the generalized polarizability of the cluster. The dynamical dipole polarizabilty $\alpha(\omega)$ has been derived by its imaginary part applying the dispersion relation. The imaginary part of the polarizability has been obtained from the experimental data on photoabsorption (Bréchignac and Connerade 1994) using the relationship (11). This scheme of calculations is rather reliable and provides a good agreement with the RPAE method (Korol et al 1995).

Using the generalized polarizability, the spectra of the polarizational bremsstrahlung

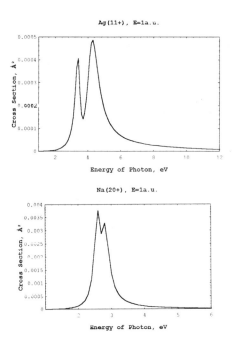

Figure 8: Polarizational BrS cross sections in collison electron with Na^{20+} and Ag^{11+} clusters. The initial energy of the electron is equal to 1 a.u.

in collision of electron with Na^{20+} and Ag^{11+} clusters have been calculated by Conner-ade and Solov'yov (1995) in the Born approximation. These results are plotted in figure 8.

This figure demonstrates that the collective giant resonances appear in the emission spectra of clusters. The shape of these resonances is similar to the profiles arising in the corresponding photoabsorption cross sections. However, the BrS and the photoab-sorption profiles have some difference. The right hand wing of the resonances shown in figure 8 is enhanced compared to the corresponding photoabsotption profiles. This enhancement arises from the dominating real part of the dynamic polarizability of the cluster, which does not contribute to the photoabsorption cross sections. We have omit-ted the ordinary BrS contribution in figure 8. This contribution is small compared with the polarizational BrS one, but nevertheless simultaneous accounting for both mecha-nisms brings some additional asymmetry to the shape of the giant resonance arising in the BrS spectrum.

The absolute value of the polarizational BrS cross sections for clusters is rather large compared with that for atoms. There are two reasons for the large value of the cross sections. First, the number of valence electrons involved in the collective motion in a cluster -N_e can be much larger then in an atom. These electons radiate coherently in the polarizational BrS process (Amusia et al 1985) and thus the cross section of the process gains a large enhancement factor- N_e^2. The second reson of the large cross sections consists in the possibility for the process to take place at relatively small collision velocity- v. Indeed, kinematically the process of the polarizational BrS occurs with a large probability if the following condition is fulfilled: $\omega R_0/v \leq 1$. Substituting the typical values $\omega \sim \omega_0 \sim 0.1$, $R_0 \sim 20$, we deduce that this condition is fulfilled, if $v \geq 2$ or $\varepsilon_1 \geq 54.4eV$. The cross section of the polarizational BrS is inversely proportional to v^2 and therefore it is larger at smaller velocities of the collision.

SUMMARY

We deduce from our calculations (i) that the polarisation mechanism dominates in the radiative spectra formed in collisions of electrons with atoms and clusters in the region of giant resonances, (ii) that an interference effect due to coupling between direct bremsstrahlung emission and polarisational radiation results in an asymmetry of the giant resonance profile as observed in emission even when the corresponding photoabsorption profile is nearly symmetrical, (iii) that polarizational BrS effect is much stronger in collisions of electrons with clusters then with atoms, (iv) that the polarizational mechanism manifest itself in the radiative electron capture process by clusters, (v) that a characteristic spectral signature of the emitted light should be searched for to indicate whether the polarization radiation process is occurring.

Finally, we want to note that the polarizational BrS problem is much broader then considered in this paper, because this kind of radiation can be emitted in any collision involed structured particles- nuclei, atoms, molecules, clusters. The number of various colliding pairs, different interaction forces between particles, kinematical conditions and the frequency ranges make this problem quite spacious and interesting.

ACKNOWLEDGMENTS

This work was supported by the grant 076 of the International Science and Technology Center and the grant JHJ 100 of the International Science Foundation.

REFERENCES

Amusia M Ya, Baltenkov A S and Gilerson V B 1977 *Pis'ma v Zh. Tehn. Fiz.* **3** 1105 (in Russian)

Amusia M Ya, Zimkina T M and Kuchiev M Yu 1982 *Zh. Tehn. Fiz.* **52** 1424 (*Sov. Phys.–Techn. Phys.* **27** 866)

Amusia M Ya, Dolmatov V K and Ivanov V K 1983 *Zh. Eksp. Teor. Fiz.* **85** 115 [*Sov. Phys. JETP* 1983 **58** 67]

M Ya Amusia, M Yu Kuchiev and A V Solov'yov 1985a *Sov.Phys.JETP* **89** 1512

M Ya Amusia, M Yu Kuchiev and A V Solov'yov 1985b *Pis'ma v Zh.Tech.Fiz.* **11** 1401

Amusia M Ya, Ivanov V K and Kupchenko V A 1989 *Z. Phys. D: At. Mol. Cl* **14** 219

Amusia M Ya 1990 *Atomic photoeffect* ed. P G Burke and H Kleinpoppen (New York: Plenum)

Amusia M Ya, Chernysheva L V and Korol A V 1990 *J. Phys. B: At. Mol. Opt. Phys.* **23** 2899

Amusia M Ya and Korol A V 1991 *J. Phys. B: At. Mol. Opt. Phys.* **24** 3251

Amusia M Ya and Korol A V 1992 *J.Phys.B: At.Mol.Opt.Phys.* **25** 2383.

Amusia M Ya and Korol A V 1994 *Phys. Lett.* **A**

Avdonina N B, Amusia M Ya, Kuchiev M Yu and Chernysheva L V 1986 *Zh. Tekn. Fiz.* **56** 246 (in Russian)

Becker U, Kerkhoff H G, Lindle D M, Kobrin I H, Ferret T A, Heimann P A, Truesdale C M and Shirley D A 1986 *Phys. Rev.* **A34** 2585

Berestetskii V B, Lifshitz E M and Pitaevskii L P 1980 *Quantum Electrodynamics* (Oxford: Pergamon)

Brack M 1993 *Rev.Mod.Phys.* **65** 677.

Bréchignac C and Connerade J P 1994 *J.Phys.B:At.Mol.Opt.Phys.* **27** 3795.

Bréchignac C, Cahuzac Ph, Carlier F and Leygnier J 1989 *Chemical Physics Lett.* **164** 433.

Chernysheva L V, Amusia M Ya, Avdonina N B and Korol A V 1989 *Preprint-1314* Ioffe Physical-Technical Institute, Leningrad

Connerade J P and Solov'yov A V 1995, XIX International Conference on the Physics of Electronic and Atomic Collisions, Abstracts, Whistler, Canada; V European Conference on Atomic and Molecular Physics, Contributed Papers, Part II, Edited by R.C.Tompson, Edinburgh, United Kingdom; XI International Conference on the Physics of Vacuum Ultraviolet, Abstracts, Tokyo; submitted to J.Phys.B:At.Mol.Opt.Phys.

de Heer W A 1993 *Rev.Mod.Phys.* **65** 611.

Ekardt W 1984 *Phys.Rev.Lett.* **52** 1925.

Ekardt W 1985a *Phys.Rev.B* **31** 1558.

Ekardt W 1985b *Phys.Rev.B* **32** 1961.

Fomichev A V, Zimkina T M, Gribovskii S A and Zhukova I I 1967 *Fiz. Tverd. Tela* **9** 1490 [*Sov. Phys. Solid State* 1967 **9** 1163] Ishii K and S Morita 1985 *Phys. Rev. A* **31** 1168.

Ivanov V K, Ipatov A N, Kharchenko V A, Zhizhin M L 1993 *Pis'ma JETPh* (in Russian) **58** 649

Ivanov V K, Ipatov A N, Kharchenko V A, Zhizhin M L 1994 *Phys.Rev.* **A50** 1459.

Kelly H P 1987 *Physica Scripta* **17** 109

Knight W D, Clemenger K, de Heer W A, Saunders W A, Chou M Y and Cohen M L 1984 *Phys.Rev.Lett.* **52** 2141.

Knight W D, Clemenger K, de Heer W A, Saunders W A 1985 *Phys.Rev.B* **31** 2539.

Korol A V, Lyalin A G, Shulakov A S and Solov'yov A V 1995a *J.Phys.B: At. Mol. Opt. Phys.* **28** L155.

Korol A V, Lyalin A G, Shulakov A S and Solov'yov A V 1995b submitted to *J.Phys.B: At. Mol. Opt. Phys.*.

Kutzner M, Zikri Altun, Kelly H P 1990 *Phys. Rev. A* **41** 3612

Richter M, Meyer M, Pahler M, Prescher T, Raven S V, Sonntag B, and Wetzel H E 1989 *Phys. Rev. A* **40** 7007

Selby K, Kresin V, Masui J, Vollmer M, de Heer W A, Scheidemann A and Knight W D 1991 *Phys.Rev.* **B43** 4565.

Solov'yov A V 1992 *Z.Phys.D: Atoms, Molecules and Clusters* **24** 5.

Tsitovich V N, edited, 1994, Polarization Radiation, Plenum.

Verkhovtseva E T, Gnatchenko E V, Zon B A, Nekipelov A A and Tkachenko A A 1990 *Zh.Eksp.Teor.Fiz.* **98** 797 (in Russian)

Zimkina T M, Shulakov A S and Brajko A P 1981 *Fiz.Tverd.Tela,* **23** 2006 (in Russian)

Zimkina T M, Shulakov A S, Brajko A P, Stepanov A P and Fomichev V A 1984, *Fiz.Tverd.Tela,* **26** 1984 (in Russian)

Zimkina T M and S A Gribovskii 1971 *J. de Physique,* Coll C-4, Sup.10 **32** C4-282.

Zon B A 1979 *Z. Exp. Teor. Phys.* **77** 44 (in Russian)

ION-ATOM COLLISIONS INVOLVING LASER-PREPARED Na ATOM TARGETS

Z. Roller-Lutz, Y. Wang, S. Bradenbrink,
H. Reihl, T.Wörmann, C. Sprengel, and H.O. Lutz

Fakultät für Physik
Universität Bielefeld
33501 Bielefeld
Germany

1. Introduction

Experimental developments, in particular improved laser-optical pumping techniques, have stimulated collision studies with electronically excited states. Such systems can be adjusted with great flexibility to the special requirements of the investigated phenomena. For example, laser-prepared collision partners can carry initial alignment and orientation and thus illuminate the role of "shape" and "rotation" of the initial atomic state.

Mainly due to its easy handling, as well as to the availability of the required laser radiation, collision studies with a Na target are at present being found at the center of attention. In particular, charge exchange in H^+ + Na collisions around 1 keV [1-13] is serving as a test case in which a quite complete understanding is reached on the basis of large-scale ab initio computer calculations.

Several aspects of the collision dynamics have been studied: total [1-6] and angle-differential cross sections [7-9], alignment effects [8,10,11], and the so-called "left-right asymmetry" [8,12,13] which is related to collisionally accumulated phases of the involved electron states. This system, also being of some technical relevance, provides a convenient model case for the study of electron dynamics in ion-atom interactions; similarly to other few-electron systems [14] it may even provide a ground for the most severe ("quantum-mechanically complete") tests of our understanding of the collision. For such an experiment, identification of the final channel (e.g., H(2p)) is essential since knowledge about *all* states is required.

In our work, the processes under investigation are

$$H^+ + Na(3s) \rightarrow H(2p) + Na^+ \tag{1}$$

(endothermic by 1.74 eV), and

$$H^+ + Na(3p) \rightarrow H(2p) + Na^+ \tag{2}$$

(exothermic by 0.37 eV).

The collision energy ranges from 1 to 5 keV, i.e., $V = v/v_e$ lies between approximately 0.3 and 0.7 (v is the collision velocity, v_e the classical orbiting velocity of the active electron); thus, the system addresses the region between the purely "quasimolecular processes" at $V \ll 1$ and the "fast collisions" at $V \gg 1$. It is well known that this regime still offers quite demanding problems for charge exchange theory.

2. Experimental method

The experiment is based on the coincident detection of scattered H particles and emitted Lyman-α radiation; on the basis of its dipole nature, the latter characterizes the final H(2p) state. Photon as well as scattered H atom detection is resolved according to polar $(\vartheta_\gamma, \vartheta_s)$ and azimuthal $(\varphi_\gamma, \varphi_s)$ angles; the coincidence requirement thus identifies the scattering plane and the photon direction for each interaction event.

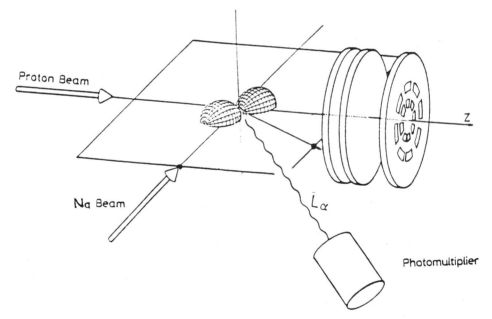

Figure 1. Experimental arrangement (schematic).

The experimental set-up is shown in figure 1. A proton beam from a Colutron ion source is accelerated to an energy in the range of 1 to 5 keV. After mass-selection in a Wien filter and collimation to better than 0.03^o in a differentially pumped drift tube, the ion beam is injected into the target chamber. The background pressure is 1 x 10^{-7} mbar. The target chamber is free of electric fields to avoid quenching of collision-produced H(2s), and the earth's magnetic field is compensated by Helmholtz coils to values smaller than 50 mG. About 25 mm before the interaction region, the ion beam is cleaned from neutrals with 3 pairs of deflector plates. In the interaction region the ion beam is crossed at right angles by a Na-atom beam. The sodium beam is produced in a two-stage oven which reduces the dimer content to less than 1 %. After skimming it is collimated to a divergence below 0.2^o. The Na-atom density in the collision area is 10^9–10^{10} cm^{-3}; it is kept constant by controlling the oven temperature. Lyman-α radiation (121.6 nm) emitted from capture-produced H(2p) is detected in a solar-blind photomultiplier. It has a MgF$_2$ window, providing a sensitivity region between approximately 115 and 200 nm; its specified quantum efficiency for Lyman-α radiation is about 15 %. It is placed at approximately the "magic angle" ($\vartheta_\gamma = 54.7^o$ relative to the ion beam axis); situated in the plane defined by the atomic and the ion beam, it views a length of ion beam around the interaction region of about 6 mm, thereby subtending a solid angle of approximately 0.3 sr. The Lyman-α photons are measured in coincidence with neutralized projectiles which are scattered through defined angles into a position-sensitive detector. This detector consists of a tandem micro-channel plate and a divided-anode array which is composed of 16 individual electrodes covering 4 azimuthal angles $\varphi_s(0, \pi/2, \pi, 3\pi/2)$ for each of 4 scattering angles ϑ_s (given in center-of-mass). The arrangement allows at the same time to continuously monitor the proper alignment of the primary beam. By choosing different distances (2.0, 2.6 and 3.6 m, respectively) between the collision region and the particle detector, scattering angles between 0.05^o and 0.4^o have been covered. The relative efficiency for particle detection in each individual anode segment is determined in a separate experiment. To this end, the proton beam is scattered on a dense N$_2$ target instead of the Na target; this gives a rather homogenous scattering profile within the angular range of interest. The efficiency determination is repeated after every few runs.

Pulses from the 16 anodes are processed separately; particular care has been taken to eliminate cross-talk. The pulses serve as stop input for 16 time-to-digital converters; the photomultiplier gives the necessary start signal. The time resolution ($\sim$ 15 ns) is limited by the finite length of ion beam viewed by the photon detector. Data processing is performed by a personal computer which also controls the experiment. The Na-beam is mechanically chopped and the signals detected in synchronisation with the chopper. Data are taken under alternating Na beam-on and beam-off conditions. All the data are taken under single-collision conditions with proper correction for background contributions as obtained from Na beam-on and beam-off measurements. One run typically takes about 20 hours of continuous measurements, and several runs are added together at one projectile energy.

When the process (2) is studied, a laser beam is injected into the scattering region, either perpendicularly to the ion- and atom- beams, or collinearly with the ion beam.

3. Charge exchange starting from the Na(3s) initial state

3.1 Definition of parameters for a complete experiment

In the case of process (1) the situation is comperatively simple (see Fig.2(a)). As is well known - provided the coherence of the process under study is not disturbed e.g. by cascades - a complete description of reaction (1) is possible on the basis of the complex excitation amplitudes $f_0, f_{\pm 1}$ leading from the initial s state to the final $p_0, p_{\pm 1}$ substates. Three real quantities are generally defined and suffice for a complete description, for example the excitation cross sections $\sigma_0 = |f_0|^2, \sigma_1 = |f_1|^2$ and the phase χ between f_0 and f_1 (the phase $f_{-1} = -f_1$ is fixed by reflection symmetry in the scattering plane, with the z-axis pointing along the incident beam direction). Various alternative schemes exist to characterize the final H(2p) state, for example, $\sigma = \sigma_0 + 2\sigma_1, \lambda = \sigma_0/\sigma$ and χ, or the state multipoles $< T_{KQ}^+ >$ (cf, e.g. Blum and Kleinpoppen [15]); the latter are related to σ, λ, χ by

$$< T_{00}^+ >= \sigma/\sqrt{3}$$

$$< T_{11}^+ >= -i\sigma[\lambda(1-\lambda)]^{\frac{1}{2}} \sin \chi = -i\sqrt{2} \, \mathrm{Im} \, (f_1 f_0^*)$$

$$< T_{20}^+ >= \sigma(1-3\lambda)/\sqrt{6} = \sqrt{\frac{2}{3}}(\sigma_1 - \sigma_0) \tag{3}$$

$$< T_{21}^+ >= -\sigma[\lambda(1-\lambda)]^{\frac{1}{2}} \cos \chi = -\sqrt{2} \, \mathrm{Re} \, (f_1 f_0^*)$$

$$< T_{22}^+ >= -\sigma(1-\lambda)/2 = -\sigma_1.$$

Only the *in-scattering-plane* part of the H(2p) wavefunction can be excited since the initial state has positive reflection symmetry; thus, an instructive picture of the H(2p) state is that of two orthogonal oscillators in the x and z directions, i.e. *in the scattering plane*, having a certain relative phase between them. This picture suggested to define an alignment angle γ, i.e. the inclination of the resultant charge cloud with respect to the quantization axis,

$$\tan 2\gamma = \frac{-2\sqrt{2} \, |f_0| \, |f_1| \, \cos \chi}{\sigma_0 - 2\sigma_1} \tag{4}$$

where [16] γ is defined modulo π. The transferred angular momentum $< L_y >$ perpendicular to the scattering plane is

$$< L_y >= \frac{-2\sqrt{2} \, |f_1| \, |f_0| \, \sin \chi}{\sigma} \tag{5}$$

Clearly, several of these quantities are redundant; three independent measurements are sufficient to determine the s-p collision dynamics completely. We are giving the results in terms of σ, λ, χ.

The experimental results will be compared to two sets of theoretical calculations (Shingal and Bransden [17] 1987, 1993, Dubois et al [18] 1993 (a), (b). Both are based on the semiclassical coupled-channels impact parameter model; 34 atomic states (Shingal) and 19 atomic states (Dubois) are used as basis.

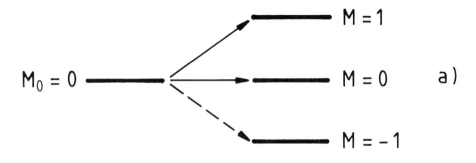

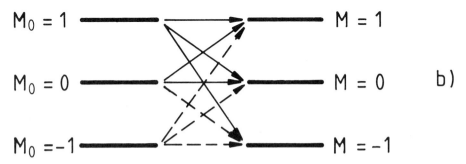

Figure 2. Schematic representation of the various excitation path ways for (a) $s \to p_{0,\pm1}$ and (b) $p_{0,\pm1} \to p_{0,\pm1}$ reactions; quantization axis in the projectile beam direction. Broken lines indicate amplitudes which are redundant in a suitably chosen geometry due to reflection symmetry (see text).

3.2. The differential H(2p) charge exchange cross section σ

The angle-differential cross sections $\sigma = \sigma_0 + 2\sigma_1$ for capture into H(2p) as well as the cascade contributions for the projectile energies 1, 2, 3.5 and 5 keV are shown in Fig. 3. The cascade contributions depend strongly on the projectile velocity and the scattering angle. At 1, 3.5 and 5 keV they are small throughout; only at 2 keV and scattering angles bigger than approximately 0.1 degrees, they give a contribution of up to 40 % to the measured H(2p) yield. After correction for cascades, the agreement between experiment and theory is quite good at all collision energies.

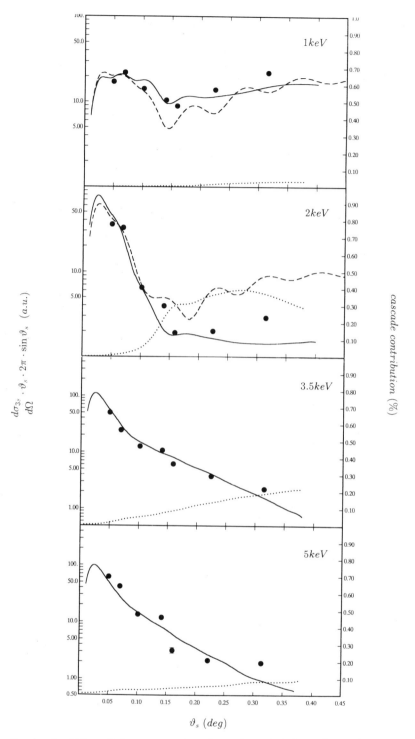

Figure 3. Angle-differential cross section (atomic units) for H$^+$+Na(3s) → H(2p) + Na$^+$ charge capture at 1,2,3.5 and 5 keV impact energy. •, experimental data (cascade corrected); full curves, theory (Shingal 1993); broken curves, theory (Dubois 1993); dotted curves, relative cascade contribution to the H(2p) yield (right-hand scale)

3.3. The coherence parameters λ and χ

The alignment parameter $\lambda = \sigma_0/\sigma$ is shown for 1 and 2 keV collision energy in Fig.4; the rather large error bars in case of 2 keV and $\varphi_s > 0.2^o$ are mainly due to the cascade correction. At $\lambda = 1/3$ one has $\sigma_0 = \sigma_1$, i.e. the H($2p_0$) and H($2p_{\pm1}$) states are equally populated, and the final state is isotropic. At both collision energies, preferential H($2p_0$) population is obtained at small scattering angles, i.e., large impact parameters which mainly determine the total cross section.

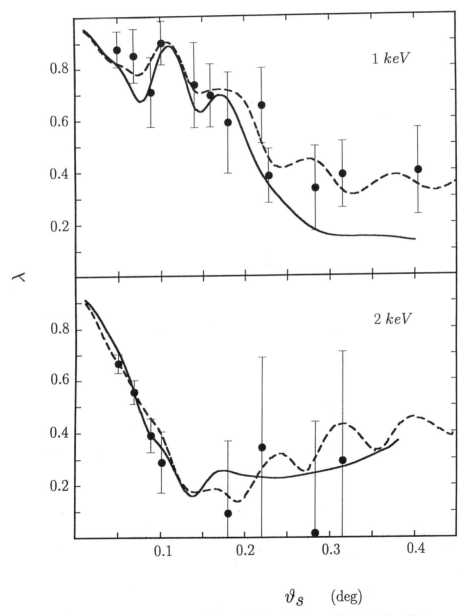

Figure 4. The alignment parameter $\lambda = \sigma_0/\sigma$ for (a) 1 and (b) 2 keV collision energy. •, experimental data (cascade corrected). Theory: full curve (Shingal 1993); broken curve (Dubois 1993).

The cosine of the phase χ between the amplitudes f_0 and f_1 leading to H(2p_0) and H(2p_1) final states, respectively, is shown in Fig. 5 for 1 and 2 keV collision energy. The agreement between experiment and theory is again quite good; in case of 2 keV and $\vartheta_s > 0.2^o$ the error bars due to the cascade correction are too large, and the data have been omitted. We note that at these fairly low energies the detailed behaviour of the phase χ cannot be predicted by general ("propensity") rules; it depends sensitively on the quasimolecular levels active during the collision. The situation is analogous to the one found in other quasi-one electron systems at low impact energies (cf. Hippler[14] and references therein); in H$^+$ + Na collisions, additional complications are caused by trajectory interference effects due to the very small scattering angles of interest [16,17]. It is quite gratifying, however, that in spite of these complexities such a good agreement between theory and experiment exists even on a quantum mechanically complete level.

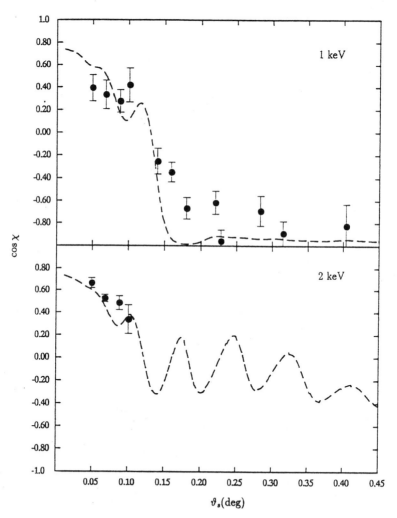

Figure 5. cos χ, with χ the phase between the amplitudes f_0 and f_1 leading to H(2p_0) and H(2p_1) final states, respectively; collision energy 1 keV (a) and 2 keV (b). •, experiment (cascade corrected). Theory: broken curve (Dubois 1993).

4. The experiment with a non-isotropic excited initial state

4.1 Remarks on the "theory of the experiment"

For process (2) the situation is considerably more complex. Three initial and three final magnetic substates have to be connected by altogether 9 complex amplitudes. (Fig. 2 b); without special symmetry considerations, this would require 17 real quantities for a complete characterization. Symmetry consideration reduces 17 real quantities for a complete characterization to nine: 5 amplitudes and 4 phases are required. Furthermore, the initial state is in general not quantum-mechanically pure and, even more, optical pumping conditions can produce a spin-polarized active electron. After the collision it couples with the angular momentum L of the final state and influences its photon decay properties. Specifically, we have recently derived the necessary relations for the case of a p-p scattered particle-photon angular correlation measurement [19] as well as for the corresponding Stokes parameters [20]. The general expression for the photon-scattered particle coincidence intensity covering spin-polarized as well as spin-unpolarized cases is given by [19]:

$$
\begin{aligned}
I = & \sum_{K=0,2} \{B_K \bar{G}_K \sum_{Q=0}^{K} (2 - \delta_{Q0}) d(\vartheta_\gamma)_{0Q}^{K} \\
& \sum_{K_0 Q_0} Re[< T(L_0)_{K_0 Q_0}^{+} \times t_{00} > A(KQ, K_0 Q_0) e^{i[Q(\varphi_\gamma - \varphi_s) + Q_0 \varphi_s]}]\} \\
& + \sum_{K=0,2} \{B_K \cdot g_{11K} \sum_{Q=0}^{K} (2 - \delta_{Q0}) d(\vartheta_\gamma)_{0Q}^{(K)} \\
& \sum_{Q'=-1}^{1} (1Q', 1Q - Q' \mid KQ) \\
& \sum_{K_0 Q_0} Re[< T(L_0)_{K_0 Q_0}^{+} \times t_{1,Q-Q'}^{+} > A(KQ', K_0 Q_0) \\
& e^{i[Q\varphi_\gamma + (Q_0 - Q')\varphi_s]}]\}
\end{aligned}
$$

(6)

$d(\vartheta_\gamma)_{KQ}$ are the elements of the rotation matrix which transform the photon detector (located at ϑ_γ, φ_γ) into the frame in which the final state multipoles are defined, and

$G_{K' k_0 K}^{Q' q_0 Q}$ are generalized perturbation coefficients given by the expression

$$
\begin{aligned}
G_{K' k_0 K}^{Q' q_0 Q} = & \\
= & (K'Q' k_0 q_0 \mid KQ) \sum_{J} (2J + 1)^2 \\
& \sqrt{(2K' + 1)(2k_0 + 1)(2K + 1)} \begin{Bmatrix} K0K \\ L\frac{1}{2}J \\ L\frac{1}{2}J \end{Bmatrix} \begin{Bmatrix} K'k_0 K \\ L\frac{1}{2}J \\ L\frac{1}{2}J \end{Bmatrix} \\
\equiv & (K'Q' k_0 q_0 \mid KQ) g_{K' k_0 K} \quad ;
\end{aligned}
$$

(7)

(|) denote Clebsch-Gordan coefficients and {−−} standard 9j-symbols.

The quantities $A(KQ, K_0Q_0)$ defined as

$$A(KQ, \; K_0Q_0) = \sum_{\substack{MM' \\ M_0M_0'}} (-1)^{L_0 - M_0 + L - M}$$

$$(LM', L - M \mid KQ)(L_0M_0', \; L_0 - M_0 \mid K_0Q_0)$$

$$\cdot f(M'M_0')f(MM_0)^* \tag{8}$$

contain all details about the collision dynamics, and $f(MM_0)$ for scattering from an initial M_0 to a final M state are the amplitudes after separation of the spin dependence. We assume that the total spin is conserved during the collision so that the spin component of the active electron *immediately after the collision* is the same as that in the initial state, and the fine interaction in the final state is negligible.

The goal of a "complete experiment" is the determination of all amplitudes $f(MM_0)$ and the corresponding phases, contained in $A(KQ, K_0Q_0)$. In contrast to the case of an isotropic initial state (cf. e.g. [9] for s-p excitation), the expression eq. (6) for the angular distribution of emitted radiation in the p-p case contains state multipoles of rank 0, 1 and 2. As a consequence (and again in contrast to the s-p case), even the sign of the phase may generally be obtained by a correlation analysis; a polarization measurement is not necessary. Still, there are at least nine independent measurements necessary. The complexity of the relation (6) suggests to employ geometries of high symmetry, and, if possible, pure initial states. Note that in order to derive the phase between transitions involving $M_0 = 0$ and $M_0 = 1$ initial states at least one experimental configuration must provide a coherent superposition of $M_0 = 0$ and $M_0 = 1$ initial states (i.e. the laser photons should propagate at an angle different from zero to the z-axis).

4.2 The differential H(2p) charge exchange cross section

In order to minimize alignment effects of the initial state, Na(3p) was optically pumped using circularly polarized light. The experimental results for the projectile energy of 1 keV are shown in fig.6; they are again compared with calculations by Shingal [17] and Dubois [18]. As expected, distinct differences are observed between capture from Na(3s) and Na(3p) targets. Note that cascade contributions are not to be neglected in case of Na(3p). Correcting the experimental data for cascades (mainly from H(3d) capture), they are in fair overall agreement with the predictions of both theories. Finally, at the collision energy of 1 keV studied here, the H(2p) capture cross section is considerably more forward-peaked in case of a Na(3p) target as compared to Na(3s).

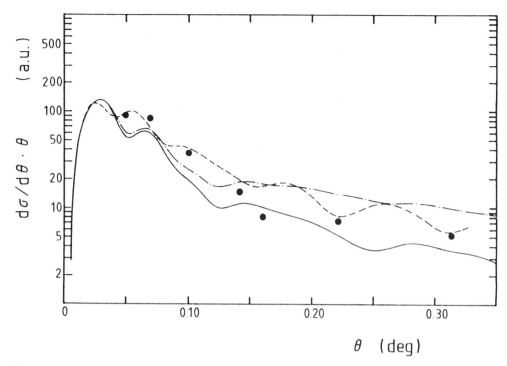

Fig. 6 Angle-differential H(2p) electron capture cross section in 1 keV H$^+$ + Na(3p) collisions. (•) Experimental data; (−) theory. The cascade contribution can be seen from (−·−) which shows Shingal's results including cascades.

4.3 The role of an initial Na(3p) state alignment

We investigate the effect of an *initial* Na(3p) state alignment on the capture cross section. An appropiate quantity is the *anisotropy parameter*

$$A = \frac{\sigma_\Sigma - \sigma_\Pi}{\sigma_\Sigma + \sigma_\Pi} \qquad (9)$$

with Σ, Π denoting the angular momentum of the initial Na(3p) orbital projected on the internuclear axis, and σ the corresponding charge exchange cross section. The impact energy dependence of A has been measured for this and similar systems [10,21−23]; a simple model involving the overlap of momentum-space wave functions has been suggested [22,23] which stresses the role of the moving frame of the final projectile state. In contrast, the present communication deals with low impact velocities V = v / v$_e$ (with v the projectile velocity and v$_e$ the classical orbiting velocity of the active electron) in which case such "electron translation effects" are small, and the system is best visualized in terms of a quasimolecule. We concentrate on the angular dependence of A at 1 keV impact energy (V ≈ 0.4). Results are shown in Figure 7. The agreement between the experiment and the numerical calculations is (particularly at small scattering angles) quite good, although at the larger scattering angles the coincidence rates are small resulting in fairly large statistical uncertainties. This agreement may be thought to be rather satisfying, however, it does not directly bring out the underlying physics.

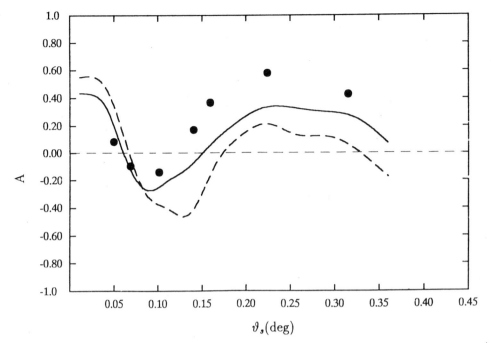

Fig. 7 Scattering angle-dependent anisotropy parameter A(2p) for 1 keV H$^+$ + Na(3p) → H(2p) + Na$^+$; Experiment: •, corrected for cascade contribution; the theoretical data (solid curve) are convoluted with the experimental angular resolution.

4.4 Left-Right Asymmetry from Circular Atomic State

The use of left and right circularly polarized laser light for preparation of oriented, excited atomic states (maximum electron angular momentum projection) prior to the collision has initiated a series of crossed-beam experiments in which the role of orientation in atomic collision dynamics, illustrated classically by the sense of rotation of the active electron around the atomic core, has been discussed [12,13]. A classical picture modeling the electron-capture process in terms of simple velocity matching, suggests that electron transfer is favored when the projectile and the active electron move in the same direction on the same side of the nucleus, as shown in Fig. 8a. The corresponding asymmetry of the intensity I is defined as $\bar{A}_c = (I^+ - I^-)/(I^+ + I^-)$; the index c indicates that in our case this asymmetry is a result of a particle-photon coincidence measurement; (+) signifies left, (-) right-handed circular polarization of the laser, injected perpendicular to the z and the atomic beam axis. Such a circular state is a coherent superposition of $M_0 = 0, \pm 1$ initial states; it can, therefore, serve to derive the above-mentioned phase between the $M_0=0$ and $M_0=1$ state, thus being the first step to a complete experiment with the Na(3p) initial state. The second step then has to involve collinear pumping of Na(3p) into $M_0=\pm 1$, an experiment which has yet to be performed.

5. Outlook

Scattering experiments involving optically pumped initial states allow great flexibility in adjusting the experimental conditions to the subject of interest. For example, we have also applied the laser optical pumping technique to the preparation of high n Rydberg states. Similar to the situation discussed in section 4.4 above, Kohring et al [24]

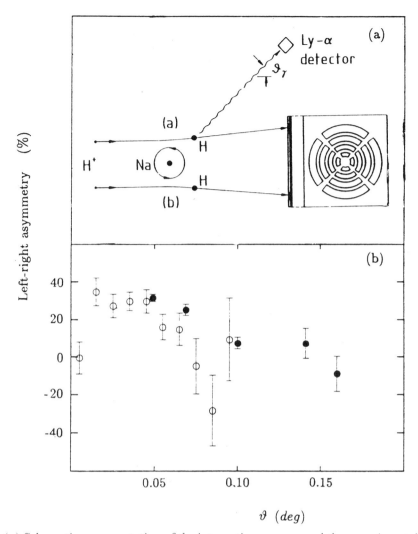

Fig.8.(a) Schematic representation of the interaction process and the experimental setup, (b) the left-right asymmetry in 1 keV $H^+ + Na(3p)$ collisions vs scattering angle ϑ

suggested that electron capture should be particulary efficient if the Rydberg electron moves parallel to the collision plane. Our experiments as well as classical-trajectory Monte Carlo (CTMC) calculations [22] show that the influence of the initial Rydberg electron alignment on the cross section depends strongly on the initial state alignment as well as the collision velocity, in agreement with the prediction of a simple model. More recently, such experiments have been performed by the Aarhaus group, in which coherent elliptic initial states have been produced, corresponding to classical ellipses of well-defined rotation and location in space. Our Classical Trajectory Monte Carlo calculations [26] of the capture probability yielded a pronounced dependence on the excentricity of the elliptical state. Strong propensity for charge exchange has been found for spatial as well as velocity matching of the ion and the captured electron. They also show [27] that the total interaction can roughly be classified into two categories. In a type-1 process, the Rydberg electron is directly captured by the projectile into a stable orbit. This mechanism is very sensitive to the spatial overlap and matching velocities of the Rydberg electron and the projectile. In contrast, in a type-2 capture process the

291

electron returns in a close passage to its own (target) nucleus after its initial atomic orbit has been perturbed by the projectile; this "double scattering" is rather insensitive to velocity matching. The analogy between these two processes and the leading terms in a strong potential treatment of charge exchange is suggestive. Further extensions of such studies are conceivable. Certainly, coincidence experiments involving such highly excited initial states are (as yet) not possibly due to the small scattering angles involved. Other developments, however, are well underway. For example, a Rydberg electron's electrostatic Coulomb energy in the nuclear field is only of the order of meV. This can be comparable to its energy in magnetic fields of laboratory strength. We have, therefore, recently performed CTMC calculations of the charge capture probability in collisions involving Rydberg states, in magnetic fields of a few Tesla. Interesting differences are found to collisions in a field-free region, regarding the total as well as the impact-parameter dependant cross sections[28].

Acknowledgement:

This work has been supported by the Deutsche Forschungsgemeinschaft in Sonderforschungsbereich 216, and by the Volkswagen-Stiftung.

References

1. V.S.Kushawaha, C.E.Burkhardt and J.J.Leventhal, Phys.Rev.Lett.**45** 1686 (1980); V.S.Kushawaha, Z.Phys.A**313** 155 (1983).
2. M.Kimura, R.E.Olson, and J.Pascale, Phys.Rev.A**26** 3113 (1982).
3. J.Allan, R.Shingal, and D.R.Flower, J.Phys.B**19** L251 (1986).
4. F.Aumayr and H.Winter, J.Phys.B**20** L803 (1987); M.Gieler, F.Aumayr, M.Hütteneder, and H.Winter, *ibid.***24** 4419 (1991).
5. K.Finck, Y.Wang, Z.Roller-Lutz, and H.O.Lutz, Phys.Rev.A**38** 6115 (1988).
6. T.Royer, D.Dowek, J.C.Houver, J.Pommier, and N.Andersen, Z.Phys.D**10** 45 (1988).
7. R.J.Allan, C.Courbin, P.Salas, and P.Wahnon, J.Phys.B **23** L461 (1990); C.Courbin, R.J.Allan, P.Salas, and P.Wahnon, *ibid.* **23** 3905 (1990).
8. C.Richter, N.Andersen, J.C.Brenot, D.Dowek, J.C.Houver, J.Salgado, and J.W.Thomsen, J.Phys.B**26**, 723 (1993).
9. Z.Roller-Lutz, K.Finck, Y.Wang, and H.O.Lutz, Phys.Lett.A**169** 173 (1992).
10. D.Dowek, J.C.Houver, C.Richter, J.Pommier, T.Royer, N.Andersen, and B.Palsdottir, Phys.Rev.Lett.**64**, 1713 (1990).
11. C.Richter, D.Dowek, J.C.Houver, and N.Andresen, J.Phys.B**23** 3925 (1990).
12. D.Dowek, J.C.Houver, C.Richter, and N.Andersen, Phys.Rev.Lett.**68**, 162 (1992).
13. Z.Roller-Lutz, Y.Wang, K.Finck, and H.O.Lutz, Phys.Rev.A**47** (1993).
14. Cf., e.g., R.Hippler and H.O.Lutz, Comments At.Mol.Phys.**28** 39 (1992).
15. K.Blum and H.Kleinpoppen, Phys.Rep.**52** 203 (1979).
16. N.Andersen, J.W.Gallagher and I.Hertel, Phys.Rep.**165** (1988) 1.
17. R.Shingal and B.H.Brandsen, J.Phys.B**20** 4851 (1987). private communication (1993)
18. A.Dubois, S.E.Nielsen and J.P.Hansen, J.Phys.B**26** 705 (1993) J.P.Hansen, S.E.Nielsen, and A.Dubois, Phys.Rev.A**46** R5331 (1992).
19. Z.Roller-Lutz, R.Höllmann, K.Blum, H.O.Lutz, J.Phys.B**27** 2008 (1994).
20. Z.Roller-Lutz, S.Knezović, Y.Wang and K.Blum, Z.Phys.D (1996)

21. F.Aumayr, M.Gieler, J.Schweinzer, H.Winter, J.P.Hansen, Phys.Rev.Lett.**68**, 3277 (1992).
22. T.Wörmann, Z.Roller-Lutz, H.O.Lutz, Phys.Rev.**A47**, R 1594 (1993).
23. S.Schipper, A.R.Schlatmann, R.Morgenstern, Phys.Lett.**A181**, 80 (1993).
24. G.A.Kohring, A.E.Wetmore, and R.E.Olsen, Phys.Rev.**A28**, 2526 (1983).
25. S.B.Hansen, T.Ehrenreich, E.Horsdal-Pedersen, K.B.MacAdem, L.Dubé, Phys.Rev.Lett.**71** 1522.
26. S.Bradenbrink, H.Reihl, Th.Wörmann, Z.Roller-Lutz, and H.O.Lutz, J.Phys.**B27**, L 391 (1994).
27. S.Bradenbrink, H.Reihl, Z.Roller-Lutz, and H.O.Lutz, J.Phys.**B28**, L 133 (1995).
28. S.Bradenbrink, H.Reihl, Z.Roller-Lutz, and H.O.Lutz, to be published.

COHERENT EXCITATION OF H(n=2) IN FEW-ELECTRON COLLISION SYSTEMS

B. SIEGMANN[1], R. HIPPLER[1], H. KLEINPOPPEN[2] and H.O. LUTZ[1]

[1] Fakultät für Physik
Universität Bielefeld
Universitätsstraße 25
D-33615 Bielefeld
Germany

[2] Atomic Physics Laboratory
University of Stirling
Stirling FK9 4LA
Scotland

Introduction

Excitation processes in simply-structured one- and two-electron systems are both of fundamental importance and of practical relevance for a number of fields. Valuable information about the dynamics of such collisions can be extracted from a measurement of so-called coherence parameters. These parameters relate to the population of energetically degenerate states and are sensitive to the relative phases between the corresponding excitation amplitudes. In collisions involving hydrogenic atoms, not only the coherence between the energetically degenerate magnetic substates (magnetic quantum number m) but also between the near-degenerate angular momentum states (angular momentum quantum number l) with the same principle quantum number n becomes accessible[1].

In the following we discuss the excitation of hydrogen atoms to the lowest excited H($n = 2$) level in collisions with rare gas atoms X (X = He, Ne, Ar) either by charge exchange excitation of incident protons

$$H^+ + X \;\rightarrow\; H(n = 2) + X^+$$

Selected Topics on Electron Physics
Edited by Campbell and Kleinpoppen, Plenum Press, New York, 1996

or by direct excitation of fast neutral H(1s) hydrogen.

$$H(1s) + X \rightarrow H(n = 2) + X$$

The excited $H(n = 2)$ atoms decay with a mean lifetime of 1.6 ns by emission of Lyman-α radiation ($\lambda = 121.6$ nm) to the H(1s) ground state,

$$H(2p) \rightarrow H(1s) + Lyman - \alpha.$$

The emitted Lyman-α photon carries the information about the collision process and, in particular, about the relative population of the various substates and the relative phases between the excitation amplitudes. This information is extracted, in the present experiment, from a polarization analysis of the emitted Lyman-α radiation.

Theoretical considerations

In atomic collision processes the interaction time is typically of the order of $\tau_{col} \approx 10^{-15} \ldots 10^{-16}$ sec and, hence significantly shorter than the time scale along which the hydrogen $H(n)$ states due to their near-degeneracy evolve; for an excited $H(n = 2)$ atom this time scale is determined by the fine-structure splitting and of the order of 10^{-11} sec. As a consequence, the hydrogenic $H(n)$ states with the same principal quantum number n become coherently excited during a collision. We may then express the wavefunction of an excited $H(n = 2)$ atom as a linear combination of H(2s) and H(2p) wavefunctions

$$\Psi(n = 2) = f_{00} |2s_0> + f_{10} |2p_0> + f_{11} |2p_1> + f_{1-1} |2p_{-1}> \tag{1}$$

where f_{lm} are complex excitation amplitudes for the corresponding $|lm>$ substates. Alternatively, the excited hydrogen atom may be described in terms of a density matrix which is more appropriate in those cases where an averaging over several collision events take place. Due to the implicit integration over the projectile scattering angle in the present experiment, the resulting *mixed* states are only partly coherent and most of the off-diagonal elements, for example, the coherence between the H($2p_0$) and H($2p_{\pm1}$) states, disappear. In the axially symmetric case the density matrix, hence, has the following simple form

$$\rho = \begin{pmatrix} \sigma_{2s} & 0 & \sigma_{sp} & 0 \\ 0 & \sigma_{2p_{-1}} & 0 & 0 \\ \sigma_{sp}^* & 0 & \sigma_{2p_0} & 0 \\ 0 & 0 & 0 & \sigma_{2p_{+1}} \end{pmatrix} \tag{2}$$

The diagonal elements of the density matrix are related to the partial cross sections σ_{lm} for excitation of the $|lm>$ substates. The relative population of the H(2p) substate is commonly expressed in terms of the alignment parameter A_{20} which characterizes the shape of the charge cloud[2]:

$$A_{20} = \frac{\sigma_{2p_1} - \sigma_{2p_0}}{\sigma_{2p}} \tag{3}$$

where $\sigma_{2p} = \sigma_{2p_0} + 2\sigma_{2p_1}$ $(\sigma_{2p_{+1}} = \sigma_{2p_{-1}} = \sigma_{2p_1})$ is the total cross section for H(2p) excitation.

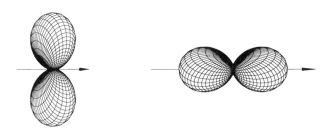

Figure 1. Schematic representation of the H(2p) charge cloud aligned perpendicular $(A_{20} = +\frac{1}{2})$ and parallel $(A_{20} = -1)$, respectively, to the beam direction.

In the present experiment, A_{20} is obtained from a polarization analysis of the emitted Lyman-α radiation in the absence of an external field. The determination of the partial cross section σ_{2s} for excitation of the H($2s_0$) state and of the only non-vanishing off-diagonal element σ_{sp} requires measurements inside an external electric field[3–5]. The so-called $s - p-$coherence σ_{sp} connects the excitation amplitudes of the H($2s_0$) and the H($2p_0$) substates. Depending on the relative phase χ between these excitation amplitudes, a shift of the centre-of-charge relative to the centre-of-mass is induced. In the axially symmetric case, only the $z-$component (chosen as the direction of the incident projectile) of the collision-induced dipole moment $\langle D \rangle$ is different from zero. It is related to the real part of the $s - p-$coherence,

$$\langle D_z \rangle = \frac{6 \cdot \Re(\sigma_{sp})}{\sigma_{tot}} = \frac{6 \cdot \sqrt{\sigma_{2s}\sigma_{2p_0}} \cdot \langle \cos \chi \rangle}{\sigma_{tot}}, \tag{4}$$

where $\sigma_{tot} = \sigma_{2s} + \sigma_{2p}$ denotes the total cross section for H($n = 2$) excitation and $\langle ... \rangle$ a weighted average over all scattering angles.

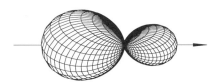

Figure 2. Schematic representation of a collision-induced dipole moment.

The imaginary part causes an inclination of the charge cloud with respect to the z-axis. It is related to the moment $\langle (\vec{L} \times \vec{A})_z \rangle$ where $\vec{L}$ and $\vec{A}$ are the angular momentum vector and the Runge-Lenz vector, respectively. In a classical picture $\langle (\vec{L} \times \vec{A})_z \rangle$ is related to the velocity vector of the excited electron at the perihelion of its Kepler orbit. For the

$z-$component we have

$$\langle(\vec{L}\times\vec{A})_z\rangle = -\frac{2\cdot\Im(\sigma_{sp})}{\sigma_{tot}} = \frac{2\cdot\sqrt{\sigma_{2s}\sigma_{2p_0}}\cdot\langle\sin\chi\rangle}{\sigma_{tot}}. \qquad (5)$$

To measure the partial cross section σ_{2s} and the $s-p-$coherence σ_{sp} it is necessary to apply an external electric field which effectively mixes the H($2s$) and H($2p_0$) states[4]. The collision-induced dipole moment $\langle D_z\rangle$ was obtained from the measured forward-backward asymmetry of the linear polarization P_1 in a longitudinal electric field (i.e. parallel or anti-parallel to the incident beam direction). Here P_1 is defined as

$$P_1 = \frac{I_\parallel - I_\perp}{I_\parallel + I_\perp} \qquad (6)$$

where $I_\parallel$ and $I_\perp$ are the light intensities with the electric field vector parallel or perpendicular to the beam direction. To determine the imaginary part of the $s-p-$coherence, i.e., the moment $\langle(\vec{L}\times\vec{A})_z\rangle$, we perform measurements in a transverse electric field and measure the circular polarization P_3 as a function of the electric field strength. Here P_3 is defined as

$$P_3 = \frac{I(-) - I(+)}{I(-) + I(+)} \qquad (7)$$

where $I(-)$ and $I(+)$ are the light intensities with negative and positive helicity, respectively.

Experimental set-up

A proton beam with kinetic energy of 1-25 keV is produced in a duoplasmatron ion-source[6,7].

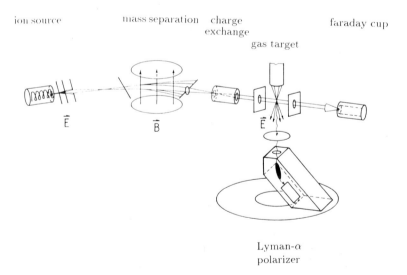

Figure 3. Experimental setup (schematic)

After mass separation in a 30° magnet the collimated beam interacts with the target gas (He, Ne or Ar). In case of a neutral projectile beam the proton beam passes through a charge exchange cell after which the unwanted H^+ contribution is removed by electrostatic deflection. The Lyman-α radiation emitted in the decay of the excited $H(n = 2)$ atoms is analyzed with a rotatable Brewster polarizer. The same device with a MgF_2 quarter-wave plate inserted in front of the Brewster polarizer was used for the circular polarization measurements.

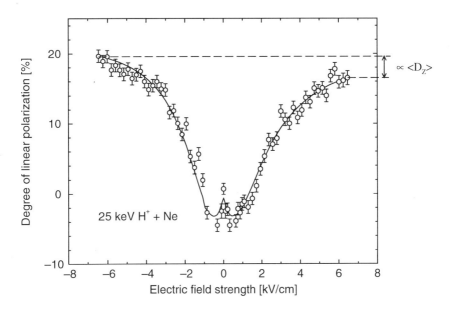

Figure 4. The measured linear polarization P_1 *versus* field strength of a longitudinal electric field for 25 keV H^+–Ne collisions.

In order to extract all relevant elements of the density matrix, we applied an external electric field to two field plates which are oriented parallel or transversal to the projectile beam. The field plates are separated from each other by 1 cm; voltages up to ±3 kV were applied to these plates resulting in an electric field strength of up to 6 kV/cm.

Figure 4 shows the measured linear polarization as a function of the applied electric field strength for 25 keV H^+–Ne collisions. The collision induced dipole moment manifests itself in a pronounced forward-backward asymmetry. A non-linear least-squares fit to the experimental data is used to extract the cross section ratio $\frac{\sigma_{2s}}{\sigma_{2p}}$, the alignment parameter A_{20}, and the real part of the $s - p$–coherence $\Re(\sigma_{sp_0})$.

Results and Discussion

The collision-induced dipole moment for H^+–He is displayed in Figure 5 a. As a general tendency, the dipole moment is small or negative at small energies and rises steeply towards higher energies with a maximum around 35 keV. The experimental data are in satisfactory agreement with theoretical calculations based on one- and two-

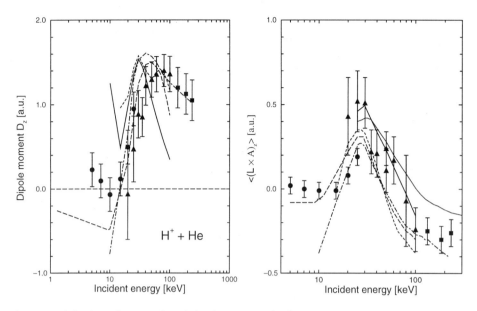

Figure 5. (a) The collision induced dipole moment $\langle D_z \rangle$ versus incident energy for H($n = 2$) excitation in H[+]–He collisions. (b) The velocity vector $\langle (\vec{L} \times \vec{A})_z \rangle$ versus incident energy for H[+] – He collisions. Experiment: Cline et al.[8] (1991, △), DeSerio et al. (1988, □); Hippler et al.[5] (1991, ●). Theory: one-electron AO[+] calculations (Jain et al.[9] 1988, ——; Shingal et al.[10] 1991, - - - -); two-electron AO-MO calculations (Kimura et al.[11] 1986, – –); two-electron AO calculations (Slim et al.[12] 1991, -·-··-) continuum-distorted wave approximation including post-collision interaction (DeSerio et al.[13] 1988, ····)

electron models[8]. Generally the characteristics of the experimental data are in better agreement with two-electron calculations than with the one-electron calculations. The positive dipole moment indicates that the captured electron is lagging behind the proton. It has a tendency of staying in between the proton and the residual He[+] ion which both provide an attraction for it. This argument seems to be supported by results for the neutral H–He collision system, where such a behavior was not observed.

A complete determination of the $s - p$–coherence additionally requires the determination of the imaginary part. Figure 5 b displays the moment $\langle (\vec{L} \times \vec{A})_z \rangle$ as a function of incident energy for H[+]–He collisions. As before for the dipole moment, the experimental data are about zero at energies up to 15 keV and increase towards larger energies with a maximum around 30 keV. Above 80 keV the moment $\langle (\vec{L} \times \vec{A})_z \rangle$ becomes negative.

A comparison of the trends for the collision-induced dipole moment in more complicated collision systems using protons as projectiles is shown in Figure 6.[14] As a general tendency, the dipole moment is small or negative at low velocities and rises towards larger energies with maxima around 10 keV for Ar and 15 keV for Ne whereas it appeared at 35 keV for He. All experimental curves seems to display similar features with minima and maxima which are shifted to larger energies with increasing binding energy of the active electron. The theoretical calculations appear to predict the correct tendencies of the dipole moment although the comparison in H[+] – Ne and H[+] – Ar collisions is limited by the small energy range covered by the calculations which still show significant discrepancies with the experimental data. It should be mentioned that the calculations were done with a minimal basis set of only five molecular states[14].

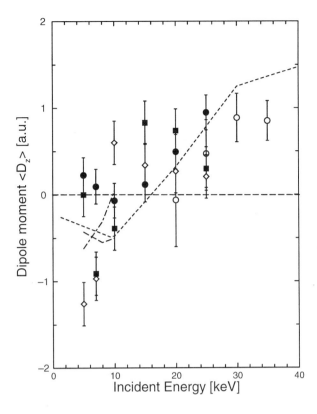

Figure 6. The collision induced dipole moment $\langle D_z \rangle$ versus incident energy for H($n = 2$) excitation in H$^+$ – He, Ne, Ar collisions. Cline et al.[8]: H$^+$ – He ($\circ$); Tepehan et al.[14]: H$^+$ – He ($\bullet$), H$^+$ – Ne ($\square$) and H$^+$ – Ar ($\diamond$); also shown are MO close-coupling calculations from Kimura for He[11] (- - -), Ne[14] (– –) and Ar[14] (-·-·-).

A different behavior of the collision-induced dipole moment is observed for the neutral H–rare-gas atom systems (Figure 7). Generally, the dipole moment is now small or negative around 10–25 keV and no maximum is noted in that range. The negative dipole moment observed in H–He collisions shows that the electron moves in front of the proton. At first glance this is consistent with the idea, that in H$^+$ – He collisions the excited electron is located between the proton and the positive target ion. In H – He collisions the neutral target atom exerts no Coulomb force on the positive proton and the electron moves in front of the proton. However, the positive dipole moment observed in H – Ne and H – Ar collisions indicate that this easy picture is not sufficient in all cases.

Conclusions

The $s - p$–coherence of the excited hydrogen atom H($n = 2$) is of fundamental importance for the understanding of the collision processes. Our results for the collision induced dipole moment of H($n = 2$) in H – rare gas collisions are generally smaller than in H$^+$ – rare gas collisions but both collision complexes show a shift of minima and

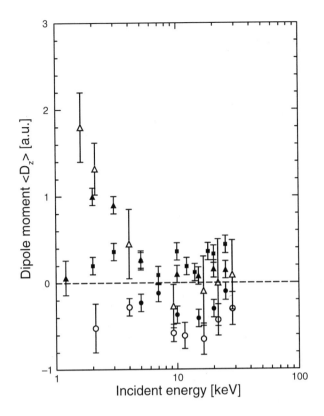

Figure 7. The collision induced dipole moment $\langle D_z \rangle$ versus incident energy for H($n = 2$) excitation in (a) H – He, (b) H – Ne and (c) H – Ar collisions: (a) ○ Krotkov et al.[15] (1980), ● Siegmann et al.[16] (1994); (b) □ Siegmann et al.[16] (1994); (c) △ Krotkov et al.[15] (1980), △ Siegmann et al.[16] (1994).

maxima towards larger projectile energies with increasing binding energy and decreasing polarisibility of the target atom. In particular, the excitation process of H($n = 2$) in H – rare gas collisions indicates that a different population of the $| \pm >$ state may result in an opposite sign of the collision induced dipole moment.

Acknowledgements

We like to thank Prof. Mineo Kimura and Prof. G.G. Tepehan for helpfull discussions. This work was supported by the Deutsche Forschungsgemeinschaft in Sonderforschungsbereich 216 *Polarisation und Korrelation in atomaren Stoßkomplexen*.

References

1. R. Hippler, J. Phys. **B26**, 1 (1993).

2. U. Fano, J. Macek, Rev. Mod. Phys. **45**, 554 (1973).

3. T.G. Eck, Phys. Rev. Letter **31**, 270 (1973).

4. A. VanWijngaarden, E. Goh, G.W.F. Drake, P.S. Farago, J. Phys. **B9**, 2017 (1976).

5. R. Hippler, O. Plotzke, W. Harbich, H. Madeheim, H. Kleinpoppen, H.O. Lutz, Z. Phys. **D18**, 16 (1991).

6. R. Hippler, M. Faust, R. Wolf, H. Kleinpoppen, H.O. Lutz, Phys. Rev. **A31**, 1339 (1985).

7. H. Madeheim, R. Hippler, H.O. Lutz, Z. Phys. **D15**, 327 (1990).

8. R.A. Cline, W.B. Westerveld, J.S. Risley, Phys. Rev. **A43**, 1611 (1991).

9. A.Jain, C.D. Lin, W. Fritsch, J. Phys. **B21**, 1545 (1988).

10. R. Shingal, C.D. Lin, J. Phys. **B24**, 963 (1991).

11. M. Kimura, C.D. Lin, Phys. Rev. **A34**, 176 (1986).

12. H.A. Slim, E.L. Heck, B.H. Bransden, D.R. Flower, J. Phys. **B24**, 1683 (1991).

13. R. DeSerio, C. Gonzales-Lepera, J.P. Gibbons, J. Burgdörfer, I.A. Sellin, Phys. Rev. **A37**, 4111 (1988).

14. G.G. Tepehan, B. Siegmann, H. Madeheim, R. Hippler, M. Kimura J. Phys. **B27**, 5527 (1994).

15. R. Krotkov, J. Stone, Phys. Rev. **A22**, 473 (1980).

16. B. Siegmann, G.G. Tepehan, R. Hippler, H. Madeheim, H. Kleinpoppen, H.O. Lutz, Z. Phys. **D30**, 223 (1994).

COHERENT EXCITATION OF HE ATOMS BY PROTON IMPACT

Bengt Skogvall and Gebhard von Oppen

Institut für Strahlungs- und Kernphysik, Technische Universität Berlin
Hardenbergstr. 36, D-10623 Berlin, Germany

1.   INTRODUCTION

Theoretical and experimental studies of excitation in ion-atom collisions have been carried out extensively for many years. Most studies were concerned with the cross section of a particular channel. However, these cross sections yield only extremely restricted information about the actual state of the atoms after the collision with the ion, as the collisionally excited states generally differ from the eigenstates of the atomic system. In particular, the center of the charge cloud of the excited electron may not coincide with that of the nucleus. This possible occurrence of electric dipole moments after collisional excitation was considered already in 1915 by N. Bohr as mentioned by W Pauli in his review article about the old quantum theory[1]. A complete description of the excitation process can be obtained by introducing excitation matrices[2]. When measuring cross sections, one is measuring the trace of the excitation matrix (the sum of the diagonal elements). The size of the matrix is determined by the energetic resolution of the experimental setup. In order to study the different diagonal elements of the excitation matrix, special experimental techniques are required. The study of polarization and angular distribution of the emitted light after the collision yield some more information, but for a complete determination of the substate population for states with angular momenta $L > 1$, sophisticated methods, like level-crossing techniques, must be employed[3].

For a complete analysis of the excitation matrix, also the off-diagonal matrix elements must be considered. These are connected to coherent excitation of different basis states. The off-diagonal matrix elements of the excitation matrix are known as coherence parameters. In this paper, we consider the coherent excitation of states with different L. These coherence parameters are related to the spatial charge distribution[4], as well as the current distribution[5] of the electronic cloud after the collision.

The former corresponds to the real part of the off-diagonal matrix elements and the latter as the imaginary part. In particular, electric dipole moments of the charge cloud are determined by the off-diagonal matrix elements of opposite-parity states. As the symmetry properties of charge and current distribution, also the symmetry of the real and imaginary part of the off-diagonal matrix elements is different: the real part is time-reversal invariant whereas the imaginary part changes sign under time-reversal. This circumstance necessitates the employment of different techniques when measuring the charge and current distribution of impact-excited states.

Theoretical and experimental studies of coherent excitation in ion-atom collisions have been carried out for more than 20 years. Most measurements were performed on atomic hydrogen. This atom has the advantage that all states with equal principal quantum number n are almost degenerate. Measurements of coherence parameters for opposite-parity states were first suggested by Eck[6] and performed by Sellin *et al.*[4] In these beam-foil measurements, quantum beats of the emitted impact-radiation were recorded and analyzed. The experimental technique involves an electric field applied to the observation region to couple levels of opposite parity. In this case, these levels can decay to the same final state and, therefore, coherence phenomena become observable. By measuring amplitude and phase of the quantum beats, the parameters of both charge and current distribution can be measured.

However, this experimental technique is less suited for investigations on ion-atom collisions where gas targets are used. The main problem arises from the condition of pulsed excitation, which must be fulfilled in quantum beat measurements[7]. But coherent excitation of opposite-parity states can be analyzed also under stationary excitation conditions. This possibility was first demonstrated by Krotkov and Stone[8]. The experimental technique was applied by Risley and coworkers[9] to investigations on p-He charge-exchange excitation of $H(n = 3)$ states and thoroughly developed further[10]. Exciting under stationary conditions entails that - strictly speaking - only population numbers of substates and coherence parameters of degenerate substates can be measured. Nevertheless, also this experimental technique allows a determination of the complete excitation matrix. By applying electric fields parallel or antiparallel to the proton beam and investigating the impact radiation as a function of the electric-field, the charge distribution of the hydrogen states excited by charge exchange could be determined[9]. By applying electric fields perpendicular to the proton beam, also the current distribution was analyzed[5].

For many years, charge and current distribution of collisionally excited states were investigated only for hydrogen atoms both experimentally and theoretically[11-13]. In many experiments protons were impinged on He, but only the collisonally excited states of hydrogen were studied. Using helium as the object of study and investigating the collisionally excited He states was not tried until Aynacioglu *et al.*[14] demonstrated that observable electric dipole moments are indeed also present in He excited by ions, in this case H_2^+. This finding was the starting point of a series of experiments by which the charge distribution of helium atoms excited by protons[15-17] and other projectiles[18,19] was studied. All experiments were performed under stationary excitation conditions and with electric fields parallel or antiparallel to the proton beam. Therefore, we obtained information only about charge distributions but not about current distributions. In this paper, the experimental techniques allowing a detailed study of proton-impact excitation of helium atoms are surveyed. Helium does not have the advantage of an almost perfect ℓ-degeneracy as hydrogen. Therefore, it is more difficult to investigate coherent excitation of opposite-parity states of helium than of hydrogen atoms. However, a unique feature of He, compared to H, is the presence of two multiplet systems. This fact can be exploited in various ways for an extremely detailed analysis of the charge distribution of collisionally excited He states, as will be shown in the following sections.

2. EXPERIMENTAL METHOD

All experiments discussed below were performed with essentially the same experimental setup. A well collimated beam of protons is directed under a gas nozzle, where He is emitted. The gas nozzle consists of a 2 mm thick micro-channel plate, through which the He gas emerges. The pressure on the micro-channel plate is approximately 5 mbar. The gas expands in the 2mm long and 25 μm wide tubes of the micro-channel plate. Thus, a directed beam of He atoms is formed. The pressure in the target region was low enough to ensure single collision conditions. An electric field in the target region is applied, either parallel or anti-parallel to the proton beam. The electric field is produced by two tubes with an inner diameter of 7 mm. The distance between the two tubes is 7 mm, thus providing a "Helmholtz" type of arrangement. In the center region a homogeneous electric field is formed. Photons emitted after the collision are detected from a direction perpendicular to the projectile beam (and the electric field). A system consisting of two lenses transports the photons onto the cathode of a photo multiplier. The transitions studied in this work are the nd-$2p$ transitions in the singlet and triplet system of He I. The wavelength of the emitted light is in the visible range. The selection of wavelength is performed using an interference filter with a typical bandwidth of 10-15 nm. The interference filter is situated in front of the cathode of the photo multiplier. Though, usually, the spectral selectivity of the interference filter is sufficiently high, in some cases the observed line is blended by other transitions. As an example, the $5d\,^3D$-$2p\,^3P$ -transition is blended by $7d\,^1D$-$2p\,^1P$. A schematic of the experimental setup is shown in reference 19.

3. ENERGIES AND EIGENSTATES OF HE I

Due to the near degeneracy of states with different angular momenta ℓ, collisional excitation of hydrogen atoms can be studied in greater detail than the collisional excitation of most other atoms, except for helium. These studies rely on the fact that the spherical eigenstates $|n, \ell, j, m\rangle$ of atomic hydrogen at zero-field are turned into parabolic eigenstates

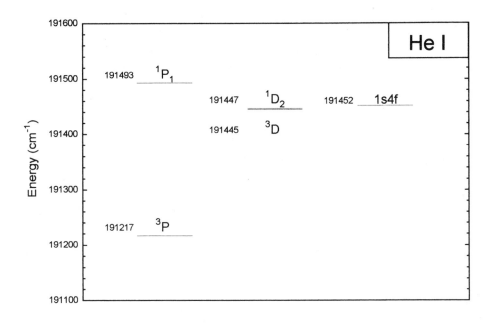

Figure 1. Fine structure of the $n=4$ shell of He I (except $1s4s\ {}^1S$ and $1s4s\ {}^3S$ level).

$|n, n_1, n_2, m\ell, m\rangle$ by electric fields of a few kV/cm or less. Though the ℓ-degeneracy of He I is not as good as for hydrogen, the energy separations of excited He I levels with different angular momenta are small enough for approaching a linear Stark splitting at electric fields which can still readily be applied experimentally. An additional advantage of helium, however, is the existence of singlet and triplet states. Therefore, helium offers various opportunities for detailed experimental studies of collisional excitation processes, which do not have a correspondence in experiments on hydrogen. In this section, we outline these peculiar features of the He I level scheme[20], which are of fundamental importance for the experimental techniques and measurements described below.

At zero-field, singlet levels (with $S = 0$) and triplet levels (with $S = 1$) of all $1sn\ell$ configurations with $\ell \leq 2$ are energetically well separated. Comparing the energy separation $\Delta E_{ST} = E(n\ {}^1L) - E(n\ {}^3L)$ of corresponding singlet and triplet levels with the magnetic fine structure splitting $\Delta E_{fs} = E(n\ {}^3L_{L+1}) - E(n\ {}^3L_{L-1})$, one finds for the ratio $r(1sn\ell) = \Delta E_{ST}/\Delta E_{fs}$ values of the order 3000 and 100 for $1snp$ and $1snd$ configurations, respectively[21]. Accordingly, the Russell-Saunders basis states $|1sn\ell, L, S, J, m\rangle$ of He I, which are eigenstates of the spin operator $\vec{S}^2 = (\vec{s}_1 + \vec{s}_2)^2$, are good approximations to the real eigenstates of He I. However, one finds $r(1sn\ell) < 1$ for all configurations with $\ell \geq 3$. Therefore, the eigenstates of the He I levels $1sn\ell\ {}^1L_L$ and ${}^3L_L$ (with total angular momentum $J = L \geq 3$) are mixtures of singlet and triplet states[20]. These mixtures give rise to strong intercombination transitions.

Another important feature of the zero-field energy levels of He I is the fact that the level order of He I within an n-shell is different for singlet and triplet system (Figure 1). Due to the large exchange interaction of the $1snp$ configuration, the energy of the ^1P level is the

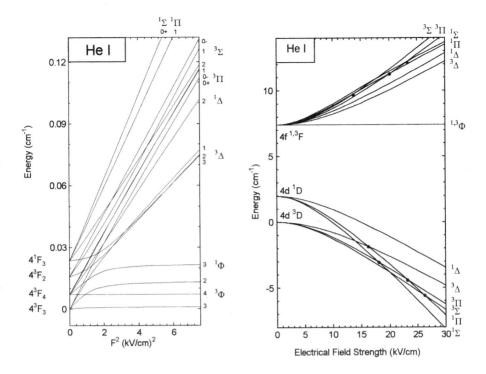

Figure 2. (left) Electric field splitting of the $1s4f$ fine structure multiplet with the levels $4\,^1F_3$ and $4\,^3F_J$, where $J=2$, 3 and 4. (The sublevels are labeled by quantum numbers $\mu=|m|$ and reflection parity if $\mu=0$. At high electric fields, the sublevels are grouped into Stark levels $^1\Lambda$ and $^3\Lambda$ with $\Lambda=|m_L|$)

Figure 3. (right) Electric-field splitting of the $1s4d$ and $1s4f$ configurations up to $F=30$ kV/cm (without magnetic fine structure). Anticrossings are marked by dots.

highest of all levels of the n-manifold, whereas the ^3P level is lower than all other $1sn\ell$ levels of the n-manifold with $\ell > 1$. Regarding the $1sn\ell$ configurations with $\ell \geq 2$, the energies of the $1sn\ell$ levels increase with increasing ℓ in the singlet as well as in the triplet system, and the singlet-triplet splitting of an $1sn\ell$ configuration is generally smaller than the energy separation from the neighboring $1sn(\ell+1)$ configuration. The energetically lowest levels of an n-manifold in both the singlet and the triplet system are the levels of the $1sns$ configurations. Both the $1sns\,^1S$ and ^3S level are energetically separated from the other levels of the manifold so well that they are little affected by the moderate electric fields applied in the experiments discussed below.

By applying an electric field F_z (z direction) each level with total angular momentum J is shifted and split in $J+1$ Stark sublevels with $\mu = |m| = 0$, 1, ... J. All sublevels with $\mu \neq 0$ are twofold degenerate. The splitting is determined by the tensor polarizabilities α_{ten} ($1sn\ell$ ^{2S+1}L) which strongly increase with n and ℓ. Along with this Stark splitting, the eigenstates of the sublevels change[22]. At weak fields, where the splitting of a triplet level reaches the order of the zero-field energy separations of the fine structure multiplet, orbital and spin angular momenta L and S are decoupled. Furthermore, considering high-ℓ configurations with $\ell \geq 3$, singlet and triplet states with $|m_L| \leq 2$ are demixed. This demixing is strongest for

Table 1. Electric-field positions of singlet-triplet anticrossings of He($1sn\ell$) states with $n \leq 5$ in kV/cm.

$^1\Lambda \times {}^3\Lambda$	$1s3d$	$1s4d$	$1s4f$	$1s5d$	$1s5f$	$1s5g$
$^1\Sigma \times {}^3\Pi$	80.87	18.13	23.29	5.74	6.60	8.90
$^1\Pi \times {}^3\Sigma$	88.83	26.31	13.72	10.01	6.50	-
$^1\Pi \times {}^3\Pi$	91.01	23.13	20.11	7.82	7.06	6.36
$^1\Pi \times {}^3\Delta$	99.55	16.33	49.23	4.71	9.72	20.54
$^1\Delta \times {}^3\Pi$	227.17	81.82	-	87.45	13.07	-

$|m_L| = 0$ sublevels and absent for $|m_L| \geq 3$ sublevels. The singlet-triplet demixing arises from the electric-field induced interaction of the high-ℓ states with the $1snd$ configuration. Due to this interaction with a configuration where singlet and triplet states are well separated at zero-field, singlet and triplet sublevels with $|m_L| \leq 2$ of the high-ℓ configuration are shifted apart and, therefore, the singlet-triplet mixing in this configuration is reduced by the electric field. The details of the Stark shifts and splittings of $1sn\ell$ configurations were calculated by D. Kaiser et al.[22] As an example, the Stark splitting of the $1s4f$ configuration is shown in Figure 2.

With increasing field strength, states with opposite parity and different angular momenta ℓ are increasingly mixed. Thereby, the atom attains electric dipole moments, which increase until complete mixing is reached and the Stark splitting becomes linear. The closely spaced high-ℓ states are mixed completely first, and then progressively also states with lower ℓ values. The level order of the Stark sublevels is strongly affected when, finally, the $\ell \geq 2$ states with $\Lambda = |m_L| \leq 1$ are significantly mixed with states of the $1snp$ configuration. Since the ^1P level is above and the ^3P level below the high-ℓ levels the $^1\Lambda$ sublevels with $\Lambda \leq 1$ are pushed downward, whereas the $^3\Lambda$ sublevels with $\Lambda \leq 1$ are pushed upward. Therefore, at large electric fields, some singlet and triplet sublevels of the $\ell \geq 3$ configuration approach again and cross each other (Figure 3). Also for the $1snd$ configurations such level crossings can be induced by moderate electric field strengths. These level crossings are turned into anticrossings by the spin-orbit interaction provided the quantum numbers m of the total angular momenta of the crossing sublevels are equal. As a result, the following anticrossings are found in the splitting schemes of most $1sn\ell$ configurations with $\ell \geq 2$ (Figure 3): $^1\Sigma_0^+ \times {}^3\Pi_0^+$, $^1\Pi_1 \times {}^3\Sigma_1$, $^1\Pi_1 \times {}^3\Pi_1$, $^1\Pi_1 \times {}^3\Delta_1$ and $^1\Delta_2 \times {}^3\Pi_2$. The electric-field positions of the anticrossings of the $1sn\ell$ configurations with $n \leq 5$ are given in Table 1. At these anticrossings, singlet and triplet states are completely mixed thereby giving rise to intercombination transitions.

4. IMPACT RADIATION IN AXIAL ELECTRIC FIELDS

For analyzing the state of He atoms immediately after a p-He collision, the collisional induced radiation emitted by He atoms in external fields can be investigated. Here, we refer to measurements, where electric fields parallel or antiparallel to a continuously running proton beam (z-axis) were applied and the scattered protons remained unobserved. Under these conditions, intensity, angular distribution and polarization of the impact radiation do not depend on a coherent excitation of different Stark eigenstates. The rotational symmetry with respect to the z-direction implies that only states with the same Zeeman-quantum number can be excited coherently. But states with $\Delta m = 0$ have different energies. Therefore, a possible initial coherence of the excitation amplitudes is averaged out during the radiative lifetimes of the excited states under stationary experimental conditions.

Under these circumstances, the population of the Stark eigenstates can be evaluated using rate equations[23]. An eigenstate $|\varepsilon, m\rangle$ with energy ε and Zeeman quantum number m is populated by direct collisional excitation and cascade feeding. The collisional excitation rate

$$\sigma(\varepsilon, m) = I_p \cdot \sigma_{\varepsilon,m} \tag{1}$$

depends on the intensity I_p of the proton beam (number of protons per m$^2 \cdot$ sec) and the excitation cross section $\sigma_{\varepsilon,m}$ of the eigenstate. Cascade feeding from an energetically higher state $|\varepsilon', m'\rangle$ with population number $\rho(\varepsilon', m')$ contributes to the population rate of $|\varepsilon, m\rangle$ with $A_{\varepsilon m}^{\varepsilon'm'} \cdot \rho(\varepsilon', m')$ where A is the appropriate transition rate. Finally, the depopulation rate $\sum_{\varepsilon',m'} A_{\varepsilon m}^{\varepsilon'm'} \rho(\varepsilon', m')$ is given by the decay-rate $\Gamma(\varepsilon, m) = \sum_{\varepsilon',m'} A_{\varepsilon m}^{\varepsilon'm'}$ of $|\varepsilon, m\rangle$ and the population number $\rho(\varepsilon, m)$ of $|\varepsilon, m\rangle$. Under stationary conditions, therefore, the following rate equation is valid:

$$\sigma(\varepsilon, m) + \sum_{\substack{\varepsilon'm' \\ (\varepsilon' > \varepsilon)}} A_{\varepsilon m}^{\varepsilon'm'} \rho(\varepsilon', m') - \Gamma(\varepsilon, m) \rho(\varepsilon, m) = 0 \tag{2}$$

The population numbers $\rho(\varepsilon, m)$ are connected with intensity, angular distribution and polarization of the emitted spectral lines[24]. Considering a transition $|\varepsilon, m\rangle \rightarrow |\varepsilon', m'\rangle$ with transition rate $A_{\varepsilon'm'}^{\varepsilon m}$ (where $\varepsilon' < \varepsilon$), the integral intensity $I_{\varepsilon'm'}^{\varepsilon m}$ (number of photons per sec) is given by

$$I_{\varepsilon'm'}^{\varepsilon m} = A_{\varepsilon'm'}^{\varepsilon m} \cdot \rho(\varepsilon, m) \cdot N_T \tag{3}$$

where N_T is the number of target atoms within the observation volume. Angular distribution and polarization depend on $\Delta m = |m - m'|$. For π transitions ($\Delta m = 0$), the emitted light is linearly polarized with a polarization vector parallel to the plane $\{z, \vec{k}\}$ spanned by the z axis and the propagation direction $\vec{k}$, and the angular distribution is given by

$$w_\pi^\parallel\left(\vartheta,\varphi\right)=\tfrac{3}{8\pi}\sin^2\vartheta \tag{4}$$

For σ transitions ($\Delta m = 1$) the emitted light is usually only partly linearly polarized. The angular distributions $w_\sigma^\parallel$ and $w_\sigma^\perp$ are given by

$$w_\sigma^\parallel\left(\vartheta,\varphi\right)=\tfrac{3}{16\pi}\cos^2\vartheta$$
$$w_\sigma^\perp\left(\vartheta,\varphi\right)=\tfrac{3}{16\pi} \tag{5}$$

These angular distributions are normalized such that

$$\int w_\pi^\parallel\left(\vartheta,\varphi\right)d\Omega=1$$
$$\int\left\{w_\sigma^\parallel\left(\vartheta,\varphi\right)+w_\sigma^\perp\left(\vartheta,\varphi\right)\right\}d\Omega=1 \tag{6}$$

Using the relations (3), (4) and (5), the intensity $I_{\varepsilon'm'}^{\varepsilon m}\left(\vartheta,\varphi,p\right)d\Omega$ of a transition $\left|\varepsilon,m\right\rangle\rightarrow\left|\varepsilon',m'\right\rangle$, where a photon with polarization $p=\parallel$ or $\perp$ is emitted in the direction $\left(\vartheta,\varphi\right)$ within a solid angle $d\Omega$ can be evaluated:

$$I_{\varepsilon'm'}^{\varepsilon m}\left(\vartheta,\varphi,p\right)d\Omega=I_{\varepsilon'm'}^{\varepsilon m}\cdot w_{\Delta m}^{p}\left(\vartheta,\varphi\right)d\Omega \tag{7}$$

With respect to the experiments considered below, we have finally to discuss the electric-field dependence of $I_{\varepsilon'm'}^{\varepsilon m}\left(\vartheta,\varphi,p\right)$. Both the transition rates $A_{\varepsilon'm'}^{\varepsilon m}\equiv A_{\varepsilon'm'}^{\varepsilon m}\left(F_z\right)$ and the excitation rates $\sigma(\varepsilon,m)=\sigma(\varepsilon,m;F_z)$ depend on the electric-field strength F_z, since the eigenstates $\left|\varepsilon,m\right\rangle$ are field-dependent as discussed in section 3. The transition rates $A_{\varepsilon'm'}^{\varepsilon m}$ are functions of the electric dipole matrix elements $\left\langle\varepsilon,m\left|e(\vec{r}_1+\vec{r}_2)\right|\varepsilon',m'\right\rangle$ and can be evaluated using standard quantum mechanics. The excitation rates, however, depend on the collision process and their evaluation is less obvious.

The functions $\sigma(\varepsilon,m;F_z)$ can be reduced to a finite set of excitation parameters using the concept of impact-excited states. If the collision process can be considered as approximately instantaneous compared with the evolution periods $\hbar/\left(\varepsilon'-\varepsilon\right)$ of the free atom, one can visualize an impact excited state

$$\left|f\right\rangle=\sum_{\varepsilon,m}\left|\varepsilon,m\right\rangle\left\langle\varepsilon,m\left|f\right.\right\rangle \tag{8}$$

found immediately after the collision, which is a coherent superposition of neighboring eigenstates of the atom. The instantaneity condition implies that a coherent excitation can reasonably be considered only for levels with

$$\varepsilon-\varepsilon'\ll\frac{\hbar}{\tau_{coll}} \tag{9}$$

For p-He collisions, where the proton velocity v_p is of the order of 1 $a.u.$, an estimate of the collision time τ_{coll} yields[14]:

$$\tau_{coll} \approx \frac{n^2}{v_p^2} \; a.u. \tag{10}$$

Using this estimate, condition (9) is fulfilled for $He(1sn\ell)$ states with the same principal quantum number n.

When considering integral excitation rates, the density-matrix formalism has to be employed for describing impact-excited states. Therefore, we introduce excitation operators σ where the diagonal matrix elements $\langle \varepsilon,m|\sigma|\varepsilon,m \rangle = \sigma(\varepsilon,m)$ are the excitation rates of the states $|\varepsilon,m\rangle$ and the non-diagonal matrix elements $\langle \varepsilon,m|\sigma|\varepsilon',m' \rangle$ are the coherence parameters of the impact-excited state. As all density operators, the excitation operator σ is hermitean and positive definite, i.e. all expectation values of σ are real and positive. Due to the instantaneity condition, the excitation operator can be assumed being independent of the strength of the external fields used in the experiments.

For discussing the properties of the excitation operators σ_n of the He atoms collisionally excited into n-shell states by protons, we use the Russell-Saunders representation with matrix elements

$$\langle n; L, m_L; S, m_S|\sigma_n|n; L', m_L'; S', m_S' \rangle \tag{11}$$

These matrix elements fulfill the following selection rules: (i) Due to Wigner's spin selection rule, only singlet states can be excited by proton impact. Therefore, only matrix elements with $S = S' = 0$ and $m_S = m_S' = 0$ are non-zero. (ii) Due to the rotational symmetry with respect to the proton beam, all matrix elements with $m_L \neq m_L'$ vanish as mentioned above. (iii) Due to reflection symmetry with respect to any plane containing the beam direction, the matrix elements do not depend on the sign of m_L, but on $\Lambda = |m_L|$. Accordingly, the excitation matrix is completely described by the following set of parameters

a) the excitation rates $\sigma(n \; ^1L, \Lambda)$ of the singlet states $|n; L, m_L; 0, 0\rangle$ and

b) the coherence parameters $\sigma(L, L'; n \; ^1\Lambda) = \langle n; L, \pm \Lambda; 0, 0|\sigma_n|n; L', \pm \Lambda; 0, 0\rangle$

The coherence parameters are usually complex. Using standard phase conventions for the Russell-Saunders states[25], the real components represent the time-reversal invariant part of the excitation matrix and are related to the charge distribution of the impact-excited state, in particular, the real components of the parameters with $\Delta L = 1$ to its electric dipole moment. The imaginary components change sign under time reversal and describe the multipole components of the current distribution. As long as only the real excitation rates

$$\sigma(\varepsilon, m) = \langle \varepsilon, m|\sigma_n|\varepsilon, m \rangle \tag{12}$$

of Stark eigenstates $|\varepsilon,m\rangle$ are measured, only the charge distribution of the impact excited state can be evaluated.

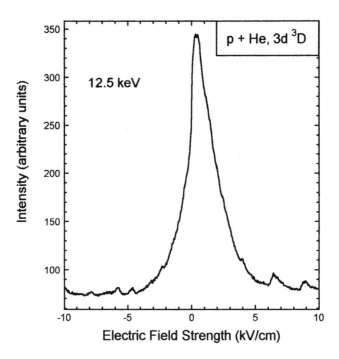

Figure 4. Intensity of the He I spectral line at $\lambda(3\ ^3D - 2\ ^3P) = 588$ nm as a function of an axial electric field for 12.5 keV proton impact.

5.   ELECTRIC-FIELD DEMIXING OF SINGLET AND TRIPLET STATES

According to Wigner´s spin conservation rule, impact excitation of He I triplet states is forbidden in collisions of bare nuclei with ground-state He-atoms. In apparent contradiction to Wigner´s rule, the emission of triplet lines, especially of lines originating from the decay of 1snd ^{3}D states had been observed after collisions of He atoms with protons[26] and α particles[27]. For demonstrating that these experimental findings do not contradict Wigner´s spin conservation rule, but are a result of singlet-triplet mixing in high-ℓ states, Aynacioglu *et al.*[28] investigated the effect of an axial electric field on the intensity of the $\lambda(3\ ^3D - 2\ ^3P) = 588$ nm line of He I induced by collisions with 10-25 keV protons.

A more recent recording[29] of the intensity I(588 nm) as a function of the strength of an axial electric field F_z for 12.5 keV proton impact is shown in Figure 4. By applying electric fields $F_z \approx 5$ kV/cm either parallel ($F_z > 0$) or antiparallel ($F_z < 0$) to the direction of the proton beam, the line intensity is reduced by about a factor 5. This dramatic intensity decrease is caused by the demixing of singlet and triplet states of the 1s$n\ell$ configurations with $\ell \geq 3$, in particular, of the 1s4f configuration.

Several conclusions can be drawn from a qualitative inspection of the signal structure. The existence of a strong field dependence of I(588 nm) already makes obvious that the 1snd ^{3}D level is mainly populated by cascade feeding, because an electric field $F_z \leq 5$ kV/cm hardly effects the eigenstates of the $1s3d\ ^3D$ level due to its small tensor polarizability[30] $\alpha_{ten}(1s3d\ ^3D)$ = 2.7 MHz/(kV/cm)^2. Cascade feeding of the $3\ ^3D$ level is most likely via primary excitation of the $1s4f$ and other $\ell \geq 3$ configurations from which the atom exclusively or at least predominantly cascades down in one or several steps to the $1s3d$ configuration. The $\ell \geq 3$ levels have tensor polarizabilities orders of magnitude larger than the $1s3d\ ^3D$ level since the polarizabilities strongly increase with increasing n and ℓ. For example[31], the polarizability $\alpha_{ten}(1s4f\ ^3F) \approx 400$ MHz/(kV/cm)^2. Therefore, by assuming cascade feeding of the $1s3d\ ^3D$ level from the $1s4f$ configuration and higher $\ell \geq 3$ levels, the strong field dependence of $I(588$ nm$)$ at weak electric fields can be explained.

The assumption that the $1s3d\ ^3D$ level is populated by cascade feeding via high-ℓ states is even more supported by the fact that the $\ell \geq 3$ states are the only He I levels where singlet and triplet states are strongly mixed at zero-electric field. Only due to this mixing, $\ell \geq 3$ states decay by one or more electric dipole transitions to both the $1s3d\ ^1D$ and the $1s3d\ ^3D$ level. Therefore, inspite of Wigner's spin conservation rule, the almost pure triplet states of the $1s3d\ ^3D$ level can be populated by proton impact. However, by applying an electric field, the singlet-triplet mixing of the $\ell \geq 3$ states is reduced and, therefore, the gateway to the triplet states is closed. As a consequence, the intensity $I(588$ nm$)$ is much lower at high electric fields than at zero-electric field.

For a more detailed analysis of the signal structure not only the excitation rates of the zero-field $\ell \geq 3$ states have to be taken into account, but also the coherence parameters. In particular, the coherence parameters are needed for explaining the asymmetry of the signal structure with respect to the sign of F_z. However, the influence of coherence parameters on $I(588$ nm$)$ is complicated by cascade feeding. Their influence becomes more transparent when direct excitation is dominant. Therefore, we postpone the discussion of coherent excitation and its influence on the collisionally induced radiation to section 6 and 7.

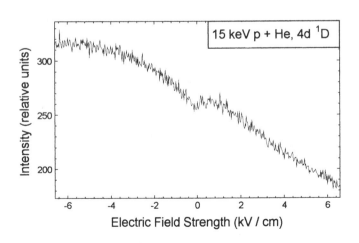

Figure 5. Intensity of the He I spectral line at $\lambda(4\ ^1D$ - $2\ ^1P)$ = 588 nm as a function of an axial electric field for 15 keV proton impact.

6. ELECTRIC-FIELD MIXING OF OPPOSITE-PARITY STATES

The charge asymmetry of impact excited states, e.g. their electric dipole moments strongly affects the electric-field dependence of the impact radiation, when the electric field is strong enough to mix opposite-parity states. A strong-mixing regime is reached when the Stark shifts change from a quadratic to a linear electric-field dependence. We exemplify the influence of this charge asymmetry by measurements on the $\lambda(1s4d\,{}^1D - 1s2p\,{}^1P) = 492$ nm line of He I excited by 5 - 20 keV proton impact[15] (Figure 5). In this case, the electric-field dependence of $I(492$ nm$)$ is mainly caused by the Stark mixing of the $1s4d\,{}^1D$ and $1s4f\,{}^1F$ states. The state vectors

$$\left|1s4d\,{}^1D, m_L; F_z\right\rangle = \cos\tfrac{\vartheta}{2}\left|1s4f\,{}^1D, m_L\right\rangle + \sin\tfrac{\vartheta}{2}\left|1s4f\,{}^1F, m_L\right\rangle \tag{13}$$

of the ^1D Stark states in an electric field F_z are linear combinations of the zero-field Stark states of the $1snd$ and $1snf$ configuration. The mixing angle is given by

$$\tan\vartheta = \frac{\left\langle 1s4f\,{}^1F, m_L \left| e(z_1 + z_2) \right| 1s4d\,{}^1D, m_L \right\rangle \cdot F_z}{\tfrac{1}{2}\left(E(1s4f) - E(1s4d\,{}^1D)\right)} \tag{14}$$

where $e(z_1 + z_2)$ is the z component of the electric-dipole operator for the 2-electron system and the energy separation is[20]

$$E(1s4f) - E(1s4d\,{}^1D) = 5 \text{ cm}^{-1} \cdot hc \tag{15}$$

The electric-dipole matrix element in the Heisenberg approximation[32] is given by

$$\begin{aligned}\left\langle 1s4f\,{}^1F, m_L \left| e(z_1 + z_2) \right| 1s4d\,{}^1D, m_L \right\rangle = \\ \frac{2}{\sqrt{5}}\left(9 - m_L^2\right) \cdot ea_B = 0.038 \text{ } kV^{-1} \cdot \left(9 - m_L^2\right) \cdot hc\end{aligned} \tag{16}$$

Accordingly, for the $m_L = 0$ state, $\tan\vartheta = 1$ is reached at $F_z \approx 7.5$ kV/cm. Due to this mixing, the atomic eigenstates acquire electric dipole moments

$$d(F_z) = \sin\vartheta \cdot \frac{2}{\sqrt{5}}\left(9 - m_L^2\right) \cdot ea_B \tag{17}$$

Since the energies of the 4 ^1D Stark states decrease with increasing field strength, He-atoms in the 4 ^1D state are attracted by electric fields, i.e. the electric-dipole moment has the same direction as the electric field. These F_z-induced electric dipole moments can be compared with the electric dipole moment of the collisionally excited state[19] as illustrated in Figure 6. Assuming a collision-induced electric dipole moment directed upstream, the excitation rates $\left\langle 1s4d\,{}^1D, m_L \left| \sigma \right| 1s4d\,{}^1D, m_L \right\rangle$ of the $1s4d\,{}^1D$ Stark states can be expected to decrease when scanning the F_z through zero-field from upstream to downstream. An oppo-

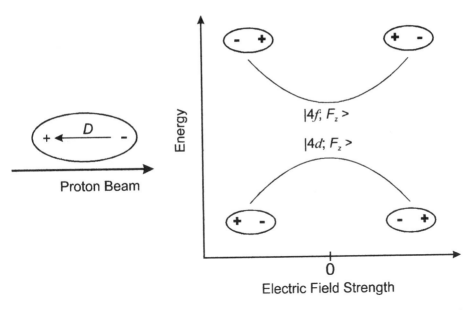

Figure 6. Electric dipole moment D of the impact-excited state relative to the beam direction (left) and of the atomic eigenstates relative to the applied field F_z (right).

site behavior is expected for the population of the $1s4f\,{}^1F$ states with $|m_L| \leq 2$, which are repelled by the electric field.

The change of the excitation rates was detected by observing the 492 nm line selected using an interference filter with a half width of about 10 nm. Therefore, both $1s4d\,{}^1D$ - $1s2p\,{}^1P$ and electric-field induced $1s4f$ - $1s2p\,{}^1P$ transitions were detected. For a detailed analysis of the recordings shown in Figure 5, also the electric-field dependence of the branching ratios of these transitions must be taken into account. However, at fields $F_z \ll 7.5$ keV/cm, the dependence of $I(492$ nm$)$ on the branching ratio is negligible. Therefore, the electric-field dependence of $I(492$ nm$)$ reflects approximately the electric-field dependence of the population rates. Obviously, the recording in Figure 5 confirms the assumption that the collision-induced electric dipole moments are directed upstream.

Though the direct decay of the $1s4f\,{}^1F$ states connected with emission of the infrared $\lambda(1s4f$ - $1s3d) \approx 1.9$ μm line was not investigated, the excitation rates of the $1s4f\,{}^1F$ Stark state can be determined by analyzing the recording of $I(588$ nm$)$ (Figure 4). As outlined in section 5, this triplet line is excited by cascade feeding of the $1s3d\,{}^3D$ level via $\ell \geq 3$ states. The dominant contribution stemming from collisional excitation of the $1s4f\,{}^1F$ state. Therefore, the asymmetry of the 588 nm recording is mainly due to the electric-field dependence of the $1s4f\,{}^1F$ excitation rates. The observed asymmetry obviously also agrees with the assumption that the collisionally induced electric dipole moment points upstream.

7.   ANTICROSSING MEASUREMENTS

The experimental techniques discussed so far are useful only for determining averaged population rates of several Zeeman and fine structure sublevels assuming the Stark splitting cannot be resolved spectroscopically. An experimental method sensitive to the excitation rate of one single Stark level which does not employ high spectral resolution is provided by anticrossing techniques[16]. As shown in section 3, anticrossings of singlet and triplet Stark levels occur for $1sn\ell$ configurations with $\ell \geq 2$. At the center of such an anticrossing, its singlet and triplet states are mixed completely. As high-ℓ states at zero-field, also the sublevels of an anticrossing decay both to singlet and triplet levels due to the singlet-triplet mixing at the anticrossing. However, since the energy separations at the anticrossing resulting from spin-orbit coupling are small compared with the exchange interaction of the $1snd$ configurations, narrow resonance-like intensity variations arise from the singlet-triplet mixing at anticrossings. Examples of these resonance-like structures are found already in the recording of Figure 4, namely at electric fields $F_z \approx$ -10.0, -7.8, -5.7, -4.7, +6.6 and +8.9 kV/cm. These measured electric field strengths can be compared with calculated electric-field positions of anticrossings (Table 1). One finds that the signals at fields $F_z < 0$ correspond to anticrossings of the $1s5d$ configuration and the signals at fields $F_z > 0$ to anticrossings of the $1s5f$ and $1s5g$ configuration. Two remarkable conclusions can be drawn immediately from these signals: (i) cascade feeding from the $1s5f$ ^1F states contributes to the population of the $1s3d$ ^3D level, and (ii) the $n = 5$ states excited by 12.5 keV proton impact have nearly the maximum possible electric dipole moments. The first conclusion is based on the very existence of the anticrossing signals, and the second conclusion on their asymmetric distribution.

Both the $1s5d$ and the $1s5f$ anticrossings can also be detected by observing the impact radiation at 403 nm corresponding to $1s5d$ ^3D - $1s2p$ ^3P transitions. According to the decay rates of the Stark states of these anticrossings, one expects that the $1s5d$ anticrossings can be detected better by observing the 403 nm line than by observing the 588 nm line. The opposite is true for the $1s5f$ anticrossings, where $1s5f$ - $1s2p$ ^3P transitions are induced only by the electric field. The recording of the 403 nm intensity as a function of F_z (Figure 7) confirms this expectation. Strong signals of the $1s5d$ anticrossings were found for $F_z < 0$, whereas only rather weak signals of $1s5f$ anticrossings were observed at electric-fields $F_z > 0$. Again, the signal structure is extremely asymmetric. Except for a small peak at the position of the $1s5d$ $^1\Sigma \times ^3\Pi$ anticrossing at $F_z = +5.7$ kV/cm, the strong $1s5d$ anticrossing signals at fields $F_z < 0$ do not have counterparts at fields $F_z > 0$.

We note that also the broad background structure of the recording (Figure 7) has a pronounced asymmetry. However, this background arises mainly from the $\lambda(1s7d$ ^1D - $1s2p$ $^1P) = 401$ nm line which was also transmitted by the interference filter used for wavelength selection. This $1s7d$ ^1D signal can be interpreted in essentially the same way as the $1s4d$ ^1D signals discussed in section 6, namely, by the electric-field mixing of the $1s7d$ ^1D states with high-ℓ states. But one has to take into account that $1s7d$ atoms can be polarized by much weaker electric fields than $1s4d$ atoms. Therefore, the variation of decay rates affects the intensity I(401 nm) and even the $1s7p$ admixture to the $1s7d$ states becomes significant at high electric fields. Due to this admixture, the $1s7d$ ^1D Stark states (with Λ=0 or 1) preferentially decay to the $1s^2$ ^1S groundstate of He I at high electric fields and, therefore, the intensity at 401 nm almost vanishes at high electric fields.

Inspite of this background structure of the 401 nm line, the anticrossing signals of the $n = 5$ configurations can be easily identified. Most surprising is the absence of $1s5d$ signals at fields $F_z > 0$. This absence can only be explained by assuming that the $n = 5$ states of He I excited by 12.5 keV-proton impact are essentially orthogonal to the $1s5d$ ^1D Stark states at the electric fields of the anticrossings, if $F_z > 0$. This assumption has to be made inspite of the fact that the $1s5d$ Stark levels are strongly excited if $F_z < 0$. This orthogonality can be understood by assuming that the impact-excited $n = 5$ states are rather pure $^1\Lambda$ states, where the excited electron is approximately in a hydrogen-like parabolic Stark state.

In view of the fact that the impact-excited states result from integral measurements, where a statistical average of collisions with different impact parameters must be considered, the conclusion that almost pure states instead of weighted mixtures are excited is surprising. In the following section, we shall try to explain this surprising result by proposing a basic mechanism mainly responsible for excitation processes at intermediate proton energies. Here, we shall discuss the experimental basis for the identification of the impact-excited states.

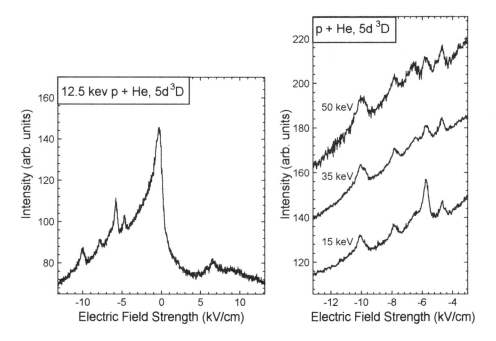

Figure 7. (left) Intensity of the $\lambda \approx 403$ nm lines of He I as a function of an axial electric field F_z for 12.5 keV proton impact. Resonance-like structures at $F_z \neq 0$ kV/cm are due to singlet-triplet anticrossings of the $1s5d$ and $1s5f$ configurations.

Figure 8. (right) Signals of singlet-triplet anticrossings of the $1s5d$ and $1s5f$ configuration measured at $\lambda \approx 403$ nm for 15, 35 and 50 keV proton impact. A constant has been added to separate the different measurements.

Since the amplitudes of the high-ℓ components of He I Stark states at high electric fields become equal to the corresponding amplitudes of the parabolic Stark states $|n; n_1, n_2, m_\ell\rangle$ of hydrogen, the orthogonality mentioned above suggests to base the analysis of the anticrossing signals on this system of basis states. The z-component d_z of the electric dipole moments of these states are[32]

$$d_z\left(n; n_1 - n_2\right) = -\tfrac{3}{2}n\left(n_1 - n_2\right)ea_B \tag{18}$$

For $n = 5$, the only parabolic Σ and Π states with $n_1 - n_2 > 0$ are $|5; 4, 0, 0\rangle$, $|5; 3, 1, 0\rangle$ and $|5; 3, 0, 1\rangle$, $|5; 2, 1, 1\rangle$ respectively. Assuming selective excitation of these basis states, the relative amplitudes of the anticrossings of $^1\Sigma$ and $^1\Pi$ Stark levels were evaluated[16,17].

The experimental results for 12.5 keV proton impact are in excellent accord with the calculated amplitudes when selective excitation of the parabolic $^1\Sigma$ and $^1\Pi$ Stark states $|5; 4, 0, 0\rangle$ and $|5; 3, 0, 1\rangle$, respectively, with the ratio $\sigma(5; 4, 0, 0)/\sigma(5; 3, 0, 1) \approx 3{:}1$ is assumed. That are the Stark states with the largest electric dipole moments. However, the experimental results disagree significantly with the calculated amplitudes when assuming selective excitation of other parabolic Stark states. Therefore, the excitation rates of these Stark states are obviously significantly smaller than $\sigma(5; 4, 0, 0)$ and $\sigma(5; 3, 0, 1)$.

The relative excitation rates of parabolic Stark states change with increasing proton energy. Recordings of the 403 nm intensity for proton energies of 35 keV and 50 keV are shown in Figure 8. Here, the signal of the $1s5d$ $^1\Sigma \times \!^3\Pi$ anticrossing has an amplitude comparable with the amplitudes of the $1s5d$ $^1\Pi \times \!^3\!\lambda$ anticrossings. Thus the ratio $\sigma(5; 4, 0, 0)/\sigma(5; 3, 0, 1)$ is reduced for 35 keV and 50 keV protons. However, also a new anticrossing signal appears at $F_z = -6.6$ kV/cm indicating that at these proton energies not only the $^1\Sigma$ Stark state $|5; 4, 0, 0\rangle$ but also the $|5; 3, 1, 0\rangle$ state is strongly excited (see Table 1 in Ref. 16, where the relative amplitudes of the $1s5f$ $^1\Sigma \times \!^3\Pi$ anticrossing at 6.6 kV/cm are tabulated).

8.   DISCUSSION OF THE EXCITATION PROCESS

Excitation of helium atoms by proton impact at intermediate velocities is one of the basic processes of atomic collision physics. Therefore, the various experimental results obtained for p-He collisions may be of significance also for other collision processes. However, we shall constrain the following discussion to p-He collisions. For this collision system, the most prominent features of collisionally excited He I states were found at proton energies of 10 - 15 keV. We emphasize the following properties of these states:
(i)   Their total electron spin is $S = 0$.
(ii) The impact excited states are strongly aligned with a predominant population of $\Lambda \leq 1$ states.
(iii) The excitation cross section of high-ℓ states have values comparable with those of low-ℓ states.

(iv) The impact-excited states have large electric dipole moments. The parabolic $^1\Sigma$ and $^1\Pi$ Stark states with the largest electric dipole moments are populated almost exclusively.

Property (i) was proved experimentally, in particular, by measurements of Aynacioglu et al.[28] and is a consequence of Wigner's spin conservation rule. Property (ii) became obvious already from measurements on the polarization of ($n\ ^1D$ -$2'P$) spectral lines where polarization degrees of about 40% were found[33]. The relative population of magnetic 4 ^1D-sublevels has been studied in detail using level-crossing techniques[34], but only at proton energies above 35 keV. Property (iii) is documented most clearly by the strong cascade feeding of the $1s3d\ ^3D$ level proved experimentally by investigations using magnetic depolarization[28] and electric-field demixing (section 5). Finally, property (iv) became apparent from the investigations on electric-field singlet-triplet anticrossings (section 7).

Only property (i) is a consequence of a well established law of collision physics. The other properties were unexpected and rather surprising, in particular, as integral excitation cross-sections are considered. For interpreting these results, the main task consists in finding an explanation for the selective excitation of parabolic Stark states with the largest electric dipole moments, that is an explanation of property (iv). The properties (ii) and (iii) are then immediate consequences of this selectivity.

Without regarding its specific state vector, it is already surprising that the impact-excited He I state is essentially independent of the impact parameter b of the collision process. Regarding a particular n-shell, only the ratio $\sigma(n; n-1, 0, 0)/\sigma(n; n-2, 0, 1)$ (or the ratio of the corresponding excitation amplitudes) is possibly dependent on b. This b-independence indicates that the final-phase evolution determines the properties of the collisionally excited n-shell states ($n \geq 3$). This conclusion can be drawn since the molecular axis of the collision system is almost parallel to the proton beam during the final phase and, hence, the impact parameter is here of little importance.

The selective excitation of extreme parabolic Stark states at intermediate proton energies can now be explained by referring to molecular orbital (MO) saddle dynamics[16]. Rost and Briggs[35] pointed out that electrons can be promoted in a symmetric two-center Coulomb potential along in-saddle sequences when the two centers recede from each other. Therefore, starting from the molecular groundstate $1s\sigma$, an electron is promoted to the parabolic states $|n; n-1, 0, 0\rangle$, and starting form the MO $2p\pi$ which can be reached by rotational coupling, the electron is promoted to the parabolic states $|n; n-2, 0, 1\rangle$. At first sight, this reasoning applies only to the symmetric system H_2^+. However, the singly excited states of HeH$^+$ resemble corresponding H_2^+ states. Also the excited electron of HeH$^+$ moves in an approximately symmetric two-center Coulomb potential. Therefore, MO saddle dynamics of the H_2^+ system may be considered also as a first approach for describing p-He collisions.

Apparently, this approach yields a satisfying explanation of the excitation of extreme parabolic Stark states. However, this MO approach should be applicable only to low-velocity collisions. But 12.5 keV protons have already the velocity $v_p = 0.7$ a.u. and experiments with higher proton energies showed that the impact excited states have still a large forward-backward asymmetry at velocities as high as $v_p \approx 2$ a.u., where the MO approach should have become invalid. A possible reason for the applicability of saddle dynamics to collisions at rather high velocities has been pointed out by one of the authors[36]. Due to the rotation of the molecular axis of the collision system, an electron can be stabilized on the Coulomb saddle during the collision time similarly as charged particles are captured in a Paul trap[37]. This mechanism works best at proton velocities of the order of 1 *a.u.* and, therefore, possibly explains the large electric dipole moments of the collisionally excited He I states found at intermediate proton velocities.

REFERENCES

1. W. Pauli: Handbuch der Physik XXIII (1926) 1 (The reference to the work of N. Bohr is on page 140)
2. O. Nedelec: Journ. Physique $\underline{27}$ (1966) 660
3. G. v. Oppen: Comments At. Mol. Phys. $\underline{15}$ (1984) 87
4. L.A. Sellin, J.R. Mowat, R.S. Peterson, P.M. Griffin, R. Laubert, H.H. Haselton: Phys. Rev. Lett. $\underline{31}$ (1973) 1335
5. C.C. Havener, N. Rouze, W.B. Westerveld, J.S. Risley: Phys. Rev. Lett. $\underline{53}$ (1984) 1049
6. T.G. Eck: Phys. Rev. Lett. $\underline{31}$ (1973) 270
7. I.A. Sellin, L. Liljeby, S. Mannervik, S. Hultberg: Phys. Rev. Lett. $\underline{42}$ (1979) 570
8. R. Krotkov, J. Stone: Phys. Rev. $\underline{A22}$ (1980) 473
9. CV.C. Havener, W.B. Westerveld, J.S. Risley, N.H. ,Tolk, J.C. Tully: Phys. Rev. Lett. $\underline{48}$ (1982) 926
10. J.R. Ashburn, R.A. Cline, P.J.M. van der Burgt, W.B. Westerveld, J. S. Risley: Phys. Rev. $\underline{A41}$ (1990) 2407
11. J. Burgdörfer: Phys. Rev. Lett. $\underline{43}$ (1979) 505
12. J. Burgdörfer: Z. Phys. A - Atoms and Nuclei - $\underline{309}$ (1983) 285
13. A. Jain, C.D. Lin, W. Fritsch: Phys. Rev. $\underline{A35}$ (1987) 3180
14. A.S. Aynacioglu, S. Heumann, G. v. Oppen: Phys. Rev. Lett. $\underline{64}$ (1990) 1879
15. S. Heumann: Thesis, TU Berlin D83 (1993)
16. S. Büttrich, G. v. Oppen: Phys. Rev. Lett. $\underline{71}$ (1993) 3778
17. G. v. Oppen, S. Büttrich: Z. Phys. D $\underline{30}$ (1994) 193
18. A.S. Aynacioglu, S. Heumann, G. v. Oppen: Supplement to Z. Phys. D $\underline{21}$ (1991) S 297
19. M. Hanselmann, A.S. Aynacioglu, G. v. Oppen: Physics Letters $\underline{A175}$ (1993) 314
20. W.C. Martin: Phys. Rev. $\underline{A36}$ (1987) 3575
21. G. v. Oppen: Physica Scripta $\underline{T26}$ (1989) 34
22. D. Kaiser, Y.-Q. Liu, G. v. Oppen: J. Phys. B $\underline{26}$ (1993) 363
23. E.W. Thomas: Excitation in Heavy Particle Collisions, Wiley-Interscience (New York, 1972)
24. E.U. Condon, G.H. Shortley: The Theory of Atomic Spectra, At The University Press (Cambridge, 1964)
25. A.R. Edmonds: angular Momentum in Quantum Mechanics, Princeton University Press (1960)
26. D. Krause, E.A. Soltysik: Phys. Rev. $\underline{175}$ (1968) 142
27. G.D. Myers, J.D. Ambrose, P.B. James, J.J. Leventhal: Phys. Rev. $\underline{A18}$ (1978) 85
28. A.S. Aynacioglu, G. v. Oppen, R. Müller: Z. Phys. D $\underline{6}$ (1987) 155
29. S. Büttrich: Thesis, TU Berlin, in preparation
30. W.D. Perschmann, G. v. Oppen, D. Szostak: Z. Phys. A $\underline{311}$ (1983) 49
31. A.S. Aynacioglu, G. v. Oppen, W.D. Perschmann, D. Szostak: Z. Phys. A $\underline{303}$ (1981) 97
32. H.A. Bethe, E.E. Salpeter: Handbuch der Physik 35, Springer Verlag (1957)
33. J. van Eck, F.J. de Heer, J. Kistenmaker: Physica $\underline{30}$ (1964) 1171
34. M. Anton, D. Detleffsen, K.-H. Schartner, A. Werner: J. Phys. B $\underline{26}$ (1993) 2005
35. J.M. Rost, J.S. Briggs: J. Phys. B $\underline{24}$ (1991) 4293
36. G. v. Oppen: Europhys. Lett. $\underline{27}$ (1994) 279
37. W. Paul, Physik. Blätter $\underline{46}$ (1990) 227

ANOTHER TYPE OF COMPLETE EXPERIMENT: ION-IMPACT INDUCED MOLECULAR COULOMB-FRAGMENTATION

U. Werner and H.O. Lutz

Fakultät für Physik
Universität Bielefeld
Universitätsstraße 25
D-33615 Bielefeld
Germany

Introduction

The ion-induced fragmentation of molecules is a process of fundamental importance in various areas of science ranging from the physics and chemistry of upper planetary atmospheres[1] to the understanding of radiation damage to biological tissue. In contrast to studies of the molecular break-up by electron bombardment[2] the ion-impact induced fragmentation has yet received comparatively little attention. Experiments in which *all* fragment ions emitted after a particular collision are detected in coincidence, can not only provide valuable information about the charge state and potential energy surfaces of the intermediate multiply-charged molecular ion, they also shed light on the excitation and fragmentation dynamics. Most work so far has concentrated on the observation of individual reaction products without further attention to the correlated behavior of the remaining fragments[3]. Exceptions can be found in dissociation studies of some diatomic molecules where the correlation of both fragments has been studied in detail[4] and in the "Coulomb explosion" studies of molecular ions[5,6]. This latter work is of particular interest since it allows a *kinematically complete* study of molecular fragmentation. The molecule under study is passed at high velocities through a thin foil where the outer electrons are stripped off. The fragments from the ensuing Coulomb explosion are detected in a position- and time-sensitive device; thereby the momentum vectors of correlated fragments can be derived. As powerful as it may be, this technique is in practice only applicable to molecular ions with kinetic energies of at least several MeV, and the data analysis is complicated by the complex interactions in the foil. Many experimental situations, however, require the handling of fairly low-energy particles; a typical experiment employs, after the dissociation, an acceleration of the charged reac-

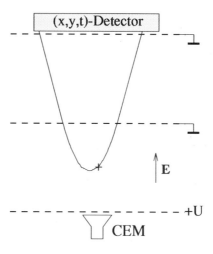

Figure 1. Geometry of the fragmentation experiment. + is the locus of the Coulomb-fragmentation, — projection of correlated fragment-ion paths.

tion products and their detection in a time-of-flight (TOF) spectrometer. In the work to be reported here we use a time- and position-sensitive detector to establish correlations between low-energy fragments from a particular molecular break-up. In case of a Coulomb-fragmentation where all fragments emerge as positive ions the conditions of a kinematically complete experiment are fulfilled. The complications caused by a dissociation foil are avoided.

Experimental set-up

Figure 1 shows the principle of the experiment. Collimated beams of H^+, He^+, O^{6+}, and O^{7+} projectiles interact with a molecular gas target. H^+ and He^+ ions were produced at the ion accelerator in Bielefeld; highly charged O^{q+} ions were provided by the electron cyclotron resonance ion source (ECRIS) of the KVI in Groningen. The slow ions and electrons generated in the collision process are separated by a weak homogeneous electric field perpendicular to the incident ion beam. Electrons are detected in a channeltron (CEM) at one side of the interaction region; positive ions are accelerated towards the position- and time sensitive multi-particle detector[7] at the other side. The ions pass a field-free time-of-flight region before they hit the detector which is based on micro-channel-plates in combination with an etched crossed-wire structure consisting of independent "wires" in x- and y-direction. The first electron registered in the channeltron serves as a start pulse for the coincidence electronics which mainly consists of a specially developed time-to-digital conversion (TDC) system. Thereby, for each positive fragment the position (x_i, y_i) on the detector and the time-of-flight t_i relative to the start electron are recorded.

The present setup allows the simultaneous measurement of all reaction channels resulting in at least one electron and one or more positive fragments. Thus relative cross sections for the production of selected ions (as H_2O^+, H^+, or O^{q+} in collisions with H_2O) and special processes (like $H_2O{\rightarrow}H^++OH^+$ and $H_2O{\rightarrow}H^++H^++O^{q+}$) can be obtained[8]; furthermore, if all fragments of a particular break-up process are detected information about the fragmentation dynamics as well as the molecular structure can be derived. In the following we will concentrate on the latter application.

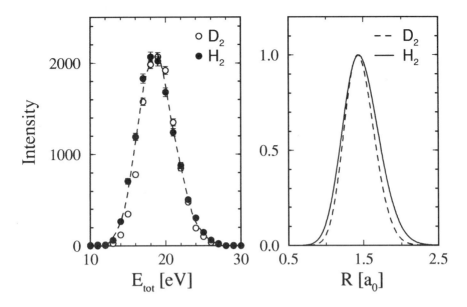

Figure 2. a: Total kinetic energy of coincident fragments in collisions of 200 keV He$^+$ on H$_2$ and D$_2$. The dashed line is the result of a Franck-Condon calculation for the Coulomb-explosion of H$_2$ convoluted with the response function of the detector system. b: Ground state wavefunctions of H$_2$ and D$_2$.

Diatomic molecules

The simplest and most frequently studied molecular target is H$_2$. Among the various reaction channels occuring in the collision processes we will concentrate on those fragmentations leading to coincident proton pairs; these are widely called 'Coulomb explosions':

$$
\begin{aligned}
\mathrm{X}^+ + \mathrm{H}_2 \;\to\;& \mathrm{X}^+ + \mathrm{H}^+ + \mathrm{H}^+ + e + e && \text{double ionization} \\
\to\;& \mathrm{X} \;+ \mathrm{H}^+ + \mathrm{H}^+ + e && \text{capture and ionization} \\
\to\;& \mathrm{X}^- + \mathrm{H}^+ + \mathrm{H}^+ && \text{double capture}
\end{aligned}
$$

with X$^+$ the projectile ion. Since the detector system is currently triggered by an electron, only the first and second process are observed. For each individual event the positions (x_i, y_i) of the correlated protons at the detector and their flight-times t_i are registered. If the geometry of the detector and the applied electric fields are known these data allow an absolute determination of the dissociation energy $\mathcal{E}$ by the use of classical mechanics[7].

Figure 2a shows the total kinetic energy distribution of coincident fragment ions from collisions of 200 keV He$^+$ with H$_2$ and D$_2$ molecules. The H$_2$ data are in good agreement with a Franck-Condon calculation[9] assuming a Coulomb explosion process. The same holds for the isoelectronic D$_2$ which shows a somewhat narrower distribution as expected by a comparison of H$_2$ and D$_2$ ground state wavefunctions (Fig. 2b). Although the exact shape of the spectrum depends also on the curvature of the H$^+$H$^+$-potential curve, the difference of the wavefunctions particularly at large bond-lengths is clearly reproduced at the low energy side of the energy distribution.

The Coulomb fragmentation of H_2 and D_2 may thus be considered as fairly well understood (at least concerning the fragmentation energy). This is not the case for more complex diatomic molecules like N_2 or CO: each fragment ion may occur in several charge states resulting in a larger number of reaction channels. However, the application of coincidence techniques as the one described here allows in many cases a clear separation of individual processes, as e.g. $CO \rightarrow C^+ + O^+$, and a calculation of the corresponding kinetic energy release. More difficult is the interpretation of the obtained energy spectra. Whereas in case of H_2 only one potential curve describes the final $H^+ + H^+$ products, there are many states of e.g. $(CO)^{2+}$ which finally result in $C^+ + O^+$. It turns out (c.f. next section) that an analysis in terms of a point-charge Coulomb explosion model is in general insufficient. Even more complicated (and also more interesting) is the Coulomb fragmentation of poly-atomic systems, a situation our detector system was built for.

Poly-atomic molecules

We studied the collision induced ionization and fragmentation of several small poly-atomic molecules. As pointed out above in general many different combinations of reaction products occur. As a first example we consider the fragmentation of CH_4. This system is fairly simple in the following sense: it consists only of two kinds of atoms which can be easily distinguished due to their large mass difference, and only one atom, namely C, may occur in different charge states. The coincidence map* (Fig. 3) gives an overview about the two-particle events detected in collisions of $742\,keV$ O^{7+} with CH_4. As a consequence of momentum conservation complete fragmentations into equally charged ions ($CH_4 \rightarrow CH_2^+ + H_2^+$ and $CH_4 \rightarrow CH_3^+ + H^+$) result in narrow structures perpendicular to the diagonal. If one or more fragments are not detected the structure is broadened due to the momentum carried away by the undetected fragments. In the case of CH_4 most channels can be separated and analyzed in great detail; in particular, cross sections for the correlated production of selected ions can be derived. Similar methods may be used for coincidences between three and more fragments, although there is no intuitively understandable graphic representation in higher dimensions. Computers, however, do not depend on a visual representation and a separation by topological algorithms is possible.

As useful as the coincidence maps and similar concepts (as e.g. the covariance map which was successfully applied in photoionization experiments by Codling and coworkers[10,11]) are for the fragmentation analysis of diatomic and certain small poly-atomic molecules one should be aware that they contain only part of the information inherent in the fragmentation dynamics. Although there are situations in which a careful analysis of the shape of these correlation structures gives information about the fragmentation energies, it is in general more reliable to derive them directly from the measured momentum vectors of individual events. Furthermore, although the coincidence map is a valuable tool for the identification and separation of reaction channels, there are often pathological situations in which it is of limited use only. In particular, if several fragments with similar masses but different charge states occur, this results in overlapping structures from different processes. An example of this kind is CF_4 where several reaction channels overlap in this way[12]. An analysis of such data would start

---

*Due to the internal crossed-wire structure of the detector coincidence data can be ordered by position on the detector. For example, a coincidence between an H^+ on the left of the detector with an C^{++} at the right results in an event at $T_R \sim 525ns$ and $T_L \sim 200ns$; for H^+ on the right and C^{++} on the left, also T_R and T_L are inverted.

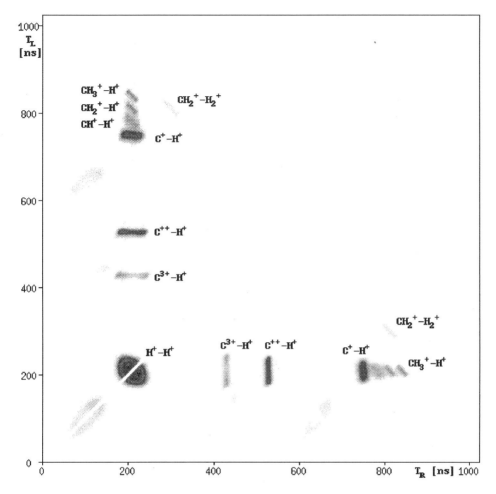

Figure 3. Coincidence map of correlated two-fragment events from collisions of 742 keV O^{7+} on CH_4. T_L and T_R are the flight times of the fragments which hit the detector at the left- and rightmost position.

with 5-fold coincidences, using the obtained information for further interpretation of the 4-fold coincidence data and so forth. However, the required deconvolution methods in higher dimensions are still a challenge. In case of overlapping complete or incomplete fragmentation channels the capabilities of the position- and time-sensitive detector may be applied: from the measured times and positions on the detector the momentum vectors of the fragments can be obtained and may be used for a check of momentum conservation[13]. This procedure is also useful to remove random coincidences from the data set.

Complete fragmentation of H_2O

Among the numerous reaction channels occuring in the collision processes we will concentrate on complete fragmentations of the type

$$X^{p+} + H_2O \rightarrow H_2O^{(q+2)+} + X^{(p-m)+} + (q+2-m)e^-$$
$$\rightarrow H^+ + H^+ + O^{q+}$$

where $q + 2 - m \geq 1$. In the experiment these events appear as 4-fold coincidences between an electron and the three positive fragment ions. In collisions with H^+ and He^+ we observed the break-up of H_2O into $H^+ + H^+ + O^+$ and $H^+ + H^+ + O^{++}$;[8] in collisions with O^{6+} and O^{7+} complete fragmentations up to $H^+ + H^+ + O^{5+}$ were separated[14].

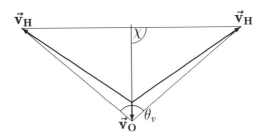

Figure 4. Definition of characteristic angles for a complete fragmentation of H_2O in velocity space.

If the time-of-flight and the position on the detector are recorded *for each fragment from a particular process* the conditions for a *kinematically complete* experiment are fulfilled. From the derived momentum vectors various parameters may be calculated which allow to analyze the dissociation dynamics. Besides the total kinetic energy of all fragments we choose the angle θ_v between the two O^{q+}-H^+ relative velocities v_{OH}, as well as the angle χ between the H^+-H^+ relative velocity v_{HH} and the velocity of the O^{q+}-ion v_O (see Fig. 4). Among these parameters the angle χ gives a first insight into the fragmentation dynamics: it may be used as an indicator whether the molecular bonds break simultaneously or in a step-wise fashion[15]. A break-up of both OH-bonds in a time short on a time scale defined by the rotation and vibration periods of the system leads to a strong angular correlation between the corresponding velocities which shows up as a narrow peak in the $\cos \chi$-distribution. In case of a two-step process the 'intact' $OH^{(q+1)+}$ subsystem may rotate around its center of mass and the correlation would be lost resulting in a uniform $\cos \chi$ distribution. Figure 5 shows measured $\cos \chi$ spectra for 100 keV He^+ and 126 keV O^{7+} impact together with a Monte-Carlo simulation of a simultaneous fragmentation into $H^+ + H^+ + O^+$ based on the MCSCF-calculations described below. The widths of experimental and calculated curves are in reasonable agreement indicating a practically simultaneous bond-breaking in the $H_2O^{3+} \rightarrow H^+ + H^+ + O^+$ fragmentation. A more detailed analysis of the simulation shows that the width of the $\cos \chi$ distribution is different for each of the considered molecular states (see below) of the intermediate H_2O^{3+} ion; thus, the remaining small deviations may be attributed to neglected states or different transition strengths. The same holds in case of all observed $H_2O^{(q+2)+} \rightarrow H^+ + H^+ + O^{q+}$ processes ($q = 1, 2$ for H^+ and He^+ impact and $q = 1 \ldots 5$ for O^{6+} and O^{7+} impact) at the collision energies studied.

Since the ion-induced fragmentation of H_2O into three positive ions may be considered as a simultaneous process, the following model is assumed: In the collision process

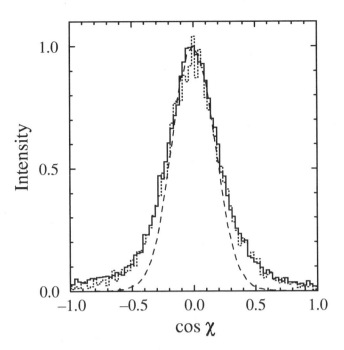

Figure 5. $\cos\chi$-distribution of $H^+ + H^+ + O^+$ from collisions of 100 keV He^+ (—) and 126 keV O^{7+} (...). (- - -) Monte-Carlo simulation based on a MCSCF-calculation assuming a simultaneous break-up of both H-O bonds.

several (ideally all) valence electrons are removed from the molecule during an interaction time which is short on the time scale defined by the rotation and vibration periods of the molecule. In this case the fragmentation dynamics is governed by the strong mutual repulsion of the generated positive ions, and the kinetic energies and emission angles may be computed by assuming Coulomb forces acting between point charges as a first approximation. Though this Coulomb explosion (CE) model is frequently used (due to its simplicity), and the energy spectra are sometimes reported to be in agreement with its predictions, this model turns out to be insufficient if energy and angular spectra are studied in more detail. First of all, in a pure Coulomb explosion process the fragmentation energy and the angular distributions should be independent of the projectile type and its energy. The shapes of the measured spectra of θ_v and of the total kinetic energy (Fig. 6), however, clearly change with projectile; in case of highly charged ions they even depend on the incident energy[14]. This is in contrast to the simple "point-charge" CE model: several competing processes which all result in three positive fragment ions must be involved to explain the observed behavior.

To account for the most important reaction channels we used the MOLPRO code[16,17] for an *ab initio* multi-configuration self-consistent field computation (MCSCF) of the nine lowest molecular states of the intermediate H_2O^{3+} and H_2O^{4+} ions[18]. The calculated potential surfaces correspond to the lowest states of the O^+ (O^{2+}) ion having a fully occupied 2s-orbital and three (two) electrons in the 2p-orbitals. For each of these surfaces the resulting total kinetic energy and angular distributions were calculated by Monte-Carlo techniques. We assumed a Franck-Condon transition from the H_2O ground state to the particular dissociating $H_2O^{(q+2)+}$ state to obtain the starting point

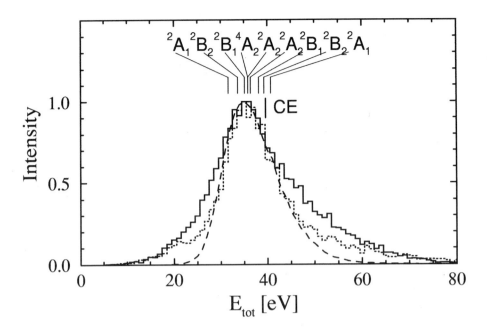

Figure 6. Total kinetic energy distribution of coincident H^+-H^+-O^+ fragments from collisions of H_2O with $250\,\text{keV}$ He^+ (—) and $92.4\,\text{keV}$ O^{6+} (...). The dashed line is a MCSCF-calculation taking into account the indicated molecular states of H_2O^{3+} which are labeled by their symmetries.

for the subsequent dissociation along the respective potential surface. The derived energy and angular distributions show a distinct dependence on the occupation of the $H_2O^{(q+2)+}$ orbitals. The simulated kinetic energy spectra were convoluted with the response function of the detector system. A major problem for a quantitative comparison of measured and theoretical data are the unknown transition strengths for the individual states. A natural approach would be a determination of these nine parameters by a fit procedure. Unfortunately, several different sets of parameters gave an equally good agreement to the experimental data. To avoid this ambiguity we assumed a transition strength proportional to $1/E_i^2$ (with E_i the excitation energy of the corresponding intermediate H_2O^{3+} state), a scaling behavior which is well known e.g. in inner shell ionization. Of course a more detailed calculation would be highly desirable. Figure 6 shows the positions of the maxima and the weighted sum of the simulated energy spectra; the χ- and θ_v-distributions (Fig. 5 and 7) were derived by a similar approach.

A comparison of the measured total kinetic energy of correlated $H^+ + H^+ + O^+$ fragments to the MCSCF-prediction shows reasonable agreement (Fig. 6): for He^+ (and H^+) the deviations at the low- and high-energy edge of the distribution may be attributed to neglected higher excited states in H_2O^{3+} (which not necessarily lead to higher released kinetic energies) and to transition strengths different to the assumed $\propto 1/E_i^2$ scaling. Note that the point-charge CE model predicts a maximum at a significantly higher energy. The spectra obtained with highly charged primary O-ions show an even better agreement with the calculated distribution. According to the classical over-barrier model excited states are expected to be less important in such "gentle" collisions in agreement with the experimental finding. A more detailed study

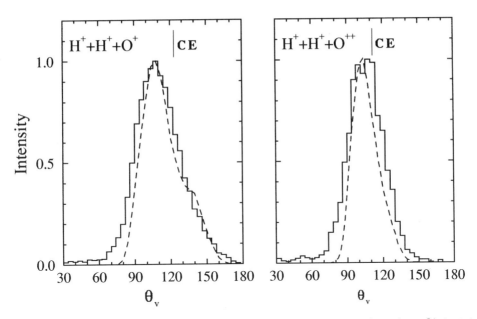

Figure 7. Distribution of θ_v for coincident $H^+ + H^+ + O^+$ (left) and $H^+ + H^+ + O^{2+}$ (right) in collisions with 250 keV He^+. - - - Monte-Carlo simulation based on MCSCF-calculation. The values calculated by a pure Coulomb-model are marked with vertical lines.

reveals characteristic differences between individual collision systems indicating changes in the transition strength to particular molecular states of the intermediate H_2O^{3+} ion.

The left part of Fig. 7 shows measured θ_v spectra for the complete fragmentation $H_2O \rightarrow H^+ + H^+ + O^+$ in comparison to simulations based on the MCSCF-calculation. In all collision systems studied the maxima of measured and calculated spectra are located at angles smaller than predicted by the CE-model. In case of H^+ and He^+ impact we observed no significant changes of the angular distribution in the studied energy regime (50–350 keV); in collisions with highly charged ions, however, variations with projectile charge and energy were observed[14]. The right part of Fig. 7 shows the corresponding results for the $H^+ + H^+ + O^{2+}$ fragmentation channel. Since the H^+H^+ repulsion is dominated by the stronger H^+O^{2+} repulsion, the θ_v spectra are shifted towards lower angles. The CE-prediction gives again too large values, though the differences are smaller here. Experiment and MCSCF-based simulations are in reasonable agreement; small systematic differences at lower angles and individual changes may be attributed to the neglect of higher excited states and to transition strengths different to the $\propto 1/E_i^2$ assumption.

Conclusion

We have studied the kinematically complete fragmentation of diatomic and small poly-atomic molecules as H_2, D_2, H_2O, and CH_4 using a newly developed time- and position-sensitive multi-particle detector system. From the measured data information on the channel-resolved multi-fragmentation as well as the dynamics of the initial ionization process can be obtained. Whereas the measured kinetic energy spectra for the complete Coulomb-fragmentations $H_2 \rightarrow H^+ + H^+$ and $D_2 \rightarrow D^+ + D^+$ are in good agreement with a point-charge Coulomb-explosion model, this model is insufficient for more complex diatomic molecules, like CO or N_2, and poly-atomic systems: there are in general several states of the fragmenting multiply-ionized intermediate ion, and the energy and angular spectra show a strong dependence on the involved electronic configuration. In case of H_2O this was confirmed by *ab initio* MCSCF-calculations of the lowest states of the $(H_2O)^{3+}$ system which are in reasonable agreement with the measured data. This finding has direct implications on the application of the Coulomb-explosion imaging (CEI) method[6] to the determination of the geometric structure of molecules: reliable data can be expected for molecules containing mainly H-atoms where all valence electrons have been removed during the collision; in case of more complex systems, a careful analysis of the role of excited states in the highly charged molecular complex is required.

Acknowledgements

We wish to thank the Groningen KVI group for the generous support during the experiments with highly charged ions, as well as for many stimulating discussions. We also acknowledge discussions with our guests from Tokyo. This work was supported by the Deutsche Forschungsgemeinschaft (DFG), the Japan Society for the Promotion of Science (JSPS), and the EU-network CHRX-CT94-0643.

References

1. K.P. Kirby, in *The Physics of Electronic and Atomic Collisions, Proceedings of XVIII ICPEAC*, edited by T. Andersen, B. Fastrup, F. Folkmann, H. Knudson and N. Andersen (AIP Press, New York, 1993), pp. 48–58.

2. see e.g.: R.N. Compton and J.N. Bardsley, in *Electron Molecule Collisions*, edited by I. Shimamura and K. Takayanagi, (Plenum Press, New York, 1984).
   T.D. Märk, in *Electron Impact Ionization*, edited by T.D. Märk and G.H. Dunn (Springer Verlag, Vienna and New York, 1985).

3. see e.g.: C.J. Latimer, Advances in Atomic, Molecular, and Optical Physics, Vol. 30, 105 (1993) and references therein.

4. see e.g: D.P. de Bruijn and J. Los, Rev. Sci. Instrum. **53**, 1020 (1982).
   A.K. Edwards, R.M. Wood, and R.L. Ezell, Phys. Rev. **A31**, 99 (1985).
   F.B. Yousif, B.G. Lindsay, and C.J. Latimer, J. Phys. **B21**, 4157 (1988).

5. D.S. Gemmell, Chem. Rev. 80, 301 (1980).

6. Z. Vager and E.P. Kanter, Nucl. Instrum. Methods **B33**, 98 (1988).

7. J. Becker, K. Beckord, U. Werner, and H.O. Lutz, Nucl. Instrum. Methods **A 337**, 409 (1994).

8. U. Werner, K. Beckord, J. Becker, H.O. Lutz, Phys. Rev. Lett. **74**, 1962 (1995).

9. K.E. McCulloh, J. Chem. Phys. **48**, 2090 (1968).

10. L.J. Frasinski, K. Codling, and P.A. Hatherly, Science **246**, 1029 (1989).

11. L.J. Frasinski, P.A. Hatherly, and K. Codling, Physics Letters **A 156**, 227 (1991).

12. U. Werner, J. Becker, and H.O. Lutz to be published.

13. K. Beckord, J. Becker, U. Werner, and H.O. Lutz, J. Phys. **B27**, L585 (1994).

14. U. Werner, K. Beckord, J. Becker, H.O. Folkerts, and H.O. Lutz, Nucl. Instrum. Methods **B98**, 385 (1995).

15. C.E.M. Strauss and P.L. Houston, J. Phys. Chem. **94**, 8751 (1990).

16. MOLPRO is an *ab initio* program written by H.J. Werner and P.J. Knowles with contributions from J. Almlöf, R. Amos, S. Elbert, K. Hampel, W. Meyer, K. Peterson, R. Pitzer and A. Stone.

17. H.J. Werner and P.J. Knowles, J. Chem. Phys. **82**, 5053 (1985).

18. K. Beckord, J. Becker, U. Werner, H.J. Werner and H.O. Lutz, to be published.

PHOTOIONIZATION DYNAMICS STUDIED BY COINCIDENCE SPECTROSCOPY

Uwe Becker and Jens Viefhaus

Fritz-Haber-Institut der Max-Planck-Gesellschaft
Faradayweg 4 - 6
D-14195 Berlin
Germany

Photoionization studies may be subdivided into two fields of interest, structure and dynamics. The first sub field is the domain of all kinds of spectroscopic investigations, in particular, high resolution ion yield studies of resonances, whereas the second sub field is concerned with electron spectrometry at different levels of differentiation. However, both field are closely linked to each other because resonance behavior in the photoionization continuum is directly affected by the dynamics of the continuum electron due to the interference between bound and free electrons above the first ionization threshold. Nevertheless, the two fields concentrate on different aspects of the photoionization process. Whereas structural studies have the ambition to achieve the best resolution available, dynamical studies depend more on the photon flux in a certain energy range and on a high degree of linear polarization of the ionizing radiation. This is because dynamical studies require the determination of more parameters than just the photoionization cross section. In order to derive these parameters unambiguously, the target and/or the products of the photoionization process have to be prepared and/or analyzed in more detail. An feasible way to do this is the spin-sensitive detection of the emitted photo-electron, a method which has been pioneered since more than a decade by Ulrich Heinzmann and is, in fact, the method which provided most of the presently known dynamical photo-ionization data beyond the photoelectron angular distributions [1]. However, there are other methods for such dynamical studies, e. g. the electron spectroscopy of polarized atoms [2] and oriented molecules [3]. Both are just starting to become more routinely used methods as spectroscopic tools for a variety of applications in dynamical studies. A third method suitable for the investigation of dynamical properties is coincidence spectroscopy, in particular angle-resolved electron-electron coincidence spectroscopy [4]. Although many groups started to use coincidence techniques in one or the other way, the group which pioneered this field with respect to dynamical photoionization studies in an indeed systematic way, was the group of Volker Schmidt at Freiburg. Starting from these bench mark experiments it will be shown in this report how new dynamical information may be gained from coincidence experiments if new techniques, such as time-of-flight (TOF) spectroscopy, are introduced [5]. The paper is subdivided into three sections: Electron-electron-coincidence studies of the double photoionization of helium, of the double photoionization of heavier rare gases, and coincidence studies of the two electron emission in resonant Auger decay.

Selected Topics on Electron Physics
Edited by Campbell and Kleinpoppen, Plenum Press, New York, 1996

DOUBLE PHOTOIONIZATION OF HELIUM

Photoionization of helium is an archetypal experiment, in particular with respect to electron correlations. This is because helium represents the only pure case of the three-body Coulomb problem which governs the photoionization behavior, especially near threshold. For a long time the study of two electron satellites in helium has been a showcase for decisive experiments concerning the theoretical interpretation of this correlated process. Satellite peaks up to a principal quantum number of n=10 could clearly be identified so far [6,7]. However, more recently the direct two-electron emission or double photoionization process became the more favored field of experimental studies of correlation processes, in particular with respect to their dynamics. Double photoionization (DPI) occurs when a single, energetic photon is absorbed by an atom or molecule resulting in the emission of two electrons. In contrast to single ionization, which can be treated in the non-resonant regime at least approximately by the independent particle model, double ionization cannot be described in this way, since it is entirely due to relaxation and electron correlation effects. Depending on the target, double ionization can account for up to about a third of the total photoionization cross section. It is of fundamental interest because it results in an ion and two escaping electrons which are subject to long-range Coulomb forces.

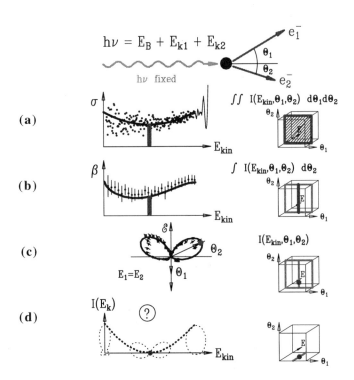

Figure 1. Schematic description of the double photoionization of helium. From the top to the bottom the different parts of this figure show examples of differential partial cross sections at a rising level of differentiation.

Figure 1 shows the double ionization process and the different steps of its investigation schematically. In contrast to normal photoelectron spectroscopy, which deals with line structures in a spectrum, double ionization gives rise to a continuum distribution of ejected electrons, the so-called shake off continuum [8], although shake off is not the only source for double ejection of electrons. Figure 1(a) shows a complete electron spectrum taken above the double ionization threshold at $h\nu = 120$ eV. The double ejection continuum with its characteristic u-shaped intensity distribution is clearly visible. This characteristic shape changes with photon energy. Such measurements are the first step beyond ion yield measurements, which give the integral cross section seperated with respect to single and double photoionization.

A further step in differentiation is the non-coincident measurement of the angular distribution of the electrons emitted in the double ionization process [8]. Figure 1(b) displays the angular distribution asymmetry parameter β of the double ejection continuum as a function of the kinetic energy of the electrons. In contrast to the intensity distribution, which is non-uniform but symmetric with respect to the half "excess energy", the β-parameter distribution can be non-symmetric. This is because energy conservation forces only the cross section to be symmetric with respect to equal energy sharing but not the differential cross section into certain space directions [9]. The most sensitive and direct way of studying this three-body Coulomb problem is by measuring the two outgoing electrons in coincidence using monochromatised synchrotron radiation. Such measurements only became feasible following technological advances in the early 90's [4,10] and since then these have provided a powerful stimulus for detailed theoretical studies [11]. Figure 1(c) shows the coincident emission pattern of electrons ejected in the double ionization of helium at different energy sharings. The patterns change dramatically with respect to the energy sharing [12]. For example, at equal energy sharing there is a node in the parallel and anti-parallel directions which reflects, in a sense, the symmetry conditions of the two outgoing electrons, whereas for unequal sharing there are in part maximal intensities at these particular positions. The information from these coincident emission patterns is very useful and provides for the first time the opportunity for a complete understanding of the two-electron emission process in terms of a rigorous theoretical description. The bottom part of this figure 1(d) shows a hypothetical coincident intensity distribution along the kinetic energy of the ejected electrons. This intensity distribution should be particularly sensitive to the dynamics of the double photoionization process because it reflects the symmetry-dependent change of patterns at the most sensitive coincidence geometries. The measurement of such complete coincident intensity distributions requires a multi-channel detection technique such as time-of-flight electron spectroscopy. This method offers the additional advantage that it is particularly suited for the detection of low energetic electrons, an aspect of special interest in double photoionization studies. Figure 2 shows a scheme of the experimental set-up being used for the first time to apply the time-of flight technique for electron-electron coincidence studies. The main disadvantage of this method is the length of the spectrometer which makes it difficult to rotate them with respect to each other, which is essential for obtaining complete angular patterns. In order to overcome this problem a small, rotatable TOF spectrometer has been added to the three fixed ones with long flight paths. In this way all possible relative positions of the different spectrometers with respect to the reference spectrometer could be obtained. First experiments were performed in the so-called anti-parallel position which is particularly sensitive to the nodal structure of the angular patterns.

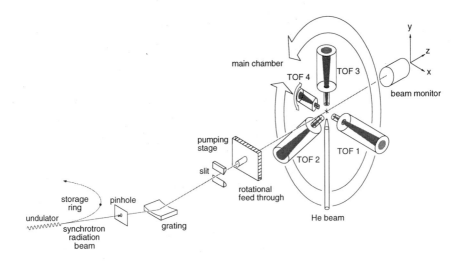

Figure 2. Scheme of the experimental set-up for TOF electron-electron coincidence spectroscopy.

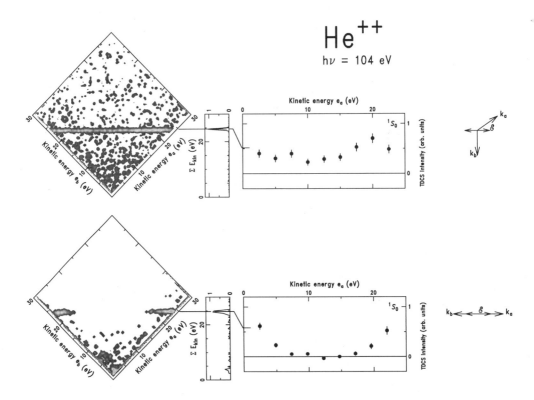

Figure 3. Two-dimensional representation of two helium electron-electron coincidence spectra recorded at two different angle combinations. The insets show the coincidence intensities along the shake off diagonal.

A TOF spectrum taken under such geometrical conditions exhibiting zero intensity in the range of equal and near equal energy sharing is shown in Figure 3. Experiments for selected non-equal energy sharing conditions have revealed a dramatic change in the energy-dependent angular patterns with photon energy [12,13]. In order to study this change in more detail, especially with respect to the continuous pattern change along kinetic energy, we have performed such an experiment in anti-parallel geometry for different photon energies.

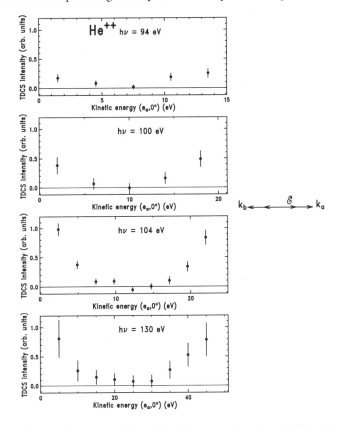

Figure 4. Anti-parallel coincidence intensities of the helium double photoionization versus kinetic energy sharing for four different photon energies.

Figure 4 shows the changing coincident intensity distributions for four different photon energies. It is clearly seen that the u-shaped distribution with its broad zero bottom becomes more and more flat towards lower photon energy. This is because the anti-symmetric part of the wave function looses importance near threshold. As a result the strong pattern variation along kinetic energy disappears with lower photon energy [13]. Such a measurement is a direct measure of the anti-symmetric component of the two-electron wave function. Further studies under other geometrical conditions are planned in the near future and will provide complete information about the dynamics of the double photoionization process.

DOUBLE PHOTOIONIZATION OF HEAVIER RARE GASES

Double photoionization studies have concentrated on helium so far. This is because a bare nucleus is left behind and no process other than direct double ionization, in which the two

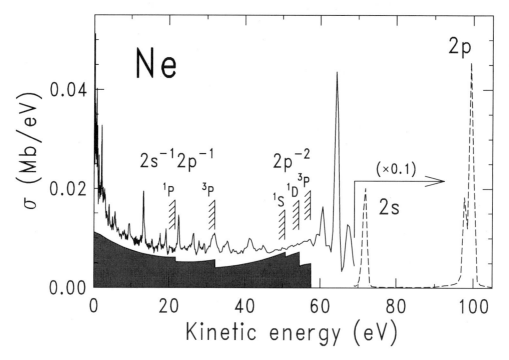

Figure 5. Complete photoelectron spectrum of the valence photoionization of neon together with the different double ionization intensities as shaded areas.

electrons escape simultaneously, can occur. In contrast, all heavier atoms are subject to other, indirect double ionization processes such as, for example, Auger decay, which also leads to the production of doubly charged ions, even in valence photoionization [14,15]. Moreover, these residual ions can be in various excited states with different symmetries. These symmetries are reflected in the wave function of the two outgoing electrons, which leads to a larger variety of angular patterns and energy sharings than for helium [16].

Figure 5 shows a complete photoelectron spectrum of neon with the direct double ionization part plotted as a shaded area [15]. The steps due to different double ionization continua are taken from a theoretical calculation because they are not exhibited in the experimental spectrum [17]. Coincidence spectroscopy, however, provides the opportunity to unravel these different double ionization continua in an unambiguous way.

This is shown in Figure 6, which represents the coincidence map of neon taken at a photon energy of 121 eV at angles of 55° and 90° with respect to the polarization vector. If one integrates along the continua, a line spectrum is obtained which represents the integrated intensity over the kinetic energy of the ejected electrons but still into certain space directions.

TWO ELECTRON EMISSION IN RESONANT AUGER DECAY

A particular interesting case of double photoionization studies with respect to simultaneous and sequential processes is the valence double photoionization of Xe. This is because of the close relationship between these valence ionization processes and the resonant Auger transitions observed in the decay of the $4d \rightarrow np$ excitations in Xe [18]. It is still unclear how many of these Auger transitions enhance open satellite channels and how many occur via non-resonantly unpopulated ionization channels [19]. Furthermore the question of the strengths

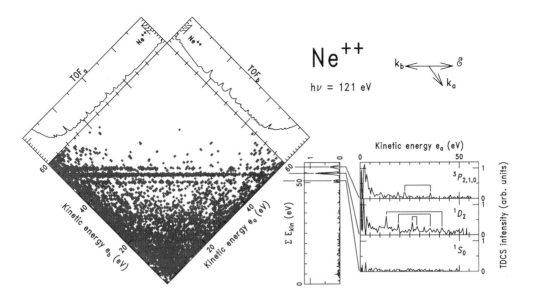

Figure 6. Two-dimensional coincidence spectrum of the valence double photoionization of neon along with the integrated double ionization intensity and final state selected spectra.

of resonant shake off transitions in the decay of these resonances has attracted considerable interest in recent years [20]. The major problem in the study of these electron spectra however, was the fact, that all the different transitions are superimposed on each other. Even high resolution spectroscopy is unable to resolve the large number of closely spaced transition lines. Note that a separation of the different double electron emission continua is, of course, not possible. The only way to disensemble these different transitions is electron-electron coincidence spectroscopy because it makes it possible to differentiate between different final ionic states of the doubly charged ion.

Figure 7 shows a non resonant coincidence spectrum of the valence photoionization of xenon. The different double photoionization continua are clearly resolved in the two-dimensional spectrum as diagonal lines. Integrating along these lines shows the discrete two-step transitions as regular line structures, but separated with respect to the final, doubly charged, ionic state. This separation allows to distinguish between otherwise indistinguishable satellite lines.

Figure 8 shows the coincident resonant Auger spectrum of the Xe $4d_{5/2} \rightarrow 6p$ excitation. This two-dimensional spectrum exhibits besides the lower states of the doubly charged ion several higher excited states. The population of these states contributes predominantly to the intensity enhancement at low kinetic energies of the decay spectrum [21]. Part of this enhancement is suspected to be simultaneous double ionization which is characterized by a continuous intensity distribution. Surprisingly even at these low energies several discrete two-step transitions exist, which contribute to the resonant double ionization yield. A quantitative analysis of the fractional intensities of these two decay modes however is not possible at the present stage of investigation. The situation is more favourable for the low lying ionic states of the doubly charged ion. Here the different ionic states can be clearly distinguished from each other making it possible to integrate along their diagonal intensity distribution. The result is displayed in the insets of Figure 8 which shows the clear predominance of two step transitions over resonant shake off. In addition, the spectra disclose that many of the resonant Auger transitions are indeed due to resonance enhancement of satellite transitions, in particular, in case of the ^1D final ionic state. This spectrum shows, that many so far unsolved problems of resonant and non resonant double photoionization can be disentangled already with very few

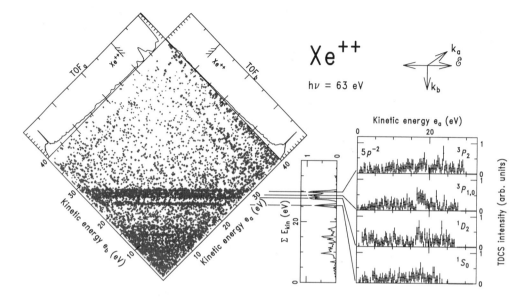

Figure 7. Coincident non-resonant xenon valence double photoionization spectrum taken below the $4d_{5/2} \rightarrow 6p$ resonance of xenon at hν = 63 eV.

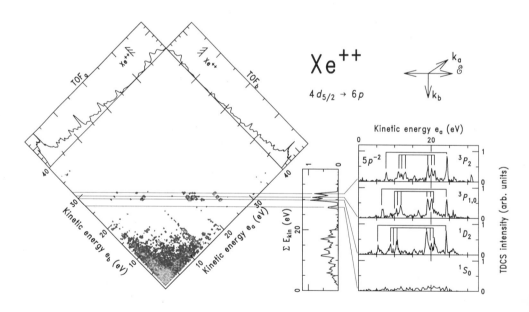

Figure 8. Coincident resonant ($4d_{5/2} \rightarrow 6p$) Auger spectrum of xenon.

coincidence spectra. Measurements at different angle positions allow to derive coincident angular patterns which contain, as in the case of helium, complete information on the photoionization process above the double ionization threshold.

This work was supported by the BMBF and in part by the EU within the HCM program.

REFERENCES

1. U. Heinzmann, and N. Cherepkov, in *VUV and soft X-ray photoionization studies*, edited by U. Becker and D. A. Shirley (Plenum Press, New York, 1996), p. 521.

2. H. Klar and H. Kleinpoppen, J. Phys. B 15:933 (1982), B. Sonntag and M. Pahler, in *New Directions in Research with Third-Generation Soft X-Ray Synchrotron Radiation Sources*, edited by A. S. Schlachter and F. J. Wuilleumier (Kluwer Academic Publishers, Maratea, Italy, 1992), p. 103.

3. E. Shigemasa, J. Adachi, M. Oura, and A. Yagishita, Phys. Rev. Lett. 74:359 (1995).

4. O. Schwarzkopf, B. Krässg, J. Elmiger, and V. Schmidt, Phys. Rev. Lett. 70:3008 (1993).

5. R. Wehlitz, J. Viefhaus, K. Wieliczek, B. Langer, S. B. Whitfield, and U. Becker, Nucl. Instrum. and Meth. in Phys. Res. B 99:257 (1995).

6. R. Wehlitz, B. Langer, N. Berrah, S. B. Whitfield, J. Viefhaus, and U. Becker, J. Phys. B 26:L783 (1993).

7. S. Svensson, S. Kikas, A. Ausmees, S. J. Osborne, S. Aksela, A. Naves de Brito, and E. Nommisk, J. Phys. B 28:L293 (1995).

8. R. Wehlitz, F. Heiser, O. Hemmers, B. Langer, A. Menzel, and U. Becker, Phys. Rev. Lett. 67:3764 (1991).

9. F. Maulbetsch, and J. Briggs, J. Phys. B 26:1679 (1993).

10. J. Mazeau, P. Selles, D. Waymel, and A. Huetz, Phys. Rev. Lett. 67:820 (1991).

11. F. Maulbetsch and J. Briggs, J. Phys. B 26:L647 (1993), C. Dal Capello, B. Joulakian, and J. Langlois, J. Physique IV C6:125 (1993), A. K. Kasansky and V. Ostrovsky, J. Phys. B 27:44 (1994), S. C. Ceraulo, R. M. Stehmann, and R. S. Berry, Phys. Rev. A 49:1730 (1994), F. Maulbetsch and J. Briggs, J. Phys. B 27:4095 (1994), M. Pont and R. Shakeshaft, Phys. Rev. A 51:R2676 (1995), A. Kasansky and V. Ostrovsky, Phys. Rev. A 51:3712 (1995).

12. O. Schwarzkopf, B. Krässig, V. Schmidt, F. Maulbetsch, and J. Briggs, J. Phys. B 27:l345 (1994).

13. P. Lablanquie, J. Mazeau, L. Andric, P. Selles, and A. Huetz, Phys. Rev. Lett. 74:2192 (1995), G. Dawber, L. Avaldi, A. G. McConkey, H. Rojas, M. A. MacDonald, and G. C. King, J. Phys. B 28:L271 (1995).

14. U. Becker, R. Wehlitz, O. Hemmers, B. Langer, and A. Menzel, Phys. Rev. Lett. 63:1054 (1989).

15. U. Becker and R. Wehlitz, J. Electr. Spectr. Rel. Phen. 67:341 (1994).

16. S. J. Schaphorst, B. Krässig, O. Schwarzkopf, N. Scherer, and V. Schmidt, J. Phys. B 28:L233 (1995)

17. S. L. Carter and H. P. Kelly, Phys. Rev. A 16:1525 (1977).

18. E. von Raven, M. Meyer, M. Pahler, and B. Sonntag, J. Electr. Spectr. Rel. Phen. 52:677 (1990), K. Okuyama, J. H. D. Eland, and K. Kimura, Phys. Rev. A 41:4930 (1990).

19. H. Aksela, S. Aksela, O.-P. Sarainen, A. Kivimäki, A. Naves de Brito, and E. Nömmiste, J. Tulkki, S. Svensson, A. Ausmees, and S. J. Osborne, Phys. Rev. A 49:R4269 (1994).

20. U. Becker, D. Szostak, M. Kupsch, H. G. Kerkhoff, B. Langer, and R. Wehlitz, J. Phys. B 22:749 (1989), P. Lablanquie and P. Morin, J. Phys. B 24:4349 (1991).

21. U. Becker and D. A. Shirley, Physica Scripta T31:56 (1990).

A NEW APPROACH TO THE COMPLETE PHOTOIONIZATION EXPERIMENT FOR STRONTIUM AND CALCIUM ATOMS

J B West
Daresbury Laboratory
Warrington WA4 4AD
UK

INTRODUCTION

It is well known that the inner shell photoionisation spectra of the alkaline earth atoms contain prominent autoionising structure, in particular the so-called "giant" resonances corresponding to the excitation of an outer p electron to an empty d-orbital. Partial collapse of the d-orbital wave function is responsible for the large overlap between the p and d wavefunctions and as a result these resonances have large peak cross sections, $\sim 10^{-15}\,cm^2$. Through configuration interaction these resonances are also quite complex, as can be seen from the absorption spectrum taken by Mansfield and Newsom (1977) for calcium, in which the 3p - 3d resonance is composed of several discrete features; similarly for strontium, also measured by Mansfield and Newsom (1981), where the structure in the 4p - 4d resonance is much more obvious. This report summarises the measurements made at the Daresbury SRS using electron spectroscopy and fluorescence spectroscopy in these resonance regions, where the aim was to measure the branching ratios for populating the excited states of the ion, and the accompanying photoelectron angular distributions. In this way further understanding of the complex nature of these resonances is to be gained. In addition these studies, when combined with the measurement of the polarisation of the fluorescence resulting from the residual ion as it decays, thereby determining its alignment, form part of a complete photoionisation experiment. For both atoms, the process being studied is the following, the example for strontium being given at the 490Å 4p - 4d resonance:

$$Sr\,(4p^6 5s^2\,^1S_0) + h\nu\,(490\text{Å}) \rightarrow Sr^*\,(4p^5 4d 5s^2\,^{1,3}P).$$

This Sr^* state can then decay by autoionisation, and it does this mainly to the ground state of

Sr⁺, but there is also a large cross section for decay to an excited state of Sr⁺ as follows:

$$Sr^* (4p^5 4d 5s^2 \ ^{1,3}P) \rightarrow Sr^+ (5s\ ^2S_{1/2}) + e^-, \text{ or } \rightarrow Sr^+ (5p\ ^2P_{1/2,3/2}) + e^-$$

The excited Sr⁺ 5p state will then decay by fluorescence to the ground state

$$Sr^+ (5p\ ^2P_{1/2,3/2}) \rightarrow Sr^+ (5s\ ^2S_{1/2}) + h\nu\ (4078\text{Å}, 4215\text{Å})$$

where the 4215Å component corresponds to decay of the $^2P_{1/2}$ level and is removed by filtering. The usefulness of the fluorescence polarisation measurement does depend on the transition above being pure, ie the excited 5p level in Sr, or the corresponding 4p level in calcium, being populated only via the mechanism shown above and not, for example, by cascade processes. The complementary electron spectrometry measurements, to detect the electron ejected in the above processes, are required to investigate this, and also to provide the information required for the complete experiment. These form the basis of this report.

EXPERIMENTAL METHOD

The experimental geometry for this work is shown in figure 1, with the photon beam incident in the y-direction. The photoelectron angular distribution measurements were made with respect to the primary E-vector component in the x-z plane, where the angle θ is the angle

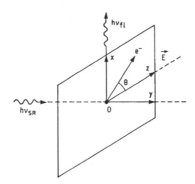

Figure 1. The experimental geometry

between the ejected photoelectron and the E-vector. The fluorescence polarisation measurement was made for photons emitted in the x-direction, in a separate experiment. The experimental apparatus for both measurements is shown schematically in figure 2, where the incident photon beam is perpendicular to the plane of the figure. Photons in the range 20 - 30 eV, with a resolution of ~50 meV and a peak flux ~10¹¹ photons/sec, were provided by a toroidal grating monochromator fitted to beamline 3 at the SRS (for details of this instrument, see West and Padmore, 1987) in the case of the calcium measurements, and by a normal incidence monochromator (see Holland et al 1989) for the strontium measurements. A capillary light guide was used to bring the photons close to the interaction region from the exit slit of the monochromator. The oven shown in figure 2 is a compact design published by Ross (1994), and gave a vapour density of ~10⁻³ torr in the interaction region. The electron spectrometer, a

150° hemispherical sector of 90 mm mean radius, was fitted with a special shield at its entrance, to prevent its entrance slit assembly being exposed to the metal vapour when moving between the $\theta = 0°$ and $90°$ positions. Particularly for strontium, this was necessary to preserve the efficiency and resolution performance of the analyser.

The angular distribution of the photoelectrons was calculated from the two measurements at the above angles, using the well known expression for the differential cross section where

$$\frac{d\sigma}{d\theta} = \frac{\sigma}{4\pi}\left[1 + \frac{\beta}{4}(3p\cos2\theta + 1)\right] \tag{1}$$

β is the angular distribution parameter and σ is the total cross section for the photoionisation channel being measured; for a derivation of this expression see West (1991). The polarisation p of the incident light was measured in a separate experiment by using helium for which

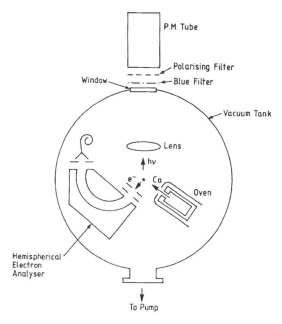

Figure 2. The experimental equipment

$\beta = 2$. The electron spectrometer was calibrated for variations in its detection efficiency with electron energy by making a series of measurements with argon gas and comparing the data to the known cross section of argon tabulated by Marr and West (1976); its angular symmetry was determined by checking the β-parameters measured for argon with those published by Holland et al (1982).

The fluorescence data were taken independently of the photoelectron measurements, and to measure the polarisation two measurements were needed, one with the transmission axis of the polarisation analyser shown in figure 2 parallel to the incoming light beam, the other with this axis perpendicular to it. The blue filter shown was chosen to exclude the $^2P_{1/2}$ component of the fluorescence, which is unpolarised and would therefore cause the value of the alignment parameter A_{20} to be too low. The alignment parameter is related to the fluorescence polarisation P_x by the expression (see, for example, Berezhko and Kabachnik, 1977)

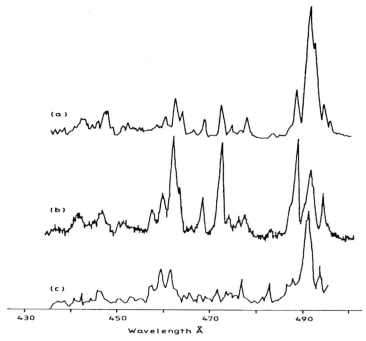

Figure 3. The singly charged ion yield (upper curve), fluorescence yield (middle curve) and 5p partial cross section (bottom curve) for Sr⁺

$$ P_x = \frac{3\alpha_2 A_{20}}{\alpha_2 A_{20} - 2} \tag{2} $$

where the coefficient α_2 is determined by the transition being measured; in this case, for the $^2P_{3/2} \rightarrow {}^2S_{1/2}$ transition $\alpha_2 = 0.5$.

RESULTS

Strontium

In figure 3 the results taken by Hamdy et al (1991) of the total fluorescence intensity for strontium in the region of the 4p -4d resonance are shown. For comparison the singly charged ion yield measurements from Nagata et al (1986), and the electron spectrometry measurements of Yagishita et al (1988), who obtained the partial cross section for populating the 5p level by measuring the intensity of the electrons which leave the Sr⁺ ion in the 5p excited state, are also shown. It is clear that this resonance is composed of at least three components, which were not assigned by Mansfield and Newsom (1981) and indicate that this resonance is indeed complex. The fluorescence spectrum is in effect part of the singly charged ion yield, and could be expected to resemble the fluorescence spectrum; the 5p partial cross section measurements should be identical to it. However, it can be seen clearly from figure 3 that although the peak positions for the three sets of data agree quite well, their relative intensities do not. This is almost

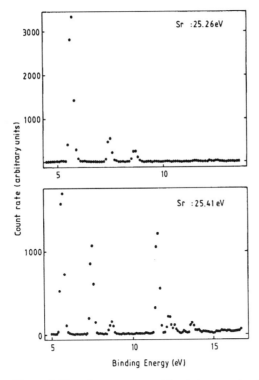

Figure 4. Photoelectron spectra at the photon energies shown

certainly due to the presence of cascade processes, the 5p level being populated not just through the mechanism outlined in the introduction, but by higher satellite states being excited and decaying to this level. It is interesting to note that Hamdy et al found that the peak at 487.9Å (25.41 eV) was completely unpolarised, which would be the result of a contribution from cascades. A photoelectron spectroscopy experiment was therefore required to resolve this issue. The data shown in figure 4 were taken at the central peak of the 4p - 4d resonance at 490.8Å (25.26 eV), in the upper part of the figure, and in the lower part is shown the electron spectrum at the 25.41 eV peak. Here the difference is clear; at 25.26 eV there are only three electron peaks, corresponding to leaving Sr⁺ in its ground state, 4d or 5p excited levels, whereas at 25.41 eV many other levels are also populated. This same result was confirmed by the much more detailed experiment of Jiménéz-Mier et al (1993). It seemed reasonable to conclude therefore that the polarisation measurement made by Hamdy et al at 25.26 eV was free from cascades, since the excited 4d level cannot decay by fluorescence to the ground state of the ion. This was an important finding, since it meant that this measurement could be used to determine the alignment of the residual ion, using the equation for P_x above, and with the additional information from the angular distribution measurement determine the photoionisation parameters. This was what Ueda et al (1993a) proceeded to do.

In figure 5 are shown the relative partial cross section for populating the Sr⁺ 5p level, and the angular distributions for the corresponding ejected electrons. The numbering of the resonance features follows that of Mansfield and Newsom (1981), and most of these features are evident in both the partial cross section and angular distribution parameter. The second strongest peak in the lower half of figure 4 corresponds to excitation to the Sr⁺ 6s level; in fact

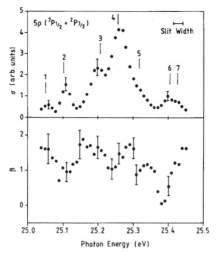

Figure 5. Data for the Sr 5p level

the branching ratio for population of this level rises to 27% at the 25.41 eV resonance, compared to 4% for the 5p level. This highlights the different nature of the component parts of the complex 4p - 4d resonance, because it indicates that the 25.41 eV resonance may have an uncollapsed Rydberg character, thereby allowing substantial overlap with the Sr⁺ 6s wavefunction, and in distinct contrast to the main resonance at 25.26 eV. It is notable also that the 5p β-parameter shows a large excursion here towards β = 0, indicating that it is dominated by a p → s transition, and in general reflecting large phase differences between outgoing s and d waves.

In order to extract the photoionization parameters, in this case the relative phase and the ratio of the dipole matrix elements corresponding to the s and d outgoing waves, it is necessary to assume LS coupling. Otherwise the $d_{3/2}$ and $d_{5/2}$ waves must be separated and from the two measurements, the alignment parameter and angular distribution parameter, there is insufficient information to do this. Ueda et al (1993a) found that in the region of the resonance at 25.26 eV, the angular distribution parameter for the 5s electrons, shown in figure 6 together with the relative partial cross section for leaving the Sr⁺ ion in its ground state, is, within experimental error, ~2. This indicates that here the two outgoing waves $p_{1/2}$ and $p_{3/2}$ are not separable and supports the assumption of LS coupling. This is certainly not the case at other resonances in the 4p - 4d complex, where the β-parameter falls to ~1.6 in some regions, but it does seem that, surprisingly for a large atom such as strontium, LS coupling can be assumed at the 25.26 eV resonance. In further support of this, in figure 7 are shown the relative intensities and β-parameters for the spin-orbit split components of the 5p doublet, where it is evident that although the ratio $^2P_{3/2}$:$^2P_{1/2}$ varies substantially over the whole region, at the peak of the 25.26 resonance it is 2:1, again as would be expected if LS coupling were valid.

Within the assumption of LS coupling, the alignment parameter A_{20} and β can be defined in terms of the dipole moments D_s and D_d, for the outgoing s and d waves respectively, and the phase difference D between them as follows:

$$A_{20} = -\frac{|D_s|^2 + |D_d|^2/10}{|D_s|^2 + |D_d|^2} \tag{3}$$

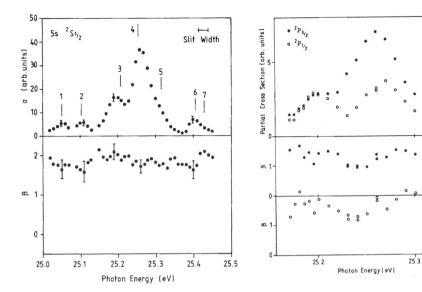

Figure 6. Data for the Sr⁺ 5s electrons

Figure 7. The Sr⁺ 5p spin-orbit components

and

$$\beta = \frac{|D_d|^2 - 2\sqrt{2}\,|D_s|\,|D_d|\cos\Delta}{|D_s|^2 + |D_d|^2} \tag{4}$$

From these equations it is possible to calculate, from the β measurement of the $^2P_{3/2}$ component and the value of A_{20} calculated from equation (2) above, the ratio $|D_s|^2/|D_d|^2$ and the phase difference Δ. Ueda et al (1993a) found the following values:

$$|D_s|^2/|D_d|^2 = 0.157 \pm 0.024 \text{ and } |\Delta| = 114 \pm 30°.$$

Calcium

The fluorescence data in the region of the 3p - 3d resonance, from Hamdy et al, are shown in figure 8, and are compared with the singly charged ion spectra taken by Sato et al (1985). Although it was no longer possible in this experiment to resolve clearly the individual components in the 3p - 3d resonance, differences are evident between the ion and fluorescence spectra. Because of this inability to resolve these peaks it was impossible to apply the same kind of analysis as used above for strontium to the calcium case, even though the assumption of LS coupling would be more reasonable for calcium. Ueda et al (1993b) made electron spectrometry measurements on calcium, and figure 9 shows the data for the cross section for populating the Ca⁺ 4p level and the angular distribution for the corresponding electrons; as before, the numbering scheme follows that of Mansfield and Newsom (1977). Two main structures in the resonance are now clearly resolved, and these features show completely different behaviour for the β-parameter. By taking wide range photoelectron spectra, Ueda et al (1993b) showed that in the calcium case there was substantial intensity in higher Rydberg levels; their data are reproduced in the following table in the form of the percentage population of the levels shown

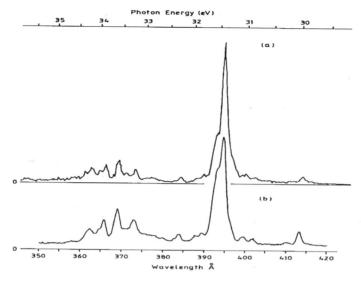

Figure 8. Comparison of the singly charged ion data (upper curve) and fluorescence data (lower curve)

at the two resonance energies.

Energy (eV)	4s	3d	4p	5s	4d	5p	nl	$3p^{-1}4s3d$
31.41	51	3.7	3.8	1.6	7.4	2.7	13	16
31.52	28	5.0	10	6.3	12	5.8	19	14

It is clear from this that cascades are a serious problem for calcium; for the 31.52 eV resonance over half the intensity is in the population of the higher states, many of which will decay to the 4p level. Using equation (4), all that can be said is that the phase difference term $\cos \Delta$ is positive in the region of the main resonance at 31.41 eV, since β is negative there.

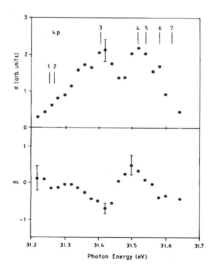

Figure 9. Data for the Ca⁺ 4p level

The solution to this difficulty is to make measurements in coincidence, ie to detect the fluorescent photon in coincidence with the photo-ejected electron which leaves the calcium ion in the 4p level. This was done recently by Beyer et al (1995), and the result of their experiment is shown in figure 10 for two electron-fluorescent photon correlation angles. In the course of

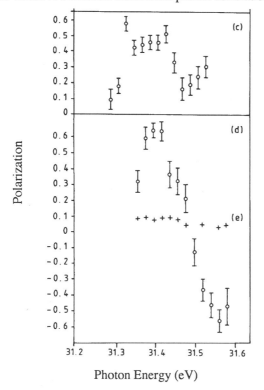

Photon Energy (eV)

Figure 10. Coincidence fluorescence polarization versus incident photon energy: (c)θ= -135°; (d) θ=-90°; (e) non-coincident measurements

the coincidence measurements values of the non-coincidence fluorescence polarization are also obtained and these, shown in the figure as curve (e), compare well with the earlier values of Hamdy et al (1991). Referring to the experimental geometry shown in figure 1, when the angle θ between the direction of the ejected electron and fluorescent photon is -90° (curve (d) on the figure), it can be seen that at 31.4 eV the polarisation value is around 0.6. This is what would be expected if LS coupling were valid, and furthermore this value is independent of the dipole moment ratio and the phase difference, ie the coincidence measurement at this angle contains no dynamic information. It can be seen that away from the resonance peak position the polarization is substantially different from 0.6, indicating that the assumption of LS coupling is invalid. The data taken at -135° can be used to calculate the photoionization parameters, although the analysis is more complicated because the polarization measurement is no longer being averaged over all electron ejection angles. With the use of statistical tensors, Berezhko et al (1978) have shown that the polarization of the fluorescent radiation is given by

$$P_x = \frac{\alpha_2 \left(3A_{20} + \sqrt{6}\, A_{22} \right)}{\alpha_2 \left(A_{20} - \sqrt{6}\, A_{22} \right)} \tag{5}$$

351

where now both the coefficients A_{22} and A_{20} contain terms which, when applied to the geometry of this experiment, include the linear and circular polarisations of the incident light (Kabachnik, 1994). As a first approximation, Beyer et al (1995) assumed that the degree of circular polarisation in the incoming light was zero. Knowing the value of the angular distribution parameter for the 4p electrons, and the polarization of the fluorescence, they found the following values for the dipole moment ratio $|D_s||D_d|$ and the phase diference $|\Delta|$:

$$|D_s||D_d| = 1.4 +0.6, -0.3, \text{ and } |\Delta| = 44.2° + 2°, -5°$$

The geometry of the monochromator and associated beamline optics are such that the expected value of the circular polarisation component of the incoming radiation is zero, since equal aperture above and below the orbit plane of the storage ring is accepted by the monochromator. However, because of small errors in the alignment of the optics, it is possible that the circular polarisation component S_3 may be as large as ± 0.6, which means that the values given above for the dipole moment ratio and the phase could lie within the ranges shown below:

$$0.74 \leq |D_s||D_d| \leq 3.6 \text{ and } 19° \leq |\Delta| \leq 46.2°$$

The way to reduce this uncertainty is to make measurements over a range of different electron-fluorescent photon correlation angles, and in particular to see if there is a difference between measurements taken at $\theta=45°$ and $135°$, referring to figure 1. Preliminary measurements of this have just been completed at the SRS and indicate that, within the limits of our experimental error, the circular polarisation component can be assumed to be zero, but it was also clear that much higher precision is needed in the experimental data before this can be firmly established.

CONCLUSION

The branching ratio and photoelectron angular distribution measurements summarised here on the photoionisation of calcium and strontium in their p - d giant resonance regions show high sensitivity to resonant structure and provide a means of investigating and determining the nature of the presently unassigned components within the p - d complex. The measurements of the angular correlation between the photo-ejected electron and the fluorescent photon which results from the decay of the excited ion are also sensitive to resonant behaviour, and such measurements have recently been completed for calcium. Using the electron angular distribution measurement from the non-coincidence experiment, the ratios of the dipole amplitudes, and their relative phases have been calculated. With a comprehensive set of coincidence measurements it should be possible to obtain enough information to calculate these parameters for the case where LS coupling is no longer valid, for example for the subsidiary resonances in the region of the 4p - 4d main resonance in strontium.

ACKNOWLEDGMENTS

It is a pleasure to acknowledge my collaborators in the above experiments, H Kleinpoppen and H-J Beyer, Stirling University, UK; K J Ross, Southampton University, UK; K Ueda, Research Institute for Scientific Measurements, Sendai, Japan; H Hamdy, Beni-Suef University, Cairo,

Egypt and N Kabachnik from Moscow State University. We are grateful for the support provided by the Engineering and Physical Sciences Research Council to run these experiments.

REFERENCES

Berezhko E G and Kabachnik N. M, *J. Phys. B: At. Mol. Phys.* **10**, 2467 (1977)

Beyer H-J, West J B, Ross K J, Ueda K, Kabachnik N, Hamdy H and Kleinpoppen H, *J. Phys. B: At. Mol. Opt. Phys.* **28**, L47 (1995).

Hamdy H, Beyer H-J, West J B, and Kleinpoppen H, *J. Phys. B: At. Mol. Opt. Phys.* **24**, 4957 (1991).

Holland D M P, Parr A C, Ederer D L, Dehmer J L and West J B *Nucl. Instrum. Methods* **195**, 331 (1982)

Holland D M P, West J B, MacDowell A A, Munro I H and Beckett A G *Nucl. Instrum. Methods* B **44**, 233 (1989)

Jiménez-Mier J, Caldwell C D, Flemming M G, Whitfield S B, and van der Meulen P, *Phys. Rev.* A **48**, 442 (1993)

Kabachnik N M, 1994 (private communication)

Mansfield M W D and Newsom G H, *Proc. R. Soc.* A **357**, 77 (1977)

Mansfield M W D and Newsom G H, *Proc. R. Soc.* A **377**, 431 (1981)

Nagata T, West J B, Hayaishii T, Itikawa Y, Itoh Y, Koizumi T, Murakami J, Sato Y, Shibata H, Yagishita A and Yoshino M, *J. Phys. B: At. Mol. Phys.* **19**, 1281 (1986)

Ross K J 1993 *Vacuum* **44**, 863

Sato Y, Hayaishi T, Itikawa Y, Itoh Y, Koizumi T, Murakami J, Nagata T, Sasaki T, Sonntag B, Yagishita A and Yoshino M, *J. Phys. B: At. Mol. Phys.* **18**, 225 (1985)

Ueda K, West J B, Ross K J, Hamdy H, Beyer H J and Kleinpoppen H, *J. Phys. B: At. Mol. Opt. Phys.* **26**, L347 (1993a)

Ueda K, West J B, Ross K J, Hamdy H, Beyer H J and Kleinpoppen H, *Phys Rev* A **48**, R863 (1993b)

West J B and Padmore H A, *Handbook on Synchrotron Radiation* Vol. II ed. G V Marr (North Holland, Amsterdam, 1987) p21

West J B, *Vacuum Ultraviolet Photoionization and Photodissociation of Molecules and Clusters* ed C-Y Ng (World Scientifiic, New Jersey, 1991) chapter 8

Yagishita A, Aksela S, Prescher Th, Meyer M, von Raven E and Sonntag B, *J. Phys. B: At. Mol. Phys.* **21**, 945 (1988)

PHOTOIONIZATION OF ATOMS NEAR INNER SHELL THRESHOLDS

M. Ya. Amusia

Institute of Physics and Astronomy, University of Aarhus
DK-8000 Aarhus, Denmark
and
A. F. Joffe Physical-Technical Institute, St. Petersburg, 194021, Russia

ABSTRACT

It is demonstrated that photoionization of atoms near inner shell thresholds is in essence a multielectron process which is determined to a large extent by interelectron interaction. The mechanisms of multiple ionization are discussed, including not only Auger-decay chains but shake-off, direct knock-out and ground-state correlations. The frequency dependence of the mean ion charge near inner shell threshold is obtained. Simple formulae are presented describing Post Collision Interaction photoelectrons' energy shift accompanying multi-step Auger-decay of an inner vacancy. Some peculiarities of an inner-shell discrete excitation are analyzed. Possible mechanisms of single charge ion production are considered. The paper emphasizes that photoionization near inner shell thresholds, particularly for heavy atoms, is an intriguing process, affected by different mechanisms of multielectron correlations.

INTRODUCTION

The interest in investigating the photoionization of atoms in the frequency region near inner shell thresholds for medium and heavy atoms is justified by the multielectron or correlative nature of this process. It is in general accompanied by forming highly charged ions and excited states of the residual ions. Although at first glance the photoionization process must become simpler while going to deeper and deeper levels due to simplification of the initial state electron wave function, this is not the case. In fact the complexity of the final state overweighs the simplicity of the initial one. Altogether the big variety of the effects accompanying the inner shells ionization makes the theory of this process rather complicated.

The experimental investigation of atomic ionization by high energy photons becomes possible nowadays because of construction of new powerful sources of electromagnetic radiation. These are giving the possibility to reach higher and higher energies.

The creation of inner vacancies is accompanied and followed not only be emitting electrons but also by emitting photons. The probability of the radiative inner vacancy decay increases very fast with the nuclear charge Z growth. Therefore radiative decay is of considerable importance for inner shells of intermediate and in particular for heavy atoms.

The inner shell ionization has in its entirety as its consequence formation of ions of different charges and electrons and photons. To calculate the probability of the given charge ion formation and electron/photon distribution is the task of the many-electron theory of this process. Although this aim is far from being achieved, we will describe in what follows some steps in the desired direction.

MANY-BODY EFFECTS IN NEAR INNER SHELL THRESHOLD PHOTO-ABSORPTION

The process of absorbing a photon whose frequency is close to intermediate and inner thresholds of medium and heavy atoms is far from being of pure single electron nature. On the contrary, it is strongly affected by a number of multielectron effects, among which the most important are the so-called rearrangement (or relaxation) effects [1]. Namely, when an inner vacancy is created, not only the photoelectron, but all other electrons are affected by the vacancy, being attracted to it. This increases the screening of the field acting upon the photoelectron thus decreasing the photoionization cross section, mainly near inner shells thresholds. With photon energy growth well above threshold this effect disappears.

The other important effect is the non-radiative or Auger decay of an inner vacancy. As a result, the field acting upon the outgoing photoelectron changes even asymptotically: instead of being $(-1/r)$ it becomes $-2/r$ or even stronger*. The growth of the attraction by the field results in the increase of the threshold cross section values.

The cross section near inner thresholds is thus determined by a competition of these two effects. It is difficult to account for these effects rigorously, but to have good estimations of their influence, two following simple approximate approaches were suggested [2].

The increase of screening due to inner vacancy i formation may be taken into account by calculating the cross section $\sigma_i(\omega)$ using the following expression:

$$\sigma_{i\tilde{f}}(\omega) \sim |< i|r|\tilde{f}>|^2 \ , \tag{1}$$

with $< i |$ and $| \tilde{f} >$ belonging to different sets of the wave functions. To be definite, the photon-electron interaction operator is presented here in the length ($\vec{r}$) form, i is the Hartree-Fock (HF) wave function of an electron in the inner shell, while $\tilde{f}$ describes the photoelectron with the energy ϵ_f moving in the self-consistent field of the residual ion with vacancy i instead of the initial atom.

The influence of the inner vacancy decay can be estimated in the same way, leading to the cross section expression given by (1) with $\tilde{\tilde{f}}$ instead of $\tilde{f}$. The function $\tilde{\tilde{f}}_q$ describes the photo-electron in the HF self-consistent field of that ion state q, which is created after Auger decay of the vacancy i:

$$\sigma_{i\tilde{f}}(\omega) \sim \sum_q \frac{\Gamma_{iq}}{\Gamma_i} |<i|r|\tilde{\tilde{f}}_q>|^2 \ . \tag{2}$$

Here it is $\Gamma_i = \Sigma_q \Gamma_{iq}$, where Γ_{iq} and Γ_i are the partial and total Auger width of the vacancy i, respectively.

For inner shells all other many-electron effects, for example those which are taken into account by the Random Phase Approximation with Exchange (RPAE) [3] are much less

---

*Atomic system of units $e = m_e = \hbar = 1$, with e and m_e being the electron charge and mass, respectively, is used throughout this paper.

important than those accounted for by (1) and (2) leading to deviation from a pure one-electron result given by σ_{if} (ω) ~ $|<i|r|f>|^2$ where f is the photo-electron's HF wave function which is calculated in the field of an atom with i electron being simply removed.

Due to the interaction between atomic electrons, the creation of a vacancy can be followed up by excitation or ionization of some other atomic electrons. These excitations when accompanying the creation of an inner vacancy are called satellites (or in some cases - shadows) of the main process which is the formation of an inner vacancy. The cross section of atomic ionization with formation of satellite states is far from being negligible. On the contrary, their contribution is considerable and must be taken into account.

MECHANISMS OF IONIZATION

In photoabsorption near and above inner shell thresholds mainly multiply charged ions are created [4]. The simplest mechanism of this is an Auger-decay chain, which is a sequence of step-by- step Auger decay of vacancies, starting with the first vacancy created in photo-ionization. If energetically allowed, these chains, which could be several, form the most probable processes. It is comparatively simple to estimate the charge of an ion formed as a consequence of these chains, in particular the highest possible degree of ionization $n_{max}^{(A)}(\omega)$.

The real experimental data demonstrates, however, considerable deviation from these simple estimations (see e.g., [5]). The role of other mechanisms is most transparent in creating highly charged ions A^{+n} with $n > n_{max}^{(A)}(\omega)$ and in forming ions with smaller charge $n < n_{min}^{A}(\omega)$, the best example of which is A^+ formation. Here $n_{min}^{(A)}(\omega)$ is the minimal amount of electrons which can be removed from an atom due to Auger-decay chains started with vacancies j created at $\omega \geq I_j$, I_j being the jth subshell ionization potential. The electrons emitted due to Auger-chain are mostly fast, their energy being high as compared to that of the vacancies created in the process.

The increase of $\sigma^{(n)}(\omega)$ - the cross section of photoionization with creation of A^{+n} ions - as compared to the corresponding cross sections $\sigma_A^{(n)}(\omega)$ describing A^{+n} production via Auger chains, is caused by the following mechanisms: shake-off, direct knock-out and ground-state correlations.

Shake-off is the result of an instant variation of the self-consistent field acting upon the outer electrons due to vacancy creation. In shake-off mainly slow electrons are emitted with almost isotropic angular distribution. The role of this mechanism is particularly important in eliminating electrons off the outer shells when the vacancy deep inside an atom is created and the primary photoelectron is a fast one. For photon energies well above thresholds the shake-off probability is frequency independent and in this ω-region this process adds some contribution to $\sigma^{(n)}(\omega) - \Delta\sigma_{so}^{(n)}(\omega)$ which has almost the same ω dependence as $\sigma_A^{(n-1)}(\omega)$ but is different for different atoms and ionized shells.

The **direct knock-out** is a process of inelastic scattering of the primary photoelectron with atomic electrons, which leads to their elimination of the atom. It is known from studies of the inelastic electron-atom/ion scattering, that the ionization cross section of this process reaches its maximum at the projectile energy of about three times higher than the ionization potential. Then the cross section decreases at high energies E as $E^{-1} \ln E$. Therefore the contribution to $\sigma^{(n)}(\omega)$ due to a direct knock-out $-\Delta\sigma_{ko}^{(n)}(\omega)$ is decreasing as a function of ω if ω is well above ionization potentials of the target atom inner shell. In this case emitted electron energies are of the order of the ionization potentials and their angular distribution is close to being isotropic.

Ground state correlations is a mechanism, which accounts for interaction of atomic electrons in the initial state of the target atom, due to which the initial state wave function deviates from being a pure HF one, it is from a simple antisymmetrized product of one-electron wave functions. Ground state correlations must be distinguished from the Auger-decay, shake-off and direct knock-out, in which the elimination of the second, third and so for electrons take place after the incoming photon is absorbed, creating an inner vacancy and a primary photoelectron. All these mechanisms are representing the effect of the interelectron interaction in the final state. Due to ground state correlations the cross section $\sigma^{(n)}(\omega)$ is corrected by the quantity $\Delta\sigma_{gs}^{(n)}(\omega)$ which as a function of ω is similar to $\Delta\sigma_{so}^{(n)}(\omega)$. Its magnitude can be, however, considerably different. The angular distribution of extra electrons emitted by this mechanism is in essence similar to that formed by the shake-off, and they are predominantly slow.

One should also specify the **quasi-Auger-decay** process in which an additional electron is eliminated via virtual creation of an inner vacancy, i.e., below the threshold of its real formation. As quasi-Auger a resonant process of ionization is considered, which is going via virtual formation of a single-vacancy state even if its decay into the final state is forbidden by energy conservation. The quasi-Auger decay mechanism is correcting $\sigma^{(n)}(\omega)$ by the magnitude $\Delta\sigma_{qa}^{(n)}(\omega)$ which is increasing when the threshold of the primarily ionizing shell is approached from below.

In a real photoionization process all mechanisms described above interfere, leading to the cross section $\sigma^{(n)}(\omega)$.

The expressions for the above mentioned corrections become considerably simpler, if the innermost shell threshold region is considered. In this case, the contribution due to shake-off is given by an expression:

$$\Delta\sigma_{SO}^{n}(\omega) = \sigma_{A,i}^{(n-1)}(\omega) \times \sum_{p \leq F} \int_{\epsilon>0} \left[\langle p|1/r|\tilde{\epsilon}\rangle / (\epsilon_p - \epsilon) \right]^2 , \tag{3}$$

where summation is performed over outer occupied electron states $p \leq F$, i is the deep ionizing shell. The continuum electron state wave function $\tilde{\epsilon}$ is calculated in the field, which is different from that $|p>$ by the extra potential $(-1/r)$. Index i in $\sigma_{A,i}^{(n-1)}(\omega)$ denotes that the Auger-chain starts with the formation of vacancy i.

The additional contribution due to direct knock-out is given by a formula

$$\Delta\sigma_{kO}^{(n)}(\omega) = \sigma^{(n-1)}(\omega) \times \sum_{\ell'=\ell\pm1} \left[1 - \exp\left(-2\tilde{Im}\ \delta_{\ell'}(\epsilon) \right) \right] , \tag{4}$$

where $\delta_{\ell'}$ is the photoelectron's elastic scattering phase shift, its energy being equal to $\epsilon = \omega - I_i$. The symbol $\sim$ above Im in (4) denotes that only that fraction of the phase shift's imaginary part $Im\ \delta_{\ell'}(\epsilon)$ is taken into account, which corresponds to ionization of any of the other atomic electrons by the photoelectron.

Ground state correlations which include inner shell electrons are small due to big separation in space and energy between the inner and the outer electrons. The role of quasi-Auger decay near inner threshold will be discussed in the next section.

It is a close correspondence between $\sigma^{(n)}(\omega)$ behaviour and that of the photoionization cross section of a given subshell $\sigma_j(\omega)$: the magnitude of $\sigma^{(n)}(\omega)$ is quite big, where $\sigma_j(\omega)$ contributes, if the elimination of $(n-1)$ electrons can be achieved via Auger-decay chains, it is in the frequency region for which n is within the limits:

$$n_{min}^{(A)}(\omega) < n < n_{max}^{(A)}(\omega) . \tag{5}$$

The extreme numbers $n_{max(min)}^{(A)}(\omega)$ of electrons removed depend upon ω as stepwise functions, their values increasing by steps at any ionization potential of an atom with energy growth - from the first ionization potential I to the highest I_{1s}. As an example, let values $n_{\max}^{(A)}$ and $n_{\min}^{(A)}$ for Ar atom are presented: $n_{\max}^{(A)} = 1$ for $\omega < 1_{2p}$, $n_{\max}^{(A)} = 2$ for $I_{2p} < \omega < I_{2s}$, $n_{\max}^{(A)} = 3$ for $I_{2s} < \omega < I_{1s}$ and $n_{\max}^{(A)} = 6$ for $\omega > I_{1s}$. As to the values of $n_{\max}^{(A)}$ they coincide with $n_{\max}^{(A)}$ for $\omega < I_{2p}$, while $n_{\min}^{(A)} = 2$ for $I_{2p} < \omega$.

The cross section $\sigma^{(n)}(\omega)$ with $n \le n_{\min}^{(A)}(\omega)$ is a direct measure of outer shell ionization contribution. For example, to produce single-charged ions A^+ an electron from an outer shell only can be removed, otherwise the process will end up creating A^{+2}, A^{+3}, etc. Therefore, in order to form A^{+n} with $n \le n_{\min}^{(A)}(\omega)$ a deeper vacancy (or even several of them) must decay radiatively finally creating a vacancy in the outer shell. The radiative decay can contribute considerably only for inner electrons starting from medium heavy atoms. This means, that processes of virtual excitations of atomic states with their subsequent decay which leads to formation of outer shell vacancies become important in creating A^{+n} with $n \le n_{\min}^{(A)}(\omega)$.

Of particular interest is the one-electron photoionization cross section $\sigma^+(\omega)$. The cross section of direct elimination of an outer electron in the vicinity of an intermediate or inner shell threshold is very small. As a function of ω it decreases fast and evenly. Any maxima or variation in $\sigma^+(\omega)$ at ω in the region of intermediate or inner shell ionization thresholds are due to virtual excitation of continuous or discrete states of these shells, which end up then by eliminating a single outer electron.

MEAN ION CHARGE

Different mechanisms of multiple ionization represent themselves in the mean ion charge $N(\omega)$, which is determined by the following relation:

$$N(\omega) = \sum_{n \ge 1} n\sigma^{(n)}(\omega) \bigg/ \sum_{n \ge 1} \sigma^{(n)}(\omega) \qquad (6)$$

This is an integral characteristic of the process, which conveniently eliminates all singularities - narrow high maxima, etc., of the cross sections $\sigma^{(n)}(\omega)$. If all but Auger-decay chains as mechanisms to produce multiple charged ions would be negligible, $N(\omega)$ as a function of ω would increase by a step of each threshold. Indeed, when ω is growing from I_{j+1} to I_j, the value $N(\omega)$ increases by a step when ω reaches the value $\omega \ge I_j$. At $\omega \ge I$ one has the following expression for $N(\omega)$:

$$N(\omega) = \frac{\bar{n}_{j+1}\sigma_{j+1}(\omega) + \bar{n}_j\sigma_j(\omega)}{\sigma_{j+1}(\omega) + \sigma_j(\omega)} \qquad (7)$$

Here the average values of the ion charge $\bar{n}_j$ are determined by the expression:

$$\bar{n}_j = \sum_q n_j^{(q)} \Gamma_j^{(q)} \big/ \Gamma_j , \qquad (8)$$

where $\Gamma_j^{(q)}$ is the partial width of the vacancy j Auger decay into the channel q. The total number of electrons eliminated from the atom when vacancy j is created and decays into channel q is equal to $n_j^{(q)}$. The total width Γ_j also includes the contribution of the radiative decay, but if this is neglected or negligible one has $\Gamma_j = \sum_q \Gamma_j^{(q)}$. It is assumed in (7) that in the vicinity of I_j the contribution of only the nearest shell - $j + 1$ - must be taken into account:

the cross sections of all other - $\sigma_{j+2}(\omega)$ and so on - are much smaller than $\sigma_{j+1}(\omega)$ at $\omega \lesssim I$. According to (7), the step in N (ω) at $\omega = I_j$ is given by the following expression

$$\Delta N(I_j) = \frac{(\bar{n}_j - \bar{n}_{j+1})\sigma_j}{\sigma_j + \sigma_{j+1}} , \qquad (9)$$

where σ_j, σ_{j+1} are the values of $\sigma_j(\omega)$ and $\sigma_{j+1}(\omega)$ at $\omega = I_j$. If $\sigma_j \gg \sigma_{j+1}$, equation (9) leads to a very simple expression

$$\Delta N(I_j) = (\bar{n}_j - \bar{n}_{j+1}) .$$

The other mechanisms of multiple ionization, namely shake-off, direct knock-out and ground state correlations affect the numbers $n_j^{(q)}$ adding to them positive, or, in principle, because of the virtual nature of some of these processes, negative corrections, which are in general frequency-dependant.

Due to non-zero Auger width of the intermediate and inner atomic levels, the function $N(\omega)$ is in fact not jumping at thresholds, but is increasing rather fast in the width region. It will be demonstrated in this section that the growth of $N(\omega)$ is noticeable even quite far below the inner or intermediate shell threshold. To prove this let us consider photoionization cross section close but below the inner shell j ionization threshold. The outer shell ionization cross section may increase in this region, but this cannot lead to the growth of $N(\omega)$ below threshold j. The latter can be achieved only if the final state, namely that which is reached at $\omega > I_j$ is created. To feel that the j-threshold is approaching, the vacancy j must be created virtually, as an intermediate state of the photoionization amplitude. It means that a process of the Auger-decay of a virtual state - i.e., of a not yet created vacancy - takes place. The analytical expression for the cross section of photoionization due to this mechanism is given by [6]:

$$\sigma_{j'}(\omega) = \frac{4\pi^2\omega}{c} \sum_{i_1 i_2} \sum_{\epsilon_k} \int_0^\infty d\epsilon |<j|\hat{d}|\epsilon_k>|^2 \frac{|<j\epsilon|V|i_1 i_2>|^2}{(\omega - I_j - \epsilon_k)^2 + \left(\frac{1}{2}\Gamma_j\right)^2} \times \qquad (10)$$

$$\times \quad \delta(\epsilon_k + \epsilon + I_{i_1 i_2} - \omega) .$$

Here the sign of quotation emphasizes that (10) is valid also below the threshold of the j vacancy formation. The vacancies i_1 and i_2 are those into which the vacancy j is decaying in the Auger-process. The matrix element $<j|\hat{d}|\epsilon_k>$ is the amplitude of the electron transition from j to ϵ_k after absorbing a photon, the interaction of which with the electron is described by the operator $\hat{d}$. The decay of the vacancy into the state $\epsilon\, i_1 i_2$ is presented in the lowest non-vanishing order of the Coulomb interelectron interaction V by the matrix element $<j\epsilon|V|i_1 i_2>$. Integration over ϵ_k includes also summation over discrete levels, which can be occupied by the primary photoelectron.

The formulae (10) can be considerably simplified, bearing in mind that for ω close to I_j the matrix element $<j|d|\epsilon_k>$ may be substituted by its threshold value and the Coulomb matrix element $<i|V|i_1 i_2>$ may be considered as energy ϵ independent in the ϵ region $\epsilon \approx \epsilon_{Aj} \equiv I_j - I_{i_1 i_2}$, where ϵ_{Aj} is the energy of the Auger-electron created in the decay of vacancy j into $i_1 i_2$. Having in mind, that Γ_j is given by the equation

$$\Gamma_j = 2\pi \sum_{i_1 i_2} |<1s \, \epsilon_{A_j} |V| i_1 i_2>|^2 , \tag{11}$$

the following expression is obtained, neglecting the contribution of discrete excitation ϵ_k:

$$\sigma_{j'}(\omega) = \sigma_j \left[\frac{1}{2} + \frac{1}{\pi} \, arctg \left(\frac{\omega - I_j}{\frac{1}{2} \Gamma_j} \right) \right] . \tag{12}$$

At ω well below I_j, expression (12) demonstrates a comparatively slow decrease with $|I_j - \omega|$ growth:

$$\sigma_{j'}(\omega) \approx \frac{\sigma_j}{2\pi} \frac{\Gamma_j}{I_j - \omega} . \tag{13}$$

Those high-lying discrete excited levels ϵ_k starting with ϵ_k^*, the spacing between which is smaller than Γ_j, can be easily taken into account substituting I_j in (12) and (13) by $I_j^* = I_j - |\epsilon_k^*|$. The contribution to (10) from several first well-separated excited levels ϵ_k is given by the following expression:

$$\Delta\sigma_{j'}(\omega) = \frac{4\pi^2\omega}{c} \sum_{k,i_1 i_2} |<j| \hat{d} |\epsilon_k>|^2 \frac{|<j\epsilon| V| i_1 i_2>|^2}{(\omega - I_j - \epsilon_k)^2 + \left(\frac{1}{2}\Gamma_j \right)^2} \tag{14}$$

where $\tilde{\epsilon}_k \approx \omega - \epsilon_k - I_{i_1 i_2}$. It is seen, that (14) decreases with decrease of ω (at $\omega < I_j$) as $(I_j + \epsilon_k - \omega)^{-2}$, i.e., much faster than (13).

Accounting for the contribution (12), the mean ion charge $N(\omega)$ for $\omega < I_j$ can be presented in the following way:

$$N(\omega) = \frac{\bar{n}_{j+1}\sigma_{j+1}(\omega) + \bar{n}_j\sigma_{j'}(\omega)}{\sigma_{j+1}(\omega) + \sigma_{j'}(\omega)} \tag{15}$$

and the increase of $N(\omega) - \Delta N(\omega)$ - can be obtained using (13):

$$\Delta N_j(\omega) = N(\omega) - N(\omega)|_{\omega = I_{j+1}}$$
$$= (\bar{n}_j - \bar{n}_{j+1}) \frac{\beta}{\epsilon + \beta} , \tag{16}$$

where $\epsilon = (I_j^* - \omega)/T_j$. According to (14), the contribution of several first well-separated discrete excitations decreases with growth of $(I_j - \omega)^2$ at $\omega < I_j$ much faster than (16) and therefore can be neglected well below the threshold j. Obviously, if the matrix elements $<j| \hat{d} |\epsilon_k>$ or corresponding oscillator strengths for these levels are known, it is not necessary to neglect in (16) the discrete level contribution. On the contrary, they can be easily accounted for.

The formulae (12-16) are much more general than the simplifying assumptions, leading to (10) and (11). Therefore (12) and (16) can be used to parametrize the observed frequency dependance of $N(\omega)$ and to derive the parameters such as the width and the jump in $\bar{n}_j$ values $(\bar{n}_j - \bar{n}_{j+1})$. Instead of Auger-decay values $\bar{n}_j^{(A)}$, those ones which account for multiple ionization via other processes such as shake-off, direct knock-out and ground state correlations - $\bar{n}_j$ can be used in (16). More precisely bearing in mind the different frequency dependencies of all these processes described above, one should substitute $\bar{n}_{j+1}$ by the value, which accounts for contributions off all processes, responsible for multiple ionization. As $\bar{n}_j$ the corresponding Auger value $\bar{n}_{.}^{(A)}$ must be used.

The expressions (13), (15) and (16) were applied recently to describe the experimentally observed dependence of $N(\omega)$ near $1s$ threshold in Ar [5]. Satisfactory agreement was achieved. It appeared that even at 30-40 eV below $1s$ threshold the contribution to $\Delta N_j(\omega)$ due to decay of virtually created $1s$ vacancy is quite noticeable.

MULTISTEP POST COLLISION INTERACTION

Near intermediate and inner shell thresholds the well-known and well-studied Post Collision Interaction [7] becomes considerably modified due to the chain of sequential Auger decay, instead of a single one. This chain is the reason why the photo-electron will finally find itself moving in the Coulomb field of the residual ion with the charge $n^{(A)}$, i.e., in the field $(-n^{(A)}/r)$ instead of $(-1/r)$, which would be without Auger-decays.

Qualitatively, multistep PCI can be understood in the following way. Each time when an extra Auger-decay from the total Auger-chain takes place, the primary photoelectron feels an instant increase of the field it is moving in, by the quantity $- 1/r$. As a result, the photoelectron loses its energy step by step.

To calculate the total energy shift, let us consider this process classically. If the Auger-chain consists of N elements, one can determine the distance which the primary photoelectron moves before 1st, 2nd....$n^{(A)}$th decay as

$$
\begin{aligned}
R_1 &= v_1 T_1 \\
r_2 &= v_2 T_2 + r_1 \\
r_t &= v_t T_t + r_{t-1} \\
r_N &= v_N T_N + r_{N-1} = \sum_{t=1}^{N} v_t T_t
\end{aligned}
\tag{17}
$$

Alteration of the field, acting upon the primary photoelectron proceeds by steps

$$
\Delta U_t = -1/r_t \ , \tag{18}
$$

with t between 1 and $n^{(A)}$. The total energy shift of the primary photoelectron is

$$
\Delta U = -\sum_{t=1}^{n^{(A)}} r_t^{-1} \tag{19}
$$

At each Auger-decay t the photoelectron loses some energy, its velocity thus becoming smaller and smaller:

$$
v_i^2 + 2\Delta U_i = v_{i+1}^2 \tag{20}
$$

and after the last step the outgoing electron energy reaches its final value v_f^2.

As an example, let us consider a two-step PCI, for which the primary photoelectron energy decrease is given by

$$\Delta U = -\frac{\Gamma_1}{v_1}\left[1 + \frac{\Gamma_2}{\Gamma_2 + \frac{v_2}{v_1}\Gamma_1}\right],$$

$$\left(\frac{v_2}{v_1}\right) = \left(1 - \frac{2\Gamma_1}{v_1^3}\right)^{1/2}.$$

(21)

Expressions (16-18) are obtained under the assumption that Auger-electrons have much higher speed than that of the primary photoelectron. If this is not the case, the interaction between the photo- and Auger-electrons must be taken into account. This is very difficult to do quantum-mechanically. But to have an estimation of the corresponding effect, classical consideration is sufficient. It means that at each decay step the energy change $\Delta U_t = -r_t^{-1}$ must be substituted by the $\Delta\tilde{U}_t$ given by the following formula:

$$\Delta\tilde{U}_t \equiv -\tilde{r}_t^{-1} \approx -(r_t^{-1} - |\boldsymbol{r}_t - \boldsymbol{v}_t^A T_t|^{-1})$$

(22)

Here $\boldsymbol{v}_t^A$ is the velocity of the Auger-electron, emitted at the t-th step of the decay, $\boldsymbol{r}_t$ is the radius-vector, which determines the photoelectron's position after the t-th step of the decay process. Neglecting for simplicity the alteration of the direction of the photo-electron's motion due to its interaction with the Auger-electrons, one has

$$\boldsymbol{r}_t = (\mathbf{v}_1/\mathbf{v}_1)r_t.$$

The term $\boldsymbol{v}_t^A T_t$ determines the distance passed by Auger-electron from the beginning of the t-th step of the Auger-decay. The second term in (21), $|\boldsymbol{r}_t - \boldsymbol{v}_t^A T_t|^{-1}$, is the potential energy of the photo-electron and Auger-electron interaction.

The modified total energy shift $\Delta\tilde{U}$ is now given by

$$\Delta\tilde{U} = -\sum_{t=1}^{N} r_t^{-1}$$

(23)

with the speeds v_t and v_{t+1} being determined by the formula:

$$v_t^2 + 2\Delta\tilde{U}_f = v_{t+1}^2.$$

(24)

Note, that if v_t^A is big enough, ΔU_t reduces to ΔU_t. On the contrary, for small v_t^A, i.e., for $v_t^A \ll v_t$, the correction ΔU_t becomes negligible.

The mutual repulsion of the Auger-electrons and the photoelectron modifies at each step of decay not only the magnitude of v_i but its direction. Therefore Eq. (21) must be corrected. But this makes the formula of the energy shift even more complicated and far from being transparent.

Multistep PCI affects considerably the Auger-electrons' energy distribution. Their total energy becomes shifted to higher values by the magnitude $(-\Delta U)$ or $(-\Delta\tilde{U})$. The widths of Auger-profiles are also increased.

PECULIARITIES IN INNER SHELL OSCILLATOR STRENGTHS

The oscillator strength of an inner shell is almost completely determined by the effective nucleus charge, which acts upon the excited electron*. As a result, the magnitudes of the oscillator strengths and their dependence upon the principle quantum number of the excited electron carry information on the electronic structure and the state of the outer atomic region.

Let us start with a one-particle approach. In its frame the oscillator strength F_{if} of an electron transition from the level i to f is given by

$$F_{if} = 2\omega_{if} |<i|r|f>|^2$$
$$\omega_{if} = \epsilon_f - \epsilon_i .$$

(25)

While the initial state wave function $\psi_i(r)$, corresponding to a vacancy i in an inner shell, can be described reasonably accurate within the pure Coulombic model with the nuclear charge Z_i, another approach should be used to determine ψ_f. This wave function is of interest at distances $r \leq r_i$, with r_i being the inner shell radius. This distance is small for the outer level wave function $\psi_f(r)$, so its radial dependence can be approximated by Cr^ℓ, ℓ being the angular momentum of the f-level. The coefficient C in front of r^ℓ in ψ_f is determined from the Coulombic wave function with the f quantum numbers, but in the nuclear field with the charge Z_f. As a result, for F_{if} as a function of Z_i and Z_f it is obtained:

$$F_{if} \sim Z_i^{-5} Z_f^5$$

(26)

Contrary to the completely stripped ions, for which $Z_i = Z_f$ and therefore F_{if} is Z-independent, for atoms with their $Z_i \gg Z_f$ the oscillator strength becomes very sensitive to Z_f and Z_i. Let us compare the F_{if} values for $1s$ shell in a noble gas atom, e.g., Ar and its alkali and alkali earth neighbours. The Z_i values are almost the same, while Z_f's are quite different. For $f = 4p$ in Ar, K and Ca the Sleter's effective charge Z_{4p} values are the following:

$$Z_{4p}^{(Ar)} = 2.2 , \quad Z_{4p}^{(K)} = 2.85 \quad \text{and} \quad Z_{4p}^{(Ca)} = 3.5 .$$

(27)

As a result F_{1s4p} in Ca is about six times bigger than in Ar. The reason for this effect is simple: there is no screening for the $4p$-excited electron from the $4s$ ones. Therefore $4p$-electron in Ca is in the field with the effective charge close to $Z_{4p}^{(Ca)} \approx 3$ while for Ar this is much smaller.

With the growth of the excitations principal quantum number the situation changes rapidly, leading to $Z_{np}^{(Ar)} \approx Z_{np}^{(K)} \approx Z_{np}^{(Ca)}$ already for $n \geq 6$. Thus the relative intensity of several first excited levels in Ca and K is considerably bigger than in Ar. This qualitative peculiarity was recently observed by comparing the intensities of $1s \to 4p$ excitations in Ar and K [8].

If an outer electron is excited, it can essentially affect the inner shell oscillator strength.

---

*The results in this section have been derived together with A. S. Baltenkov and G. I. Zhuravleva.

For instance, if the $3p$ electron in Ar goes into $4s$ level, the intensities of $1s$ - np transitions become similar to that in K. The excitation of the outer $4s$ electron in K into n_1s or n_1p levels leads to a prominent increase of oscillator strengths of all $1s$ - np levels, up to the $n' \leq n_1$: for all of them the effective charge $Z_{n'}^{(K^*)}$ will be close to $Z_{4p}^{(K)}$. This dependence upon Z_f and n_f is a quite general feature. It has a simple physical foundation and must manifest itself in any neighbouring atoms and not only in the innermost shell.

If the nuclear field could be screened by outer electrons completely, leading to $Z_f \approx 0$, the discrete excitations would disappear. While this is impossible for isolated atoms, where Z_f is not less than one, this can take place for atoms imbedded in metals, where the excited levels of inner-shell electrons can be eliminated.

Of course the simple approximation, which leads to (26) is crude. This is reflected also in the big difference between results of the oscillator strength calculations in the length (r) and velocity (∇) forms. Therefore calculations were performed in a much more reliable model, describing the creation of the $1s$-vacancy in the Generalized Random-Phase Approximation with Exchange frame, which takes into account the comparatively fast Auger-decay of $1s$-vacancy. The details of this method, used in these calculations, can be found in [9]. However, its essence is in taking into account that the excited electron feels the field of a double vacancy (or even more complicated) state, which is created after $1s$-vacancy Auger-decay. As a result of this decay, different final two-vacancy (and more complicated) states are formed, thus leading to different fields acting upon the excited electron. Therefore an average oscillator strength must be introduced, if the final states are not fixed in the experiment. This average oscillator strength is determined by the equation

$$\overline{F}_{if} = \sum_{jk} F_{i\tilde{f}(jk)} \Gamma_{i,jk} / \Gamma_i ,\qquad(28)$$

where $\Gamma_{i,jk}$ is the partial width of the inner vacancy i Auger decay into the state with two vacancies j and k. For simplicity, the next steps of the decay with formation of more than two vacancies were neglected. The oscillator strengths $F_{i\tilde{f}(jk)}$ are calculated using formula (26), in which, however, the excited electron wave function $\tilde{f}$ is found in the field of an ion, having vacancies j and k. The total Auger-width Γ_i is determined by the equation

$$\Gamma_i = \sum_{jk} \Gamma_{i,jk} .\qquad(29)$$

All possible jk HF states were taken into account and the wavefunctions of Auger-electrons were calculated in the self-consistent HF field of an ion with two vacancies jk. It appeared, that the oscillator strengths $F_{i\tilde{f}(jk)}$ weakly depend upon the jk configuration. The internal accuracy and consistency of calculations were controlled by performing them using two forms for the operator describing the interaction of an electron with electromagnetic field, namely the length r and velocity - $i \nabla/\omega c$. For precise wavefunctions the results with r or - $i \nabla/\omega c$ must be the same. A reasonable coincidence, within several percent, was achieved in our calculations. The results of them, performed using the computing code (ATOM) [10], are presented in the Table:

Oscillator strengths _F_ of 1s electron

	Frozen Core			Relaxed Core		
Level	K	Ar	Ratio	K	Ar	Ratio
1s - 4p	1.69	0.76	2.22	4.26	3.02	1.41
1s - 5p	0.34	0.26	1.31	1.25	1.13	1.11
1s - 6p	0.14	0.12	1.17	0.54	0.54	1.00
1s - 7p	0.075	0.067	1.12	0.29	0.29	1.00

It is seen from the Table that the magnitude of oscillator strengths depend strongly upon whether or not the 1s-vacancy decay is taken into account, the latter effect increasing the _F_-values by a factor 3-4. It confirms the conclusion already made in investigating the photoionization of 1s-electrons in Ar near their threshold [2] that multielectron effects such as vacancy decay are extremely important in this process. As is seen, these effects are also strongly emphasized in the ratios, particularly in their difference for 1s - 4p and higher _n_ transitions. Presented above simple model leading to (26) overestimates the value of the effective charge acting upon the excited electron. The data in the Table being results of calculations accounting for all possible physical effects which can alter the oscillator strengths, must be considered as upper and lower limits to the forthcoming results of absolute measurements. The recent observations [8] are confirming that the oscillator strength of the 1s - 4p transition in _K_ is considerably bigger than in Ar, but not giving absolute values for the ratio.

SINGLE-CHARGED ION PRODUCTION

Single-charged ions A^+ are produced only when an outer electron is eliminated off an atom. This can be accompanied by emission of photons. The cross section of direct outer electron elimination in the frequency region near the inner shell ionization threshold is small and a smooth function of ω. The contribution of indirect ionization, i.e., that which proceeds via the virtual excitation of other atomic electrons is considerable, has a strong ω dependence, and can therefore noticeably affect the A^+ production cross section $\sigma^+(\omega)$. Recent experiments demonstrated [5] that $\sigma^+(\omega)$ as a function of ω has a maximum in the vicinity of the 1s-threshold in Ar. Such a variation does not exist in the direct outer shell one electron ionization cross section. It can come only if the virtual excitations of 1s are included. The virtual excitation of all but 1s electrons is leading to the frequency variation of $\sigma^+(\omega)$ near their own thresholds. Therefore at $\omega \approx I_{1s}$ only the virtual excitation of 1s electrons must be taken into account. This is achieved when the dipole operator r in the photoionization cross section is substituted by $\boldsymbol{r}$, given by the following equation:

$$\boldsymbol{r} = r\left(1 + \frac{\alpha^{(1s)}(\omega)}{r^3}\right) \tag{30}$$

where $\alpha^{(1s)}(\omega)$ is the dipole dynamical polarizability of the 1s-shell.

The action of an inner shell (in particular 1s) upon the outer one can be described in GRPAE frame, by taking into account the interaction of the outer and inner shells only. It was done recently for Ar* and it appeared that this effect for $\omega > I_{1s}$ is quite small. The discrete excitations, mainly the $1s$ - $4p$ level are manifested in Ar$^+$ yield by a maximum, but a quite small.

More effective is the production of Ar$^+$ via the PCI - Post Collision Interaction. The latter prominently decreases the photoelectron's energy near the threshold of its creation. The PCI energy shift can be bigger than the photoelectron's energy, thus leading to its recapture into a discrete level in the field of that ion which is created by the Auger-decay. One should note, however, that most probable is the recapture to highly excited levels, which then can decay via autoionization by emitting mainly slow electrons from the outer shell. Therefore the capture cross section is bigger than that of A^+ formation. This difference is particularly considerable for the innermost shells. For them only those Auger-decay channels can contribute to A^+ formation which are connected with creation of vacancies in the outer shell: otherwise ions A^{+n}, $n \geq 2$ will be formed. This explains why although A yield can have a maxima in the vicinity of the inner shell threshold, the cross section for A^+ formation is there small.

CONCLUDING REMARKS

A number of processes connected with the elimination of an electron from an inner atomic shell was discussed in this paper. Although the estimation of their probability can be done relatively simply within the frames of existing theoretical approaches, the accurate calculations to be performed require accounting for multielectron correlations of several outgoing electrons formed in ionization process as well as those which are virtually excited in the intermediate states. This important problem is quite far from being solved. On this way of studying of the role of interelectron interaction in atoms the experimental data are particularly important. They will come and are already coming from new sources of continuous spectrum electromagnetic radiation, which are already operating and under construction. The increase of the photon energy and simultaneously of the flux intensity is opening the possibility to study the threshold region of the heaviest atoms' innermost shells and to consider in details more and more sophisticated process which accompany the primary creation of a deep vacancy.

Of importance and interest are energy and angular distributions of secondary particles, emitted after photon absorption, namely electrons and photons. While that particles which are leaving an atom due to Auger - or radiative vacancy decay have characteristic energies the other mechanisms are supplying continuous spectrum particles. For electrons these are shake-off, direct knock-out and quasi-Auger. The continuous spectrum electrons are emitted in the course of direct multiple ionization when it proceeds via the ground state correlation. Another source of such electrons is the two- and multielectron Auger-decay, in which the transition energy is distributed between at least two outgoing electrons.

Of interest is the process of continuous spectrum photon emission or Internal Bremsstrahlung, as well as the vacancy decays with simultaneous emission of an electron and photon. Here as an example can serve the so-called Radiative Auger-decay.

These processes can proceed directly, for instance as the usual Bremsstrahlung of an electron in the statical field of an atom or via virtual excitation of some other atomic electrons. In the latter case strong collective effects can manifest themselves, if several atomic electrons are virtually excited coherently.

Entirely, it is a lot what can be done in the studies of the inner shell photoionization of atoms, which step by step is becoming an important area of the research in the whole domain of photo processes studies.

---

*Considered together with A. S. Baltenkov and G. I. Zuravleva

REFERENCES:

[1]   M. Ya. Amusia, in *Atomic Physics 5*, Plenum Press, pp 537-565, 1977.

[2]   M. Ya. Amusia, V. K. Ivanov and V. A. Kupchenko, J. Phys. B: At. Mol. Phys. **14**, L667-671 (1981).

[3]   M. Ya. Amusia, *Atomic Photoeffect*, Plenum Press, New York - London, 1990.

[4]   H. Tawara, T. Hayaishi, T. Koizumi, T. Matsuo, K. Shima and A. Yagishita, J. Phys. B: At. Mol. Opt. Phys. **25**, 1476 (1992).

[5]   J. Doppelfeld, N. Anders, B. Esser, F. von Busch, H. Scherer and S. Zinz, J. Phys. B: At. Mol. Opt. Phys. **26**, 445 (1993).

[6]   M. Ya. Amusia, Phys. Lett. A **183**, 201-204 (1993).

[7]   M. Yu. Kuchiev and S. A. Sheinerman, Sov. Phys. Uspekhi **32**, 569-587 (1989), in Russian.

[8]   Y. Azuma, H. G. Berry, T. Le Brun et al., 1994, private communication.

[9]   M. Ya. Amusia, V. K. Ivanov, S. A. Sheinerman and S. I. Sheftel, Sov. Phys. JETP **78**, 910-923 (1980)

[10]   M. Ya. Amusia and L. V. Chernysheva, *Computation of Atomic Structure and Processes*, Adam Hilger, IOP Publishing, 1996, in press.

THRESHOLD PHOTODOUBLE IONISATION IN ATOMS AND MOLECULES

George C. King and Grant Dawber

Department of Physics and Astronomy
University of Manchester
Manchester, M13 9PL
UK

INTRODUCTION

The process of single photon double ionisation is of particular interest because it is a means of studying the dynamics of two outgoing electrons in the field of a doubly charged ion: the fundamental, three body Coulomb problem. These two electrons are highly correlated and so photodouble ionisation measurements are a sensitive way of studying these correlations. Photodouble ionisation may occur through a direct process where the two photoelectrons are emitted simultaneously, or indirectly - for example via an excited singly charged state, which subsequently decays, by autoionisation, to an energetically lower lying doubly charged state. So far, less attention has been paid to these indirect processes as they constitute an extra degree of complexity to the overall photodouble ionisation process. Interestingly, however, these can have much larger cross-sections than those for the direct process, and so are beginning to receive increased attention[1,2]. Most of the work on the study of correlations in photodouble ionisation has centred on atoms rather than molecules: the additional molecular motions are expected to complicate the basic three body Coulomb problem. In fact in the molecular case it is the *spectroscopy* of the doubly charged states created that has formed the centre of interest[3].

The process of single photoionisation may also be dominated by electron correlation effects. This is the case, for example, when the incident photon causes the excitation of two valence electrons: one is ejected and the other is promoted to an unfilled orbital, giving rise to so called 'satellite' states. This process can be viewed as a case of aborted photodouble

Selected Topics on Electron Physics
Edited by Campbell and Kleinpoppen, Plenum Press, New York, 1996

ionisation, in which two correlated electrons initially recede from the core, but evolve to a final state in which only one electron escapes. Again, these singly charged states may be formed either directly or indirectly. In this case the indirect route proceeds via neutral doubly excited states (resonances).

The correlation effects noted above will be most important when the outgoing electrons are moving very slowly in the field of the residual ion. There they dominate the dynamical behaviour of the electrons and the consequent evolution of the cross section as the available excess energy increases. This occurs in near threshold photoionisation where the energy of the incident photon is just greater than the single or double ionisation onset. The nearly zero energy (threshold) photoelectrons that result can be very effectively detected using the penetrating field technique, developed by Cvejanovic and Read[4]. This technique offers both high detection efficiency and sensitivity; the threshold electrons can be collected over $\sim 4\pi$ Sr with an energy resolution of a few meV. The high detection efficiency is of particular importance because of the low cross-sections for the formation of singly charged satellite states near threshold. The high energy resolution means that it is possible to resolve many of the features observed in the Threshold PhotoElectron Spectra (TPES), including, for example, rotational structure in the molecular case[5].

In an analogous manner, the most direct way of studying the threshold photodouble ionisation process is to detect, simultaneously, the two outgoing threshold electrons in a coincidence measurement. Such a technique, Threshold PhotoElectrons COincidence spectroscopy (TPEsCO), has been developed by Hall et al[6] and applied to wide variety of atomic and molecular targets.

Whilst the TPES and TPEsCO techniques afford high detection efficiency and energy resolution, all angular information is lost as a result. Information about the angular distribution of one of the two electrons that are ejected following photodouble ionisation can be obtained in a Threshold PhotoElectron PhotoElectron COincidence measurement (TPEPECO). Ultimately, in order to learn about the angular distribution of both of these electrons, a PhotoElectron PhotoElectron COincidence measurement (PEPECO) can be performed. The present chapter describes the application of the above techniques and some of the physical insights into photodouble ionisation that have been obtained.

EXPERIMENTAL ARRANGEMENTS

A generic form of the apparatus is presented in figure 1. Although various forms of the apparatus have been used[6,7,8], only one version - which embodies all of the main points - will be discussed. It comprises a pair of identical photoelectron energy analysers, mounted in a plane perpendicular to the photon beam, which is directed out of the page. The photon beam comes from the Daresbury SRS TGM beamline, and is tunable over the energy range 10-150eV, with an energy resolution that varies from 10 to 150meV, depending on the energy. The analysers are able to rotate, independently, about the photon beam axis. Each analyser is made up of a lens system, electron energy analyser and detector. The lens system has been specially designed to collect and focus very low energy electrons from the interaction region to the entrance of the energy analyser. The analyser used in these studies was a 127^0 Cylindrical Deflector Analyser (CDA). The signal transimitted by the CDA is detected using a Channel Electron Multiplier (CEM).

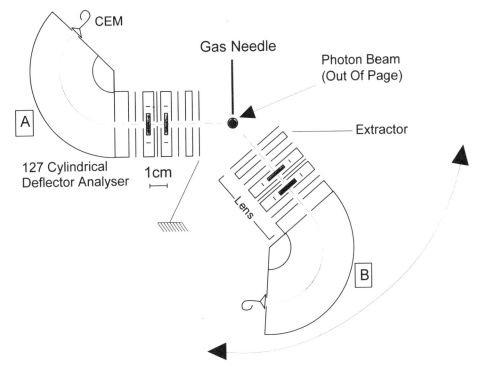

Figure 1. Schematic diagram of the apparatus used in the present studies.

Each analyser is capable of working in either the threshold mode, in which threshold electrons are collected using the penetrating field technique, or conventionally, in which low (non-zero) energy electrons are collected over a small solid angle. In the threshold mode a high voltage (~100V) is applied to an extraction electrode, and the resulting field penetrates through an earthed screening electrode into the interaction region. The effect of this field is to collect preferentially nearly zero energy electrons - of energy $\Delta E < 20 meV$ - irrespective of their initial direction, whilst discriminating against electrons of higher energy. In this case the CDA filters the threshold signal from the high energy 'tail' arising from energetic electrons ejected in the direction of the analyser. A threshold photoelectron spectrum (TPES) can then be obtained by scanning the photon energy over the range of interest and measuring the yield of threshold photoelectrons. One of the great strengths of this technique is that the widths of peaks in such a spectrum depend only on the available photon energy resolution, and can therefore be as narrow as a few meV using very high resolution radiation[6].

If the two analysers are placed 180^0 apart, and the penetrating field technique is used in both analysers, then the threshold photoelectron signal will be divided between them according to their relative efficiencies, with ~2π Sr into each analyser. If the photon energy is scanned across a photodouble ionisation threshold, then two nearly zero energy electrons will be ejected. If these two threshold electrons drift in approximately opposite directions, then one electron will be detected in each analyser. By performing a coincidence measurement between the signals collected in each analyser, using standard timing electronics, these photodouble ionisation events can be distinguished from, for example, single photoionisation events. The yield of true coincidences as a function of the photon energy is called a threshold photoelectron's coincidence spectrum (TPEsCO). The analogy between the TPES spectrum and its double ionisation counterpart, TPEsCO, is illustrated in figure 2.

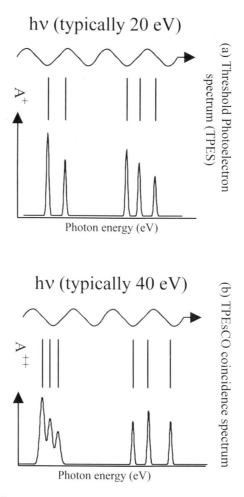

Figure 2. Schematic disgram illustrating the analogy between the use of the TPES in photoelectron spectroscopy (a) and the TPEsCO technique in photodouble ionisation (b).

If one analyser is tuned away from threshold, so that only electrons from a small solid angle are collected, then angularly differential TPEPECO measurements can be performed. Such a coincidence measurement, in which the other analyser collects threshold electrons now over 4πSr, corresponds to a Doubly Differential Cross Section (DDCS) for the case of extreme energy sharing between the two electrons. By taking DDCS measurements as the position of the angularly differential analyser is varied with respect to the polarisation vector of the light, then the asymmetry parameter for photodouble ionisation, β, can be obtained, for this case of extreme energy sharing.

Finally by taking both analysers away from threshold, so that now both are angularly differential, a PEPECO measurement can be performed. The yield of true coincidences corresponds to the Triple Differential Cross Section (TDCS), and represent a measurement of the photodouble ionisation process at its most fundamental level.

THRESHOLD PHOTOELECTRON STUDIES IN HELIUM

Helium is the archetypal atomic system for probing the effects of electron correlation in photodouble ionisation because of the absence of structure in the residual ion core.

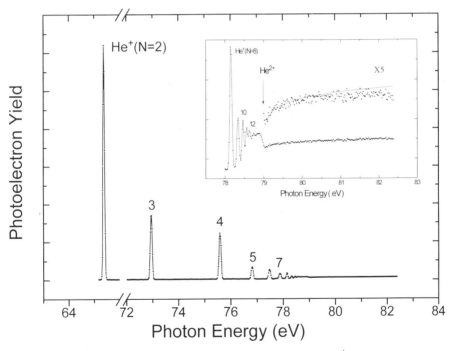

Figure 3. Threshold photoelectron spectrum of helium in the region of the $He^+(N=2\rightarrow\infty)$ satellite states, including the region around the double ionisation threshold.

An example of a TPES spectrum obtained in helium[9] in the region of the double ionisation potential is shown in figure 3. Below the double ionisation potential the structure corresponds to satellite formation while above it the yield corresponds to direct photodouble ionisation. In the spectrum of figure 3, satellite states of He^+ up to $n = 12$ are clearly visible.

Higher levels merge into an almost flat shoulder which falls off rapidly reaching a sharp minimum at the double ionisation potential, to which the satellites converge. This singularity has the form of a strongly asymmetric cusp and can be seen more clearly in inset of the figure.

Both regions, either side of the double ionisation potential, are dominated by electron correlations and there are several points of particular interest. Below the double ionisation potential, the intensity of the satellites can be seen to decrease in a non-uniform way. This can be explained as being due to the effect doubly excited neutral states, which interfere with the direct cross sections. Studies of such satellite states in helium[10,11] and the other rare gases[12,13,14,15] indicate that the indirect processes do indeed dominate their cross-sections.

The model of Wannier[16] makes predictions about various aspects of the two electron escape problem and, for example, predicts that the cross-section for photodouble ionisation, σ, varies as E^m with m equal to 1.056. Since the yield of only threshold electrons (of energy between zero and ΔE) is measured in the TPES spectrum of figure 3, this corresponds to a partial cross-section which may be written,

$$\sigma_{partial} \propto E^{m-1} \propto E^{0.056} \tag{1}$$

where E is the excess energy above the double ionisation threshold. Since m-1 rather than m is being measured this is a very sensitive test of theory. Above the double ionisation potential the form of the TPES can be compared with the Wannier model. The full curve shown in the inset of figure 3 is a fit of the data, over the energy range 0.08 <E< 0.48 eV for the Wannier exponent m=1.056 and then extrapolated to higher energies. The data are consistent with this value of m and give the range of validity of the threshold law of about 2 eV. Also of interest in figure 3, is the symmetry of the cusp about the double ionisation potential. The physics of the two-electron pair either side of the double ionisation potential is expected to be essentially the same, being dominated by their mutual correlations. The only difference is the sign of their total energy, being negative below and positive above. Theory would therefore suggest a symmetric shape for the cusp, and experimental investigation of the detailed shape of this cusp is of continuing interest.

THRESHOLD PHOTOELECTRONS COINCIDENCE SPECTROSCOPY FOR PHOTODOUBLE IONISATION

TPEsCO Studies In The Rare Gases

The process of photodouble ionisation involves the solution of a three-body Coulomb problem where the boundary conditions of the two electrons have to be included. As noted above, the physical concepts governing this process were first discussed in the pioneering work of Wannier. Further theoretical analysis of the electrons in the Wannier geometry has

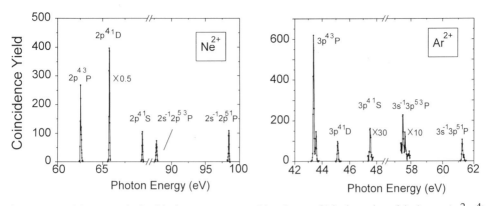

Figure 4. TPEsCO spectra obtained in the rare gases neon (a) and argon (b) in the region of the lowest (ns^2np^4) and first excited $(nsnp^5)$ double ionisation thresholds.

shown that the the cross section depends upon the symmetry of the wavefunction of the correlated electron pair. The rare gases, with electronic configuration ns^2np^6, are ideal for a study of this selectivity since they have several doubly charged states of different symmetry. For example, removal of two of the valence p electrons leads to ns^2np^4 ^3P, ^1D and ^1S states, whilst removal of one s and one p electron gives $nsnp^5$ ^3P and ^1P states.

TPEsCO spectra have been measured[17] for all the rare gases from neon to xenon in an energy range that includes the doubly charged states np^4,(^3P, ^1D and ^1S) and with the exception of xenon, the $nsnp^5$ (^3P and ^1P) states as well. Examples of TPEsCO spectra, obtained in neon and argon are shown in figures 4 (a) and (b) respectively. The ion state energies deduced from these spectra are in agreement with the known spectroscopic values. The first important point of note in the spectra of figure 4 is that all the possible states are observed. Following the arguments of Greene and Rau[18], the two electron wavefunction can be classed as either *favoured* or *unfavoured*, depending on the symmetry. From the theoretical work of Huetz *et al*[19], for the rare gases from neon to xenon, this gives the final ns^2np^4 ^3P state of the ion as *favoured*, followed by ^1D and ^1S in decreasing *favouredness*. Similarly the $nsnp^5$ ^1P state is favoured over $nsnp^5$ ^3P. The TPEsCO observations are in general accord with this propensity rule, the only exception being in neon, where the $Ne^{2+}2s^22p^4$ ^1D peak has the largest intensity (this peak is divided by two in figure 4). So far this anomalous behaviour has not been explained.

Indirect excitation of doubly charged states could also influence the intensities of the peaks observed in the TPEsCO spectra. This would be analogous to the excitation of singly charged ions via neutral state resonances as noted above. However in the TPEsCO spectra, only satellite states within about 20 meV of the thresholds can contribute to the observed intensities. In the case of neon, for example, a high resolution investigation of the satellites has shown[15] that all three doubly charged states of Ne appear to be free from any intense satellite structure. Therefore the cross-sections deduced from the peaks in the TPEsCO spectrum should correspond to those for the direct process.

TPEsCO Studies In Molecules

Doubly charged states of molecules (dications) are a significant constituent in many important environments, for example plasma and atmospheric physics[20]. Despite a wide range of experimental techniques being applied to their study, until very recently only either low accuracy data over a broad range of energies, or else highly accurate data over a very narrow energy range, was available. On the other hand, advances in theoretical modelling of these doubly charged species had lead to the prediction of many bound states for several molecules. Although the energy required to form these states is always greater than that for dissociation, they are often quasi-stable as a result of a potential barrier that prevents dissociation, until the molecule falls apart by quantum tunneling through the barrier. In many cases this potential barrier can be several electron volts deep, and may support many vibrational levels. It is in the study of these systems that the TPEsCO method can perhaps be most usefully applied. The main advantage of the technique being that the overall energy resolution is just that of the incident photon beam, which can be several tens of meV. This is often sufficient to resolve the vibrational structure of many dicationic states. Furthermore, the TPEsCO signal is obtained independently from the subsequent fate of the doubly charged state.

An example of a TPEsCO spectrum, obtained[21] in O_2 is shown in figure 5. This shows a long progression of 19 vibrational levels that are assigned to the O_2^{2+} $^1\Sigma_g^+$ ground

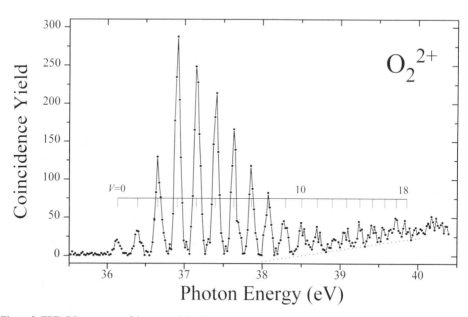

Figure 5. TPEsCO spectrum of the ground dicationic state of oxygen. This state supports 19 vibrational levels, which are indicated on the figure.

doubly charged state. The peak widths in this spectrum represent an overall instrumental energy resolution of 80 meV which corresponds to the energy spread of the photon beam.

The first peak in the spectrum is the $v=0$ level and the double ionisation potential of O_2 represented by this level is located at 36.13±.02eV. Interestingly this value is nearly 0.5eV lower than the previously accepted value, a difference explainable by the relatively low intensity of the first two vibrational levels. The energies of the 19 vibrational levels have been fitted to those of an anharmonic oscillator to obtain values for the molecular parameters ω_e and $\omega_e x_e$. This gave values of 0.276±0.001eV and 3.5±0.2 meV respectively, which compare favourably with the more recent theoretical values[22,23]. Also of note is the excitation of very high vibrational levels with appreciable intensity, which lie well outside of the Franck-Condon region for direct excitation. This suggests again the presence of indirect processes and the excitation of these vibrational levels via high lying singly charged states. The TPEsCO technique has been applied to a number of diatomic molecules[20,3]. to reveal a wealth of new information. It has also been applied to triatomic and polyatomic molecules[8] although here it is more difficult to resolve the vibrational structure with the available photon energy resolution and flux. Nevertheless, the first observation of a vibrational series belonging to a polyatomic dicationic state has been recently reported, for the case of photodouble ionisation of benzene[3]. The situation is now improving with the availability of third generation synchrotron radiation sources, where the available photon energy resolution approaches the natural vibrational widths for many dicationic states.

NEAR-THRESHOLD ANGULAR DISTRIBUTION MEASUREMENTS OF PHOTODOUBLE IONISATION

More detailed information regarding the physics of the photodouble ionisation process can be obtained by looking at the angular distribution of one of the two outgoing photoelectrons. In this case the cross section is differential both in energy E_1 and angle θ_1[1] of one of the two electrons (the other electron has an energy $I^{2+}-E_1$, and is integrated over all angles) and is referred to as a DDCS. One aim is to make these measurements as close to threshold as possible where the angular correlation effects are expected to be strongest. In this region experimental results can be compared with the predictions of different competing models. In the model of Huetz *et al*[19], which is based on the Wannier model, the value of the asymmetry parameter β, which describes the angular behaviour of the first electron, increases from a value of -1 at the double ionisation threshold itself, and rises rapidly to a value near zero at E~1.5eV before levelling off. The precise shape is governed by a Gaussian

---

[1]All angles are measured with respect to the direction of the major polarisation ellipse of the incident radiation.

correlation function which has a width $\theta_{1/2}=\theta_0E^{1/4}$, where E is the excess energy and θ_0 is an angular constant. A β value of -1 indicates a strong degree of correlation, with preferential ejection at 90^0, whilst a value of 2 indicates preferential ejection at 0^0. The β value may take on all values in between these two extremes with, for example, a value of zero indicating an isotropic angular distrubtion.

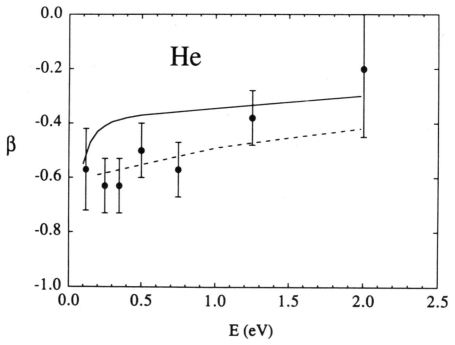

Figure 6. β for photodouble ionisation of helium, for the case of extreme energy sharing. The full curve shows the calculations of Maulbetsch and Briggs[25], where one electron has twice the energy of the other. The dashed curve shows the calculations of Kazansky and Ostrovsky[24].

More recent calculations by Kazansky and Ostrovsky[24], using an extension of the Wannier mechanism, have predicted a less dramatic evolution of β with excess energy, in which the value of the parameter θ_0 itself evolves with excess energy. Very recently the first *ab initio* calculations for near threshold photodouble ionisation of helium were made by.Maulbetsch and Briggs[25]. Their results confirmed the dominant role of electron correlation in the final state. Their result further demonstrated that, contrary to the Wannier situation, in the first few eV above threshold the behaviour of β is strongly dependent on the energy partitioning between the two electrons. The strongest manifestation of electron correlation was found to occur for equal energy sharing which produces the most negative values of β.

The asymmetry parameter β for the case of extreme energy sharing was measured in the region 0.2 < E < 2.0 eV for the cases of helium and neon, using the TPEPECO method mentioned above[26,27]. In both cases the final doubly charged ion was left with $^1S^e$ symmetry. The results for the case of helium are shown in figure 6. The error bars give an estimate of the reproducibility of the measurements. The values of β, which are expected to exhibit the weakest degree of correlation, were found to have a constant value of approximately -0.6, or else were slowly rising, over this region. These measurements implied that the degree of angular correlation between the two outgoing electrons was underestimated in the model of Huetz *et al*, as β was more negative than predicted. The *ab-initio* model of Maulbetsch and Briggs also appears to underestimate the degree of correlation. A better agreement was obtained with the results of Kazansky and Ostrovsky, using their 'extended Wannier ridge' model, mentioned above, although their results were obtained for the condition of equal energy sharing between the two outgoing electrons. Assuming the validity of the Wannier model - that the energy sharing plays no role in the angular distributions for the photodouble ionisation process - then their results would be applicable to the measured data.

NEAR THRESHOLD TDCS FOR PHOTODOUBLE IONISATION OF HELIUM

The most basic, and consequently most informative, level of the photodouble ionisation process come in the form of the Triple Differential Cross Section (TDCS). In this case both of the two outgoing photoelectrons are detected in small solid angles, and the measurement is

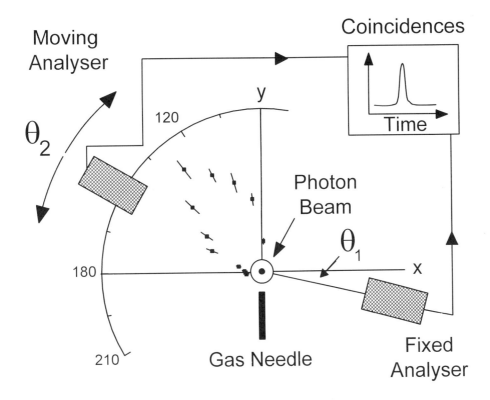

Figure 7. Schematic diagram of the conventional PEPECO measurement, where one analyser is fixed at some angle θ_1, and the other analyser is rotated about the photon beam over its angular range.

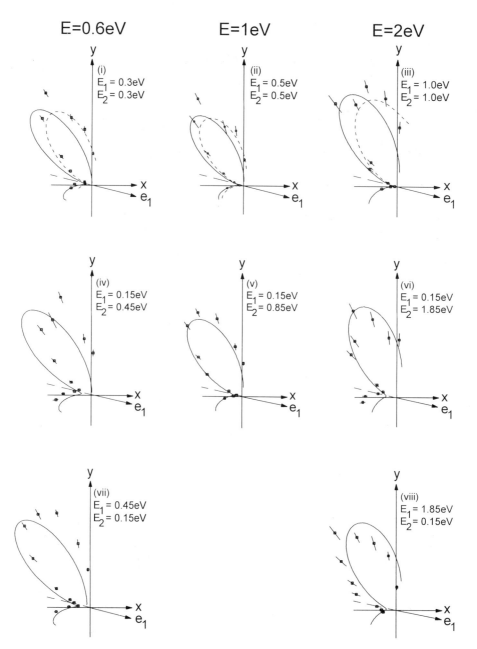

Figure 8. TDCS for photodouble ionisation of helium, for the (E_1, E_2) energy sharings indicated. The broken curves represent the Wannier based model of Huetz *et al*[19], whilst the full curves show the calculations of Maulbetsch and Briggs[29,30].

differential in the two angles θ_1 and θ_2, as well as the energy E_1 of one of the two electrons (the other has energy $E_2=E-E_1$, where E is the excess energy). Measurements of the TDCS, using PEPECO methods, have only been attempted in the past few years, on account of the very low cross sections involved. The measurement of a TDCS is usually done by mapping out the number of true coincidences recorded at an angle θ_2 for a fixed position θ_1 of the other analyser. This process is illustrated in figure 7.

TDCS have been measured in the case of photodouble ionisation of helium[28], for the case of equal and unequal energy sharing of the two electrons. The results, for the various (E_1,E_2) energy sharings indicated, are presented in figure 8. In these measurements one electron was detected at an angle of $\theta_1=-12.4^0$, whilst the other electron was detected over the range $88^0<\theta_2<188^0$. The TDCS are characterised by lobes which have a node at the configuration $\theta_1-\theta_2=180^0$, which, in the Wannier based model of Huetz et al[19], arises from the unfavoured nature of the final doubly charged state. In the case of equal energy sharing between the two outgoing electrons, i.e $E_1=E_2=E/2$, then the predictions of both Huetz et al, and the ab-initio model of Maulbetsch and Briggs[29,30] can be applied. In the case of unequal energy sharing between the two electrons, $E_1 \neq E_2$, only the ab-initio theory has been compared to the measured data.

One of the assumptions of the Wannier model is that the total cross section for photodouble ionisation has no dependence upon the way in which the two electrons share the available excess energy. If this independence in energy sharing were to be valid at the level of the TDCS, then the Wannier model should be equally valid for the TDCS from any one column of the figure (corresponding to $E=E_1+E_2$=constant). On the other hand, in the Wannier based model of Huetz et al, the TDCS contains contributions from one term, a_g, which is symmetric with respect to the exchange $E_1 \leftrightarrow E_2$ - giving the familiar Wannier energy sharing independence - and a second, asymmetric term, a_u;

$$TDCS \propto | \, a_g(\cos\theta_1+\cos\theta_2) + a_u(\cos\theta_1-\cos\theta_2) \, |^2 \qquad (2)$$

In the near threshold region this second term is negligible in the model, on account of its energy dependence in the near threshold region. The squared amplitude of the a_g term can be equated with the Gaussian correlation function, referred to above, so that the TDCS reduces to a product between this correlation term and another term which describes the interaction between the two electrons and the incident radiation;

$$TDCS \propto \exp[-4\ln2 \,(180^0-\theta_{12})^2 / \theta_{1/2}^2] \, (\cos\theta_1+\cos\theta_2)^2 \qquad (3)$$

where $\theta_{12}=\theta_1-\theta_2$ is the interelectron angle. This equation in principle should apply exactly to the case of equal energy sharing, i.e where $a_u=0$. For unequal energy sharing then the strength of the departure of the measured TDCS from their equal energy sharing counterparts should provide a measure of the importance of this asymmetric term for near threshold photodouble ionisation. This is most apparent in the region around $\theta_{12}=180^0$, where a *node* is observed in the equal energy sharing case, and a *near-node* is observed in the unequal energy sharing case. The ab-initio results of Maulbetsch and Briggs can be seen to give a reasonable agreement with the measured data for E=1 and 2eV. However, at the lowest excess energy studied, E=0.6eV, there is a considerable discrepancy.

CONCLUSIONS

The threshold photoelctron spectroscopy technique (TPES) has been shown to be a very sensitive probe of single ionisation processes, where the inherent properties of the penetrating field technique - high energy resolution and sensitivity - have been used to great effect. The extension of this technique to the study of doubly charged states of both atoms and molecules is provided by the TPEsCO method. This method has not only provided insights into the role of electron correlations in the photodouble ionisation of atoms, but has also revealed the existence and properties of many doubly charged molecular states. The energy resolution acheived has enabled these states to be studied at the vibrational level, and has provided important clues into their mechanism of formation.

Measurements of the asymmetry parameter for photodouble ionisation of helium and neon have revealed the strength of the electron correlations in the region near threshold. These correlations have been investigated at the most fundamental level of the TDCS for photodouble ionisation. As expected in the Wannier mechanism, the general shapes of the measured TDCS were found to be the similar for different (E_1,E_2) sharings. In the Wannier based model of Huetz *et al*[19], this would be due to the dominance of the symmetric part of the cross section. On the other hand, the departure from the energy independence picture from the Wannier model could be clearly seen around the nodes in the TDCS at $\theta_{12}=180^0$, whuch can be interpreted as being due to the asymmetric term (see (2) and (3)).

All of these techniques would clearly benefit from higher flux, higher energy resolution synchrotron radiation sources, which hold the promise of a great deal of new and interesting physics.

References

1. B.Krässig, O.Schwarkopf and V.Schmidt *J.Phys.B: At.Mol.Opt.Phys* **26**:2589 (1993)
2. G.Dawber, L.Avaldi, A.G.McConkey, H.Rojas, N.Gulley, M.A.MacDonald, G.C.King and R.I.Hall XIX ICPEAC, *to be published in Can.J.Phys* (1996)
3. R.I. Hall, L. Avaldi, G. Dawber, A.G. McConkey, M.A. MacDonald, and G.C. King, *Chem.Phys.* **187**:125 (1994).
4. S. Cvejanovic and F.H. Read *J.Phys.B:At.Mol.Opt.Phys.* **7**:1180 (1974).
5. K.Ellis, R.I.Hall, L.Avaldi, G.Dawber, A.G.McConkey, L.Andric and G.C.King *J.Phys.B: At.Mol.Opt.Phys.* **27**:3415 (1994).
6. R.I. Hall, A.G. McConkey, K. Ellis, G. Dawber, L. Avaldi, M.A. MacDonald and G.C. King, *Meas.Sci.Technol.* **3**:316 (1992)
7. R.I.Hall, L.Avaldi, G.Dawber, P.M.Rutter, M.A.MacDonald and G.C.King *J.Phys.B: At.Mol.Opt.Phys* **22**:3205 (1989)
8. A.G. McConkey, G. Dawber, L. Avaldi, M.A. MacDonald, G.C. King, and R.I. Hall. *J.Phys.B: At.Mol.Opt.Phys.* **27**:271 (1994).

9. R.I. Hall, L. Avaldi, G. Dawber, M. Zubek, K. Ellis and G.C. King *J.Phys.B: At.Mol.Opt.Phys.* **24**:115 (1991).
10. M. Zubek, G.C. King, P.M. Rutter and F.H. Read *J.Phys.B: At.Mol.Opt.Phys.* **23**:3411 (1989).
11. M. Zubek, G. Dawber, R.I. Hall, L. Avaldi, K. Ellis and G.C. King *J.Phys.B: At.Mol.Opt.Phys.* **24**:L337 (1991).
12. R.I. Hall, L. Avaldi, G. Dawber, P.M. Rutter, M.A. MacDonald and G.C. King, *J.Phys.B: At.Mol.Opt.Phys.* **22**:3205 (1989)
13. R.I. Hall, L. Avaldi, G. Dawber, M. Zubek, and G.C. King, *J.Phys.B: At.Mol.Opt.Phys.* **23**:4469 (1990).
14. R.I. Hall, G. Dawber, K Ellis, M. Zubek, L. Avaldi, and G.C. King, *J.Phys.B: At.Mol.Opt.Phys.* **24**:4133 (1991).
15. R.I. Hall, A.G.McConkey, L. Avaldi, M.A. MacDonald and G.C. King, *J.Phys.B: Atom.Molec.Opt.Phys.* **25**:377 (1992).
16. G.H. Wannier, *Phys.Rev.* **90**:817 (1953).
17. R.I. Hall, G. Dawber, A. McConkey M.A. MacDonald and G.C. King, *Z.Phys.D.* **23**:377 (1992).
18. C.H. Greene and A.R.P. Rau *Phys.Rev.Letts.* **48**:533 (1982).
19. A. Huetz, P. Selles, D. Waymel and J. Mazeau *J.Phys.B: Atom.Molec.Opt.Phys.* **24**:1917 (1991).
20. G. Dawber, A.G.. McConkey, L. Avaldi, M.A. MacDonald, G.C. King, and R.I. Hall *J.Phys.B: At.Mol.Opt.Phys.* **27**:2191 (1994)
21. R.I. Hall, G. Dawber, A.G. McConkey M.A. MacDonald and G.C. King *Phys.Rev.Letts.* **68**:2751 (1992).
22. L.G.M. Pettersson and M.Larsson *J.Chem.Phys.* **94**:818 (1991)
23. M.W.Wong, R.H.Nobes, W.J.Bouma and L.Radom *J.Chem.Phys.* **91**:2971 (1989)
24. A.K.Kazansky and V.N.Ostrovsky *J.Phys.B:At.Mol.Opt.Phys* **26**:2231 (1993)
25. F.Maulbetsch and J.S.Briggs *Phys.Rev.Letts.* **68**:2004 (1992)
26. G.Dawber, R.I.Hall, A.G.McConkey, M.A.MacDonald and G.C.King *J.Phys.B:At.Mol.Opt.Phys* **27**:L341 (1994)
27. R.I.Hall, G.Dawber, A.G.McConkey, M.A.MacDonald and G.C.King *J.Phys.B:At.Mol.Opt.Phys* **26**:L653 (1993)
28. G.Dawber, L.Avaldi, A.G.McConkey, H.Rojas, M.A.MacDonald and G.C.King *J.Phys.B:At. Mol.Opt. Phys* **28**:L271 (1995)
29. F.Maulbetsch and J.S.Briggs *J.Phys.B:At.Mol.Opt.Phys* **26**:L647 (1993)
30. F.Maulbetsch and J.S.Briggs *J.Phys.B:At.Mol.Opt.Phys* **27**:4095 (1994)

PHOTOIONIZATION OF POLARIZED ATOMS: APPLICATIONS TO FREE ATOMS AND FERROMAGNETS

N. A. Cherepkov*

Johannes Gutenberg-Universität
Institut für Physik
55099 Mainz, Germany

INTRODUCTION

Discussion of connections between different phenomena observed in seemingly different situations usually helps to better understanding of the physics of underlying processes. Some time ago Farago[1] discussed analogies and contrasts between light polarization and electron spin polarization. He showed that though in both cases the same Stokes vector formalism[2] can be applied, the analogies between them have rather limited validity. We will discuss here the applicability of equations derived for the description of photoionization of free polarized atoms[3,4] to photoemission from ferromagnets[5,6]. We show that qualitative features of photoemission from core levels of ferromagnets are correctly reproduced by equations derived for free polarized atoms, while for a quantitative description one should take into account also the solid state effects. Recently[7] it was shown that the qualitative analysis of angle- and spin-resolved photoemission from core levels of nonmagnetic solids (in that case from Cu) can be also performed using the equations derived for free atoms[8].

Originally photoionization of polarized atoms[3] was considered as a way to perform "complete" or "perfect" experiment[9], that is the experiment from which one can extract all theoretical parameters necessary for a complete description of the process in a given approximation, in the case of photoionization in the electric-dipole approximation. Photoionization process is rather simple, therefore the first complete experiment in the electric-dipole approximation has been performed already relatively long ago[10] by measuring the angular distribution of photoelectrons with defined spin polarization ejected from unpolarized atoms. According to the theory[8], one can extract no more than five parameters from this kind of experiment, and determine from them three dipole matrix elements and two phase shift differences necessary for theoretical description of photoionization process. But in a general case of an open shell atom, when neither in the initial state nor in the final ionic state the total angular momentum J is equal to zero, five parameters is not sufficient for a complete experiment[11]. One of possible ways to overcome this problem is to polarize atoms in the initial state, which can be done with open shell atoms, and to measure the angular distributions of photoelectrons, which are characterized now by more than five parameters[3].

*Permanent address: State Academy of Aerospace Instrumentation, 190 000 St. Petersburg, Russia.

In a general case the angular distribution of photoelectrons ejected from polarized atoms has so complicated structure that it hardly can be used for extracting parameters or for some other analysis. There are several ways to simplify it. The most obvious one is to choose a particular experimental geometry when some terms disappear. The other way is to determine differences between two angular distributions corresponding to different directions of either the incoming light polarization, or the initial target polarization. Then many terms in the angular distributions cancel. The other advantage of defining these differences is the fact that for definite geometries of experiment they have characteristic values (for example, zeroes) independent of photon energy which can be used for qualitative analysis of the target characteristics. For example, one can define from them the direction and degree of atomic polarization[4]. And these differences play an important role now in investigations of magnetic splitting of core levels in ferromagnets[5,6]. Investigated is a differences between photoelectron intensities for two opposite (or mutually perpendicular) directions of sample magnetization which is called magnetic dichroism in angular resolved photoemission. This difference originates mainly from the fact that the hole state in ferromagnets is polarized, and therefore it can be described using the equations derived for free polarized atoms.

PHOTOIONIZATION OF POLARIZED ATOMS

We will imply in the following that the electromagnetic field is weak, that photon energy is low enough for the electric-dipole approximation to be applied, and that after absorption of one photon only one electron is ejected. We will also imply that an ensemble of atoms with the total angular momentum J and its projection M_J is polarized (aligned or oriented) in some direction $\mathbf{n}$, and is described by an incoherent superposition of JM_J sublevels, i.e. that this ensemble is axially symmetric[12]. Then the corresponding atomic density matrix is diagonal. The process under consideration is characterized by three vectors. One of them is used to define the coordinate system (in our case the unit vector $\mathbf{q}$ in the direction of the photon beam), therefore the angular distribution can be presented as a double expansion in spherical functions of two other directions, the atomic polarization $\mathbf{n}$, and the photoelectron momentum $\boldsymbol{\kappa}$ (both are unit vectors), which are arbitrary[4]

$$I^J(\boldsymbol{\kappa}, \mathbf{n}) = \sigma_{nl}^J(\omega)\sqrt{3[J]} \sum_{k,L} \sum_N C_{kLN}^J \rho_{N0}^{\mathbf{n}} \sum_{x,M} \sum_{M_N} Y_{LM}^*(\hat{\mathbf{k}}) Y_{NM_N}^*(\hat{\mathbf{n}}) \rho_{kx}^{\gamma} \begin{pmatrix} k & L & N \\ x & M & M_N \end{pmatrix} \quad (1)$$

Here σ_{nl}^J is a photoionization cross section of nl-subshell at photon energy ω, $[J] \equiv 2J+1$, $\rho_{N0}^{\mathbf{n}}$ and ρ_{kx}^{γ} are state multipoles[12] characterizing the polarization of atoms and of the photon beam, respectively. Atoms are said to be oriented if $\rho_{10}^{\mathbf{n}} \neq 0$ (when populations of sublevels with projections M_J and $-M_J$ are not equal), and aligned if $\rho_{10}^{\mathbf{n}} = 0$, $\rho_{20}^{\mathbf{n}} \neq 0$ (sublevels with the projections M_J and $-M_J$ are equally populated). Eq. (1) is valid for any atom or ion, and the parameters C_{kLN}^J depend on subshell and the coupling scheme (LS, jj, or intermediate). The particular expressions for these parameters are given in[4]. When the number of independent parameters C_{kLN}^J is not larger than five, they can be expressed through the spin polarization parameters A, γ, and η, and the angular asymmetry parameter β[13].

Equation (1) is very convenient for investigations since a geometrical and a dynamical parts in it are separated. The dynamical part, that is dependence on the photon energy, is given by the cross section and by the parameters C_{kLN}^J which are proportional to products of the dipole matrix elements and sine or cosine function of the phase shift differences. The geometrical part, that is dependence on the geometry of experiment, is given by the spherical functions. Similar expression has been recently obtained and investigated in autoionization resonances by Baier et al[14]. It is important to mention that the parameters C_{kLN}^J, like the angular asymmetry parameter β, have the same sign and nearly the same magnitude for two fine-structure components corresponding to $j = l + 1/2$ and $j = l - 1/2$, contrary to the spin polarization parameters which have the opposite signs for two fine-structure components and different magnitudes[8].

All the summations in (1) are restricted, $k \leq 2$, $N \leq 2J$, L is even and $L \leq 2l_{max}$ where l_{max} is the largest orbital angular momentum of photoelectrons. Nevertheless, the number of terms there can be rather large. If there are more parameters in (1) than the number of dipole matrix elements and phase shift differences necessary for theoretical description of the process, these parameters are not independent, and there are some relations between them. It is much more convenient to present (1) in a vector form using the explicit expressions for the spherical functions as it was done for unpolarized atoms[8], but then the equation becomes too long. In two particular cases of a pure linear and a pure circular light polarization they are presented in[4] restricting by terms with $N \leq 2$.

As an example, consider the most frequently used geometry of experiment shown in Fig. 1a when two light beams are counterpropagating and photoelectrons are detected in the plane perpendicular to the light beams. The first linearly polarized light beam produces excited atoms aligned in the direction $\mathbf{n}$ which can be rotated, and the second light beam with polarization vector $\mathbf{e}$ ionizes them. We will restrict our consideration here by $np_{3/2}$-subshells, and by the simplest case when spin-orbit interaction in continuous spectrum is neglected which usually is a good approximation except for the region of autoionization resonances[8,14]. Then only three parameters are necessary to characterize the process, two reduced dipole matrix elements d_s and d_d (see[8] for definitions) corresponding to the transitions from np to s and d continuum states, respectively, and one phase shift difference between two continua. It means that for a complete experiment in this approximation one needs to extract three parameters from experimental data. For the geometry of experiment presented in Fig. 1a the general equation (1) gives

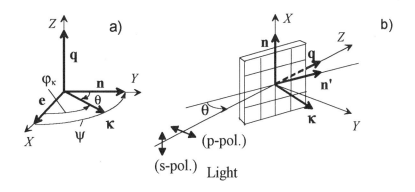

Fig. 1. Different geometries of experiment (see text for details).

$$I(\psi, \varphi_k) = \frac{\sigma(\omega)}{2\pi}[a_0 + a_1\cos 2\psi + a_2\cos 2\varphi_k + a_3\cos 2(\psi - \varphi_k) + a_4\cos(2\psi - 4\varphi_k)]$$
(2)

where

$$a_0 = \left(1 + \tfrac{1}{4}\beta\right)\left(\rho_{00}^{\mathbf{n}} - \tfrac{1}{4}\rho_{20}^{\mathbf{n}}\right), \quad a_1 = -\tfrac{3}{4}\left(1 + \tfrac{1}{4}\beta - \tfrac{9}{8}W\right)\rho_{20}^{\mathbf{n}},$$
(3)

$$a_2 = \tfrac{3}{4}\beta\left(\rho_{00}^{\mathbf{n}} - \tfrac{1}{4}\rho_{20}^{\mathbf{n}}\right), \quad a_3 = -\tfrac{9}{16}\beta\rho_{20}^{\mathbf{n}}, \quad a_4 = -\tfrac{27}{32}W\rho_{20}^{\mathbf{n}}, \quad W = d_d^2/(d_s^2 + d_d^2),$$

β is the angular asymmetry parameter, and the state multipoles characterizing the aligned excited state in our particular case are $\rho_{00}^{\mathbf{n}} = 1/2$, $\rho_{20}^{\mathbf{n}} = -1/2$. Since there are three parameters in (2), σ, β, and W, which can be extracted from experimental data, the measurement of the angular distribution (2) constitute a complete experiment in the approximation accepted here. Instead of W one can use the spin polarization parameter A^J, $A^{3/2} = 3W/4 - 1/2$. Eq. (2) (as well as eq. (1)) is equally applied in autoionization resonances and outside them. In the former case the resonance behaviour is incorporated into the parameters σ, β, and W (or into the parameters σ_{nl}^J and C_{kLN}^J in (1)). The orientation vector $\rho_{10}^{\mathbf{n}}$ does not enter eq. (2) (it will appear when the first light beam will be circularly polarized and $\mathbf{n}$ will be parallel to $\mathbf{q}$), therefore one can not distinguish between the states aligned or oriented along $\mathbf{n}$. For example, for the aligned state with $m_j = 1/2$ and $m_j = -1/2$ sublevels equally populated, provided $m_j = \pm 3/2$ sublevels are empty, the result will be the same as for the oriented state with only one $m_j = 1/2$ sublevel populated.

There were rather many experimental investigations using the geometry of experiment shown in Fig. 1a (see[15-19] and references therein). We consider here only a recent measurement[19] for Li excited state $1s^22p(^2P_{3/2})$ in the region of autoionization resonance $1s2p^2(^2D)$. Since the cross section in this resonance is several orders of magnitude larger than outside the resonance, one has $d_d \gg d_s$ and as a consequence $\beta \cong W \cong 1$ so that there is no need to perform atomic calculations. Fig. 2 shows the angular distribution obtained from eq. (2) with these limiting values of the parameters which is in a good agreement with the experimental data of Pahler et al[19]. More detailed description of theoretical results for Li atoms and comparison with the experiment is given in[20]. To illustrate the dependence of the angular distributions on values of the parameters β and W, we show in Fig. 3 several examples calculated for the parameters obtained earlier for $6s^26p$ initial state of Tl atom[8] and with the state multipoles given above. Dotted and dot-dashed curves in this figure correspond to the peaks of $6s6p^2$ (^2D) and (^2S) autoionization resonances, respectively, and demonstrate a typical behaviour of angular distributions in autoionization resonances with these configurations. In general, if the cross section in a strong resonance is dominated by contribution of only one continuum channel, the angular distribution has a characteristic behaviour which enables one to identify it easily[19]. Other examples of variations of the angular distributions across the autoionization resonances were given in[14,16].

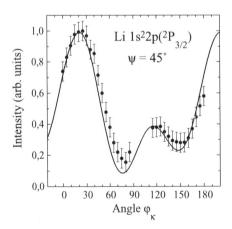

Fig. 2. Angular distribution of photoelectrons from aligned excited $1s^22p(^2P_{3/2})$ state of Li as a function of the angle φ_κ for $\psi = 45°$ (see Fig. 1a) calculated using Eq. (2) with $\beta = W = 1$ (full curve), and measured by Pahler et al[19] (points), in arbitrary units.

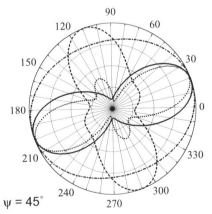

Fig. 3. Polar plot of the angular distributions (2) calculated with different values of the parameters β and W as a function of φ_κ for $\psi = 45°$ (see Fig. 1a). Full curve: $\beta = 1.36$, $W = 0.33$; dotted curve: $\beta = 1$, $W = 1$; dashed curve: $\beta = -0.57$, $W = 0.28$; dot-dashed curve: $\beta = 0.14$, $W = 0.01$.

If autoionization resonances with $J = 3/2$ and $J = 5/2$ are resolved (as in photoionization of excited Na atoms[18]), the $d_{3/2}$ and $d_{5/2}$ continua are strongly different because each of these resonances interacts only with that continuum channel which has the same total angular momentum J, and one could not use the approximation accepted above. Then there are five independent parameters defining the process, and simple eqs. (3) for the a_i coefficients are not valid. They can be expressed through the parameters C_{kLN}^J. Much more complicated is the case of photoionization of polarized Ar^* ($3p^54p\ ^3D_3$) excited state in the photon energy region between the $3p^5(^2P_{3/2})$ and $3p^5(^2P_{1/2})$ ionization thresholds investigated in[21]. Here there are six open channels in continuous spectrum, and consequently 11 parameters to be determined from experiment. The presence of autoionization resonances in this region helps in defining the amplitudes of different channels.

DIFFERENCES OF ANGULAR DISTRIBUTIONS

To obtain more specific information on the target atoms one can consider a difference between two angular distributions corresponding either to different light polarization or to different polarization of the target atoms. One can define

i) a difference between angular distributions obtained with left and right circularly polarized light which is called Circular Dichroism in the Angular Distribution (CDAD)[22,23,13];

ii) a difference between angular distributions obtained with linearly polarized light of two mutually perpendicular polarizations which is called Linear Dichroism in the Angular Distribution (LDAD)[24,25];

iii) a difference obtained with light of a fixed polarization for two opposite directions of atomic polarization, which for the first time has been defined in investigations of photoemission from ferromagnets, and was called Magnetic Dichroism in the Angular Distribution (MDAD)[26]; MDAD is different from zero for circularly polarized, linearly polarized, and unpolarized light[5,27];

iv) a difference obtained with light of a fixed polarization for two mutually perpendicular directions of atomic polarization, this difference has no special name yet[4,28].

There are two reasons to define these differences. First, some terms in these differences cancel, so that when the angular distribution itself has too complicated structure, the difference is usually much simpler. It does not contain the constant term and therefore has sharper variation with the angle. Second, the differences usually have characteristic values, zeroes in particular, for a definite geometry which can be used to define unknown direction and degree of polarization of atoms. The detailed description of these differences and some numerical examples were given in[4]. We present here the only available result of CDAD measurement[29] for polarized excited $7p_{3/2}$ state of Cs for the geometry of experiment shown in Fig. 1a. The measured CDAD intensity was normalized according to the equation

$$S_{CDAD}(\theta) = \frac{I^+(\theta) - I^-(\theta)}{\frac{1}{2}[I(0) + I(\pi/2)]} = \frac{3\eta^{3/2}\rho_{20}^n \sin 2\theta}{\left(\rho_{00}^n - \frac{1}{4}\rho_{20}^n\right)\left(1 + \frac{1}{4}\beta\right)} \tag{4}$$

where $I^+(\theta)$ and $I^-(\theta)$ are the angular distributions obtained with left and right circularly polarized light, respectively, θ is the angle between $\mathbf{n}$ and $\boldsymbol{\kappa}$, $\eta^{3/2}$ is defined as follows[8]

$$\eta^{3/2} = -\frac{1}{2}\eta^{1/2} = \frac{3}{2\sqrt{2}}[d_s d_d \sin(\delta_d - \delta_s)]/\left(d_s^2 + d_d^2\right) \tag{5}$$

and the state multipoles have been defined above. Fig. 4 shows the experimental data[29] compared to calculations[30] according to eq. (4). The magnitude of experimental CDAD signal is considerably smaller which is connected with the existence of unresolved hyperfine splitting of the $7p_{3/2}$ state. The hyperfine coupling reduces the alignment initially produced in

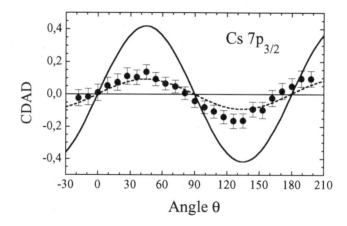

Fig.4. Experimental[29] (circles) and theoretical[30] (full curve) CDAD values for the $7p_{3/2}$ level of Cs normalized according to eq. (5). Dashed curve shows the theoretical result corrected for hyperfine depolarization (From Cuéllar et al[29]).

the excitation step. The inclusion of hyperfine depolarization in the long-time limit[31] (which is applicable when ionization occurs over a time period which is much longer than the time of precession of J about the total angular momentum $F=J+I$ where I is the nuclear spin) leads to much better agreement between theory and experiment.

There is a possibility to define the degree of alignment of target atoms (or the degree of their depolarization) directly from experiment by measuring the difference which was denotedd as $I_\perp$ in[4]. In particular, for the geometry of experiment shown in Fig.1a and for $p_{3/2}$-subshells it follows from eq. (2) for the corresponding normalized value

$$A_\perp^J(0) = [I(0,0) - I(\pi/2, 0)]/[I(0,0) + I(\pi/2, 0)] = -3\rho_{20}^n/(4\rho_{00}^n - \rho_{20}^n). \tag{6}$$

This value depends neither on photon energy nor on dipole matrix elements or phase shifts and is completely defined by the state multipoles, that is by alignment of atoms. In the absence of depolarization $A_\perp^{3/2}(0) = 3/5$, due to depolarization this value will be smaller. So, measurement of $A_\perp$ enables one to define the state multipole ρ_{20}^n directly from experiment.

Integrating CDAD or LDAD over the electron ejection angles, we arrive at circular or linear dichroism (CD or LD). Measurements of CD and LD can be used to obtain two independent parameters from photoionization of polarized atoms[32].

PHOTOEMISSION FROM FERROMAGNETS

The theory discussed above can be applied not only to free atoms but also to atoms adsorbed at a surface[23], which is rather evident, and to photoemission from ferromagnets, which is less evident. Consider the photoemission from core levels of ferromagnets. If in the initial state the ionized subshell was closed, in the final state there will be a hole characterized by the quantum numbers nlj with j being the total angular momentum. Due to exchange interaction with the open nd or nf valence shell in ferromagnets each nlj state will be energetically split into components with a given projection m_j. They will be characterized by the quantum numbers $nljm_j$ and therefore will be polarized. This characterization of the hole states is justified when the spin-orbit splitting is larger that the exchange one which usually takes place. We will not consider explicitly the exchange interaction between the hole state and an open subshell, as well as we will disregard the total angular momentum of the outer shell because it should not be a good quantum number in solids. The latter

Table 1. State multipoles ρ_{N0}^{n} for magnetic sublevels of $np_{1/2}$ and $np_{3/2}$ hole states.

m_j	$j = 1/2$		$j = 3/2$			
	1/2	-1/2	3/2	1/2	-1/2	-3/2
ρ_{00}^{n}	$\frac{1}{\sqrt{2}}$	$\frac{1}{\sqrt{2}}$	$\frac{1}{2}$	$\frac{1}{2}$	$\frac{1}{2}$	$\frac{1}{2}$
ρ_{10}^{n}	$\frac{1}{\sqrt{2}}$	$-\frac{1}{\sqrt{2}}$	$\frac{3}{2\sqrt{5}}$	$\frac{1}{2\sqrt{5}}$	$-\frac{1}{2\sqrt{5}}$	$-\frac{3}{2\sqrt{5}}$
ρ_{20}^{n}	0	0	$\frac{1}{2}$	$-\frac{1}{2}$	$-\frac{1}{2}$	$\frac{1}{2}$
ρ_{30}^{n}	0	0	$\frac{1}{2\sqrt{5}}$	$-\frac{3}{2\sqrt{5}}$	$\frac{3}{2\sqrt{5}}$	$-\frac{1}{2\sqrt{5}}$

approximation is used in practically all papers devoted to this problem. So, the ferromagnetic nature of the solid is described in our model by introducing the energy splitting of core hole magnetic sublevels. The magnitude of this splitting does not influence the angular dependence and therefore can be considered simply as an empirical parameter. This is fully analogous to the fact that the spin polarization of photoelectrons ejected from unpolarized atoms does not depend on the strength of spin-orbit interaction provided the spin-orbit splitting of atomic levels has been resolved[8].

It can be easily shown that the photoionization of a closed subshell provided the final ionic state is polarized, is described by the same equations as the photoionization of a one-electron subshell with the same orbital and total angular momenta which is initially polarized. Therefore for the description of photoemission from ferromagnets in the atomic model introduced above one can use eq. (1) implying that the polarized one-electron subshell nlj is ionized. The state multipoles corresponding to each magnetic sublevel of np subshells are given in Table 1.

In investigations of ferromagnets it was discovered experimentally that the shape of core levels photoelectron spectra changed when the direction of magnetisation is reversed. The effect was called magnetic dichroism in angular resolved photoemission (MDAD), and was observed for both circularly[26] (CMDAD) and linearly[28,33] (LMDAD) polarised light. These effects are evidently connected with the local magnetic field acting upon the substrate atoms, and with the spin-orbit and exchange splitting of core levels. The atomic model described above is able to give a qualitative explanation of the effects of magnetic dichroism observed experimentally[5].

A typical experimental geometry used for observation of MDAD is shown in Fig. 1b. Sample is magnetized in the direction $\mathbf{n}$ (or $\mathbf{n'}$) parallel to the surface. The incoming light beam is moving in the direction $\mathbf{q}$ which makes an angle θ with the surface. Electrons are collected in the direction of the surface normal ($\mathbf{\kappa}$) which makes the influence of the surface minimal. Also it is important to have relatively high photoelectron energy in order to reduce the influence of scattering processes in the bulk. Suppose that light is linearly polarized perpendicular to the direction of magnetization, $\mathbf{e} \perp \mathbf{n}$ (p-polarized light). Then from eq. (1) it follows

$$I_{LMDAD}^{j} = I^{j}(\mathbf{\kappa}, \mathbf{n}) - I^{j}(\mathbf{\kappa}, -\mathbf{n}) = \frac{1}{2\pi}\sigma_{nl}^{j}(\omega)\rho_{10}^{n}\sin 2\theta \begin{cases} 2\sqrt{5}\,\eta^{3/2}, & j = 3/2, \\ -\sqrt{2}\,\eta^{1/2}, & j = 1/2, \end{cases} \qquad (7)$$

where the parameters η^{j} have been defined in (5). Here only the state multipoles ρ_{10}^{n} are different for different magnetic sublevels. Their signs coincide with the signs of projections m_j of corresponding magnetic sublevels (see Table 1). Therefore LMDAD should change the

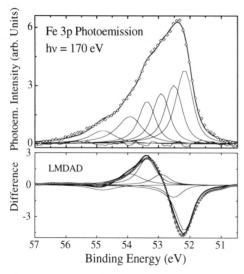

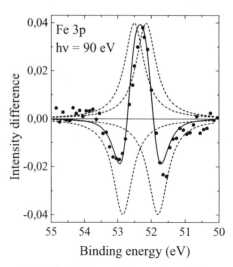

Fig. 5. Upper panel: fitting of Fe 3p photoemission spectrum from unmagnetized Fe(100) (after integral background subtraction). Bottom panel: experimental LMDAD spectrum (points) along with the result of its fit (full curve) by m_j components from the upper panel multiplied by ρ_{10}^{n}. (From Rossi et al[34]).

Fig. 6. The difference between two photo-electron spectra corresponding to two directions of magnetization **n** and **n'** in Fig. 1b for s-polarized light. Points: experiment[28], full curve: simulation by four Lorentzians shown by dashed curves. (From Cherepkov[5]).

sign within $np_{3/2}$ and $np_{1/2}$ multiplets. An examaple of application of eq. (7) is presented in Fig. 5. The upper panel shows the experimental photoemission spectrum from 3p core level of Fe for unmagnetized sample (points), and the result of the fit of this spectrum by six Lorentzians corresponding to six magnetic sublevels[34]. The bottom panel shows the experimental LMDAD spectrum (points) and the corresponding fit based on the results of the upper panel. Namely, each peak in the upper panel was multiplied by the corresponding state multipole ρ_{10}^{n} from Table 1 in accord with eq. (7). In this procedure it was implied that the ordering of $3p_{1/2}$ magnetic sublevels is reversed as compared to $3p_{3/2}$ sublevels, that is $m_j= 1/2$ sublevel is lower than $m_j= -1/2$ one. The reversed ordering of $2p_{1/2}$ magnetic sublevels was already known from previous analysis of CMDAD spectrum[35]. The result of the fit is in a good agreement with experiment, which can be considered as a proof of the assumptions made in this fit. On the other hand, within the atomic model the intensities of magnetic sublevels with m_j and $-m_j$ should be equal, while they are strongly different in the fit shown in Fig. 5 (similar results have been obtained in[36] for 2p subshell of Fe). It demonstrates that the solid state effects are also important in producing the MDAD spectra[35,37].

Another important value is a difference between photoelectron spectra obtained with s-polarized light when the direction of magnetization is changed from **n** to **n'** (see Fig. 1b)

$$I_{\perp}^{j} = I^{j}(\kappa, \mathbf{n}) - I^{j}(\kappa, \mathbf{n'}) = -\tfrac{3}{8\pi}\sigma_{nl}^{j}(\omega)\sqrt{2j+1}\,\rho_{20}^{\mathbf{n}}(1-\beta/2). \qquad (8)$$

The variation of this value within a given multiplet is completely defined by the alignment tensor ρ_{20}^{n} because all other values are constant. $np_{1/2}$ state could not be aligned, therefore only $np_{3/2}$ multiplet contributes to (8). According to the atomic model, the components with $|m_j|=1/2$ and $|m_j|=3/2$ give the contributions of opposite sign and of the same magnitude. Fig. 6 shows a simple simulation[5] of the results observed in[28] for the $3p_{3/2}$ multiplet of Fe by four equidistant Lorentzians of equal magnitudes and widths, and of the signs defined by the state multipoles ρ_{20}^{n}. The agreement with experiment is rather good, and from the figure one can define the positions of four magnetic sublevels. Actually the magnitudes of Lorentzians should be nonequal as it follows from Fig. 5 which can improve agreement with the experiment.

Recently[33,36] also angle- and spin-resolved photoelectron spectra were measured for Fe core levels. In the case of free atoms this kind of experiment could not give something new because already angle-resolved photoemission is sufficient for a complete experiment. In the case of ferromagnets where not only atomic but also solid state properties are probed this kind of experiment is informative. Theoretical description of angle- and spin-resolved photoemission from ferromagnets within the atomic model was given in[6]. For the spin-resolved photoemission the third spherical function containing the direction of photoelectron spin $\mathbf{s}$ appears in the expansion (1). An analysis of this rather complicated equation for the geometry of experiment shown in Fig. 1b shows that the most clear separation of contributions of different magnetic sublevels of $np_{1/2}$ and $np_{3/2}$ hole states is achieved with linearly polarized light, therefore we consider here only this particular case. For p-polarized light ($\mathbf{e} \perp \mathbf{n}$) the following expressions are obtained[6]

$$I_{1/2} = \frac{1}{4\pi\sqrt{2}}\sigma_{np}^{1/2}(\omega)(\rho_{00}^{n} \mp \rho_{10}^{n})\left(1+\beta-\frac{3}{2}\beta\sin^2\theta \pm \eta^{1/2}\sin 2\theta\right) \tag{9}$$

$$I_{3/2} = \frac{1}{4\pi}\sigma_{np}^{3/2}(\omega)\left[\left(\rho_{00}^{n}+\frac{1}{2}\rho_{20}^{n} \pm \frac{2}{\sqrt{5}}\rho_{10}^{n} \pm \frac{3}{2\sqrt{5}}\rho_{30}^{n}\right)\left(1+\beta-\frac{3}{2}\beta\sin^2\theta\right)\pm\right.$$

$$\left.\left(\rho_{00}^{n}+2\rho_{20}^{n} \pm \sqrt{5}\,\rho_{10}^{n}\right)\eta^{3/2}\sin 2\theta\right] \tag{10}$$

where the upper and lower signs refer to $\mathbf{s}\|\mathbf{n}$ and $\mathbf{s}\|(-\mathbf{n})$, respectively. One can easily check that for some magnetic sublevels the linear combinations of state multipoles in (9) and (10) are equal to zero, that is not all magnetic sublevels contribute to the photoelectron spectrum as it is shown in Table 2. Unity in this table means that the intensity of a given sublevel in the spin-resolved spectrum is the same as in the spin-unresolved one. It is worth while to stress that the ratios in Table 2 are independent of the angle θ. If spin-unresolved spectrum for a given magnetization is fitted by lorentzian peaks, then the spin-resolved spectrum is obtained from it by multiplying the fitted intensities by the coefficients given in Table 2 as it has been done already in[34]. So, the $np_{3/2}$ spectrum should have a double peak structure which is in accord with the experimental observation presented in Fig. 3 in[33]. The positions of two peaks on both curves in the lower part of this figure which corresponds to $\mathbf{s}\|\mathbf{n}$ are shifted to lower binding energies as compared to two peaks on both curves in the upper part of this figure which corresponds to $\mathbf{s}\|(-\mathbf{n})$, again in accord with the results of Table 2. But there is one

Table 2. The ratios of intensities of different magnetic sublevels of $np_{1/2}$ and $np_{3/2}$ hole states in spin-resolved spectra to that in spin-unresolved spectra for the geometry of experiment shown in Fig. 1b and linearly polarized light.

m_j	$j=1/2$		$j=3/2$			
	1/2	-1/2	3/2	1/2	-1/2	-3/2
$\mathbf{e}\perp\mathbf{n}$, $\mathbf{s}\|\mathbf{n}$	0	1	1	0	1	0
$\mathbf{e}\perp\mathbf{n}$, $\mathbf{s}\|(-\mathbf{n})$	1	0	0	1	0	1
$\mathbf{e}\|\mathbf{n}$, $\mathbf{s}\|\mathbf{n}$	1	0	0	1	0	0
$\mathbf{e}\|\mathbf{n}$, $\mathbf{s}\|(-\mathbf{n})$	0	1	0	0	1	0

peculiarity of experimental spectra which is not explained by the atomic model, namely, that the intensities of peaks with s||n are essentially higher than that with s||(-n). Analysis of new experimental data for the 2p subshell of Fe using the atomic model is given in[36].

For s-polarized light (e||n) the analogous consideration gives

$$I_{1/2} = \frac{1}{4\pi\sqrt{2}}\sigma_{np}^{1/2}(\omega)(\rho_{00}^{n} \pm \rho_{10}^{n})\left(1 - \tfrac{1}{2}\beta\right), \qquad (11)$$

$$I_{3/2} = \frac{1}{4\pi}\sigma_{np}^{3/2}(\omega)\left(\rho_{00}^{n} - \rho_{20}^{n} \pm \frac{1}{\sqrt{5}}\rho_{10}^{n} \mp \frac{3}{\sqrt{5}}\rho_{30}^{n}\right)\left(1 - \tfrac{1}{2}\beta\right), \qquad (12)$$

where the upper and lower signs refer again to the cases when s||n and s||(-n), respectively. Relative contributions of magnetic sublevels in this case are also given in Table 2. Now only one of magnetic sublevels of $np_{3/2}$ hole state contributes which is favorable for determination of its energy position. There are no experimental observations for s-polarized light.

CONCLUSIONS

We have shown that the general foramlism developed for the description of photoionization of free polarized atoms has actually much broader applicability and can be successfuly used, in particular, for the description of photoemission from core levels of ferromagnets. Though in this case it could not explain all the details of the observed spectra, it correctly reproduces the main peculiarities and shows that the phenomenon of magnetic dichroism in the angular resolved photoemission from core levels of ferromagnets has essentially an atomic origin. The atomic model can be successfully used also for the description of spin-resolved photoemission from ferromagnets. Comparison of experimental spectra with the predictions of the atomic model enables one to separate the effects having the solid state origin from those having the atomic origin. It is interesting to mention, for example, that the appearance of CMDAD in a geometry when the photon beam and the sample magnetization are perpendicular (n⊥q in Fig. 1b) which was called "forbidden geometry" in[38], is not forbidden in the atomic model[5], so that this effect has also essentially the atomic origin. Within the atomic model the photoemission from feromagnets is characterized by the same parameters (the angular asymmetry parameter β, the spin polarization parametsrs η and A) as the photoionization of free atoms. This clearly demonstrates the fundamental importance of the atomic physics in understanding the properties of matter.

The author acknowledges the hospitality of the Johannes Gutenberg-Universität Mainz extended to him during the work on this paper, and the Deutsche Forschungsgemeinschaft for the financial support through SFB 252.

REFERENCES

1. P.S.Farago, Analogies and contrasts between optical and electron spin polarization, *Comments At. Mol. Phys.* 6: 99 (1977).
2. P.S.Farago, Electron spin polarization, *Rep. Prog. Phys.* 34: 1055 (1971).
3. H.Klar, and H.Kleinpoppen, Angular distribution of photoelectrons from polarized atoms exposed to polarized radiation, *J. Phys. B* 15: 933 (1982).
4. N.A.Cherepkov, V.V.Kuznetsov, and V.A.Verbitskii, Photoionization of polarized atoms, *J. Phys. B* 28: 1221 (1995).
5. N.A.Cherepkov, Origin of magnetic dichroism in angular resolved photoemission from ferromagnets, *Phys. Rev. B* 50: 13813 (1994).
6. N.A.Cherepkov, and V.V.Kuznetsov, Angle and spin resolved photoemission from ferromagnets, *J.Phys.: Condensed Matter*, in press (1996).

7. Ch.Roth, F.U.Hillebrecht, W.G.Park, H.B.Rose, and E.Kisker, Spin polarization in Cu core-level photoemission with linearly polarized soft X rays, *Phys. Rev. Lett.* 73: 1963 (1994).
8. N.A.Cherepkov, Spin polarization of atomic and molecular photoelectrons, *Adv. At. Mol. Phys.* 19: 395 (1983).
9. J.Kessler, The "perfect" photoionization experiment, *Comments At. Mol. Phys.* 10: 47 (1981).
10. U.Heinzmann, Experimental determination of the phase shift difference of continuum wavefunctions describing the photoionisation process of xenon atoms II, *J. Phys. B* 13: 4367 (1980).
11. N.A.Cherepkov, On the complete photoionization experiment, *in*: "Proc. Int. Workshop on Photoioniza-tion 1992", U.Becker and U.Heinzmann, eds., AMS Press, New York (1993).
12. K.Blum, "Density Matrix Theory and Applications", Plenum, New York (1981).
13. N.A.Cherepkov, and V.V.Kuznetsov, Optical activity of polarized atoms, *J. Phys. B* 22: L 405 (1989).
14. S.Baier, A.N.Grum-Grzhimailo, and N.M.Kabachnik, Angular distribution of photoelectrons in resonant photoionization of polarized atoms, *J. Phys. B* 27: 3363 (1994).
15. R.N.Compton, J.A.D.Stockdale, C.D.Cooper, X.Tang, and P.Lambropoulos, Photoelectron angular distributions from multiphoton ionization of cesium atoms, *Phys. Rev. A* 30: 1766 (1983).
16. J.S.Keller, J.E.Hunter, and R.S.Berry, Path dependence in resonant multiphoton excitation to autoioni-zing states of barium, *Phys. Rev. A* 43: 2270 (1991).
17. C.Kerling, N.Böwering, and U.Heinzmann, Photoelectron angular distributions from laser-excited aligned Yb atoms ionized by vacuum ultraviolet radiation, *J. Phys. B* 23: L629 (1990).
18. S.Baier, M.Schulze, H.Staiger, P.Zimmermann, C.Lorenz, M.Pahler, J.Rüder, B.Sonntag, J.T.Costello, and L.Kiernan, Investigation of Na $2p^53s3p$ resonances using angular resolved photoelectron spectroscopy of laser-aligned sodium atoms, *J. Phys. B* 27: 1341 (1994).
19. M.Pahler, C.Lorenz, E.v.Raven, J.Rüder, B.Sonntag, S.Baier, B.R.Müller, M.Schulze, H.Staiger, P.Zimmermann, and N.M.Kabachnik, Angle-dependent photoelectron spectroscopy of laser-aligned atoms: Li, *Phys. Rev. Lett.* 68: 2285 (1992).
20. N.A.Cherepkov, L.V.Chernysheva, V.V.Kuznetsov, and S.K.Semenov, Photoionization of polarized excited Li atoms, *J. Electron Spectrosc. Relat. Phenom.*, to be published.
21. S.Shohl, D.Klar, N.A.Cherepkov, S.Baier, K.Ueda, and H.Hotop, Photoionization of polarized Ar*(4p J=3) atoms near threshold, to be published.
22. R.Parzynski, Circular dichroism of angular distributions of photoelectrons emitted from heavy polarized alkali atoms, *Acta Phys. Pol. A* 57: 49 (1980).
23. R.L.Dubs, S.N.Dixit, and V.McKoy, Circular dichroism in photoelectron angular distribution from adsorbed atoms, *Phys. Rev. B* 32: 8389 (1985).
24. N.A.Cherepkov, and G.Schönhense, Linear dichroism in photoemission from oriented molecules, *Europhys. Lett.* 24: 79 (1993).
25. N.A.Cherepkov, Circular and linear dichroism in atomic and molecular photoelectron emission, *in*: "Electronic and Atomic Collisions (Invited Papers of ICPEAC XVII)", I.E.McCarthy and M.C. Standage, eds., Hilger, Bristol (1992).
26. L.Baumgarten, C.M.Schneider, H.Petersen, F.Schäfers, and J.Kirschner, Magnetic X-ray dichroism in core-level photoemission from ferromagnets, *Phys. Rev. Lett.* 65: 492 (1990).
27. M.Getzlaff, Ch.Ostertag, G.H.Fecher, N.A.Cherepkov, and G.Schönhense, Magnetic dichroism in photoemission with unpoplarized light, *Phys. Rev. Lett.* 73:3030 (1994).
28. Ch.Roth, H.B.Rose, F.U.Hillebrecht, and E.Kisker, Magnetic linear dichroism in soft X-ray core level photoemission from iron, *Solid State Commun.* 86: 647 (1993).
29. L.E.Cuéllar, C.S.Feigerle, H.S.Carman, Jr., and R.N.Compton, Circular dichroism in photoelectron angular distributions for the $7P_{3/2}$ level of cesium, *Phys. Rev. A* 43: 6437 (1991).
30. R.L.Dubs, S.N.Dixit, and V.McKoy, unpublished.
31. C.H.Greene, and R.N.Zare, Photofragment alignment and orientation, *Ann. Rev. Phys. Chem.* 33: 119 (1982).
32. S.V.Bobashev, A.Yu.Elizarov, V.K.Prilipko, and N.A.Cherepkov, Linear and circular dichroism in two-step photoionization of Ba atoms, *Laser Physics* 3: 751 (1993).
33. Ch.Roth, F.U.Hillebrecht, H.B.Rose, and E.Kisker, Linear magnetic dichroism in angular resolved Fe $3p$ core level photoemission, *Phys. Rev. Lett.* 70: 3479 (1993).
34. G.Rossi, F.Sirotti, N.A.Cherepkov, F.Combet Farnoux, and G.Panaccione, 3p fine structure of ferromag-netic Fe and Co from photoemisson with linearly polarized light, *Solid State Commun.* 90: 557 (1994).
35. H.Ebert, L.Baumgarten, C.M.Schneider, and J.Kirschner, Polarization dependence of the 2p-core-level photoemission spectra of Fe, *Phys. Rev. B* 44: 4406 (1991).
36. F.U.Hillebrecht, Ch.Roth, H.B.Rose, W.G.Park, E.Kisker, and N.A.Cherepkov, Magnetic linear dichroism in spin-resolved Fe 2p photoemission, *Phys. Rev. B*, submitted.
37. G. van der Laan, Angular resolved linear and circular dichroism in core-level photoemission of metallic systems, *Phys. Rev. B* 51: 240 (1995).
38. C.M.Schneider, D.Venus, and J.Kirschner, Strong X-ray magnetic circular dichroism in a "forbidden geometry" observed via photoemission, *Phys. Rev. B* 45: 5041 (1992).

'INTENSE-FIELD MANY-BODY S-MATRIX THEORY' AND MECHANISM OF LASER INDUCED DOUBLE IONIZATION OF HELIUM

F.H.M. FAISAL and A. BECKER
Fakultät für Physik
Universität Bielefeld
D-33615 Bielefeld, Germany

1. Introduction

Surprisingly large signals for direct double ionization of He by intense laser light have been observed in a series of recent experiments [1-6]. The most dramatic observation of this effect is in the latest experiment by Walker et.at. [6], who measured the double ionization signal over a remarkable range of twelve orders of magnitude covering an intensity domain between 10^{14} W/cm^2 and 10^{16} W/cm^2 at a wavelength of 780 nm. Perhaps the most unexpected aspect of this result is that the observed double ionization signal *below* the experimental 'saturation intensity' (i.e. the intensity at which the initially present neutral target atoms in the interaction volume is depleted) is several orders of magnitude *greater* than that expected for the single ionization of the already once ionized He atoms! The most interesting question is: what is the mechanism behind this anomalously large signal?

The first thing to note is that the process of interest is an extremely non-linear one: at least some 50 photons at an wavelength of 780 nm must be absorbed, by a single He atom, to overcome the unperturbed double ionization threshold of He at 79.02 eV. In the research area of intense field atomic dynamics there are other examples that appeared to be equally surprising at the time when they were initially predicted theoretically or observed experimentally. To name only two well-known examples among them: the above-threshold-ionization (ATI) and the adiabatic stabilization (AS) phenomena. The existence of ATI was discovered by Agostini et. al. [7] and soon confirmed by Muller et al [8]; it was predicted within the so-called KFR-theory [9-11], initially at a time long before the advent of strong enough laser fields necessary for its observation. 'AS' was initially predicted using the so-called high frequency model [12,13] and subsequently by *exact numerical* solution of the Schrödinger equation for the H atom (3D) in intense laser fields[14-17]; a first experimental indication of its existence has been recently found by de Boer et.al. [18]. In the former case (ATI) a whole sequence of peaks is observed in the energy spectrum of the ejected electron, that lie above the usual Einstein photoeffect peak nearest to the ionization threshold; in the latter case (AS) the ionization probability per unit time *decreases* instead of increasing, with *increasing* intensity. These phenomena are now understood to be due to the highly non-linear nature of the atom-field interaction; at the same time they can be both analyzed in terms of the independent electron hypothesis for this interaction. Generally, this has been done within a *single active electron* (SAE) assumption for non-hydrogenic systems and exactly only for the hydrogen atom. Recent observations of the laser induced direct double ionization of He introduces a *new* element in intense field physics, namely *electron-electron correlation*, and with it a possible break-down of the active single electron hypothesis for this fundamentally two-electron process. An analysis of the process in He atom is of theoretical interest not only because it involves the simplest of the real two-electron atomic systems available but also because a detailed understanding of the process in this system is probably essential for a proper understanding of the response of many-electron systems (ranging from complex atoms to solids) to intense laser fields under similar conditions. The theoretical challenge posed by the observations is to simultaneously account for the *combined* influence of the high non-linearity of the field interaction *and* the Coulomb correlation on the process.

2. Two Heuristic Pictures

Before proceeding further let us briefly consider two recently proposed qualitative pictures for the double ionization of He in intense laser fields: (i) the so-called 'shake-off' picture, and (ii) the 'quasi-static tunneling and 'e-2e' like rescattering' picture. According to the first picture, proposed by Fittinghoff et.al.[1], at first one of the two electrons absorbs the laser photons and reaches the continuum very rapidly while the other electron, being unable to adjust to the resulting rapid change in the effective charge of the core, will be 'shaken off ' the atom. According to the second picture, suggested first by Corkum [19], at first one of the electrons tunnels out of the atom with zero initial energy, due to a supposed quasi-static influence of the strong electric field of the laser, then it behaves classically as a free electron in the laser field and absorbs the maximum possible classical energy, $E_{max.} = 3.2\ U_p$, where $U_p = I/4\omega^2$ is the socalled 'quiver energy', from the field during the first half-cycle and propagate away from the core. But in the next half-cycle, as the direction of the field will change, the electron will return near to the core and 'kick off' the second electron by a 'e-2e'-like scattering process from the ground state of He^+ ion.

Note the following characteristic differences between these two pictures or models. The first model does not make qualitative distinction between a linear and a circular polarizations of the laser field. The second model, in contrast, gives specific importance to the linear polarization for the assumed effective reencounter with the core electron, and thus predicts significant double ionization for the linear but *not* for the circular polarization case. A second important difference is that mechanism (i) does not imply any specific 'threshold intensity' for the double ionization of Helium, but mechanism (ii) predicts a threshold (or cut off) intensity for double ionization corresponding to $E_{max.}=3.2\ Up =E_B= 54.4$ eV, where E_B is the binding energy of the He^+ core. A third difference concerns the frequency and time dependence; mechanism (i) does not restrict itself to a specific range of the frequency. It requires, however, a short enough ejection time of the first electron (of the inverse order of the binding frequency of the He^+ ionic core). Mechanism (ii), on the other hand, appears to be restricted to the lower range of available laser frequencies, in view of the requirement of quasi-static tunneling of the first electron, but the over all time scale involved could be larger in this model, e.g. of the order of the inverse photon frequency, than in the 'shake off' picture.

The experiments done with linear- and circular polarizations [1-6] showed strongly reduced signals for the circular polarization case. It also seemed that a heuristic combination of the so-called ADK tunneling rate [20] and the known 'e-2e' cross sections [21] could be used to fit [19] the initially measured rates of double ionization of He [13,15-17]. These facts generated much interest in favor of the second mechanism. But with the latest precision measurements of double ionization of He by Walker et.al. [18], that extended dramatically the observed range of signals by as much as 6-orders of magnitude (made possible by the development of kHz repetition rates of the 780 nm laser) and covered an intensity range of more than an order of magnitude, revealed *no* sign of any cut-off of the double ionization signal for intensities decisively below the 'cut-off intensity' predicted by the second model. Thus at present neither model (i) nor model (ii) can satisfactorily explain the observed signals of the double ionization of He in intense laser fields.

An exact numerical simulation of double ionization of a two-electron atom in intense laser fields poses formidable practical difficulties in view of the fact that the fundamental Schrödinger equation of the system consists of a (six + one)-dimensional partial differential equation; the large size of the space-time grid over which the wave function needs to be propagated in the configurational space (and time), in order to obtain experimentally relevant information, remains as yet practically untractable. More over, numerical simulations, despite their great virtues, are often rather poor in obtaining physical insights into the *mechanism* behind a new phenomenon. In view of these, in this paper we present a two-electron diagrammatic perturbation theory, appropriate for intense fields, and apply the leading terms of the associated S-matrix series to analyze the double ionization process under experimentally relevant conditions. Use of Feynman diagrammatic technique will prove to be particularly convenient in view of the complexity of the two-electron S-matrix series. The results of numerical calculations will be compared with the recent experimental data, that will allow us to identify the dominant mechanism behind the laser induced direct double ionization process.

3. Intense-Field Many-Body S-Matrix Theory (IMST): Leading Diagrams for Double Ionization of a Two-Electron Atom

In this section we develop an 'intense-field many-body S-matrix theory (IMST)'. It will permit us to take *both* the highly non-linear electron-photon interaction as well as the electron-electron correlation into account, simultaneously. The usual S-matrix perturbation theory for multiphoton ionization of atoms is based on the expansion of the wavefunction in terms of the amplitude of the light field (or equivalently, of the vector potential) and it is known to break down for intensities above 10^{13} W/cm^2 for the usual laser frequencies (e.g., p.51[22]). The form of the IMST developed below is particularly interesting for the present problem, but it can be equally well applied to a host of other problems, due to its great flexibility. To be concrete, we shall explicity write it down it terms of a two-electron atomic system, but this will contain all the essential steps for the generalization to the many-electron atomic systems in a straight forward manner.

The Schrödinger equation of the system "He atom + laser field" (we use, unless stated explicitly otherwise, Hartree a.u., $|e|=m=h=1$) is:

$$i d/dt \, \Psi(\mathbf{r}_1,\mathbf{r}_2;t) = H(t) \, \Psi(\mathbf{r}_1,\mathbf{r}_2;t) \tag{1}$$

where,

$$H(t) = H_i^0 + V_i(t) \tag{2}$$

is the total Hamiltonian of the system, and

$$H_i^0 = 1/2 \, \mathbf{p}_1^2 + 1/2 \, \mathbf{p}_2^2 - z/r_1 - z/r_2 + 1/r_{12} \qquad (z = 2) \tag{3}$$

is the Hamiltonian of the unperturbed He atom. The associated 'initial state interaction' is, therefore:

$$V_i(t) = H(t) - H_i^0 = v_1(t) + v_2(t)$$

where,

$$v_j(t) = -1/c p_j.A(t) + 1/2c^2 A(t)^2, \qquad (j=1,2) \tag{4}$$

The total Green's function, $G(t,t')$, of the system is defined by

$$[id/dt - H(t)] G(r_1,r_1';r_2,r_2';t,t') = \delta(t-t') \, \delta(r_1-r_1') \, \delta(r_2-r_2') \tag{5}$$

With the help of $G(t,t')$, a formal solution of Eq.(1), evolving from an arbitrary initial state

$$\Phi_i^0(t) = \Phi_i^0(r_1,r_2)e^{-iE_i t},$$

where,

$$H_i^0 \, \Phi_i^0(r_1,r_2) = E_i \, \Phi_i^0(r_1,r_2), \tag{6}$$

can be written down as:

$$|\Psi_i(t)> = |\Phi_i^0(t)> + S\,dt' \, G(t,t')V_i(t') |\Phi_i^0(t')> \tag{7}$$

(The validity of Eq.(7) can be verified at once by operating with $[id/dt - H(t)]$ on the left and using the definition (5) of $G(t,t')$.)

For the present purpose we shall choose the final reference state as the momentum states observed by the detectors and given by the appropriate asymptotic wavefunction (in general, Coulombic) describing the state of the two ejected electrons (momenta, k_a and k_b). Thus, for example, we may split the total Hamiltonian as:

$$H(t) = H_f^0 + V_f(t) \tag{8}$$

with

$$H_f^0 = 1/2 \, p_1^2 + 1/2 \, p_2^2 - z/r_1 - z/r_2 \tag{9}$$

$$V_f(t) = 1/r_{12} + v_1(t) + v_2(t) \tag{10}$$

where the v's are given by Eq.(4). (We note, parenthetically, that other choices of the final state reference Hamiltonian, which include partly or wholely the electron-electron correlation in the final reference state directy, can also be introduced in the present S-matrix theory without any formal difficulty).

Thus, we split $G(t,t')$, as:

$$G(t,t') = G_f^0(t,t') + S\,dt_1 \, G_f^0(t,t_1)V_f(t_1)G(t_1,t'), \tag{11}$$

where $G_f^0(t,t')$ is the Green's function associated with H_f^0, so that on substituting (11) in (7) we get,

$$|\Psi_i(t)> = |\Phi_i^0(t)> + S\,dt' \, G_f^0(t,t')V_i(t') |\Phi_i^0(t')>$$
$$+ S\,dt' \, S\,dt_1 \, G_f^0(t,t_1)V_f(t_1)G(t_1,t')V_i(t')|\Phi_i^0(t')> \tag{12}$$

400

Finally, we can expand the total G also in terms of *any* other G^0 and U^0 (for the present choice, see Eqs.(18,19) below) as:

$$G(t,t') = G^0(t,t') + Sdt_1 \, G^0(t,t_1)U^0(t_1)G^0(t_1,t')$$

$$+ \, Sdt_2 \, Sdt_1 \, G^0(t,t_2)U^0(t_2)G^0(t_2,t_1)U^0(t_1)G^0(t_1,t') + \ldots \ldots \quad (13)$$

Thus, substituting (13) in (12) we get a general and highly flexible expansion of the total wavefunction in the form:

$$|\Psi_i(t)> = |\Phi^0_i(t)>$$

$$+ \, S^t dt' \, G^0_f(t,t')V_i(t')|\Phi^0_i(t')>$$

$$+ \, S^t dt_1 \, Sdt' \, G^0_f(t,t_1)V_f(t_1)G^0(t_1,t')V_i(t')|\Phi^0_i(t')>$$

$$+ \, S^t dt_2 \, Sdt_1 \, Sdt' G^0_f(t,t_2)V_f(t_2)G^0(t_2,t_1)U^0(t_1)G^0(t_1,t')V_i(t')|\Phi^0_i(t')> + \ldots \ldots \quad (14)$$

Projecting on to an eigenstate associated with the final reference Hamiltonian H^0_f, we get the desired S-matrix series:

$$S(t,-00) = <\Phi^0_f(t)|\Psi_i(t)>$$

$$= S_0(t) + S_1(t) + S_2(t) + S_3(t) \ldots \ldots \ldots \ldots \ldots$$

with

$$S_0(t) = <\Phi^0_f(t)|\Phi^0_i(t)> \, ,$$

$$S_1(t) = -iS^t dt' <\Phi^0_f(t')|V_i(t')|\Phi^0_i(t')> \, ,$$

$$S_2(t) = -iS^t dt_1 \, S \, dt' <\Phi^0_f(t_1)|V_f(t_1)G^0(t_1,t')V_i(t')|\Phi^0_i(t')> \, ,$$

$$S_3(t) = -iS^t dt_2 \, Sdt_1 \, Sdt' <\Phi^0_f(t_2)|V_f(t_2)G^0(t_2,t_1)U^0(t_1)G^0(t_1,t')V_i(t')|\Phi^0_i(t')> \, ,$$

$$\ldots \ldots \ldots \ldots \ldots \ldots \ldots \quad (15)$$

From the S-matrix series, Eq. (15), we observe that the first term, S_0, is merely a non-singular overlap between the initial and the final reference states and is negligible asymptotically, compared to the second and the successive terms (which all contain the singular delta function associated with the over all energy conservation, asymptotically at large times, at the end of the transition process). Hence the leading contributing terms in this expansion are the second term, S_1, and the third term, S_2. (To the formally interested reader we may point out that an 'ordering' of the series arises, in terms of the potential U^0, only from the fourth term onward; hence the second and the third terms *together* form the non-trivial 'leading term' of the series, in the strict sense).

This is a new form of the expansion of the S-matrix series and is much more flexible than the well-known conventional expansion of the S-matrix either in the 'prior' (initial state interaction) or in the 'post' (final state interaction) form. The flexibility of the present expansion arises from the fact that the reference states (the initial state and the final state) do *not* have to belong to the same reference Hamiltonian and yet *any* desired propagator,

G^0, can be introduced in the *intermediate* states of the expansion. Our result, Eq.(15), in fact, contains the usual 'post' and 'prior' forms of the transition amplitude as special cases of the intermediate propagator and interaction, G^0 and U^0. Thus, for example, if we choose:

(i) $G^0=G^0_i$ and $U^0=V_i$ in Eq.(15) we reproduce the conventional 'prior' form of the S-matrix series, while the choice:

(i) $G^0=G^0_f$ and $U^0=V_f$ in Eq.(15), reproduces the conventional time-reversed or the 'post' form of the S-matrix series. It is also easily seen that the choice:

(iii) $G^0=G^0_f$ =Volkov Green's function (see, Eq.(19 or 20), below) and $U^0=V_f$ leads to the generalization of the well-known so-called KFR-theory [9-11] to two-electron atoms. In fact, the above statements hold *mutatis mutandis* for laser interactions with *many*-electron atoms.

For the investigation of the intense-field double ionization process the above forms are found not to be very useful; consequently we propose here to choose the intermediate propagator G^0 to represent the propagation of one electron in the Volkov state and the other electron in the complete set of states of the residual ion. Thus, explicitly, we define $G^0(t,t')$ by:

$$\{id/dt -H^0(t)\}G^0(\mathbf{r_1},\mathbf{r_2},t; \mathbf{r_1}',\mathbf{r_2}',t') = \delta(t-t') \, \delta \, (\mathbf{r_1} -\mathbf{r_1}') \, \delta(\mathbf{r_2} -\mathbf{r_2}') \qquad (16)$$

with, $\quad H^0 (t) = H_{ion}(\mathbf{r_2}) +[1/2(\mathbf{p_1} -1/c\mathbf{A}(t))^2] \qquad (17)$

and the associated interaction

$$U^0(t) = H(t)-H^0(t) = 1/r_{12} + [-z/r_1 -1/c \, \mathbf{p_2}.\mathbf{A}(t) + (1/2c^2)\mathbf{A}(t)^2] \qquad (18)$$

where, $H_{ion}(\mathbf{r_2})$ is the unperturbed Hamiltonian of the He$^+$ ion and $[1/2(\mathbf{p_1} -1/c\mathbf{A}(t))^2]$, that of the free electron interacting with the laser field. The propagator G^0 can be written down explicitly by solving Eq.(16) using standard techniques as:

$$G^0(\mathbf{r_1},\mathbf{r_2},t; \mathbf{r_1}',\mathbf{r_2}',t')= -i\theta(t-t')S_j \, (2\pi)^{-3} \, S_k \, \phi_j^c(\mathbf{r_2},t)\phi^v(\mathbf{k},\mathbf{r_1},t)\phi_j^{c*}(\mathbf{r_2}',t')\phi^{v*}(\mathbf{k},\mathbf{r_1}',t') \qquad (19)$$

where $\phi_j^c(\mathbf{r_2},t)$ and $\phi^v(\mathbf{k},\mathbf{r_1},t)$ are the jth eigenstate of the ion and the Volkov state of the free electron of momentum $\mathbf{k}$, respectively. For the usual linearly polarized monochromatic radiation field, (19) has the explicit Floquet representation (c.f. [22-24]):

$$G^0(\mathbf{r_1},\mathbf{r_2},t; \mathbf{r_1}',\mathbf{r_2}',t') = -i \, \theta(t-t')S_N \, S_n \, \exp\{-iN\omega t-i(k^2/2+E_j+U_p-n\omega)(t-t')\}$$

$$.S_j \, (2\pi)^{-3} \, S_k \, [\phi_j^c \, (\mathbf{r_2})\phi^0(\mathbf{k}, \mathbf{r_1})J_{n-N}(\alpha_0.\mathbf{k};U_p/2\omega). \, J_n(\alpha_0.\mathbf{k};U_p/2\omega)\phi_j^c \, {}^*(\mathbf{r_2}')\phi^{0*}(\mathbf{k}, \mathbf{r_1}') \qquad (20)$$

where J(a;b)'s are the generalized Bessel functions of two arguments (e.g. [22,chap.1;23]); $\phi^0(\mathbf{k},\mathbf{r_1})$ is the plane wave state (momentum $\mathbf{k}$), and E_j is the jth eigen-energy of the He$^+$ ion.

In view of the analytical complexity of the intense-field two-electron S-matrix theory, it is most convenient to analyse diagrammatically the leading contributions arising from the expansion (15):

i) As already noted, the term S_0 does not contribute to the rate of the process and therefore will not be considered further below.

ii) The term S_1 gives rise to two diagrams that contribute only in the case of allowed single photon transitions via the interaction v_1 or v_2 (Eq. (4)). This term, therefore, *exactly* reproduces the weak field *one*-photon transition amplitude according to the well-known first order perturbation theory (and the associated Fermi golden rule), for example, for the ionization of He by weak synchrotron radiation. In the case of two- or multi-photon transitions, this term vanishes identically. The simple Feynman diagram associated with this term can be represented as in Fig.1.

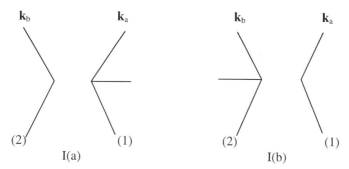

Fig.1. Time is assumed to run from below upward. The two Feynman diagrams I(a), and I(b), arising from the first term S_1 of the IMST for the one-photon ionization of a two-electron atom. (1) and (2) denote the two initially bound electrons; $\mathbf{k}_a$ and $\mathbf{k}_b$ correspond to the final momentum states of the two free electrons.

iii) The term S_2 includes the leading diagrams for the *non-linear* processes. This term, in fact, brings in the electron-electron correlation for the first time in the expansion (15). There are in fact nine diagrams in this case which are shown in Fig.2 and discussed below. Note that the first four Feynman diagrams, II(a-d), are disjoint diagrams and, therefore, can not contribute significantly for simultaneous double ionization process that requires a direct interaction joinig both the electron lines. The fifth and the sixth diagrams, II(e) and II(f), are of the latter kind.

Consider first the diagram II(e) which shows that the first electron interacts with the photon field and goes into an intermediate Volkov state of momentum $\mathbf{k}$, while at the same time, the second electron propagates in the eigenstates, $\{j\}$, of the He$^+$ ion. (The joint propagation in the intermediate states is governed by, as indicated, the Green's function G^0). This is a major conjoint diagram. During the propagation, the first electron shares energy with the second electron through the Coulomb correlation, $1/r_{12}$, until both of them have enough energy to escape together (with the respective momenta $\mathbf{k}_a$ and $\mathbf{k}_b$). We note that the (minor) diagram II(f) corresponds to the unlikely event in which the first electron interacts with the photon field but the second electron goes into the intermediate Volkov state and may be neglected here after.

The diagrams II'(a,b,c) appear explicitly in S_2 if the initial sate wavefunction is considered to be a product of two independent particle orbitals. Thus they do *not* arise if the initial state correlation is included in the (initial) wavefunction of the He atom, as may be assumed in general.

We may thus tentatively assume, from diagrammatic considerations alone, that the diagram II(e) is qualitatively a dominant diagram for the simultaneous double-escape process for a two-electron atom subjected to intense lascr ficlds. Below we analyse further the diagram II(e) and bring it into a form suitable for actual calculations. To this end we use the Floquet representation for the Green's function $G^0(t,t')$, Eq.(17), carry out the double integrations over t' and t_1 (by suitably employing partial integrations and noting the vanishing surface terms) and simplify to obtain:

Diag. II(e) = $-2\pi i \, \Sigma_N (E_a+E_b+E_B -N\omega) \, T^{(N)}$,   (21)

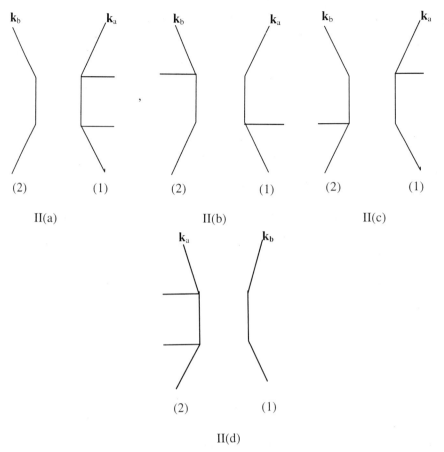

Fig.2. Time is assumed implicitly to run from below upward. The four disjoint Feynman diagrams, II(a-d), for double ionization of He, that occur in the second leading second term S_2 of the IMST. Note that the diagrams II(c) and II(d) arise simply from the diagrams II(a) and II(b) by exchanging the first interaction with the bound electron (1) by that with the bound electron (2). $\mathbf{k}_a$ and $\mathbf{k}_b$ correspond to the final momentum states of the two ionized electrons.

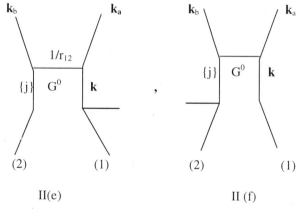

Fig.2. (contd.) The two conjoint Feynman diagrams for double ionization of He in an intense laser field. Note that while the (major) diagram II(e) describes the more likely event of electron (1) to go over into the intermediate Volkov state (momentum $\mathbf{k}$) by interacting with the field, the diagram II(f) correspons to the less likely event of electron (1) to go over into the Volkov state when merly electron (2) interacts with the field.

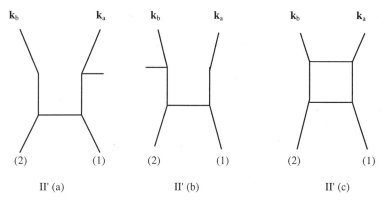

Fig.2. (contd.) The three additional diagrams in S_2 involving the initial state correlation, which arise when the initial state is given by the product of independent one-electron states but *not* if the initial wavefunction is chosen to be a correlated wavefunction.

with, $T^{(N)} = \Sigma_n (2\pi)^{-3} S_k S_j \{ <\phi^c(\mathbf{k}_a,\mathbf{r}_1) \phi^c(\mathbf{k}_b,\mathbf{r}_2)|1/r_{12}| \phi_j^c(\mathbf{r}_2) \phi^0(\mathbf{k},\mathbf{r}_1)>$

$.(U_p-n\omega)(k^2/2+E_j+E_B+U_p-n\omega+i0)^{-1}$

$.J_{n-N}(\alpha_0.\mathbf{k});U_p/2\omega). J_n(\alpha_0.\mathbf{k};U_p/2\omega) <\phi_j^c(\mathbf{r}_2)\phi^0(\mathbf{k},\mathbf{r}_1)|\Phi_i^0(\mathbf{r}_1,\mathbf{r}_2)>\}$ (22)

where, E_a and E_b are the final kinetic energies of the two outgoing electrons, E_B is the binding energy of the He atom, and the angular brackets denote integrations with respect to the coordinates.

All the physically interesting quantities, such as the angular- and energy distributions of the ejected electrons, as well as the integrated total probability of double ionization, can be now readily expressed in terms of the basic expressions of the N-photon T-matrix elements (19) above. Thus, the five-fold angular correlation distribution, corresponding to one electron emering with the momentum $\mathbf{k}_a$ and at the same time the other electron is emerging with the momentum $\mathbf{k}_b$ is given by:

$dW / dE_b d\Omega_a d\Omega_b = (2\pi)^{-5} \Sigma_N |T^{(N)}|^2 k_a k_b$ (23)

with $T^{(N)}$ defined by Eq.(22); $k_a = (2(N\omega-E_B-k_b^2/2))^{1/2}$ is a real momentum determined by the energy conservation; $d\Omega_a$ and $d\Omega_b$ are the elements of the solid angles of ejection of the two electrons.

The total probability per unit time of double ionization per atom is obtained by integrations over the energy dE_b and the solid angles $d\Omega_a$ and $d\Omega_b$.:

$\Gamma = \Sigma_N S S S dE_b d\Omega_a d\Omega_b (2\pi)^{-5} |T^{(N)}|^2 k_a k_b$ (24)

4. Numerical Estimates and Comparison with Experiments

The integration of the double ionization T-matrix elements (22), is non-trivial. Thus, on carrying out the six-fold integrations over the coordinates of the two electrons in the expression (22), we are left with three integrations over the intermediate momentum, **k,** of the Volkov state, an intermediate summation (integration) over the virtual states, {j}, of the

405

He$^+$ ion, and a double infinite sum over the virtual photon numbers, n. Furthermore, after squaring this amplitude in the final expression, Eq.(24), for the total rate Γ, we must carry out the five-fold integrations over the solid angles of the two outgoing electrons and the energy of the second electron, E_b. Finally, an infinite sum over the actually absorbed final photon numbers, N, starting from the minimum number, N_0 (determined by the threshold), has to be carried out. These many-fold integrations (8-fold) and infinite summations (3-fold) present a formidable task; we present below the first estimates of the total rates obtained by using a combination of analytical and numerical methods along with a number of simplifying approximations in the calculation. Specifically, we have evaluated the k-integration using the pole-approximation; two other integrations analytically and the rest numerically. The contribution from the intermediate ground state of the ion dominates greatly over that from the excited intermediate states. The latter contribution has been estimated by the closure technique (e.g.[22,§4.5]), and is found to be negligible. The initial state of the He atom is assumed to be the well-known simple Hylleraas wavefunction, and plane waves of momentum $\mathbf{k}_a$ and $\mathbf{k}_b$, instead of Coulomb waves, have been used in the final state to simplify the present calculationton and the possibility of taking more sophisticated initial and/or final states, e.g. with effective charges (e.g. [25]), is left open for the future investigations.

In Fig.3 we present the experimental and theoretical rates of laser induced ionizations of He. The arbitrary scale of the experimental signals has been fixed by matching the double ionization data point at the saturation intensity, $I=0.8 \times 10^{15}$ W/cm^2, with that of the present

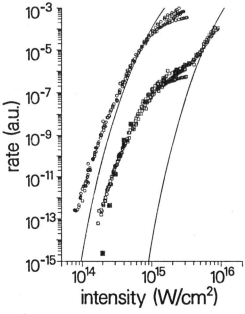

Fig.3. The experimental data of single ionization signal of He [6] (open circles), experimental data for double ionization of He [6] (open squares) and the present theory for one-atom double ionization rates of He (solid squares). The experimental data in arbitrary scale have been matched at one point to our theoretical value i.e. at the saturation intensity, $I=0.8 \times 10^{15}$ W/cm^2 (without any shift of the intensity scale). The left hand curve is the one-atom ADK-rates [20] for the single ionization of He and the right hand curve is that for the ionization of He$^+$; both the ADK curves are shifted in intensity scale by a factor of 1.4. The theoretical rates are not averaged with respect to the spatial distribution of the laser beam. The laser wavelength corresponds to 780 nm. Observe that there is *no* 'intensity threshold' for the double ionization rates in the data or in the present theory.

calculation at that intensity. The rates are compared for three different processes, namely (i) *single* ionization of He: experimental data (open circles) by Walker et.al. [6] and calculation (left hand curve) with ADK-model [20], (ii) *single* ionization of He$^+$ ion (right hand curve): calculation with ADK-model, and finally (iii) *double* ionization of He: experimental data (open squares) by Walker et.al. [6], and theoretical results (solid squares) from the present calculations. Variations of the rates of the processes are shown as functions of the intensity at a fixed laser wavelength of 780 nm. Comparisons of single ionization data with the ADK model is made to provide a qualitative orientation; numerical simulations for He$^+$ and for He, using the SAE-approximation, have been also made and shown to provide very good agreement for the single ionization in these cases[e.g. 6 and refs. therein]. The main interest here is that for the double ionization. A comparison of the experimental double-ionization rates (open squares) for neutral helium with that of the single-ionization of He$^+$ ion (right hand curve) shows a *many* orders of magnitude greater rates for the double ionization, in the intensity range between about 10^{14} W/cm^2 to about 10^{15} W/cm^2 (that is in the range below the saturation intensity). Note the excellent order of magnitude agreement of the present theoretical estimates (solid squares) with the data for double ionization of He (open squares). This is the first time such an agreement has been obtained between the data and the prediction of a theory of laser induced double ionization of He. Of particular interest is the *absence* of any 'intensity-threshold', both in the data and in the present theory, below the saturation intensity.

This is a counter-intuitive result, in view of the fact that the threshold of single ionization of He$^+$ ion is only about 54.4 eV, while that of the double ionization of He atom is about 79 eV. The experiment [6] and the present theory clearly suggest that it is *very* much more probable to transfer over 79 eV and doubly ionize the He atom than to transfer over only 54.4 eV and singly ionize the already once-ionized He atom (i.e. He$^+$-ion)! This remarkable fact underlines the important role of *e-e correlation* in the process.

5. Mechanism of Direct Double Ionization of He in Intense Laser Fields

In view of the excellent order of magnitude agreement of the predictions of the present theory with the experimental data for double ionization of He at the sub-saturation intensities, we are in a position to identify the main mechanism behind the anomalous laser induced double ionization process. As discussed above this contribution for the double ionization of He atoms is calculated here from the leading conjoint Feynman diagram II(e). A direct reading of this diagram provides a vivid picture of the associated mechanism for the process. This diagram distinguishes itself as being the major diagram in the leading term for the double-escape process, that involves electron-electron correlation explicitly in the final interaction. Reading from below upward (i.e. in the assumed direction of the flow of time) we see that at first one of the two electrons interacts with the laser field and absorbs a large number of photons and propagates in the Volkov state while the other electron propagates into the ionic states (mainly in the intermediate ground state). At this intermediate stage of the process the electron-electron correlation enters explicitly in action and mediates the transfer of the photon energy absorbed by one electron to the other electron until both of them get enough energy to escape together from the binding force of the nucleus. Thus, the e-e interaction (correlation) in the *final* stage of the process plays here a crucial dynamical role, for otherwise the electron that absorbs the large amount of photon energy from the field first, would escape alone *and* abort the double escape process.

We may note, parenthetically, that an explanation of the double ionization process in terms of tunneling alone would also fail unless the electron-electron correlation, in the final stage, in that mechanism could be included. And, by construction, all types of 'single active electron' models, which have been very useful for the single ionization [e.g.6], remain out side the scope of a simultaneous two-electron process.

The present mechanism is reminiscent of the Corkum-model of (e-2e)-like ionization of the core by rescattering of the free electron on a classical trajectory in a laser field, discussed in sec.2 above. But note the following difference between them: the mechanism of double ionization presented here permits absorption of photon-energy from the field with varying degrees of probability at *all* intensities, i.e. there is no 'threshold-intensity' for the quantum mechanical process. This is because the intermediate state energy associated with the momentum, **k**, of the Volkov electron may lie even *below* the two-electron ionization threshold and yet can cause ionization, through the *off-shell* (e-2e)-like amplitudes for simultaneous electron-photon ionization of the intermediate ionic states (including the ground state) (c.f. Eq.(19)). More over, there is no sharp restriction to the maximum kinetic energy of the intermediate Volkov electron. In contrast, as indicated above, the simple classical (e-2e)-like model is restricted, at a given intensity, by a maximum possible energy that an ejected electron can gain from the field on returning to the core.

Our analysis shows that the direct double ionization process is a prototype of a 'cooperative' quantum phenomenon at its most rudimentary form: two-particle joint escape by energy sharing through the e-e *correlation*. Thus, the high probability of the process, makes the laser induced double ionization of He, compared to the single ionization of He$^+$ at the same frequency and (sub-saturation) intensities, one of the largest effects of Coulomb correlation known in atomic physics.

6. Conclusions

To conclude, we have developed an *ab initio* intense-field many electron diagrammatic S-matrix perturbation theory and applied it to analyze the recently observed anomalously large probability of laser induced double ionization of He atom. The double ionization process is investigated using the leading terms of the new S-matrix theory from which a dominant diagram for the process is identified. Results of the priliminary evaluation of this diagram, II(e), are compared with the experimental data. Excellent orders of magnitude agreement is found for the first time between a theory and the experiment. Thus, the process is shown to be a quantum mechanical one involving energy sharing between the electrons *through* Coulomb correlation to facilitate the double escape. The mechanism suggested by the diagram is one in which the energy due to the absorption of a large number of photons by one electron is transferred to the other electron through the $1/r_{12}$ interaction in the *final* stage, until both electrons get enough energy to escape jointly. This is reminiscent of the Corkum-model of 'e-2e'-like ionization of the ion-core by collision of the classically free electron in the electromagnetic field. However, unlike that simple model which predicts a pesudo threshold at an intensity corresponding to the maximum energy of the classical electron on return to the nucleus, $3.17U_p$= ionization-energy of the ion, the present quantum model, and the experimental data, do *not* show such a threshold. Finally, the present analysis supports the view that the double ionization probability for He, below the saturation intensity, (which is observed to be *many* orders of magnitude greater than the single ioization probability of the already once ionized He atom, at the same frequency and intensity) is one of the largest effects of Coulomb correlation known in atomic physics.

7. Acknowledgment

We are thankful to L.F.DiMauro and P.Agostini for kindly making available to us their published data, on the single- and double ionization of He, in the numerical form, and for the stimulating correspondence on the results. We also take this opportunity to thank K.Kondo for kindly making available their data. This work was partially supported by the Deutsche Forschungsgemeinschaft, Bonn.

8. References

1. D.N.Fittinghoff, P.R.Bolton, B.Chang, and K.C.Kulander, Phys.Rev.Lett.69, 2642 (1992)
2. J.Peatross, B.Buerke and D.D.Meyerhofer, Phys.Rev. A47, 1517 (1993)
3. B.Walker, E.Mevel, B. Yang, P.Breger, J.P.Chambaret, A.Antonetti, L.F.DiMauro and P.Agostini, Phys.Rev. A48, R895 (1993)
4. K.Kondo, A.Sagisaka, T.Tomida, Y.Nabekawa and S.Watanabe, Phys.Rev.A48, R2531 (1993)
5. D.N.Fittinghoff, P.R.Bolton, B.Chang and K.C.Kulander, Phys.Rev. A49, 2174 (1994)
6. B.Waker, B.Sheehy, L.F.DiMauro, P.Agostini, K.J.Schafer and K.C.Kulander, Phys.Rev. Lett. 73, 1227 (1994)
7. P.Agostini, F.Fabre, G.Mainfray, G.Petite and N.K.Rahman, Phys.Rev.Lett. 42, 1127 (1979)
8. P.Kruit, J.Kimman and M.J. van der Wiel, J.Phys. B14, L597 (1981)
9. L.V.Keldysh, Zh.Ekps.Theor.Phys. 47, 1945 (1964) [Sov.Phys. JETP 20, 1307 (1965)]
10. F.H.M.Faisal, J.Phys.B5, L89 (1973)
11. H.R.Reiss, Phys.Rev. A22,1786 (1980)
12. J.I.Gersten and M.H.Mittleman, J. Phys. B9, 2561 (1976)
13. M.Gavrila and J.M.Kaminski, Phys.Rev. 52, 612 (1984); M.Pont and M.Gavrila, Phys.Rev.Lett. 65, 2362 (1990)
14. K.Kulander, K.J.Schafer and J.L.Krause, Phys.Rev.Lett. 66, 2601 (1991)
15. M.Dörr, R.M.Potvliege, D.Proulx and R.Shakeshaft, Phys.Rev. A43, 3729 (1991)
16. L.Dimou and F.H.M.Faisal, Phys.Lett A171, 211 (1992); Phys.Rev. A46, 4442 (1992); Laser Phys.3, 440 (1993); Acta.Phys.Pol. A86, 201 (1994); Phys.Rev. A49, 4564 (1994)
17. R.M.Potvliege and H.G.Smith, Phys.Rev. A48, R46 (1993)
18. M.P.deBoer, J.H.Hoogenraad, R.B.Vrijen, L.D.Noordam and G.H.Muller, Phys.Rev.Lett. 71, 3263 (1993)
19. P.B.Corkum, Phys.Rev.Lett. 71,1994 (1993)
20. M.V.Ammosov, N.B.Delone and V.P.Krainov, Zh.Eksp.Teor. Fiz. 91, 2008 (1986) [Sov. Phys. JETP 64, 1191 (1986]
21. H.Tawara and T.Koto, At. Data and Mol. Data Tables 36, 167 (1987)
22. F.H.M.Faisal, *Theory of Multiphoton Processes*, Plenum Press, N.Y. (1987), Chap.1
23. F.H.M.Faisal, *Floquet Green's Function Method for Radiative Electron Scattering and Multiphoton Ionization in a Strong Laser Field,* Computational Physics Reports, Vol.9, No.2 , North Holland, Amsterdam (1989)
24. A.Becker and F.H.M.Faisal, Phys.Rev. A50, 3256 (1994)
25. S. Jetzke and F.H.M.Faisal, J.Phys.B25, 1543 (1992)
26. M.R.Cervenan and N.R.Isenor, Opt. Commun. 13, 175 (1975)

ANISOTROPIES AND LAMB SHIFTS IN HYDROGENIC IONS: TWENTY YEARS OF PROGRESS

G. W. F. Drake, and A. van Wijngaarden

Department of Physics
University of Windsor
Windsor, Ontario, N9B 3P4
Canada

INTRODUCTION

If a hydrogen atom (or other hydrogenic ion) is prepared in the metastable $2s_{1/2}$ state and then subjected to an external electric field, the induced Ly-α radiation to the ground state displays a rich diversity of quantum beats, anisotropies, and angular asymmetries resulting from a variety of interference effects. Over the past 22 years, these have been exploited at the University of Windsor to obtain fundamental measurements of the Lamb shift, decay rates, and the relativistic magnetic dipole (M1) transition amplitude. Table 1 summarizes the key developments since the anisotropy method of measuring the Lamb shift was first proposed in 1973.[1] Peter Farago's interest and enthusiasm played an important role during the early development stages.

This article first summarizes the basic principles needed to understand the interference phenomena, and then discusses the results that have been obtained and their significance.

THEORY OF RADIATION INTERFERENCE

The interference effects discussed here are made experimentally accessible by the near degeneracy of the $2s_{1/2}$ and $2p_{1/2}$ states. In one-electron Dirac theory these states are exactly degenerate, but when the quantum electrodynamic effects of electron self-energy and vacuum polarization are included, the energy of the $2s_{1/2}$ state rises about 1047 MHz above the $2p_{1/2}$ state. This Lamb shift is about 10 times larger than the $2p_{1/2}$ level width, and about one tenth of the $2p_{1/2}$–$2p_{3/2}$ fine structure splitting.

In the absence of external fields, the $2s_{1/2}$ state decays to the ground state primarily by spontaneous two-photon emission with a slow rate of $8.22938Z^6$ s^{-1},[2] where Z is the nuclear charge. There is also a relativistic magnetic dipole (M1) mechanism, but it is still slower by a factor of $\sim 10^{-4}$.[3] However, owing to the small size of the Lamb shift, an external electric field of ~ 100 V/cm is sufficient to produce about a 20% mixing with

Selected Topics on Electron Physics
Edited by Campbell and Kleinpoppen, Plenum Press, New York, 1996

Table 1. History of Anisotropy Measurements.

Authors	Results
G. W. F. Drake and R. B. Grimley, *Phys. Rev. A* 8:157 (1973)	Method proposed
A. van Wijngaarden, G. W. F. Drake, and P. S. Farago, *Phys. Rev. Lett.* 33:4 (1974)	Lamb shift of H: 1059 ± 13 MHz
G. W. F. Drake, A. van Wijngaarden, and P. S. Farago in "Electron and Photon Interactions with Atoms" H. Kleinpoppen and M. R. C. McDowell eds., Plenum, New York (1976)	Lamb shifts H: 1057.3 ± 0.9 MHz D: 1058.7 ± 1.1 MHz
G. W. F. Drake and C.-P. Lin, *Phys. Rev. A* 14:1296 (1976)	Anisotropy theory
A. van Wijngaarden, E. Goh, G. W. F. Drake, and P. S. Farago, *J. Phys. B* 9:2017 (1976)	Quantum beats in H
G. W. F. Drake, *J. Phys. B* 10:775 (1977)	Quantum beats theory
A. van Wijngaarden and G. W. F. Drake, *Phys. Rev. A* 17:1366 (1978)	Lamb shift of D: 1059.36 ± 0.16 MHz
G. W. F. Drake, *Phys. Rev. Lett.* 40:1705 (1978)	E1-M1 interference proposal.
G. W. F. Drake, S. P. Goldman, and A. van Wijngaarden, *Phys. Rev. A* 20:1299 (1979)	Lamb shift of He$^+$: 14040.2 ± 2.9 MHz
A. van Wijngaarden and G. W. F. Drake, *Phys. Rev. A* 25:400 (1982)	E1-M1 interference in He$^+$: $\bar{M} = -(0.262 \pm 0.69)\alpha^2 e\hbar/mc$
A. van Wijngaarden, R. Helbing, J. Patel, and G. W. F. Drake, *Phys. Rev. A* 25:862 (1982)	Lifetime of He$^+$(2p): $(0.988 \pm 0.013) \times 10^{-10}$ s
G. W. F. Drake, J. Patel, and A. van Wijngaarden, *Phys. Rev. A* 28:3340 (1983)	Lifetime of He$^+$(2p): $(0.9992 \pm 0.0026) \times 10^{-10}$ s
A. van Wijngaarden, J. Patel, and G. W. F. Drake, *Phys. Rev. A* 33:312 (1986)	E1-M1 interference in He$^+$: $\bar{M} = -(0.273 \pm 0.031)\alpha^2 e\hbar/mc$
J. Patel, A. van Wijngaarden, and G. W. F. Drake, *Phys. Rev. A* 36:5130 (1987)	Lamb shift of He$^+$: 14041.9 ± 1.5 MHz
G. W. F. Drake, J. Patel, and A. van Wijngaarden, *Phys. Rev. Lett.* 60:1002 (1988)	Lamb shift of He$^+$: 14042.22 ± 0.35 MHz
A. van Wijngaarden, J. Kwela, and G. W. F. Drake, *Phys. Rev. A* 43:3325 (1991)	Lamb shift of He$^+$: 14042.52 ± 0.16 MHz
A. van Wijngaarden, J. Kwela, and G. W. F. Drake, *Phys. Rev. A* 46:113 (1992)	Lifetime of He$^+$(2p): $(0.997\,17 \pm 0.000\,75) \times 10^{-10}$ s

the $2p_{1/2}$ state, leading to rapid electric dipole (E1) transitions to the ground state. There is also a smaller ($\sim$2%) mixing with the $2p_{3/2}$ state, which also decays to the ground state by either E1 or M2 radiation. It is the interference between the two main E1 decay channels that is primarily responsible for the rotational asymmetries with

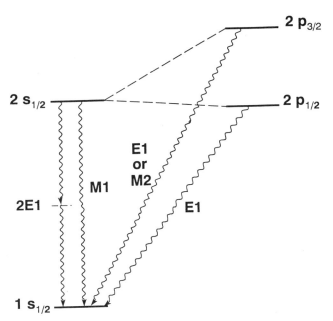

Figure 1. Summary of radiative processes from the $2s_{1/2}$ state. The dashed lines represent electric field mixing with the $2p_{1/2}$ and $2p_{3/2}$ state.

respect to the external electric field direction. The various processes are summarized in Fig. 1.

Theory of Spontaneous Radiation

From Fermi's Golden Rule, the spontaneous transition rate w_{if} from an initial state i to a final state f is

$$w_{if} = \frac{2\pi}{\hbar} |\langle f | V_{\text{int}} | i \rangle|^2 \rho_f , \tag{1}$$

where V_{int} is the interaction energy operator coupling the atom to the field, and ρ_f is the number of final states per unit energy interval in a normalization volume $\mathcal{V}$. The basic parameters characterizing the emitted photon are its frequency ω, its polarization vector $\hat{\varepsilon}$, and its propagation vector $\mathbf{k}$ ($|\mathbf{k}| = \omega/c$). Then the number of photon states of polarization $\hat{\varepsilon}$ per unit energy and solid angle $d\Omega$ in the normalization volume $\mathcal{V}$ is

$$\rho_f = \frac{\mathcal{V} k^2 \, d\Omega}{(2\pi)^3 \hbar c} , \tag{2}$$

and in relativistic Dirac notation, the interaction energy is

$$V_{\text{int}} = e\boldsymbol{\alpha} \cdot \mathbf{A}^* . \tag{3}$$

If the photon vector potential $\mathbf{A}$ is normalized to a field energy of $\hbar\omega$ per unit volume, then

$$\mathbf{A} = \frac{1}{k} \left(\frac{2\pi\hbar\omega}{\mathcal{V}} \right) \hat{\varepsilon} e^{i\mathbf{k}\cdot\mathbf{r}} . \tag{4}$$

Collecting terms, Eq. (1) becomes

$$w_{if}\,d\Omega = \left(\frac{e^2 k}{2\pi\hbar}\right)|\langle f \mid \boldsymbol{\alpha}\cdot\hat{\boldsymbol{\varepsilon}}e^{-i\mathbf{k}\cdot\mathbf{r}} \mid i\rangle|^2\,d\omega\,. \tag{5}$$

per unit time. In the nonrelativistic limit, $\boldsymbol{\alpha} \to \mathbf{p}/mc$, $e^{-i\mathbf{k}\cdot\mathbf{r}} \approx 1$, and Eq. (5) becomes the familiar dipole velocity form of the transition rate.

The $2s_{1/2}$ state can decay by spontaneous M1, and the electric field induced E1 and M2 processes. Keeping these terms in the standard multipole expansion for $e^{-i\mathbf{k}\cdot\mathbf{r}}$ yields [4]

$$e^{-i\mathbf{k}\cdot\mathbf{r}} = \left(\frac{3}{8\pi}\right)^{1/2}\sum_M\left\{\varepsilon_M\mathbf{a}_{1,M}^{(1)*} + i[\hat{\mathbf{k}}\times\hat{\boldsymbol{\varepsilon}}]_M\mathbf{a}_{1,M}^{(0)*} + i\sqrt{10/3}[\hat{\mathbf{k}},\,\hat{\mathbf{k}}\times\hat{\boldsymbol{\varepsilon}}]_{2,M}\mathbf{a}_{2,M}^{(1)*}\right\}\,, \tag{6}$$

where $\varepsilon_{\pm 1} = \mp(\varepsilon_x \pm \varepsilon_y)/\sqrt{2}$, $\varepsilon_0 = \varepsilon_z$, and the notation $[\mathbf{a},\mathbf{b}]_{2,M}$ denotes a vector coupled product to form an irreducible tensor of rank 2 and component M. The $\mathbf{a}_{L,M}^{(\lambda)}$ are multipole operators with $\lambda = 1$ for electric multipoles and $\lambda = 0$ for magnetic multipoles. In the transverse (Coulomb) gauge, they are given by

$$\mathbf{a}_{L,M}^{(1)} = \left(\frac{L}{2L+1}\right)^{1/2}g_{L+1}(kr)\mathbf{Y}_{L\,L+1}^M(\hat{\mathbf{r}}) + \left(\frac{L+1}{2L+1}\right)^{1/2}g_{L-1}(kr)\mathbf{Y}_{L\,L-1}^M(\hat{\mathbf{r}})\,, \tag{7}$$

$$\mathbf{a}_{L,M}^{(0)} = g_L(kr)\mathbf{Y}_{L\,L}^M(\hat{\mathbf{r}})\,, \tag{8}$$

where the $\mathbf{Y}_{J\ell}^M(\hat{\mathbf{r}})$ are vector spherical harmonics defined by

$$\mathbf{Y}_{J\ell}^M(\hat{\mathbf{r}}) = \sum_{m,q}Y_\ell^m\hat{\mathbf{e}}_q(\hat{\mathbf{r}})\langle\ell\,1\,m\,q \mid J\,M\rangle\,, \tag{9}$$

and $g_L(kr)$ is related to the spherical Bessel functions by

$$g_l(kr) = 4\pi i^L j_L(kr)\,, \tag{10}$$

$$j_L(z) = \frac{z^L}{(2L+1)!!}\left[1 - \frac{z^2/2}{1!(2L+3)} + \frac{(z^2/2)^2}{2!(2L+3)(2L+5)} - \cdots\right]\,. \tag{11}$$

Since $kr = \omega r/c$ is small for low-Z atoms, the long wavelength approximation in which only the leading term of Eq. (11) is retained is usually sufficient. However, as discussed below, the second term is the leading contribution to the relativistic M1 process. For ordinary allowed transitions, the nonrelativistic limits are

$$e\boldsymbol{\alpha}\cdot\mathbf{a}_{1,M}^{(1)} \longrightarrow e\sqrt{2}\Phi_{1,M}\,, \qquad e\boldsymbol{\alpha}\cdot\mathbf{a}_{1,M}^{(0)} \longrightarrow i(\nabla\Phi_{1,M})\cdot\left(\frac{e\mathbf{L}}{mc\sqrt{2}} + \sqrt{2}\boldsymbol{\mu}\right)\,,$$

$$e\boldsymbol{\alpha}\cdot\mathbf{a}_{2,M}^{(0)} \longrightarrow i(\nabla\Phi_{2,M})\cdot\left(\frac{e\mathbf{L}}{mc\sqrt{6}} + \sqrt{3/2}\boldsymbol{\mu}\right)\,,$$

where

$$\Phi_{L,M} = g_L(kr)Y_L^M(\hat{\mathbf{r}})\,, \qquad \mathbf{L} = \mathbf{r}\times\mathbf{p}\,, \qquad \boldsymbol{\mu} = \tfrac{1}{2}e\alpha a_0\boldsymbol{\sigma}\,,$$

and $\boldsymbol{\mu}$ is the magnetic moment operator.

For the single-photon $2s_{1/2}$–$1s_{1/2}$ transition, only the M1 term contributes, but the matrix element of $\mathbf{a}_{1,M}^{(0)}$ vanishes in lowest order since it becomes proportional to the overlap integral. The inclusion of relativistic corrections of $O(\alpha^2 Z^2)$ and finite

414

wavelength (retardation) corrections of $O[(\omega r/c)^2]$ gives the effective nonrelativistic magnetic moment transition operator [3]

$$M_{1,M} = \mu_B \sigma_M \left[1 - \frac{2p^2}{3m^2c^2} - \frac{1}{6}\left(\frac{\omega r}{c}\right)^2 + \frac{Ze^2}{3mc^2r} \right], \qquad (12)$$

where $\mu_B = e\hbar/2mc = \frac{1}{2}\alpha a_0$ is the Bohr magneton. With hydrogenic wave functions, the matrix element M of M_{10} is

$$M \equiv \langle 1s_{1/2} \mid M_{10} \mid 2s_{1/2}\rangle = -\left(\frac{8\alpha^2 Z^2}{81\sqrt{2}}\right) e\alpha a_0. \qquad (13)$$

After summing over polarizations and integrating over angles, the final decay rate is

$$\begin{aligned} w(2s_{1/2} \to 1s_{1/2}) &= 4k^3|M|^2/\hbar \\ &= \frac{\alpha^9 Z^{10}}{972} \text{ a.u.} \\ &= 2.496 \times 10^{-6}\,\text{s}^{-1} \quad \text{for H } (Z = 1). \end{aligned} \qquad (14)$$

Even though this is much less than the 2E1 rate, the M1 amplitude still produces observable interference effects in the Stark quenching of the $2s_{1/2}$ state.

Theory of Electric Field Quenching

The electric dipole operator $\hat{\boldsymbol{\varepsilon}}\cdot\mathbf{r}$ does not connect directly the $2s_{1/2}$ and $1s_{1/2}$ states. However, it does connect indirectly in the presence of an external D.C. electric field $\mathbf{E}$ $(\mathcal{E} = |\mathbf{E}|)$ through perturbative mixing with the $2p_{1/2}$ and $2p_{3/2}$ states. The effective operator for combined M1 and field-induced E1 radiation is

$$D = e\hat{\boldsymbol{\varepsilon}}\cdot\mathbf{r} \sum_{j,\mu} \frac{|2p_{j,\mu}\rangle\langle 2p_{j,\mu}|\mathbf{E}\cdot\mathbf{r}}{E(2s_{1/2}) - E(2p_j) + i\Gamma_p/2} + M(\hat{\mathbf{k}} \times \hat{\boldsymbol{\varepsilon}})\cdot\boldsymbol{\sigma}, \qquad (15)$$

where Γ_p is the level width. If the initial $2s_{1/2}$ state is in a linear superposition of spin states with amplitudes a_{μ_i}, then the emitted intensity summed over final states is

$$I \propto \sum_{\substack{\mu_i,\mu_i' \\ \mu_f}} \langle 2s_{1/2}, \mu_i \mid D^\dagger \mid 1s_{1/2}, \mu_f\rangle\langle 1s_{1/2}, \mu_f \mid D \mid 2s_{1/2}, \mu_i'\rangle a_{\mu_i}^* a_{\mu_i}'. \qquad (16)$$

To simplify the analysis, it is convenient to define coefficients

$$\begin{aligned} V_j &= \frac{\mathcal{E}}{3}\frac{\langle 1s \mid z \mid 2p\rangle\langle 2p \mid z \mid 1s\rangle}{E(2s_{1/2}) - E(2p_j) + i\Gamma_p/2}, \\ \bar{M} &= \langle 1s \mid M \mid 2s\rangle, \end{aligned}$$

which depend on $j = \frac{1}{2}, \frac{3}{2}$ only through the energy denominators in V_j. Then D can be expressed as a spin matrix operator

$$\mathbf{D} = \mathbf{1}(V_{1/2} + 2V_{3/2})(\hat{\boldsymbol{\varepsilon}}\cdot\hat{\mathbf{E}}) + \boldsymbol{\sigma}\cdot[i(V_{1/2} - V_{3/2} - M_{3/2})(\hat{\boldsymbol{\varepsilon}} \times \hat{\mathbf{E}}) + \bar{M}(\hat{\mathbf{k}} \times \hat{\boldsymbol{\varepsilon}})], \qquad (17)$$

where $M_{3/2}$ denotes the small M2 contribution from Eq. (6) analogous to $V_{3/2}$. If in addition the initial state is described by a density matrix $\rho = \frac{1}{2}(\mathbf{1} + \boldsymbol{\sigma}\cdot\mathbf{P})$, where $\mathbf{P}$ is the spin polarization vector, and the results summed over two orthogonal photon polarization vectors $\hat{\boldsymbol{\varepsilon}}_1$ and $\hat{\boldsymbol{\varepsilon}}_2$ perpendicular to $\hat{\mathbf{k}}$, then

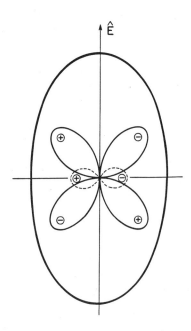

Figure 2. Intensity distributions of electric field quenching processes. The outer oval is the main Lamb shift term I_L. The clover-leaf pattern is I_Γ, and the inner left-right asymmetry is I_M.

$$I \propto \sum_{\hat{\varepsilon}_1\,\hat{\varepsilon}_2} \mathrm{Tr}\,[\rho \mathbf{D}^\dagger \mathbf{D}]$$
$$= I_\mathrm{L} + I_\Gamma + I_\mathrm{M} + I_\mathrm{small}\,, \tag{18}$$

where

$$I_\mathrm{L} = |V_{1/2} + 2V_{3/2}|^2[1 - (\hat{\mathbf{k}}\cdot\hat{\mathbf{E}})^2] + |V_{1/2} - V_{3/2} + M_{3/2}|^2[1 + (\hat{\mathbf{k}}\cdot\hat{\mathbf{E}})^2]\,, \tag{19}$$

$$I_\Gamma = -6\,\mathrm{Im}\,(V_{1/2}^* V_{3/2})(\hat{\mathbf{k}}\cdot\hat{\mathbf{E}})(\mathbf{P}\cdot\hat{\mathbf{k}} \times \hat{\mathbf{E}})\,, \tag{20}$$

$$I_\mathrm{M} = 4\bar{M}\,\mathrm{Re}\,(V_{1/2} + \tfrac{1}{2}V_{3/2})(\mathbf{P}\cdot\hat{\mathbf{k}} \times \hat{\mathbf{E}})\,, \tag{21}$$

$$I_\mathrm{small} = 2|\bar{M}|^2 + 4\bar{M}\,\mathrm{Im}\,(V_{1/2} - V_{3/2})(\hat{\mathbf{k}}\cdot\hat{\mathbf{E}})\,. \tag{22}$$

I_L is the main Lamb shift anisotropy producing the outer oval in Fig. 2. In the limit $|V_{1/2}| \gg |V_{3/2}|$, the Lamb shift is much less than $|E(2p_{3/2}) - E(2s_{1/2})|$, and the two terms in Eq. (19) sum to an isotropic distribution. The anisotropy, defined by $R = (I_\parallel - I_\perp)/(I_\parallel + I_\perp)$, is therefore approximately equal to the ratio of the Lamb shift to $E(2p_{3/2}) - E(2s_{1/2})$; i.e. $R \simeq 0.15$ for hydrogen, and decreases slowly with Z for hydrogenic ions.

The term I_Γ is proportional to the level width Γ_p of the $2p$ state, and produces the clover-leaf intensity distribution in Fig. 2. This term is proportional to $|\mathbf{P}|$, and so the initial state must be spin-polarized in order to observe it. For 100% polarization, the corresponding asymmetry $A = (I_{135^\circ} - I_{45^\circ})/(I_{135^\circ} + I_{45^\circ})$ is about a factor of ten smaller than R.

Finally, the term I_M can be used to measure the relativistic M1 matrix element $\bar{M}$. It produces the small left-right asymmetry in Fig. 2, and requires a spin-polarized beam to be observed. All of these effects have been measured to high precision, as further discussed below. The $|\bar{M}|^2$ term in I_small is much too small to be oberved, and the remaining small term in Eq. (22) averages to zero under a 180° reversal.

The contributions to the main I_L term have a simple physical interpretation. The first term in Eq. (19) proportional to $|V_{1/2}+2V_{3/2}|^2$ comes from transitions with $\Delta m = 0$, and the sum over photon polarizations gives the factor $\sum_{\hat{\varepsilon}} |\hat{\varepsilon} \cdot \hat{\mathbf{E}}|^2 = 1 - (\hat{\mathbf{k}} \cdot \hat{\mathbf{E}})^2$. The second term proportional to $|V_{1/2} - V_{3/2}|^2$ comes from transitions with $\Delta m = \pm 1$, and the sum over photon polarizations gives the factor $\sum_{\hat{\varepsilon}} |\hat{\varepsilon} \times \hat{\mathbf{E}}|^2 = 1 + (\hat{\mathbf{k}} \cdot \hat{\mathbf{E}})^2$. Since the radiation from a pure $j = \frac{1}{2} \to \frac{1}{2}$ is clearly isotropic, the anisotropy is due to the relatively smaller electric field mixing with the $2p_{3/2}$ state.

There is a simple geometrical connection between the anisotropy R of the emitted radiation and its polarization P in a particular direction $\mathbf{k}$. For an unpolarized beam, the only direction defined in space is the direction of $\mathbf{E}$, assumed to point in the z-direction. This is therefore an axis of symmetry, and the two polarization components $I_z^{(x)}$ and $I_z^{(y)}$ of $I_z \equiv I_{\parallel}$ must be equal. Thus $P = 0$ in the z-direction. Consider now $I_{\perp}$ in, say, the x-direction. Its polarization components are $I_x^{(y)}$ and $I_x^{(z)}$, with $I_x^{(y)} = I_z^{(y)}$ in the long wavelength approximation (i.e. $e^{i\mathbf{k}\cdot\mathbf{r}} = 1$). The only two free parameters are then $I^{(x)} = I^{(y)}$ and $I^{(z)}$, and one can write

$$R = \frac{I_{\parallel} - I_{\perp}}{I_{\parallel} + I_{\perp}} = \frac{I^{(x)} - I^{(z)}}{3I^{(x)} + I^{(z)}}, \tag{23}$$

$$P = \frac{I^{(x)} - I^{(z)}}{I^{(x)} + I^{(z)}} = \frac{2R}{1 - R}. \tag{24}$$

In fact, P was measured before R by Ott, Kauppila, and Fite,[5] but for experimental reasons R can be measured much more accurately.

For purposes of comparing theory and experiment, it is convenient to define the ratio

$$\zeta = \frac{V_{1/2}}{V_{3/2}} = \frac{\Delta_{\mathrm{L}} + i\Gamma_P/2}{\Delta_{\mathrm{F}} + i\Gamma_P/2}, \tag{25}$$

where $\Delta_{\mathrm{L}} = E(2s_{1/2}) - E(2p_{1/2})$ and $\Delta_{\mathrm{F}} = E(2s_{1/2}) - E(2p_{3/2})$. Then in the limit of weak fields,

$$R = -\frac{3\,\mathrm{Re}\,(\zeta) + \frac{3}{2}|\zeta|^2}{2 - \mathrm{Re}\,(\zeta) + \frac{7}{2}|\zeta|^2}. \tag{26}$$

Except for several small corrections, this displays the connection between a measured value of R and the Lamb shift Δ_{L}. The small corrections not included arise from finite field effects, mixing with np states having $n \neq 2$, final state perturbations, relativistic corrections, and the M2 contribution in Eq. (19). For weak fields, the largest of these is the M2 correction, which for He^+ decreases R by 65.4 parts per million (ppm). A detailed discussion of these corrections can be found in Ref. 6.

EXPERIMENTAL TECHNIQUE

A schematic diagram of the apparatus used to measure the various quenching asymmetries is shown in Fig. 3. The main features are as follows. A 134.2 keV beam of He^+ ions enters from the right, and passes through a gas cell containing N_2 where a substantial fraction ($\sim$5%) are excited to the metastable $2s_{1/2}$ state. After passing through a magnetic lens and various prequenching fields to remove long-lived Rydberg states, the beam enters the main quenching region with the geometry of a quadrupole. However, unlike the usual quadrupole geometry where diagonally opposite rods are held

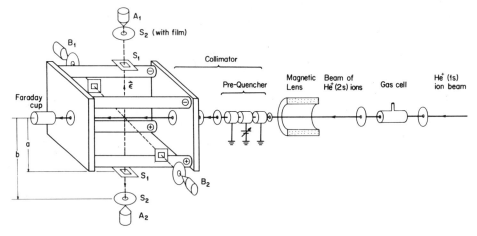

Figure 3. Schematic diagram of the quenching apparatus (from Ref. 7).

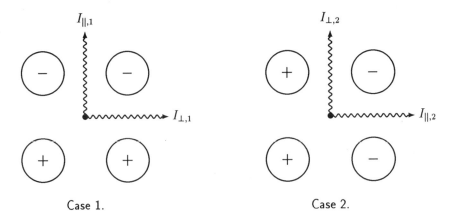

Figure 4. Cross section through the quenching cell for two electric field orientations, showing the beam at the center and the four quadrupole rods. For Case 1, the field points up, and for Case 2 it points to the right. The roles of the counters are interchanged for the two cases.

at the same potential, here the diagonally opposite rods have opposite potentials $+V$ and $-V$. Hence there is a net electric field at the center which is nearly constant across the beam diameter. Finally, the beam current of about 8 μA is monitored by a Faraday cup.

The quenching cell is conceptually the same as a parallel plate capacitor, but the quadrupole design is essential to a high precision experiment because it preserves the four-fold geometrical symmetry of the observation directions, and it allows open observation from all four sides. As shown in Fig. 4, the electric field can be rotated in steps of 90° simply by switching potentials on the rods. This feature is also essential because it allows the analysis of the data to be done in such a way that the relative sensitivities of the photon detectors cancel, along with many other systematic errors. With intensities defined as in Fig. 4, the quantity

$$\frac{I_\parallel}{I_\perp} = \left(\frac{I_{\parallel,1}}{I_{\perp,1}} \frac{I_{\parallel,2}}{I_{\perp,2}} \right)^{1/2} \tag{27}$$

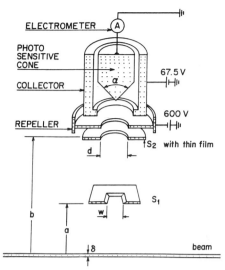

Figure 5. Detail of the photon detection system (from Ref. 7).

is independent of the relative sensitivities. A further averaging over the "down" and "left" directions removes the sensitivity to beam deflections by the quenching field.

An important lesson learned from this work is that, due to dead-time effects and other nonlinearities, photon counting techniques become inadequate at count rates higher than a few thousand s^{-1}. Since effective count rates far higher than this are required in order to accumulate adequate statistics, the photon flux is measured instead by collecting (without amplification) the photoelectron current emitted from the large surface area of metal cones coated with MgF_2. The photoelectron current is then measured directly with a high-precision Keithley electrometer to within a few parts per million. A schematic drawing of the detector design is shown in Fig. 5. Further details can be found in Refs. 6 and 8.

Quantum Beats

If a deliberate effort is made to switch on the quenching field as rapidly as possible as the beam enters the quenching cell, then, as shown in Fig. 6 for hydrogen, pronounced quantum beats can be observed as a function of position along the beam.[10,9] With a beam velocity of 4.79×10^8 cm/s, The main oscillations correspond to the Lamb shift frequency, and the rapid oscillations correspond to the $2s_{1/2}$–$2p_{3/2}$ frequency. The solid curve represents the theoretical prediction based on a numerical integration of the time-dependent Schrödinger equation, including fringing field effects around the entrance hole.[9] The rapid switching was achieved by coating the otherwise insulating end plate with a resistive thin carbon film.[10] The current distribution in the resistive coating is such that surfaces of equal potential are perpendicular to the end plate, and the field inside a small hole in a plate of zero thickness is just twice what it would be in the absence of the hole.

The quantum beats shown in Fig. 6 are among the best resolved that have ever been observed in such systems. However, for high-precision anisotropy measurements, they are an undesirable feature since the fringing field must be precisely known. For anisotropy measurements, care must be taken to switch the field on over a large distance so that the atoms enter nearly adiabatically.

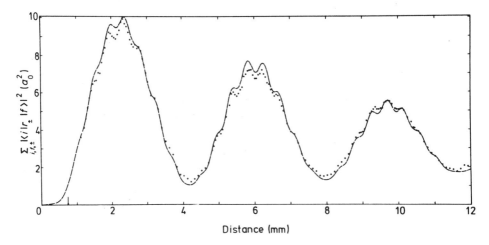

Figure 6. Quantum beats in the electric field quenching of the $2s_{1/2}$ state of hydrogen. The data points are the observed intensities, and the solid curve is the theoretical prediction, including instrumental broadening and fringing field effects (from Ref. 9).

RESULTS

The Lamb Shift Anisotropy

Early measurements of R for hydrogen and deuterium[11-13] (see Table 1) yielded results for the Lamb shift in agreement with theory, but they were not sufficiently accurate to provide an improved value for the Lamb shift. However, the Lamb shift in He$^+$ is much less accurately known by direct microwave resonance techniques. Although a recent measurement by Dewey and Dunford[14] has improved the precision by that means to ±86 ppm, this is still far short of the ±9 ppm accuracy available for hydrogen.[15,16] The problem with the measurements in hydrogen is that there is a significant correction due to the finite nuclear size, and the two available measurements of the proton radius (0.805 fm and 0.862 fm) do not agree with each other. It is not clear which value should be used in comparing theory with experiment. A precise measurement of the Lamb shift in He$^+$ is therefore of considerable value, not only because the ^{4}He radius is precisely known to be 1.673 ± 0.001 fm from the muonic helium Lamb shift (see however Ref. 6 for additional discussion), but also because it is more sensitive to higher order relativistic and two-loop binding corrections which increase in proportion to Z^5 or Z^6.

The He$^+$ Lamb shift obtained by the anisotropy method is 14042.52 ± 0.16 MHz.[6] At the time the measurement was performed, this was in excellent agreement with theory, and is still the most accurate value available. However, as shown in Fig. 7, the recent calculation of two-loop binding corrections has significantly changed the picture. The terms in the Lamb shift can be represented in the standard form (see, e.g. Drake[19])

$$
\Delta_{\rm L} = \frac{\alpha^3 Z^4}{\pi n^3} \Big\{ A_{40} + A_{41} \ln(Z\alpha)^{-2} + A_{50} Z\alpha + (Z\alpha)^2 \Big[A_{60} + A_{61} \ln(Z\alpha)^{-2}
$$
$$
+ A_{62} \ln^2(Z\alpha)^{-2} \Big] + (Z\alpha)^3 \Big[A_{70} + A_{71} \ln(Z\alpha)^{-2} \Big]
$$
$$
+ O(Z^4\alpha^4) + \frac{\alpha}{\pi} \Big[B_{40} + B_{50} Z\alpha + O(Z^2\alpha^2) \Big] \Big\}
\tag{28}
$$

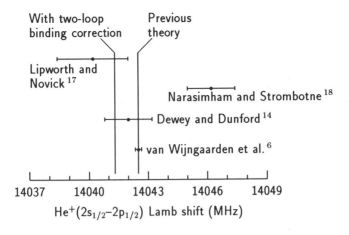

Figure 7. Comparison of theory and experiment for the He⁺ Lamb shift.

in atomic units. The A_{ij} series of terms representing the one-loop Lamb shift is accurately known from the calculations by Mohr[20] and Pachucki.[21] The B_{ij} terms are the two-loop corrections. The value of the leading term has long been known to be $B_{40} = 0.53894\cdots$ for s-states. The B_{50} term is the two-loop binding correction recently calculated by Pachucki,[22] and by Eides and Shelyuto.[23] It turns out to have the suprisingly large value $B_{50} = -21.4 \pm 0.01$. This accounts for the difference of -1.181 MHz between the two theoretical values in Fig. 7. The reason for the now rather substantial disagreement between theory and experiment shown in the figure is not clear. A repetition of the anisotropy measurement, and an independent confirmation of the experimental value by another method is clearly needed. By way of comparison, inclusion of the B_{50} term for the Lamb shift in hydrogen brings theory and experiment into agreement, provided that the larger value for the proton radius (0.862 fm) is used (see Weitz *et al.*[24] for further discussion).

The 2p Level Width

As is evident from Fig. 2, the directions of observation must be rotated by 45° relative to the quenching field direction in order to measure the clover-leaf pattern corresponding to the term I_Γ in Eq. (18). This can be accomplished simply by grounding a pair of diagonally opposite rods in the quenching cell (see Fig. 4). Also, the beam must be prepared in a spin-polarized state since the effect is proportional to $(\mathbf{P} \cdot \hat{\mathbf{k}} \times \hat{\mathbf{E}})$. The spin polarization is achieved by application of a magnetic field in the prequenching region of Fig. 3. For the case of He⁺, the magnetic substates $2s_{1/2,-1/2}$ and $2p_{1/2,1/2}$ cross near 7.4 KG. A small electric field is then sufficient to rapidly depopulate the $2s_{1/2,-1/2}$ state, while the $2s_{1/2,1/2}$ state remains long-lived. Essentially 100% polarization is thereby achieved. The apparatus is otherwise identical to the Lamb shift experiment.

For convenience, the measured 45°–135° asymmetry A can be expressed in terms of Γ_p by

$$A = \frac{3\Gamma_p |\mathbf{P}|(\delta_L - \Delta_F)}{4\Delta_F^2 - 2\Delta_F\Delta_L + \Delta_L^2 + 9\Gamma_p^2/4}, \tag{29}$$

and the result interpreted as a measurement of the lifetime $\tau_p = 1/(2\pi\Gamma_p)$. The measurement yields $\tau_p = (0.997\,17 \pm 0.00075) \times 10^{-10}$ s.[8] This agrees with, but is considerably more accurate than the beam-foil result of $(0.98 \pm 0.05) \times 10^{-10}$ s.[25]

The theoretical decay rate for the $2p_{1/2}$ state of a hydrogenic ion with nuclear charge Z and reduced electron mass $|mu = mM/(m+M)$ is

$$\gamma = 4\pi c R_\infty (1 - \mu/M)[1 + (Z-1)\mu/M]2[1 - 2\ln(8/9)\alpha^2 Z^2](2/3)^8 \alpha^3 Z^4, \qquad (30)$$

which includes corrections for the reduced mass, nuclear motion, and the leading relativistic term. Using $R_\infty = 109\,737.315\,709(18)$ cm-1, $\alpha^{-1} = 137.0359895(61)$, and $\mu/M = 1.370\,745\,62(3)$, the theoretical decay rate is $\gamma = 1.003\,118 \times 10^{10}$ s^{-1}, or $\tau = 0.996\,891 \times 10^{-10}$ s. The good agreement with the measured value at the $\pm 0.75\%$ level provides the best available test of basic radiation theory for a fundamental one-electron atomic system.

The most accurate previous measurement of an atomic lifetime is for the $2\,^2P$–$2\,^2S$ transition in neutral lithium with a claimed accuracy of $\pm 0.15\%$.[26] However, their result disagrees with recent high precision calculations[27] by five standard deviations, and has not been confirmed by other recent measurements approaching the same level of accuracy.[28]

The Relativistic Magnetic Dipole Matrix Element

The relativistic M1 process has a long and interesting history (see Cheng et al.[29] for a recent discussion). It provides the dominant decay mechanism for the metastable $1s2s\,^3S_1$ state of helium-like ions where two-photon processes are spin-forbidden, but it has never been observed directly in a one-electron system where the two-photon process dominates. However, it can be observed indirectly through the I_M interference term in Eq. (18). After some algebra, the left-right asymmetry in Fig. 2 for a 100% polarized beam can be expressed in the form

$$A_{lr} = 2\Lambda \left[\frac{1 + \frac{1}{2}|\zeta|^2 \Delta_F/\Delta_L}{1 + \mathrm{Re}\,(\zeta) + \frac{5}{2}|\zeta|^2 + \Lambda^2} \right], \qquad (31)$$

so that $A_{lr} \simeq 2\Lambda$, where

$$\Lambda = -\frac{243\sqrt{2}Z^2 \Delta_L \bar{M}}{128 e^2 a_0^2 \mathcal{E}} = -\left(\frac{\text{spontaneous M1 rate}}{\text{induced E1 rate}} \right)^{1/2}. \qquad (32)$$

Since $\mathcal{E}$ appears in the denominator of Eq. (32), the effect is enhanced by making $\mathcal{E}$ as small as possible while still maintaining an adequate signal strength. With $\mathcal{E} = 43.63$ V/cm, the measured asymmetry is $A_{lr} = (2.935 \pm 0.337) \times 10^{-4}$. Th implies a relativistic M1 matrix element of $\bar{M} = -(0.2725 \pm 0.0313)\alpha^2 e\hbar/mc$,[30] in good agreement with the theoretical value $-0.2794\alpha^2 e\hbar/mc$. The result provides a direct test of the relativistic correction terms shown in Eq. (12), including the overall sign.

DISCUSSION

The major question that remains unanswered is the reason for the discrepancy between theory and experiment for the Lamb shift of He$^+$ illustrated in Fig. 7. The good agreement between theory and experiment for the other two distributions I_Γ and I_M can be regarded as rigorous tests of the equipment and method of analysis. However, they do not eliminate all possible sources of systematic error. The disagreement for the Lamb shift corresponds to a slight excess of intensity in the parallel direction if the theoretical value is taken as correct. A possible source of systematic arror in fact arises from stray charged particles accelerated by the quenching field into the detectors. To avoid this, the noise was defined as the quenching signal still remaining when the metastables were removed from the beam by prequenching. Also, the detectors were protected with windows of thin aluminum films, and the application of magnetic fields. However, mechanical strains in the windows may have introduced a polarization sensitivity which would bias the results. The possibility of an error from this effect is being further investigated.

A new high-precision measurement of the Lamb shift anisotropy for hydrogen is currently in progress. This will provide a direct comparison between H and He$^+$ by the same method of measurement. Subsequently, the He$^+$ experiment will be repeated using improved technology now available. The major source of error in traditional resonance measurements of the Lamb shift is the large level width of the $2p_{1/2}$ state relative to the Lamb shift, and the necessity of understanding the line profile in great detail. It is therefore important to have a completely independent method of measurement. The anisotropy method is largely independent of the level width, and provides a viable alternative to traditional microwave resonance techniques.

ACKNOWLEDGEMENTS

It is a pleasure to acknowledge the stimulation provided by Peter Farago during his visits to the University of Windsor in 1973 to 1976. This work was supported by the Natural Sciences and Engineering Research Council of Canada.

REFERENCES

1. G. W. F. Drake and R. B. Grimley, Photon-scattering theory of the quenching of hydrogenic metastables, *Phys. Rev. A* 8:157 (1973).
2. G. W. F. Drake, Spontaneous two-photon decay rates in hydrogenlike and heliumlike ions, *Phys. Rev. A* 34:2871 (1986).
3. G. W. F. Drake, Theory of relativistic magnetic dipole transitions: lifetime of the metastable $2\,^3S$ state of the helium-like ions, *Phys. Rev. A* 3:908 (1971).
4. D. M. Brink and G. R. Satchler, "Angular Momentum," Clarendon Press, Oxford (1968).
5. W. R. Ott, W. E. Kauppila, and W. L. Fite, Polarization of Lyman-α radiation emitted in electron collisions with hydrogen atoms and molecules, *Phys. Rev. A* 1:1089 (1970).
6. A. van Wijngaarden, J. Kwela, and G. W. F. Drake, Measurement of the $n = 2$ Lamb shift in He$^+$ by the anisotropy method, *Phys. Rev. A* 43:3325 (1991).
7. J. Patel, A. van Wijngaarden, and G. W. F. Drake, Improved anisotropy measurement of the Lamb shift in He$^+$, *Phys. Rev. A* 36:5130 (1987).
8. G. W. F. Drake, J. Kwela, and A. van Wijngaarden, He$^+$ $2p$ lifetime by a quenching-asymmetry measurement, *Phys. Rev. A* 46:113 (1992).
9. G. W. F. Drake, Influence of fringing fields on the field-induced quantum beats of atomic hydrogen, *J. Phys. B* 10:775 (1977).
10. A. van Wijngaarden, E. Goh, G. W. F. Drake, and P. S. Farago, Quantum beats in the electric-field quenching of metastable hydrogen, *J. Phys. B* 9:2017 (1976).

11. A. van Wijngaarden, G. W. F. Drake, and P. S. Farago, New method for Lamb shift measurements, *Phys. Rev. Lett.* 33:4 (1974).

12. G. W. F. Drake, A. van Wijngaarden, and P. S. Farago, On the anisotropy of the quenching radiation from metastable hydrogen and deuterium atoms, *in:* "Electron and Photon Interactions with Atoms", H. Kleinpoppen and M. R. C. McDowell, eds. Plenum, New York (1976).

13. A. van Wijngaarden and G. W. F. Drake, Deuterium Lamb shift via quenching-radiation anisotropy measurements, *Phys. Rev. A* 17:1366 (1978).

14. M. S. Dewey and R. W. Dunford, Lamb shift in singly ionized helium, *Phys. Rev. Lett.* 60:2014 (1988).

15. S. R. Lundeen and F. M. Pipkin, Separated oscillatory field measurement of the Lamb shift in H, $n = 2$, *Metrologia*, 22:9 (1986).

16. E. W. Hagley and F. M. Pipkin, Separated oscillatory field measurement of hydrogen $2S_{1/2}-2P_{3/2}$ fine structure interval, *Phys. Rev. Lett.* 72:1172 (1994).

17. E. Lipworth and R. Novick, Fine structure of singly ionized helium, *Phys. Rev.* 108:1434 (1957).

18. M. Narasimham and R. Strombotne, Lamb shift in singly ionized helium, *Phys. Rev. A* 4:14 (1971).

19. G. W. F. Drake, Quantum electrodynamic effects in few-electron atomic systems, *Adv. At. Mol. Phys.* 18:399 (1982).

20. P. Mohr, Self-energy correction to one-electron energy levels in a strong Coulomb field, *Phys. Rev. A* 46:4421 (1992).

21. K. Pachucki, Higher-order binding corrections to the Lamb shift, *Ann. Phys. (N.Y.)* 226:1 (1993).

22. K. Pachucki, Complete two-loop binding corrections to the Lamb shift, *Phys. Rev. Lett.* 72:3154 (1994).

23. M. I. Eides and V. A. Shelyuto, Corrections of order $\alpha^2(Z\alpha)^5$ to the hyperfine splitting and the Lamb shift, *Phys. Rev. A* 52:954 (1995).

24. M. Weitz, A. Huber, F. Schmidt-Kaler, D. Leibfried, W. Vassen, C. Zimmermann, K. Pachucki, and T. W. Hänsch, Precision measurement of the $1S$ Lamb shift in atomic hydrogen and deuterium by frequency comparison, *Phys. Rev. A* 52:2664 (1995).

25. L. Lundin, H. Oona, W. S. Bickel, and I. Martinson, Mean lives of the $2p$, $3p$, $4p$, $5p$, and $6p$ ^2P levels in He II, Phys. Scr. 2:213 (1970).

26. A. Gaupp, P. Kuske, and H. J. Andrä, Accurate lifetime measurements of the lowest $^2P_{1/2}$ states in neutral lithium and sodium, *Phys. Rev. A* 26:3351 (1982).

27. Z.-C. Yan and G. W. F. Drake, Theoretical lithium $2\,^2S \rightarrow 2\,^2P$ and $2\,^2P \rightarrow 3\,^2D$ oscillator strengths, *Phys. Rev. A* 52:R4316 (1995).

28. U. Volz and H. Schmoranzer, in "Proceedings of the 5th International Colloquium on Atomic Spectra and Oscillator Strengths for Astrophysical and Laboratory Plasmas", *Phys. Scr.* T-series (1996).

29. S. Cheng, R. W. Dunford, C. J. Liu, and B. J. Zabransky, M1 decay of the $2\,^3S_1$ state in heliumlike krypton, *Phys. Rev. A* 49:2347 (1994).

30. A. van Wijngaarden, J. Patel, and G. W. F. Drake, Asymmetry measurement of the relativistic magnetic-dipole matrix element in He$^+$, *Phys. Rev. A* 33:312 (1986).

DYNAMIC SPIN-ORBIT EFFECTS
IN SPONTANEOUS E1 TRANSITIONS

M. Brieger [1] and H.A. Schuessler [2]

[1]Institut für Technische Physik, DLR, Pfaffenwaldring 38-40,
 D-70569 Stuttgart, Germany
[2]Department of Physics, Texas A & M University, College Station,
 TX 77843, U.S.A.

INTRODUCTION

Despite undoubtedly many impressing successes our ability to comprehensively under-stand and explain nature in all its facets will remain limited as long as we have to resort to simplifying models which inherently can only deliver what has been built into. Despite their shortcomings, however, these models often have proven beneficial as guidelines and eye-openers to deeper insight and ensuing progress. One of these simplifications is the notion that dynamic features with strongly differing time scales can be treated independently. Presumably the most successful example of these adiabatic approximations is the one devised by Born and Oppenheimer[1]. Extended to the *adiabatic approximation* it describes the expectation values of diatomic molecules very accurately, except for the vicinity of curve-crossings. In atomic physics the notion of *a priori* fine-structure eigenstates was very successful in interpreting spectra. This success supported the perception that quantum systems exist in eigenstates in the first place. As a consequence, transient phenomena such as quantum interference effects were explained by a superposition of eigenstates. In this article, however, we offer an alternative approach and explanation. We will demonstrate that by rigorously solving the time-dependent Schroedinger equation in interaction representation with spin-orbit *and* electric dipole interac-tion treated *simultaneously* finer details of the dynamics of spin-orbit coupling in radiatively decaying states can be described. The treatment is being kept in the most general form in order to make it applicable to any interaction of the scalar-product type like, for instance, the hyperfine interaction. This way, it also can be easily extended to include, for example, the Zeeman effect[10]. Besides providing a physically illustrative interpretation of quantum inter-ference effects such as zero-field fine-structure beats in spontaneous E1 transitions and, at the same time, reproducing the generally accepted initial wave function after impulsive, e.g. beam-foil, excitation[2-7] our theory gives a dynamical explanation for the existence of stationary fine-structure eigenstates and in this regard provides new aspects for the occurrence of sensitized fluorescence[8,9]. Furthermore, it predicts several new phenomena. Due to space constraints, unfortunately, we can only present a sketch of the new ideas here. A comprehensive account on the new theory will be given elsewhere[10].

THEORY

Spin-orbit interaction[11] is a well established subject. Yet its dynamical implications apparently have not been fully realized. For pedagogical reasons let us first discuss these by using the following scenario: When it comes to dealing with strongly perturbed excited systems like in the early stage after electron capture[12] which is the final physical process in the beam-foil excitation of neutrals[2-7,13-22], no attempt has been made as yet to treat spin-orbit and electromagnetic interaction on the same footing. Instead, the establishment of a superposition of fine-structure eigenstates before any electromagnetic decay has been assumed as part of the whole process which is expressed by the generally accepted initial wave function[2-7]. This proves to be ingeniously correct but prevents insight into the spin-orbit dynamics as we will show. During the very short collision time in the order of 10^{-14} sec the individual potentials are extremely deformed under the transient cylindrical symmetry. This gives rise to a strong mixing of the orbital angular momenta in the collision compound. Therefore, as the collision partners recede and spherical symmetry is being restored the most affected electron not only has to readjust its orbit, but may have to readjust its spin-orbit interaction as well. Depending on how spin-sublevels are initially being populated this adjustment may either lead to fine-structure eigenstates or to what we call *precession states* which, when collectively in-phase, are known as coherent states.

It is this dynamical behavior that our new approach is addressing by treating spin-orbit *and* electric dipole interaction *simultaneously* using the outset of nonrelativistic time-dependent perturbation theory in interaction representation. By expanding the wave function of a one-electron system plus radiation field with respect to the wave functions of the free radiation field and hydrogenic wave functions, i.e with uncoupled spin and orbital angular momentum, for the atomic part and taking into account the spin-orbit interaction only between levels of the same principal quantum number a system of coupled differential equations of first order is obtained for the amplitudes. In the case of *spontaneous* electric dipole transitions this system of generally higher order breaks up into the following consecutively and rigorously solvable sets of pairs of coupled differential equations of first order

$$i\hbar \frac{\partial}{\partial t} a_{\beta l' M_j' - m_s; 1/2, +m_s; \lambda}(t) = \left[H^{SO}_{\beta M_j' - m_s; \beta M_j' - m_s} - \frac{i}{2}\Gamma_{\beta l' M_j' - m_s} \right] a_{\beta l' M_j' - m_s; 1/2, +m_s; \lambda}(t) +$$

$$+ H^{SO}_{\beta M_j' - m_s; \beta M_j' + m_s} a_{\beta l' M_j' + m_s; 1/2, -m_s; \lambda}(t) + \qquad (1)$$

$$+ i\frac{eA_o}{c} \sum_{\alpha l M_J - m_s} \omega_{\alpha l, \beta l'} \sqrt{\frac{\hbar}{2\omega_\lambda}} <\beta\, l' M_J' - m_s | \epsilon_\lambda^* \cdot r | \alpha\, l M_J - m_s > a_{\alpha l M_J - m_s; 1/2, +m_s; \mu}(t) e^{-i(\omega_{\alpha l, \beta l'} - \omega_\lambda)t}$$

As a matter of fact, Eq. (1) is the concise representation of the two-dimensional system of coupled differential equations of first order, as $m_s = \pm 1/2$, and with $M_J = m_l + m_s$ being a good quantum number under spin-orbit interaction. Eq. (1) holds for any step in a cascade. In this special example the decaying level $|\beta, l', M'_J - m_s, 1/2, +m_S >$ is being fed from higher levels $|\alpha, l, M_J - m_S, 1/2, +m_S >$ through the spontaneous emission of a photon with polarization ϵ_λ, while the latter were fed from yet even higher levels through emission of a photon ϵ_μ. In order to obtain the proper behavior of a decaying level[23] the equations had to be extended phenomenologically by the imaginary diagonal element $-i\Gamma(\beta, l', M'_J - m_s)/2$. This, however, is common practice[3-5,24,27-31] and it can also be shown that the phenomenological equations follow consistently from a fully quantum mechanical treatment[25,26,32] of the vacuum electromagnetic field. Here, however, we take this a step further in that we introduce an individual lifetime width for each spin-sublevel as the labeling indicates. This will provide the flexibility necessary in cases which we will encounter in $l \rightarrow l$ -1 transitions for $l \geq 2$ due to spin-orbit dynamics and, hence, call "dynamical". Although such a measure may seem unfamiliar if not "unnatural" we will see that selection rules for electric dipole transitions make it a necessary one. For the time being we note that the "natural" situation can be obtained any time by making the lifetime widths equal. An important key to the understanding of the spin-orbit dynamics is the operator itself. Like any scalar product-type operator acting under spherical symmetry, it can be decomposed into spherical tensor operators of which two are off-diagonal and act "ping-pong"-like while the third is diagonal, and g is the radially dependent coupling strength

$$H^{SO} = g(\mathbf{l} \cdot \mathbf{s}) = g[-l_{+1}s_{-1} - l_{-1}s_{+1} + l_o s_o] \tag{2}$$

Any wave function $|\alpha, l, M_J - m_s; 1/2, m_S >$ to be acted on by the coupling "ping-pong" part under conservation of $M_J = m_l + m_S$ must have the flexibility of being transformed into either of the two wave functions $|\alpha, l, M_J \mp 1/2; 1/2, \pm 1/2 >$, respectively. In the set of $2(2l+1)$ possible wave functions there are two, however, that do not have this flexibility because they have both spin *and* orbital angular momentum stretched out with maximal components into the same direction: $|\alpha, l, \pm l; 1/2, \pm 1/2 >$. As these levels which we call *stretched states* for distinction, are only affected diagonally by the spin-orbit interaction they are fine-structure eigenstates *per se* and can equally well be represented by $|\alpha, l, 1/2; J_+, \pm J_+ >$ where $J_\pm = l \pm 1/2$. As a matter of fact we note that ^2S states are *stretched states*. It is the different response to the spin-orbit operator that makes the dynamical behavior of the *stretched states* different: Since the off-diagonal matrix elements which couple spin-orbit interacting levels vanish for these states their dynamical behavior is governed by the following differential equation of *first* order

$$i\hbar \frac{\partial}{\partial t} a_{\beta l', \pm l'; 1/2, \pm 1/2, \lambda}(t) = \left[H^{SO}_{\beta, l', \pm l'; \beta, l', \pm l'} - \frac{i}{2}\Gamma_{\beta, l', \pm l'} \right] a_{\beta l', \pm l'; 1/2, \pm 1/2, \lambda}(t) + \tag{3}$$

$$+ i\frac{eA_o}{c} \sum_{\alpha l M_J} \omega_{\alpha l, \beta l'} \sqrt{\frac{\hbar}{2\omega_\lambda}} <\beta\, l', \pm l' |\epsilon^*_\lambda \cdot \mathbf{r} | \alpha\, lM_J \mp 1/2 > a_{\alpha l M_J \mp 1/2; 1/2, \pm 1/2, \mu}(t)\, e^{-i(\omega_{\alpha l, \beta l'} - \omega_\lambda)t}$$

Accordingly, the solution has only *one* time exponential making the population in a *stretched state* quasi-stationary.

In the case of the spin-orbit interacting spin-sublevels, on the other hand, the two-dimensional system of coupled linear differential equations of first order can be transformed into an inhomogeneous linear differential equation of *second* order for one of the amplitudes as given in Eq. (4).

$$\frac{\partial^2}{\partial t^2} a_{\beta l' M_J - m_s; 1/2, +m_s, \lambda}(t) + \frac{\partial}{\partial t} a_{\beta l' M_J - m_s; 1/2, +m_s, \lambda}(t) \frac{i}{\hbar} \left\{ H^{SO}_{\beta M_J - m_s; \beta M_J - m_s} - \frac{i}{2}\Gamma_{\beta l' M_J - m_s} + \right.$$

$$\left. + H^{SO}_{\beta M_J + m_s; \beta M_J + m_s} - \frac{i}{2}\Gamma_{\beta l' M_J + m_s} \right\} + a_{\beta l' M_J - m_s; 1/2, +m_s, \lambda}(t) \frac{1}{\hbar^2}\left\{ |H^{SO}_{\beta M_J - m_s; \beta M_J - m_s}|^2 - \right.$$

$$\left. - \left(H^{SO}_{\beta M_J - m_s; \beta M_J - m_s} - \frac{i}{2}\Gamma_{\beta l' M_J - m_s} \right)\left(H^{SO}_{\beta M_J + m_s; \beta M_J + m_s} - \frac{i}{2}\Gamma_{\beta l' M_J + m_s} \right) \right\} = \tag{4}$$

$$= i\frac{eA_o}{\hbar c} \sum_{\alpha, l, M_J} \frac{\omega_{\alpha l, \beta l'}}{\sqrt{2\hbar\omega_\lambda}} e^{-i(\omega_{\alpha l, \beta l'} - \omega_\lambda)t} \left\{ <\beta l' M_J' - m_s | \epsilon^*_\lambda \cdot \mathbf{r} | \alpha l M_J - m_s > \left[\left(H^{SO}_{\beta M_J + m_s; \beta M_J + m_s} - \right. \right. \right.$$

$$\left. - \frac{i}{2}\Gamma_{\beta l' M_J + m_s} - \hbar(\omega_{\alpha l, \beta l'} - \omega_\lambda) \right) a_{\alpha l M_J - m_s; 1/2, +m_s, \mu}(t) - i\hbar\frac{\partial}{\partial t} a_{\alpha l M_J - m_s; 1/2, +m_s, \mu}(t) \right] -$$

$$\left. - <\beta l' M_J' + m_s | \epsilon^*_\lambda \cdot \mathbf{r} | \alpha l M_J + m_s > H^{SO}_{\beta M_J - m_s; \beta M_J + m_s} a_{\alpha l M_J + m_s; 1/2, -m_s, \mu}(t) \right\}$$

For the *stretched states* Eq. (4) turns out to be the time differentiated form of Eq. (3). This is not a surprise because both equations are rooted in Eq. (1). By the way of writing out the equations so far we have tacitly assumed that the matrix elements are constant and refer to a spherically symmetric potential. This always holds for photon-induced excitations. In the case of electron capture which we keep discussing, time zero in the initial conditions for the solution of Eqs. (3) and (4) thus has to be an instant after the electron transfer when the collision partners have receded far enough for the potential to be again spherical and constant, at least in first approximation. Because of the very short collision time mentioned above this is not critical and we can assume that up to this time zero no radiative decay has taken place yet. Hence, at time zero we have $A_o = 0$ which makes Eq. (4) a homogeneous one. Introducing the initial conditions

$$a_{\alpha l M_J - m_s; 1/2; +m_s; o}(t=0) = B_{\alpha l M_J - m_s} \tag{5}$$

the solutions to Eq. (4) in this case read as follows

$$a_{\alpha l M_J - m_s; 1/2, +m_s; o}(t) = \tag{6}$$

$$\sum_{j=1}^{2} (-1)^j \frac{[i\hbar X_{j+1}^{(\alpha)} - H_{\alpha M_J - m_s; \alpha M_J - m_s}^{SO} + \frac{i}{2} \Gamma_{\alpha l M_J - m_s}] B_{\alpha l M_J - m_s} - H_{\alpha M_J - m_s; \alpha M_J + m_s}^{SO} B_{\alpha l M_J + m_s}}{i\hbar(X_-^{(\alpha)} - X_+^{(\alpha)})} e^{X_j^{(\alpha)} t}$$

where $j = 1, 2$ represent $J_- = l - 1/2$, $J_+ = l + 1/2$, respectively, as do the indexes $(-)$, $(+)$ and

$$X_j^{(\alpha)} = -\frac{\Gamma_{\alpha l M_J - m_s} + \Gamma_{\alpha l M_J + m_s}}{4\hbar} - \frac{i}{2\hbar} \left\{ H_{\alpha M_J - m_s; \alpha M_J - m_s}^{SO} + H_{\alpha M_J + m_s; \alpha M_J + m_s}^{SO} + \right. \tag{7}$$

$$\left. + (-1)^j \sqrt{\left(H_{\alpha M_J - m_s; \alpha M_J - m_s}^{SO} - H_{\alpha M_J + m_s; \alpha M_J + m_s}^{SO} - \frac{i}{2}[\Gamma_{\alpha l M_J - m_s} - \Gamma_{\alpha l M_J + m_s}]\right)^2 + 4|H_{\alpha M_J - m_s; \alpha M_J + m_s}^{SO}|^2} \right\}$$

Reducing the indices to the most significant one we introduce the following abbreviations

$$A_\alpha = \frac{\Gamma_{\alpha l M_J - m_s} + \Gamma_{\alpha l M_J + m_s}}{2\hbar}, \qquad a_\alpha = \frac{\Gamma_{\alpha l M_J - m_s} - \Gamma_{\alpha l M_J + m_s}}{2\hbar} \tag{8}$$

$$B_\alpha = \frac{H_{\alpha M_J - m_s; \alpha M_J - m_s}^{SO} + H_{\alpha M_J + m_s; \alpha M_J + m_s}^{SO}}{\hbar}, \qquad b_\alpha = \frac{H_{\alpha M_J - m_s; \alpha M_J - m_s}^{SO} - H_{\alpha M_J + m_s; \alpha M_J + m_s}^{SO}}{\hbar}$$

$$c_\alpha = 2 \frac{H_{\alpha M_J - m_s; \alpha M_J + m_s}^{SO}}{\hbar}, \qquad \tan \varrho_\alpha = \frac{-2 a_\alpha b_\alpha}{b_\alpha^2 - a_\alpha^2 + |c_\alpha|^2}, \qquad \tau_\alpha^2 = (b_\alpha^2 - a_\alpha^2 + |c_\alpha|^2)^2 + 4 a_\alpha^2 b_\alpha^2$$

which allow us to write Eq. (7) more compactly

$$X_j^{(\alpha)} = -\frac{1}{2}\left(A_\alpha - (-1)^j \sqrt{\tau_\alpha} \sin(\varrho_\alpha/2) + i\left[B_\alpha + (-1)^j \sqrt{\tau_\alpha} \cos(\varrho_\alpha/2)\right]\right) \tag{9}$$

With the spin-orbit matrix elements inserted B_α turns out to be the coupling strength. In the case of "dynamical" behavior the second imaginary part represents the effective fine-structure

428

splitting. It is reduced from the "natural" amount due to the lifetime difference in "dynamically" interacting sublevels. This is also seen in the second real part which, as $-\pi \le \varrho_\alpha \le 0$, causes for the M_J levels involved a lifetime reduction in the J_- state and a prolongation in the J_+ state.

For "naturally" behaving levels, i.e. when the spin-orbit interacting spin-sublevels have equal lifetimes, $a_\alpha = \varrho_\alpha = 0$, and Eqs. (7) and (9) simplify to

$$X_J^{(\alpha,n)} = -\frac{\Gamma_{\alpha l}}{2\hbar} - \frac{i}{4} g_{\alpha l} \hbar \left\{-1+(-1)^{l+J-1/2}(2l+1)\right\} = -\frac{\Gamma_{\alpha l}}{2\hbar} - i\,\Omega_{\alpha lJ} \qquad (10)$$

where $\Omega_{\alpha lJ}$ denotes the respective fine-structure level shifts for $J_\pm = l \pm 1/2$.

Being the solution to a homogeneous differential equation of *second* order the amplitude of Eq. (6) usually contains *two* different time exponentials, of course. This is a very important feature that results from the dynamics of the spin-orbit interaction and has not been produced by any theory[2-7,24-29,32-36] so far. However, for very specific initial populations, as will be explained below, one of the time exponentials vanishes at a time, but only in the "natural" case. In the so-called "dynamical" case which we encounter in $^2D \rightarrow \,^2P$ transitions, the different lifetimes in the spin-orbit interacting sublevels prevent the atomic wave function from being normalizable and representable in closed form, as we will explain below. On the other hand, $^2P \rightarrow \,^2S$ transitions always show a "natural" behavior. As a consequence, the forefactors of $B_{\alpha,l,M\pm m}$ in Eq. (6) for the amplitudes can be expressed by 3j-symbols, and the atomic part of the wave function is obtained in closed form

$$\Psi_o^{(n)}(t) = \sum_{\alpha,l,m_s} \sum_{J,M_J} B_{\alpha l M_J - m_s}\,(-1)^{l-1/2+M_J}\,\sqrt{2J+1} \begin{pmatrix} l & 1/2 & J \\ M_J - m_s & m_s & -M_J \end{pmatrix} |\alpha,l,1/2\,;J,M_J> \cdot$$
$$\cdot\, e^{-[\Gamma_{\alpha l}/2\hbar + i\Omega_{\alpha,l,J}]t} \qquad (11)$$

This is exactly the type of coherent wave function that had been postulated[2] and broadly used[3-7] for interpreting the observed quantum interference phenomena after beam-foil excitation[2-7,12-21] i.e. fine-structure quantum beats: Macek[2,5], in an effort to explain these, concentrated on the imaginary time exponential in his ingenious choice, but was not very specific about the basis set. Other authors either followed[6] his approach or recognized from simple reasoning[3,4,7] that the LS coupling scheme should apply. This led them to a wave function where only the initial amplitudes $B_{\alpha l M-m}$ were missing as compared with Eq. (11). Their presence, however, is very crucial: If they are proportional to one of two appropriate pairs of 3j-symbols, the *Clebsch-Gordan coefficients*, which are known to diagonalize the spin-orbit operator the summation over m_s does generate a fine-structure eigenstate, indeed.

From the very beginnings[34] the occurrence of quantum interference effects in transitions from coherently excited states was always explained along the same lines: Due to the composed nature of the coherent wave function with two or more different time exponentials of Schroedinger-type eigenfunctions as in Eq. (11) the probability for a transition from such a coherent state into a single lower state always shows cross-terms in the imaginary time exponentials giving rise to intensity modulations in the emitted radiation. In order to illustrate this rather abstract, only mathematically based explanation it became common practice[27,35-37] to resort to an analogy between wave mechanics and optics by invoking Young's double slit experiment. This also holds for the time-integrated companion of the quantum beat, the level-crossing[27,38]. In the analogy, two coherently populated, nondegenerate levels are mimicking the slits. The question, however, whether a coherent wave function could represent something physically real, remained unanswered. In hindsight, from the point of view of this article, there is no doubt about this failure: That we arrived with Eq. (11) at the same type of wave function that in the past only has been obtained by an expansion with respect to fine-structure eigenfunctions is owed to the unforeseeable fact that the *two* different time exponentials in the amplitudes of Eq. (6) rearrange in building the wave function of Eq. (11) from the uncoupled hydrogenic wave functions in such a way that it seems as if an expansion with respect to single time exponential, Schroedinger-type fine-structure eigenfunctions had been the starting point as usual. This starting point proves to be inappropriate for representing the dynamics of spin-orbit interaction as we will show below.

Hitherto only guessed, the true meaning of this coherent wave function and a clear and illustrative picture of the dynamics involved is obtained when we look at the populations in the spin-sublevels of the initial state. From the absolute square of the amplitudes in Eq. (6) we deduce for "naturally" behaving *precession states* in an αl state

$$
|a^{(n)}_{\alpha l M_J - m_s; 1/2, + m_s; 0}(t)|^2 = e^{-\frac{\Gamma_{\alpha l}}{\hbar} t} \left\{ B^2_{\alpha l M_J - m_s} - \frac{\sqrt{(2l+1)^2 - (2M_J)^2}}{2l+1} [1 - \cos(\Omega_{\alpha lJ, \alpha lJ'} t)] \cdot \right.
$$

$$
\left. \cdot \left[\frac{\sqrt{(2l+1)^2 - (2M_J)^2}}{2(2l+1)} (B^2_{\alpha l M_J - m_s} - B^2_{\alpha l M_J + m_s}) - (-1)^{1/2 - m_s} \frac{2M_J}{2l+1} B_{\alpha l M_J - m_s} B_{\alpha l M_J + m_s} \right] \right\}
$$

(12)

It can be easily recognized from Eq. (12) that each spin-orbit interacting sublevel periodically exchanges part of its population with its counterpart, thus keeping the total population quasi-stationary. The amount of population that is being exchanged with the frequency equivalent $\Delta\Omega_{\alpha lJ}$ of the fine-structure splitting depends on the inclination $2M_J$ of the possible total angular momenta with respect to the quantization axis and on the initial populations of the spin-sublevels. For $M_J = \pm J_+$ no population is exchanged and we have a positively or negatively *stretched state*, respectively. For smaller inclinations the amount of population exchanged increases with $[1 - (2M_J/(2l+1))^2]$ reaching a maximum for $M_J = \pm 1/2$. At the same time the character of the spin-sublevels in terms of the fine-structure eigenfunctions changes from pure $|\alpha, l, 1/2; J_+, \pm J_+ >$ to a mixture containing $(1 \pm 2M_J/(2l+1))/2$ of $|\alpha, l, 1/2; J_\pm, M_J >$, respectively, and arrives at the maximal possible equivalence of the two fine-structure eigenfunctions i.e. the "slits", for $M_J = \pm 1/2$ as well. In either of these configurations the maximal quantum beat modulation can be expected, but they must not be equally populated or oscillating in-phase. The stationary i.e. the not exchanged part of the population determines the time averaged spin polarization as will be shown below. Eq. (12) expresses illustratively how the spin-orbit operator of Eq. (2) acts: While the "ping-pong" part is responsible for the exchange the diagonal part determines the stationary amount of population in each *precession state*.

As the sublevels are identified particularly by their magnetic quantum numbers which represent the projections of the orbital and spin angular momenta, respectively, onto the quantization axis, the exchange of population according to Eq. (12) can be interpreted as how quantum mechanics express the classical precession of the interacting angular momenta about their possible resultants. That is why the states that are made up from these population ex-changing sublevels have been called *precession states*. Although it had been intuitively concluded earlier[39] that coherent wave functions are related to precessing angular momenta, it has not been proven before. If the various *precession states* are differently populated and/or precess collectively with suitable phase relations due to e.g. impulsive excitation we have what is commonly known as a coherent state and quantum beats will be observable.

The most important result[10] is, however, that the exchange of population always dis-appears whenever the initial amplitudes of the spin-sublevels are proportional to the Clebsch-Gordan coefficients as has been mentioned in context with Eq. (11) and is easily verifiable in Eq. (12). This disappearance does not mean that the exchange back and forth between the spin-sublevels has stopped. Rather, it just means that taking and giving has become balanced between the two sublevels at any time making either population look stationary. Since the *Clebsch-Gordan coefficients* are the elements of a transformation matrix that diagonalizes the

430

spin-orbit interaction, *our result demonstrates that diagonalization covers only one aspect of the physical reality because it selects from the wide range of possible initial sublevel amplitudes only the two, so-to-say quantized ratios of sublevel amplitudes that generate stationary fine-structure eigenstates. It also demonstrates that an expansion with respect to eigenfunctions obtained from the diagonalization cannot fully describe the physical reality.* Therefore, the most general solution to the problem of describing a system under spin-orbit interaction are the *precession states*, indeed. Their unbalanced precession can be interpreted as a virtual nutation making necessary more than one angular momentum for describing the state. Furthermore, we will show below that even though in *precession states* neither the total angular momentum nor the energy is sharply defined, any transition, be it between *precession states* or into or from fine-structure eigenstates, may take place through any of the available fine-structure decay channels. This is giving rise to the occurrence of the known fine-structure lines despite the uncertainty in energy and angular momentum. The reason for this to happen is that the photon can carry away only a definite amount of angular momentum. Also, if there is more than one decay channel open from an excited fine-structure eigenstate into a lower doublet state the system will make a spontaneous transition into a lower *precession state* and not into either one of the eigenstates. This is related to the change of coupling strength in the transition and the adjustments the electron has to make.

From the populations given in Eq. (12) we can easily determine the dynamical spin polarization produced by the precessing angular momenta in the excited *precession state*

$$
P_{\alpha l M_J}(t) = \frac{B^2_{\alpha l M_J - 1/2} - B^2_{\alpha l M_J + 1/2}}{B^2_{\alpha l M_J - 1/2} + B^2_{\alpha l M_J + 1/2}} - \sqrt{1 - \left(\frac{2M_J}{2l+1}\right)^2} \; [1 - \cos(\Delta\Omega_{\alpha l J} t)] \cdot
$$

$$
\cdot \left[\sqrt{1 - \left(\frac{2M_J}{2l+1}\right)^2} \frac{B^2_{\alpha l M_J - 1/2} - B^2_{\alpha l M_J + 1/2}}{B^2_{\alpha l M_J - 1/2} + B^2_{\alpha l M_J + 1/2}} - 2 \frac{2M_J}{2l+1} \frac{B_{\alpha l M_J - 1/2} B_{\alpha l M_J + 1/2}}{B^2_{\alpha l M_J - 1/2} + B^2_{\alpha l M_J + 1/2}} \right]
$$

(13)

If the lifetime is long compared to the period of precession the time-averaged dynamical spin polarization is obtained from Eq. (13) as

$$
< P_{\alpha l M_J}(t) >_t =
$$

(14)

$$
= \left(\frac{2M_J}{2l+1}\right)^2 \frac{B^2_{\alpha l M_J - 1/2} - B^2_{\alpha l M_J + 1/2}}{B^2_{\alpha l M_J - 1/2} + B^2_{\alpha l M_J + 1/2}} + 2 \frac{2M_J}{2l+1} \sqrt{1 - \left(\frac{2M_J}{2l+1}\right)^2} \frac{B_{\alpha l M_J - 1/2} B_{\alpha l M_J + 1/2}}{B^2_{\alpha l M_J - 1/2} + B^2_{\alpha l M_J + 1/2}}
$$

Inserting the eigenstates generating initial amplitudes into Eqs. (13) or (14) yields the time-independent spin polarization of the fine-structure eigenstates $|\alpha, l, 1/2; J_{\pm} M_J>$

$$
P_{\alpha l 1/2 J_{\pm} M_J} = \pm \frac{2M_J}{2l+1}
$$

(15)

According to Eq. (15) the J_+ and J_- eigenstate components in the wave function of a *precession state* have opposite spin polarizations. Therefore the spin polarization in a *precession state* is always smaller than in an eigenstate[10].

For producing a fine-structure eigenstate, as we saw before, a delicate balance of the sublevel populations has to be generated and maintained during the lifetime. If this balance is destroyed, for instance due to the rearrangement after inelastic collisions, *precession states* will be generated and sensitized fluorescence[8,9] emitted. This phenomenon occurs when in spite of an excitation via a selected, well resolved fine-structure or hyperfine-structure transition into e.g. *Rb 11d $^2D_{5/2}$* by Doppler-free two-photon excitation the "forbidden" line *11d $^2D_{3/2}$ → 5p $^2P_{1/2}$* is observed with increasing intensity as a function of the *Rb* vapor pressure[40] i.e. collision frequency. The effect is also observed as a function of an added amount of gas[8,9]. Although these thermal collisions are slower by some orders of magnitude, they are nevertheless much faster then the radiative decay in atoms. However, as the collisions take place randomly no phase relation exists between the collision-induced *precession states* and only the "forbidden" line but no beats can be observed.

All these clues which will be corroborated by further results farther below indicate that "quantum interference" effects in atoms like quantum beats are generated by the dynamics of angular momentum coupling. This interpretation is beyond the scope of the density matrix formalism[26,28,30,31,36,41] because its current form is based on the expansion with respect to fine-structure eigenfunctions which we have proven to be not capable of describing the dynamics of spin-orbit coupling.

The coherent wave function of Eq. (11) which we prefer to call the wave function of a *spin-orbit state* for distinction, is actually composed of two dynamically different types of wave functions: Due to the summation over M_J the quasi-stationary wave functions $|\alpha,l,1/2;J_+,\pm J_+\rangle$ of the *stretched states* are present for $M_J = \pm J_+$ and the wave functions of the *precession states* for $|M_J| < J_+$. How much each of these contribute to the total wave function of the *spin-orbit state* depends on their initial populations. In an unspecific excitation process it may even happen with a finite probability that for some $M_J \neq \pm J_+$ the initial populations in the spin-sublevels meet the criterion for generating fine-structure eigenstates.

After all what we have discussed so far our conclusion is that the unique fingerprints of precessing angular momenta mandate a totally new dynamical perception of the electromagnetic decay. This will be shown next. Figure 1 depicts a modified level scheme of a *d-p-s* cascade where the spin-orbit interacting sublevels pertaining to the same M_J are indicated by back and forth swinging arrows. The most common type of cascade, with converging magnetic sublevels, has been chosen because all the peculiarities with respect to differing lifetimes in spin-sublevels only occur in *l → l -1* transitions for $l \geq 2$ as will be discussed below. In *l → l+1* transitions the system always behaves "naturally". The figure shows in detail how spin-orbit dynamics and selection rules affect the individual transition probabilities. Postponing the gene-

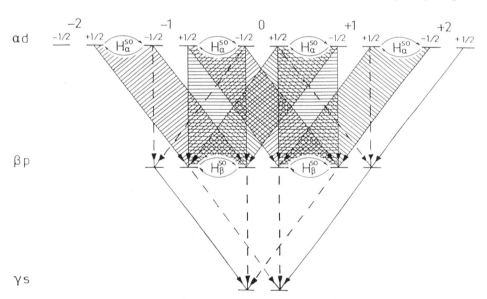

Figure 1. Modified levels scheme of a *d-p-s* cascade. Spin-orbit interacting sublevels pertaining to the same M_J are indicated by swinging arrows. Hatching marks transitions that occur between upper and lower *precession states*, dashed arrows indicate transitions between an upper *precession state* and a lower *stretched state* while solid arrows represent transitions between *stretched states*. The initial *d*-population has been assumed to be spin-up.

ral formulae until later we will first discuss some features encountered in the simpler case of the $^2P \rightarrow {}^2S$ transitions : Depending on the value of M_J in the ^2P state and with ^2S, as mentioned earlier, being a *stretched state* , these transitions take place either between *precession states* and *stretched states* (for $| M_J | < 3/2$, dahed arrows) or among *stretched states* (for $| M_J | = 3/2$, solid arrows). In either case selection rules for electric dipole transitions mandate that the spin component be conserved. This is always fulfilled for the stationary i.e. spatially fixed spins in transitions between *stretched states* and the transition probabilities comply with the "Golden Rule". Also, there is a total transfer of the individual spin polarization in this case. This does not hold true for the other type of transitions, neither with respect to the "Golden Rule" nor with respect to the spin polarization. With the precessing spin in the upper ^2P state the transition probability becomes dependent on the relative orientational matching with the spatially fixed spin orientation in the lower ^2S state. For giving an example we have calculated the probability for the transition from inititially assumed spin-up and equally populated sublevel sets, i.e. *precession states,* into a negatively *stretched state*. This applies also to ^2S and yields a *dynamically reduced transition probability* which, expressed in general terms, reads

$$| \overline{a}_{\beta l', -l', 1/2, -1/2} |^2 = \frac{e^2 A_o^2}{6 \pi \hbar c^5} \delta_{l', l-1} \frac{(2l'+2)(2l'+4)}{2l'+3} \left\{ \begin{matrix} 1 & l & l'+1 \\ 1 & l' & l \end{matrix} \right\}^2 | B_{\alpha l m} |^2 | < \beta \, l' \| r \| \alpha \, l > |^2 \cdot$$

$$\cdot \, \omega_{\alpha l, \beta l'}^2 \frac{\omega_{\alpha l, \beta l'} + \frac{1}{2}(\Omega_{\alpha l}^+ + \Omega_{\alpha l}^-) - \Omega_{\beta l'}^+}{\Gamma_{\alpha l} / \hbar} \cdot \frac{\left(\dfrac{\Omega_{\alpha l}^+ - \Omega_{\alpha l}^-}{\Gamma_{\alpha l} / \hbar} \right)^2}{1 + \left(\dfrac{\Omega_{\alpha l}^+ - \Omega_{\alpha l}^-}{\Gamma_{\alpha l} / \hbar} \right)^2} \tag{16}$$

Including the density of final states of the free radiation field Eq. (16) is obtained from the absolutely squared amplitude of Eq. (17) below after integrating over all emission frequencies, directions, and polarizations. Holding for times long compared with the lifetime it represents the asymptotic transition probability for $t \rightarrow \infty$. Apart from the absolutely squared transition matrix element and a correction to the dominant ($\omega_{\alpha l, \beta l'}$)^3 term it features the reduction factor with the ($\hbar \Delta \Omega_{\alpha l J}$ / $\Gamma_{\alpha l}$)^2 terms. This reduction factor is a measure of how often within a lifetime the system has the chance to decay into the respective *stretched state* : For $\hbar \Delta \Omega_{\alpha l J}$ $<<\Gamma_{\alpha l}$ the transition probability tends toward zero because the electric dipole decay is much faster than the precession can swing the spin from the up- into the down-orientation. In the opposite case the reduction factor tends toward unity.

This, however, is only one face of the coin. The transition probability into the postively *stretched state* compensates the reduction term making the overall transition probability into the ^2S state comply with Fermi's "Golden Rule". This holds true generally independent of the initial spin orientation. All of this can be easily understood from Eq. (13) which tells us that the precession in an upper ^2P *precession state* makes the spin bounce back and forth between the two spatially fixed orientations of the lower ^2S *stretched states*. Whenever the spin starts matching either of the allowed orientations the enhanced population makes the transition more probable by the same amount. The equality of the transition probabilities from both spin-sublevels is the essential criterion determining the behavior in these *precession states* as "natural". From Fig. 1 we also learn that if the two ^2P *precession states* were equally populated *and* precessing in-phase any linearly polarized detection parallel or perpendicular to the quantization axis would not observe any modulation because the contributions from the respective sublevels to either polarization are exactly out of phase. Thus, even with equally populated sublevel sets quantum beats would be observable provided the respective *precession states* do not precess in-phase and such a state can be prepared.

Continuing our qualitative discussion of Fig. 1 we now will learn that there are two types of $^2D \to {}^2P$ transitions: One type we encounter there are transitions between *precession states* which are enhanced in Fig. 1 by hatching. This is to indicate that in these transitions both spin-sublevels in each upper *precession state* have the same transition probability into the lower *precession state,* though not necessarily simultaneously. Besides the *precession states,* however, the lower ^2P *spin-orbit state* also comprises his two *stretched states.* Due to the conservation of the spin-orientation in electric dipole transitions only one of the upper spin-sublevels can populate either one of these radiatively. This is indicated in Fig. 1 by dashed arrows. Thus, for only one of the two spin-sublevels in each upper *precession state* involved an additional decay channel is open. This, however, must result in a different i.e. shorter lifetime and has been accounted for by allowing for different lifetimes of the spin-orbit interacting sublevels. This necessary additional flexibility is now being justified belatedly. The additional radiative population sink makes the precession unbalanced under all circumstances. This matters especially in cases where the balance counts, namely for the fine-structure eigenstates. Thus, for an initially prepared pure ^2D_j fine-structure eigenstate our theory predicts that it will develop into a *precession state* in the course of its life. According to Eq. (9), the radiatively unbalanced precession not only leads to a reduced fine-structure splitting but also to a shorter / longer lifetime of the $J_+ = l + 1/2$ / $J_- = l - 1/2$ eigenstate component, respectively, as compared with the "natural" case. All the phenomena, however, pertaining to differing lifetimes in spin-orbit interacting sublevels are predicted to occur only in $l \to l - 1$ transitions for $l \geq 2$. In $l \to l+1$ transitions they cannot occur because, due to selection rules, *precession states* can only decay into *precession states* but not into *stretched states* in this case and, therefore, we always have "natural" behavior.

The general case from which above results were deduced and which holds true for any state is obtained by solving Eq. (4) with the preceding states' amplitudes, Eq. (6), inserted

$$a_{\beta l'Mj - m_s;1/2,+m_s;\lambda}(t) = i \frac{eA_o}{c} \sum_{j,k=1}^{2} \sum_{\alpha l M_J} \frac{\hbar \omega_{\alpha l,\beta l'}}{\sqrt{2\hbar\omega_\lambda}} \cdot \frac{(-1)^k M_{\beta l'Mj - m_s;\alpha l M_J;\lambda}^{(j,k)}}{i\hbar(X_j^{(\alpha)} - X_k^{(\beta)}) + \hbar(\omega_{\alpha l,\beta l'} - \omega_\lambda)} \cdot$$

$$\cdot \left[e^{[X_j^{(\alpha)} - i(\omega_{\alpha l,\beta l'} - \omega_\lambda)]t} - e^{X_k^{(\beta)}t} \right] \quad (17)$$

Here, according to Eq. (7), $X_j^{(\alpha)}$ and $X_k^{(\beta)}$ contain the dynamical spin-orbit parameters of the initial and intermediate/final states, respectively, and, in shortened notation, $M_{\beta,\alpha}^{(j,k)}$ is the

$$M_{\beta l'Mj - m_s;\alpha l M_J,\lambda}^{(j,k)} = \frac{i\hbar X_{k+1}^{(\beta)} - H_{\beta Mj - m_s;\beta Mj - m_s}^{SO} + \frac{i}{2}\Gamma_{\beta l'Mj - m_s}}{i\hbar(X_-^{(\beta)} - X_+^{(\beta)})} \cdot \quad (18)$$

$$\cdot <\beta,l',M_J' - m_s,1/2,+m_s | \boldsymbol{\epsilon}_\lambda^* \cdot \boldsymbol{r} | \alpha,l,M_J - m_s,1/2,+m_s > C_j^{(\alpha)}(+m_s) -$$

$$- \frac{H_{\beta Mj - m_s;\beta Mj + m_s}^{SO}}{i\hbar(X_-^{(\beta)} - X_+^{(\beta)})} <\beta,l',M_J' + m_s,1/2,-m_s | \boldsymbol{\epsilon}_\lambda^* \cdot \boldsymbol{r} | \alpha,l,M_J + m_s,1/2,-m_s > C_j^{(\alpha)}(-m_s)$$

transition matrix element in which $C_j^{(\alpha)}(m_s)$ are the factors in front of the time exponentials in Eq. (6). Although superfluous, the individual transition matrix elements between spin-sublevels in Eq. (18) display the respective spin function for clarity reasons. *Only* in the case of "natural" behavior in the upper *and* lower levels the factors in front of the individual transition matrix elements as well as the analog expressions in $C_j^{(\alpha)}(m_s)$ can be related to 3j-symbols leading to the fine-structure E1 transition matrix element as found in the following atomic wave function of the intermediate state after the first step in a cascade which also would be the final one if $\Gamma_{\beta l'} = 0$.

$$\Psi_{\beta,l',M_J}^{(n)}(t) = i\frac{eA_o}{c}\sum_{\alpha l J M_J}\sum_{J'}\sum_{\tilde{m}_s}(-1)^{l-1/2+M_J}\sqrt{2J+1}\begin{pmatrix}l & 1/2 & J \\ M_J-\tilde{m}_s & \tilde{m}_s & -M_J\end{pmatrix}B_{\alpha l M_J-\tilde{m}_s}\frac{\hbar\omega_{\alpha l,\beta l'}}{\sqrt{2\hbar\omega_\lambda}}\cdot$$

$$\cdot\frac{<\beta,l',1/2;J',M_J'|\epsilon_\lambda^*\cdot r|\alpha,l,1/2;J,M_J>}{\hbar(\Omega_{\alpha lJ}+\omega_{\alpha l,\beta l'}-\Omega_{\beta l'J'}-\omega_\lambda)-\frac{i}{2}\Delta\Gamma_{\alpha l,\beta l'}}|\beta,l',1/2;J',M_J'>\cdot$$

$$\cdot\left\{e^{-[\Gamma_{\alpha l}/2\hbar+i(\Omega_{\alpha lJ}+\omega_{\alpha l,\beta l'}-\omega_\lambda)]t}-e^{-[\Gamma_{\beta l'}/2\hbar+i\Omega_{\beta l'J'}]t}\right\}\tag{19}$$

Eq. (19) has some remarkable features: If the initial population in the spin-sublevels is prepared such as to make the upper *precession state* a stationary fine-structure eigenstate the summation over $\tilde{m}_s$ of the squares of *Clebsch-Gordan coefficients* produces a Kronecker symbol that removes the sum over J and M_J by letting survive only the term involving this specific fine-structure eigenstate. The summation over J', however, is being left unaffected. As a consequence, even if the initial upper state is an eigenstate the transitions lead into lower *precession states* as long as all decay channels are open as mentioned earlier. This is being supported by the denominator under the fine-structure transition matrix element in Eq. (19) which provides the decay channels on the frequency scale as will be shown below. Only if selection rules prevent the fine-structure transition matrix element from keeping open both decay channels the transition ends in a lower eigenstate. Thus, the transition from a ^2S state into the states ^2P_J, $M_J = \pm 1/2$ is always resulting in lower *precession states*. For $M_J = \pm 3/2$, on the other hand, we have the transitions into the *stretched states* of $^2P_{3/2}$ which are eigenstates. The former process is the origin for the occurrence of Λ-type[41] quantum beats in $^2S \rightarrow {}^2P$ transitions where the short excitation pulse only serves to set time zero for the start of the decay of all excited atoms.

The expansion coefficients in the wave function of Eq. (19), if taken in the conventional sense as population amplitudes for the coherent fine-structure states, are similar to those first given by Weisskopf and Wigner[33] in connection with their work on natural line widths and cascading transitions and later used by Macek[2] as a starting point when dealing with cascades as well. $\Delta\Gamma_{\alpha l,\beta l'} = \Gamma_{\alpha l} - \Gamma_{\beta l'}$ is typical[33] of this process. However, since the spin-orbit interaction had not been included explicitly the initial amplitudes and the additional fine-structure shifts are missing. The fine-structure shifts, as we saw, in both the upper and lower states are essential for accounting for the spin-orbit dynamics in cascading transitions. This can be understood when we take a look at the transition probability. This quantity is obtained in our treatment by absolutely squaring the amplitude of Eq. (17) and multiplying it with the density of final states of the free radiation field[42]

$$|a_{\beta l'M_J-m_s;1/2,+m_s;\lambda}(t)|^2\varrho(\omega_\lambda)d\omega_\lambda d\Omega =\tag{20}$$

$$=\frac{e^2A_o^2}{16\hbar\pi^3c^5}\sum_{j;k=1}^{2}\sum_{\tilde{j};\tilde{k}=1}^{2}\sum_{\alpha lM_J}\sum_{\tilde{\alpha}\tilde{l}\tilde{M}_J}\frac{\omega_{\alpha l,\beta l'}\,\omega_{\tilde{\alpha}\tilde{l},\beta l'}\,\omega_\lambda\,(-1)^{k+\tilde{k}}\,M_{\beta l'M_J-m_s;\alpha lM_J;\lambda}^{(j,k)}\,M_{\beta l'M_J-m_s;\tilde{\alpha}\tilde{l}\tilde{M}_J;\lambda}^{(\tilde{j},\tilde{k})*}}{\left[i(X_j^{(\alpha)}-X_k^{(\beta)})+(\omega_{\alpha l,\beta l'}-\omega_\lambda)\right]\left[i(X_j^{(\tilde{\alpha})}-X_k^{(\beta)})+(\omega_{\tilde{\alpha}\tilde{l},\beta l'}-\omega_\lambda)\right]^*}\cdot$$

$$\cdot\left[e^{[X_j^{(\alpha)}-i(\omega_{\alpha l,\beta l'}-\omega_\lambda)]t}-e^{X_k^{(\beta)}t}\right]\left[e^{[X_j^{(\tilde{\alpha})}-i(\omega_{\tilde{\alpha}\tilde{l},\beta l'}-\omega_\lambda)]t}-e^{X_k^{(\beta)}t}\right]^* d\omega_\lambda d\Omega$$

In Eq. (20) the third line is responsible for the time behavior. Starting with this one we recognize that it contains three types of terms: First and third, exponentials with arguments containing terms of the type

$$\left(X_j^{(\alpha)} + X_j^{(\tilde{\alpha})*}\right)t = -\frac{1}{2}\left(A_\alpha + A_{\tilde{\alpha}} - (-1)^j\sqrt{\tau_\alpha}\,\sin\,(\varrho_\alpha/2) - (-1)^{\tilde{j}}\sqrt{\tau_{\tilde{\alpha}}}\,\sin\,(\varrho_{\tilde{\alpha}}/2)\right] +$$

$$+ i\left[B_\alpha - B_{\tilde{\alpha}} + (-1)^j\sqrt{\tau_\alpha}\cos\,(\varrho_\alpha/2) - (-1)^{\tilde{j}}\sqrt{\tau_{\tilde{\alpha}}}\cos\,(\varrho_{\tilde{\alpha}}/2)\right]\right)t \tag{21}$$

and its complex conjugate where for the excited states $\alpha = \tilde{\alpha}$ or $\alpha \neq \tilde{\alpha}$ while for the lower state $\alpha = \tilde{\alpha} = \beta$ and where, using Eqs. (8,9) for the general case, the first term of the real part represents the arithmetic average of all lifetime widths involved while the second one plays a role only in the "dynamical" case when differing lifetime widths in spin-sublevels matter. For $\alpha = \tilde{\alpha} (= \beta)$ the latter term either cancels (for $j \neq \tilde{j}$) as does the imaginary part (for $j = \tilde{j}$) or adds up (for $j = \tilde{j}$) while the imaginary part (for $j \neq \tilde{j}$) becomes equal to the effective spin-orbit splitting producing either V- or Λ-type[41] (see below) fine-structure quantum beats due to precession. The second type of terms shows exponentials with arguments containing

$$\left(X_j^{(\alpha)} + X_k^{(\beta)*}\right)t = -\frac{1}{2}\left(A_\alpha + A_\beta - (-1)^j\sqrt{\tau_\alpha}\,\sin\,(\varrho_\alpha/2) - (-1)^k\sqrt{\tau_\beta}\,\sin\,(\varrho_\beta/2)\right. +$$

$$+ i\left[B_\alpha - B_\beta + (-1)^j\sqrt{\tau_\alpha}\cos\,(\varrho_\alpha/2) - (-1)^k\sqrt{\tau_\beta}\cos\,(\varrho_\beta/2)\right]\right)t \tag{22}$$

and its complex conjugate and couples the time evolutions of the upper and lower states. The second imaginary part in Eq. (22) represents the difference of the effective fine-structure split-tings of the upper and lower states involved, so the total imaginary time exponentials would oscillate with the mismatch between the actual transition frequency ω_λ and the resonance frequency $\omega_{\alpha,l,J;\beta,l',J'}$. As the transition probability of Eq. (20) contains summations over all possible decay channels resonance can be achieved in only one channel at best leaving the other channels heavily oscillating. These channels, however, as we will see when discussing the second term in Eq. (20) describing the spectral dependences, cannot contribute much. Nevertheless, a more comprehensive physical insight can be obtained by separating the different contributions: As B_α and B_β represent the respective coupling strengths which strongly increase with the depth in the potential, and the expressions with the square roots over τ_α and τ_β, respectively, the effective fine-structure splittings, the imaginary time exponentials could also be decomposed into superimposed modulations referring to, respectively, the abrupt change of the spin-orbit coupling strength in the transition, the difference or sum of the precession frequencies in the upper and lower levels, and the mismatch between the transition frequency ω_λ and the hydrogenic resonance frequency $\omega_{\alpha l,\beta l'}$. The first contribution is of nonadiabatic origin and causes building-up oscillations, the second one arises because as soon as population is building up in the lower levels their orbital angular momenta couple with the spin and start precessing. Depending on the direction of this precession relative to the one in the upper levels the spin orientations either run out of phase giving rise to a "low" frequency modulation, as is the case for precessions of the same direction, or, in the case of opposite directions, a "high" frequency modulation occurs. These modulations, however, are *not* detectable: Due to the jittering of the carrier frequency they are smeared out under the line profile. Upon frequency integration under the line profile they contribute[10] equally to the modulations caused by the precessions in the upper and lower *precession states* as given by Eq. (21). Depending on whether it is a transition from a *precession state* into an eigenstate or vice versa V-type[41] or Λ-type[41] quantum beats, respectively, are observed. In transitions between *precession states* a superposition of both is present. Fine-structure quantum beats, if not observed in Rydberg states, are usually in the THz region and hard to detect. This is not true in the presence of hyperfine interaction which makes the electronic total angular momenta along with the nuclear spin precess about the resulting atomic total angular momenta in the upper and lower states, respectively. Given the smallness of the hyperfine interaction, the fast precession of the *electronic* spin and orbital angular momentum about the *electronic* total angular momenta is practically fully decoupled from the slow global precession about the *atomic* total angular momenta. As, however, in this case the conservation of the components of electronic *and* nuclear spins is mandatory in electric dipole transitions it is the final slowest wheel in this transmission that determines the overall time behavior: While the matching of the electronic spins occurs, as described above, very often it takes a while for the nuclear spin to loop around and make the transition possible again. Because of this time scale which is enlarged by a factor of about 10^3 hyperfine-structure quantum beats are easily detectable.

Turning now to the second line in Eq. (20) where energy conservation determines the spectral distribution of the radiation emitted in the various transitions we decompose the relevant part into what is given in Eq. (23)

$$\frac{\omega_\lambda M^{(j,k)}_{\beta l'M_j-m_s;\alpha lM_j,\lambda}\, M^{(j,\tilde{k})*}_{\beta l'M_j-m_s;\tilde{\alpha}\tilde{l}\tilde{M}_j,\lambda}}{[i(X_j^{(\alpha)}-X_k^{(\beta)})+(\omega_{\alpha l,\beta l'}-\omega_\lambda)][i(X_j^{(\tilde{\alpha})}-X_k^{(\beta)})+(\omega_{\tilde{\alpha}\tilde{l},\beta l'}-\omega_\lambda)]^*}= \tag{23}$$

$$=\frac{\omega_\lambda\left\{\dfrac{\Delta\Gamma_{\alpha ljM_j,\beta l'kM_j}}{2\hbar}\cdot\dfrac{\Delta\Gamma_{\tilde{\alpha}\tilde{l}j\tilde{M}_j,\beta l'\tilde{k}M_j}}{2\hbar}+(\omega_{\alpha ljM_j,\beta l'kM_j}-\omega_\lambda)(\omega_{\tilde{\alpha}\tilde{l}j\tilde{M}_j,\beta l'\tilde{k}M_j}-\omega_\lambda)\right\}\Re\{M_{\beta\alpha}M^*_{\beta\tilde{\alpha}}\}}{\left\{\left(\dfrac{\Delta\Gamma_{\alpha ljM_j,\beta l'kM_j}}{2\hbar}\right)^2+(\omega_{\alpha ljM_j,\beta l'kM_j}-\omega_\lambda)^2\right\}\left\{\left(\dfrac{\Delta\Gamma_{\tilde{\alpha}\tilde{l}j\tilde{M}_j,\beta l'\tilde{k}M_j}}{2\hbar}\right)^2+(\omega_{\tilde{\alpha}\tilde{l}j\tilde{M}_j,\beta l'\tilde{k}M_j}-\omega_\lambda)^2\right\}}+$$

$$+\frac{\omega_\lambda\left\{\dfrac{\Delta\Gamma_{\tilde{\alpha}\tilde{l}j\tilde{M}_j,\beta l'\tilde{k}M_j}}{2\hbar}(\omega_{\alpha ljM_j,\beta l'kM_j}-\omega_\lambda)-(\omega_{\tilde{\alpha}\tilde{l}j\tilde{M}_j,\beta l'\tilde{k}M_j}-\omega_\lambda)\dfrac{\Delta\Gamma_{\alpha ljM_j,\beta l'kM_j}}{2\hbar}\right\}\Im\{M_{\beta\alpha}M^*_{\beta\tilde{\alpha}}\}}{\left\{\left(\dfrac{\Delta\Gamma_{\alpha ljM_j,\beta l'kM_j}}{2\hbar}\right)^2+(\omega_{\alpha ljM_j,\beta l'kM_j}-\omega_\lambda)^2\right\}\left\{\left(\dfrac{\Delta\Gamma_{\tilde{\alpha}\tilde{l}j\tilde{M}_j,\beta l'\tilde{k}M_j}}{2\hbar}\right)^2+(\omega_{\tilde{\alpha}\tilde{l}j\tilde{M}_j,\beta l'\tilde{k}M_j}-\omega_\lambda)^2\right\}}$$

Here $X_j^{(\alpha)}$ and $X_k^{(\beta)}$ have been expressed in analogy to Eq. (10) but with indicating the general "dynamical" case of Eq. (9) by the dependence on the orientation through the additional index M_j. Concentrating on the pure frequency expressions we make use of an expansion to partial fractions. In order to simplify the notation we contract the strings of indices into representative singles. Thus, the first frequency part in Eq. (23) can be represented by

$$\frac{\omega_\lambda[\gamma_\nu\gamma_{\tilde{\nu}}+(\omega_\nu-\omega_\lambda)(\omega_{\tilde{\nu}}-\omega_\lambda)]}{[\gamma_\nu^2+(\omega_\nu-\omega_\lambda)^2][\gamma_{\tilde{\nu}}^2+(\omega_{\tilde{\nu}}-\omega_\lambda)^2]}= \tag{24}$$

$$=\frac{\omega_\nu(\omega_{\tilde{\nu}}-\omega_\nu)-\gamma_\nu(\gamma_\nu+\gamma_{\tilde{\nu}})}{(\gamma_\nu+\gamma_{\tilde{\nu}})^2+(\omega_{\tilde{\nu}}-\omega_\nu)^2}\cdot\frac{\omega_\nu-\omega_\lambda}{\gamma_\nu^2+(\omega_\nu-\omega_\lambda)^2}+\frac{\omega_\nu\gamma_{\tilde{\nu}}+\gamma_\nu\omega_{\tilde{\nu}}}{(\gamma_\nu+\gamma_{\tilde{\nu}})^2+(\omega_{\tilde{\nu}}-\omega_\nu)^2}\cdot\frac{\gamma_\nu}{\gamma_\nu^2+(\omega_\nu-\omega_\lambda)^2}+$$

$$+\frac{\omega_{\tilde{\nu}}(\omega_\nu-\omega_{\tilde{\nu}})-\gamma_{\tilde{\nu}}(\gamma_{\tilde{\nu}}+\gamma_\nu)}{(\gamma_{\tilde{\nu}}+\gamma_\nu)^2+(\omega_\nu-\omega_{\tilde{\nu}})^2}\cdot\frac{\omega_{\tilde{\nu}}-\omega_\lambda}{\gamma_{\tilde{\nu}}^2+(\omega_{\tilde{\nu}}-\omega_\lambda)^2}+\frac{\omega_{\tilde{\nu}}\gamma_\nu+\gamma_{\tilde{\nu}}\omega_\nu}{(\gamma_{\tilde{\nu}}+\gamma_\nu)^2+(\omega_\nu-\omega_{\tilde{\nu}})^2}\cdot\frac{\gamma_{\tilde{\nu}}}{\gamma_{\tilde{\nu}}^2+(\omega_{\tilde{\nu}}-\omega_\lambda)^2}$$

It is obvious from Eq. (23) that the first frequency expression is even with respect to the interchange of the indices. And so is the substitute expression of Eq. (24), of course, as well as the real part of the product of matrix elements in Eq. (23). The imaginary part of this product and the second frequency expression are odd under this operation as is the corresponding substitute expression of Eq. (25)

$$\frac{\omega_\lambda[\gamma_{\tilde{v}}(\omega_v-\omega_\lambda)-(\omega_{\tilde{v}}-\omega_\lambda)\gamma_v]}{[\gamma_v^2+(\omega_v-\omega_\lambda)^2][\gamma_{\tilde{v}}^2+(\omega_{\tilde{v}}-\omega_\lambda)^2]}=\tag{25}$$

$$=\frac{\omega_v\gamma_{\tilde{v}}+\gamma_v\omega_{\tilde{v}}}{(\gamma_v+\gamma_{\tilde{v}})^2+(\omega_{\tilde{v}}-\omega_v)^2}\cdot\frac{\omega_v-\omega_\lambda}{\gamma_v^2+(\omega_v-\omega_\lambda)^2}-\frac{\omega_v(\omega_{\tilde{v}}-\omega_v)-\gamma_v(\gamma_v+\gamma_{\tilde{v}})}{(\gamma_v+\gamma_{\tilde{v}})^2+(\omega_{\tilde{v}}-\omega_v)^2}\cdot\frac{\gamma_v}{\gamma_v^2+(\omega_v-\omega_\lambda)^2}-$$

$$-\frac{\omega_{\tilde{v}}\gamma_v+\gamma_{\tilde{v}}\omega_v}{(\gamma_{\tilde{v}}+\gamma_v)^2+(\omega_v-\omega_{\tilde{v}})^2}\cdot\frac{\omega_{\tilde{v}}-\omega_\lambda}{\gamma_{\tilde{v}}^2+(\omega_{\tilde{v}}-\omega_\lambda)^2}+\frac{\omega_{\tilde{v}}(\omega_v-\omega_{\tilde{v}})-\gamma_{\tilde{v}}(\gamma_{\tilde{v}}+\gamma_v)}{(\gamma_{\tilde{v}}+\gamma_v)^2+(\omega_v-\omega_{\tilde{v}})^2}\cdot\frac{\gamma_{\tilde{v}}}{\gamma_{\tilde{v}}^2+(\omega_{\tilde{v}}-\omega_\lambda)^2}$$

According to Eqs. (24,25) the emission lineshape consists always of a superposition of dispersion and Lorentzian curves. In the case of transitions between fine-structure eigenstates straight and tilded indices are equal and we have to deal with only the first part of Eq. (23). Only in this case the lineshape is almost pure Lorentzian, the dispersion component being typically of the order of 10^{-7}. It can be shown[10] that Fermi's "Golden Rule" emerges from Eq. (20) for transitions between fine-structure eigenstates after integrating over frequencies, directions and polarizations. Depending on the magnitude of the imaginary part of the product of matrix elements the dispersion component can become quite substantial in beating transitions giving rise to a Fano-type[44] profile. Thus, the characteristically asymmetric frequency distribution may not only be a fingerprint of the interference between a discrete autoionizing state and the continuum[44] but of coherences in general. On the overall transition probability, however, which is obtained by integrating over all emission frequencies, directions, and polarizations the dispersion component has no influence. As the reduced transition probability of Eq. (20) is the result of this procedure the reduction factor is obtained by summing up the forefactors of the Lorentzian components for all possible decay channels.

CONCLUSION

Our new approach has proven capable of reconciling consistently all observed phenomena, be it quantum beats of any type or just "normal" transitions between fine-structure eigenstates, within one theory. Most importantly, without resorting to mystifying "quantum interference" the origin of quantum beats can be explained in physical terms as due to a dynamical response to selection rules of in-phase precessing angular momenta. Our theory also tells us that the stationary i.e. time-independent eigenstates are only one out of many possibilities of how the system can accommodate energy. Any average energy between the limiting fine-structure eigenstates is possible, it only depends on how the initial populations of the spin-substates have been prepared. In this respect only the duration of the excitation process matters, not the way it was accomplished. Therefore, any pulsed excitation, be it by beam-foil, photons, ion or electron impact[45], generates phase related *precession states* provided there is a phase relation in the groundstate and the length of an energetically quasi-resonant pulse is shorter than the period of the precessions so that the concomitant Fourier transform of its pulse length provides the necessary spread about the mean energy. As a matter of fact, we have demonstrated that the classical principles of angular momentum coupling are not obscured by "wave mechanics" but clearly make themselves known through quantum beats.

ACKNOWLEDGEMENT

We wish to thank Professor H. Opower for encouragement and support.

References:

1. M. Born and R. Oppenheimer, *Ann. Physik* 84:457 (1927).
2. J. Macek, *Phys. Rev.* 23:1 (1969); *Phys.Rev.* A1:618 (1970).
3. H.J. Andrä, *Phys. Rev. Lett.* 25:325 (1970); *Phys. Rev.* A2:2200 (1970); *Nucl. Instr.* Methods 90:343 (1970); *Physica Scripta* 9:257 (1974).
4. I.A. Sellin, J.A. Biggerstaff, and P.M. Griffin, *Phys. Rev.* A2:423 (1970).
5. J. Macek and D.H. Jaecks, *Phys. Rev.* A4:2288 (1971).
6. D.J. Burns and W.H. Hancock, *Phys. Rev. Lett.* 27:370 (1971); *J. Opt. Soc. Am.* 63: 241 (1973).
7. H.G. Berry and J.L. Subtil, *Phys. Rev. Lett.* 27:1103 (1971); H.G. Berry, J.L. Subtil, and M. Carré, *J. de Phys.* 33:947 (1972).
8. D.A. McGillis and L. Krause, *Phys. Rev.* 153:44 (1967)
9. A. Gallagher, *Phys. Rev.* 172:88 (1968)
10. M. Brieger, to be published
11. O. Haxel, J.H.D. Jensen, and H.E.Suess, *Z.Physik* 128:295 (1950); *Ergebn. exakt. Naturwiss.* 26:244 (1952).
12. C.D. Liu and J. Macek, *Phys. Rev.* A35:5005 (1987).
13. S. Bashkin and G. Beauchemin, *Can. J. Phys.* 44:1603 (1966); S. Bashkin, *Appl. Opt.* 7:2341 (1968).
14. S. Bickel and S. Bashkin, *Phys. Rev.* 162:12 (1967); W.S. Bickel, *Appl. Opt.* 7:2367 (1968).
15. E.L. Chupp, L.W. Dotchiu, and D.J. Pegg, *Phys. Rev.* 175:44 (1968).
16. I.A. Sellin, C.D. Moak, P.M. Griffin, and J.A. Biggerstaff, *Phys. Rev.* 184:56 (1969).
17. D.J. Lynch, C.W. Drake, M.J. Alguard, and C.E. Fairchild, *Phys. Rev. Lett.* 26:1211 (1971).
18. W.N. Lennard, R.M. Sills, and W. Whaling, *Phys. Rev.* A6:884 (1972); W.N. Lennard and C.L. Cocke, *Nucl. Instr. Methods* 110:137 (1973).
19. P. Dobberstein, H.J. Andrä, W. Wittmann, and H.H. Bukow, *Z. Physik* 257:272 (1972); W. Wittmann, K. Tillmann, H.J. Andrä, and P. Dobberstein, *ibid.*:279 (1972); J. Wangler, L. Henke, W. Wittmann, H.J. Plöhn, and H.J. Andrä, *Z. Physik* A299:23 (1981); B. Becker and H.J. Andrä, *Nucl. Instr Methods* B9:650 (1985).
20. H.G. Berry, L.J. Curtis, and J.L. Subtil, *J. Opt. Soc. Am.* 62:771 (1972); H.G. Berry, J.L. Subtil, E.H. Pinnington, H.J. Andrä, W. Wittmann, and A. Gaupp, *Phys, Rev.* A7:1609 (1973).
21. M. Dufay, *Nucl. Instr. Methods* 110:79 (1973).
22. I. Martinson, *Physica Scripta* 9:281 (1974).
23. J.M. Blatt and V.F. Weisskopf. "Theoretical Nuclear Physics," John Wiley and Sons, New York (1963), pp. 412 - 422
24. W.E. Lamb, Jr. and R.C. Retherford, *Phys. Rev.* 79:549 (1950); W.E. Lamb, Jr., *Phys. Rev.* 85:259 (1952).
25. J.P. Barrat, *J. Phys. Rad.* 20:541 (1959).
26. J.P. Barrat and C. Cohen-Tannoudji, *J. Phys. Rad.* 22:329 (1961); C. Cohen-Tannoudji *Ann. Physique* 7:423 (1962).
27. P.A. Franken, *Phys. Rev.* 121:508 (1961).
28. O. Nedelec, *J. de Phys.* 27:660 (1966).
29. M.P. Silverman and F.M. Pipkin, *J. Phys.* B5:1844 (1972).
30. M. Sargent III., M.O. Scully, and W.E. Lamb, Jr. "Laser Physics," Addison-Wesley Publishing Co., New York (1974, 6th printing 1993) p. 23
31. P. Mestre and M. Sargent III., "Elements of Quantum Optics," Springer Verlag Berlin (1990) p. 102
32. M.P. Silverman and F.M. Pipkin, *J. Phys.* B5:2236 (1972).
33. V. Weisskopf and E. Wigner, *Z. Physik* 63:54 (1930); ibid. 65:18 (1930).
34. G. Breit, *Rev. Mod. Phys.* 4:504 (1932); ibid. 5:91 (1933).
35. W.W Chow, M.O. Scully, and J.O. Stoner, *Phys. Rev.* A11:1380 (1975)
36. S. Haroche, in "Topics in Applied Physics," Vol. 13, K. Shimoda, ed., Springer Verlag, Heidelberg (1976) p. 256
37. G. Leuchs and H. Walther, *Z. Physik* A293:93 (1979).
38. F.D. Colegrove, P.A. Franken, R.R. Lewis, and R.H. Sands, *Phys. Rev. Lett.* 3:420 (1959).
39. T.W. Ducas, M.G. Littman, and M.L. Zimmerman, *Phys. Rev. Lett.* 35:1752 (1975).
40. M. Brieger and H.A. Schuessler, unpublished results
41. W. Lange and J. Mlynek, *Phys. Rev. Lett.* 40:1373 (1978); J. Mlynek and W. Lange, *Opt. Comm.* 30:337 (1979); H. Harde, H. Burggraf, J. Mlynek, and W. Lange, *Opt. Lett.* 6:290 (1981); J. Mlynek, W. Lange, H. Harde, and H. Burggraf, *Phys. Rev.* A24:1099 (1981)
42. W. Heitler, "The Quantum Theory of Radiation," Oxford University Press, 3rd editon, Oxford (1954) p.176
43. J. Mlynek, K.H. Drake, G. Kersten, D. Fröhlich, and W. Lange, *Opt. Lett.* 6:87 (1981)
44. U. Fano, *Phys. Rev.* 124:1866 (1961)
45. T. Hadeishi and W.A. Nierenberg, *Phys. Rev. Lett.* 14:891 (1965)

THE SPECTROSCOPY OF ATMOSPHERIC MOLECULES PROBED BY SYNCHROTRON RADIATION AND ELECTRON IMPACT

N J Mason, J A Davies and J M Gingell
Department of Physics and Astronomy
University College London
Gower Street
London WC1E 6BT

INTRODUCTION

Accurate quantitative analysis of the composition and dynamics of the atmosphere is crucial in our assessment of the quality of the environment and our judgement of the success of those measures taken to reduce pollution. Many of the techniques used to analyse the environment are based upon our knowledge of the molecular spectroscopy of atmospheric species. However there remain many atmospheric molecules for which data is, at best, fragmentary and often is completely absent.

The absorption spectrum of a molecule provides a unique fingerprint covering a wide spectral range. Rotational states in the fundamental vibrational bands are dominant in the microwave or far infrared region. Rotation-vibration transitions in the fundamental vibration bands cover the mid infrared range while electronic transitions account for the observed spectra in both the visible and UV wavelengths. To understand the chemistry, dynamics and molecular concentrations within the troposphere and stratosphere it is therefore necessary to study the whole spectral region. For example, while UV spectroscopy is of particular importance to the environmental problem of ozone depletion, infrared spectroscopy is essential in studies of the greenhouse effect and global warming.

To date, most experimental studies probing molecular spectroscopy have involved optical absorption measurements, but these are limited to providing information on those transitions obeying electron dipole selection rules. In contrast the probing of 'forbidden transitions' requires the study of interaction of electrons with molecular targets. Forbidden transitions produce long lived excited molecular states which are amongst the most chemically reactive species in the atmosphere. Such states are produced by photo-dissociation, recombination and collisional excitation in the stratosphere and ionosphere and are also responsible for the emissions in airglow and auroral processes yet there have been few experimental studies of the spectroscopy of such forbidden states.

In this review recent experimental results using synchrotron radiation and electron impact spectroscopy to study electronic, vibronic and fragmentation of atmospheric molecules will be presented. Particular emphasis will be placed upon UV and infrared spectroscopy of stratospheric and tropospheric molecules, but in addition the role of vibronically excited or

Selected Topics on Electron Physics
Edited by Campbell and Kleinpoppen, Plenum Press, New York, 1996

'hot' molecular species in atmospheric processes will be assessed and the need for a concentrated programme on the spectroscopy of atmospheric molecules will be emphasised.

UV SPECTROSCOPY AND SOLAR RADIATION

The solar spectrum reaching the Earth's surface (figure 1) is composed of a wide range of frequencies that are characteristic of both the solar emission and atmospheric composition since much of the solar radiation is absorbed in its passage through the atmosphere. Except for some weak absorption by O_2 and some absorption by ozone (the Chappuis bands) little absorption occurs in the visible portion of the solar spectrum, however it is well known that most of the solar UV is absorbed and is unable to penetrate to the ground, indeed were this not so life on the Earth would not have developed.

Our understanding of the mechanisms of UV absorption in the terrestrial atmosphere and how these may be altered by climatic/environmental change (eg ozone depletion by CFC and nitric oxides) therefore relies upon accurate knowledge of the UV spectroscopy of atmospheric molecules.

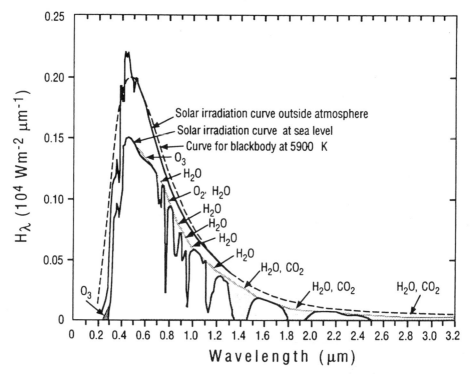

Figure 1. Spectral distribution of incident solar radiation outside the atmosphere and at sea level. The major absorption bands of some of the important atmospheric gases are indicated [from Handbook of Geophysics and Space Environments, McGraw Hill, New York, 1965].

Ozone is the most important molecule in the terrestrial atmosphere for absorption of solar UV. At 300 nm ozone begins to absorb solar radiation thereby dissociating into O_2+O and continues to do so until $\lambda \approx 200$ nm. The result of this absorption is a virtual absence of solar UV radiation at the Earth's surface between 200 and 300 nm. Below 200 nm molecular oxygen absorbs effectively with subsequent production of O atoms. Below 150 nm the photons are energetic enough to be able to ionise the molecular targets.

In principal, measurements of UV absorption cross sections would appear to be straight forward with photo-absorption cross sections being capable of measurement to accuracies of a few per cent. Many of the experimental studies of photoabsorption have involved a simple measurement of the attenuation of a beam of photons traversing a gas filled absorption cell. The light source may be a discharge lamp or tunable radiation from a synchrotron facility. The intensity of light reaching a photon detector with (I) and without (I_o) gas present in the cell is recorded and the photoabsorption cross section (σ_{pa}) is then determined by

$$I = I_o \exp\left(-\sigma_{pa}\ N\ \chi\right)$$

where N is the number density of the gas and χ the length of the cell. This so called Beer-Lambert Law is strictly only valid if (1) monochromatic light is used such that there is no change in effective absorption as the radiation passes through the gas and (2) σ_{pa} is independent of pressure and temperature of the gas. The first condition is simply fulfilled by using a grating to disperse and select the incident radiation, but the latter condition provides a serious systematic error in several experiments.

Photoabsorption measurements of intense spectral lines with very narrow band widths are prone to large errors due to an effect known as 'line saturation'. This effect arises from the finite optical resolution of the spectrometer and can be minimised if the spectrometer band width is made much narrower than the natural line width of the spectral line (Shobatake, 1992), however this is often not possible in practise. 'Line saturation' becomes evident when the sample pressure is varied, producing changes in the measured photoabsorption cross sections. The effect is most severe at high pressures and can be reduced by either using a very low pressure or using a range of pressures and extrapolating to a lower pressure. However, there are very large experimental errors attached to these low pressure measurements. A detailed quantitative analysis and theoretical investigation of the 'line saturation' effect is provided by Chan et al (1991).

Conversely, electron impact spectroscopy may be used to measure photoabsorption cross sections free of line saturation since electron impact excitation is a non-resonant process. If the incident electron energy T, is sufficiently large (T > 100 eV) and the scattering angle small, $\theta \approx 0°$, the electric field induced at the site of the molecule by the passing electron is very similar to that which would be caused by an incident photon pulse. The electric field acts most strongly with the transition electric dipole of the molecule such that electric dipole or optically allowed transitions are predominantly excited, hence the inelastic electron scattering process simulates the photoabsorption process.

It is possible to relate the electron energy-loss spectrum $I(E_L)$ obtained at $\theta \approx 0°$ and high incident energy $T \gg E_L$ (the energy loss energy) to the differential oscillator strength (DOS) df/dE by

$$df/dE_L \propto \{E_L/R\}\ \Delta\theta\{Ln(1+(\Delta\theta/\gamma)^2)\}^{-1}I(E_L)$$

where

$$\gamma^2 + (E_L/2T)^2\ (1-E_L/T)^{-1}$$

and $\Delta\theta$ is the detector acceptance angle and R is one Rydberg. The oscillator strength for a particular absorption band located between two energies E_1 and E_2 may then be obtained by the integration

$$f = \int_{E_1}^{E_2} \left(\frac{df}{dE_L}\right) dE_L$$

It is possible to normalise this relative distribution to an absolute value using either an atomic line transition (mixing the molecular target with a known amount of helium) or to an existing optical measurement. Alternatively, the TRK sumrule may be invoked to provide a calibration free of any apparatus function [Chan et al (1991)]. Invoking the TRK sumrule the high energy loss (E_L > 100 eV) portion of low resolution relative oscillator strength spectra is fitted to a curve of the form

$$\frac{df}{dE_L} = A \ E_L^{-X} + B \ E_L^{-Y} + C \ E_L^{-Z}$$

where x, y, z, are fixed values of the target and A, B, C are the best fit parameters. The area under the fitted portion of the oscillator strength is then calculated as a percentage of the whole spectrum and must equal the number of valence electrons (eg 18 for O_3, 19 for OClO) plus a calculable contribution due to Pauli excluded transitions from inner shell to the already occupied Valence shell orbitals. This electron impact energy loss technique has been used to produce DOS and thence optical absorption cross sections for many molecular targets. [eg Chan et al (1991)] with df/dE related to σ_{pa} by;

$$\frac{df}{dE} \ (eV^{-1}) = 9.11 X 10^{-3} \ \sigma_{pa}(10^{-18} \ cm^2)$$

Recent optical absorption cross sections for ozone measured using synchrotron radiation [Gingell et al (1995) and figure 2(a)] have been compared with electron impact energy loss DOS data [Davies et al (1995)] figure 2(b). While there is good agreement at long wavelengths (eg the Hartley band) at shorter wavelengths, some discrepancies remain, although the most recent synchrotron data is in much closer agreement with the electron energy loss data than previous optical measurements [Tanaka et al (1953)]. In the recent synchrotron data it has also been possible to observe well resolved Rydberg structures between 9 and 11 eV for the first time.

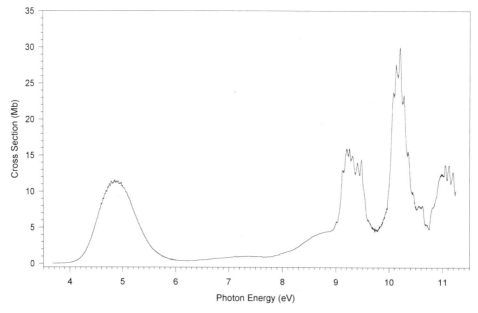

Figure 2(a). The photoabsorption cross section for ozone obtained using synchrotron radiation [Gingell et al (1995)]

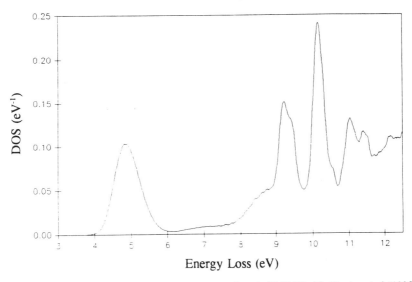

Figure 2(b). DOS spectra for ozone in the energy region, 4 eV$\leq$E$\leq$12 eV, [Davies et al (1995)]

Table One shows those molecular targets for which photoabsorption data and electron energy loss data have been compared; many atmospheric compounds have yet to be studied by electron spectroscopy and some have only tentative photoabsorption cross sections assigned. Further investigations of electron impact and synchrotron derived photoabsorption cross sections are therefore urgently needed.

FORBIDDEN TRANSITIONS

Electron impact spectroscopy is a very powerful technique for probing dipole forbidden transitions. At small angles ($\theta < 10°$) and high incident energies (> 100 eV) electron scattering obeys electron dipole selection rules. At low impact energies and large scattering

---

TABLE ONE

Photo absorption cross sections of Atmospheric Molecules

Well established standards

O_2, N_2, CO, CO_2, N_2O, NO, O_3, H_2O,

Known to tolerable degree of accuracy

CH_4, SO_2, NO_2, simple CFC's, SF_6, H_2S, C_2H_4, C_2H_6,

Poorly determined or unknown

N_2O_5, SO, OClO, OBrO, Cl_2O_3, ClOO, OH, SO_3,

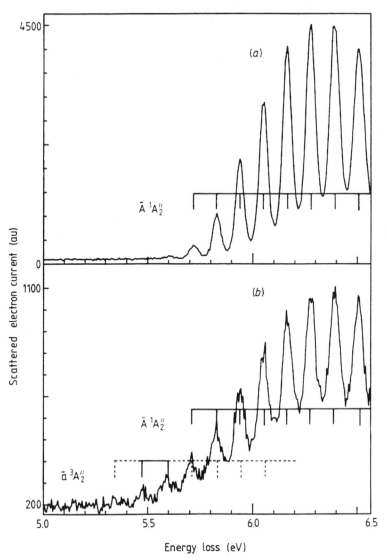

Figure 3. Electron energy loss spectra of ammonia at 15 eV impact energy; (a) scattering angle 15°, (b) scattering angle 50°. The resolution is 28 meV for both spectra. (Furlan et al (1987).

angles ($\theta > 90°$) structure observed in energy loss spectra will be due to forbidden transitions. By selective adjustment of scattering parameters towards large momentum transfer it is possible to probe those electronic states forbidden optically, but which may be excited by collision processes in the atmosphere. Figure 3 reveals energy loss spectra for ammonia at an incident energy of 15 eV and two scattering angles 15° and 50°. At the higher angle the lowest vibronic states of the forbidden triplet a $^3A_2$" are revealed in addition to the allowed A $^1A_2$" vibrational series. Crucial questions in atmospheric chemistry ask which electronic states are bound?, what are their adiabatic energies?, and how long lived (hence how reactive) are they? Even for the simplest molecules such as ozone there remain fundamental questions as to the excitation energies of these lowest excited states which are responsible for the weakly absorbing (forbidden) Chappuis and Wulf bands. For more complex radical species such as OClO, OBrO and NO_3, information is almost totally lacking.

Recently, trapped electron spectroscopy has been used to probe the excited states of stratospheric compounds ozone and OClO. This well established method [Keenan et al (1982)] has previously been used to probe low lying optically forbidden electronic states of many molecular species (eg CH_4, SiH_4, CF_4 and SiF_4 [Curtis and Walker (1989)].

Gingell et al have, for the first time, revealed the excitation of the forbidden states of ozone by direct electron impact (figure 4). The near threshold cross section for electronic excitation between 6 and 9 eV is very high and indicates that there are triplet states accessible in this region, where the optical cross section is very low (figure 2). At lower excitation energies bands of forbidden transitions have been observed from which the excitation energy of the $^1A_2$ state was determined to be around 1.65 eV, in good agreement with recent calculations of Borowski et al (1995). A second band is excited around 1.25 eV, in close proximity to the $^3A_2$, $^3B_2$ and $^3B_1$ states, but this preliminary experiment has insufficient resolution to separate these three states.

Further experiments to determine the adiabatic energies of atmospheric species are needed, a few possible molecular targets are listed in table 2.

INFRARED SPECTROSCOPY

Within the Earth's atmosphere molecules selectively absorb infrared radiation emitted by the Earth's surface and therefore act as a partial thermal blanket raising the Earth's surface temperature to a life sustaining 15°C. This blanketing is known as the natural greenhouse

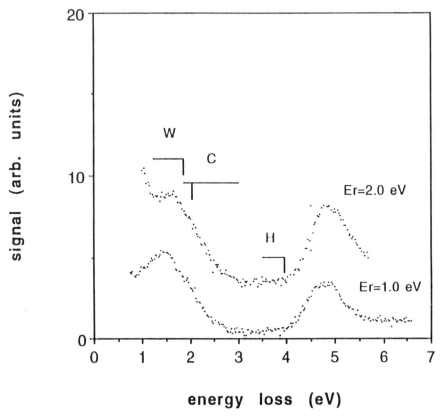

Figure 4. Scattered electron spectra recorded at residual electron energies of 2.0 and 1.0 eV respectively and showing excitation of the low-lying optically-forbidden states. [Davies et al (1995)].

TABLE 2

Atmospheric Molecules whose lowest lying (forbidden) states require characterization.

<u>Stratospheric:</u>

O_3, $OClO$, $OBrO$, H_2O_2, HNO_3, N_2O_5, NO_3,
All CFC's and replacements, HCFC's, HFC's

<u>Troposphere:</u>

HF, H_2S, SO, SO_3
Polyaromatic hydrocarbons (PAH's), Olefins

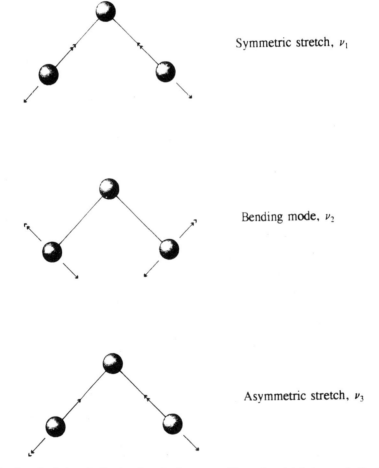

Symmetric stretch, ν_1

Bending mode, ν_2

Asymmetric stretch, ν_3

Figure 5. The three fundamental vibrational modes for ozone. The end nuclei first move in the direction of the single-headed arrow and then in the direction of the double-headed arrow.

effect and the atmospheric gases absorbing terrestrial infrared radiation are known as greenhouse gases. Nitrogen and oxygen are poor greenhouse gases neither absorbing or re-emitting infrared radiation, it is the water vapour and carbon dioxide in the terrestrial atmosphere that dominate the natural greenhouse effect. The increase in industrialisation has led to a 25% increase in CO_2 concentrations in the atmosphere since 1700, such that this increase has led to much speculation of global warming and as the role of other pollutants (eg N_2O, CH_4) is increasing the terrestrial greenhouse effect has become a topical area of environmental research.

Infrared spectroscopy is concerned with the vibrational excitation of molecules. The number of fundamental vibrational frequencies for a molecule containing N atoms is equal to 3N-6 for a non-linear molecule and 3N-5 for a linear molecule. Of these fundamental vibrational modes, N-1 are bond stretching motions while the remaining vibrations are bending motions. For example, ozone being a triatomic has three allowed vibrational modes, two of which are stretching modes and the other a bending mode (figure 5). The symmetric stretch mode has the highest frequency out of the symmetric stretch modes and is labelled υ_1, the next highest symmetric frequency belongs to the bending mode which is labelled υ_2 while the asymmetric stretch is labelled υ_3.

As well as containing fundamental (υ_1, υ_2. ...) and overtone ($n\upsilon_1$, $n\upsilon_2$, ... where n = 2, 3...) bands, the vibrational spectra for triatomics and larger molecules may also contain combination bands and difference bands. Combination bands arising from the addition of two or more fundamental frequencies or overtones eg $\upsilon_1 + 2\upsilon_2$, $3\upsilon_1 + \upsilon_2 + 2\upsilon_3$, are allowed although their intensities are usually small. Difference bands eg $2\upsilon_1 - \upsilon_3$, $\upsilon_1 + 2\upsilon_2 - \upsilon_3$, may also be observed in vibrational spectra although their intensities are very small.

In order to be infrared active (ie observable by I-R optical spectroscopy), there must be a change in the electric dipole of the molecule during the vibration. For a non-linear triatomic such as ozone, all three vibrational modes are I-R active whereas the symmetric stretch mode of a diatomic or linear triatomic molecule does not involve a dipole change and so is I-R inactive. Therefore symmetric molecules such as H_2 and O_2 are infrared inactive.

However, infrared inactive modes can be studied using electron impact, indeed electron spectroscopy has the advantage that it may be used to observe all possible vibrational modes. At low incident energies (T < 10 eV) and large scattering angles ($\theta > 20°$), the vibronic modes excited by electron impact are often quite different from those excited by photons, this may, in part, be due to the formation of resonances. In electron-molecule scattering, negative-ion resonances are formed by the temporary attachment of the incident electron to the target for between 10 and 1000 times longer than the collision period. Subsequent detachment of the electron leaves the target in an excited vibronic state.

The study of radiative transfer within the troposphere and species such as N_2O, CH_4, O_3 and the CFCs relies upon an accurate knowledge of their infrared spectroscopy. The combination of electron spectroscopy and optical methods is proving invaluable in determining the vibrational spectroscopy of such molecules. Figure 6 shows six electron energy loss spectra of ozone at a constant scattering angle ($\theta = 90°$) and a range of incident energies between E_i 3.5 and 7 eV. The relative intensities of the three vibrational peaks with respect to the elastic peak $E_i = 0$ eV) reach a maximum at $E_i = 4$ eV indicative of resonance formation.

Figure 7 reveals three vibrational spectra obtained at three scattering angles $\theta = 60$, $80°$ and $100°$, using a constant incident energy T = 6 eV. The contributions of the possible vibrational modes to each of the three vibrational peaks observed are shown in Table 3. It is found that the three energy-loss features are dominated by the vibrational modes υ_1, $2\upsilon_3$ and $3\upsilon_3$. Hence the two stretch modes appear to dominate in electron impact. In contrast, optical excitation is dominated by υ_3, $\upsilon_1 + \upsilon_3$, υ_2 and υ_1 vibrational modes. Such differences therefore reveal the advantage of electron spectroscopy in probing infrared inactive states, states that may, however, be excited by atom molecule, molecule - molecule collisions within the tropospheric system.

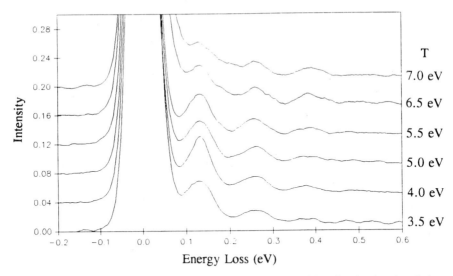

Figure 6. Electron energy loss spectra of ozone showing the intensities of the vibrational peaks relative to the elastic peak as a function of incident energy, T, with a constant scattering angle, θ=90°.

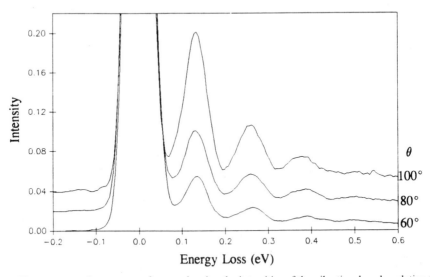

Figure 7. Electron energy loss spectra of ozone showing the intensities of the vibrational peaks relative to the elastic peak as a function of scattering angle θ, with a constant incident energy T = 6 eV.

TABLE 3

Vibrational features in ozone

	1st	2nd	3rd
Energy loss (meV)	130	260	380
Major contributions of the	60% 1_0^1	30% 3_0^2	35% 3_0^3
vibrational modes	20% 3_0^1	255% 1_0^2	20% $1_0^1 3_0^2$
observed vibrational peaks	15% 2_0^1	20% $1_0^1 3_0^1$	20% $1_0^1 2_0^1 3_0^1$
T = 6eV, θ = 60°			
Ratio to elastic peak	4.6%	1.6%	0.6%
DCS (10^{-18} cm²/sr)	7.1	2.5	0.9
T = 6eV, θ = 80°			
Ratio to elastic peak	7.1%	2.8%	1.2%
DCS (10^{-18} cm²/sr)	6.9	2.7	1.2
T = 6eV, θ = 100°			
Ratio to elastic peak	15.1%	5.8%	2.5%
DCS (10^{-18} cm²/sr)	10.0	3.8	1.6

SPECTROSCOPY OF EXCITED TARGETS

Any molecule present in a terrestrial medium is excited, the rotational modes of all molecules being sufficiently low as to be excited thermally by their environment. Therefore, in understanding any collision process it is strictly necessary to consider the target as an ensemble of rotational states. Until the development of high resolution laser spectroscopy, such refinements were impossible and all collision studies were, in reality, state insensitive.

Vibrational levels may also be populated by thermal processes, the population of an upper level of a molecule being given by the well known Boltzmann distribution

$$N_R = \frac{g_R \, e^{-E_R/kT}}{\sum g_R \, e^{-E_R/kT}}$$

where N_R is the percentage population, g_R the statistical weights, E_R the energy relative to the ground state, k Boltzmann's constant and T, the ambient temperature of the surrounding environment.

For simple diatomic molecules E_R is high (> 150 meV) and hence at room temperature (T = 288K) only a small percentage of the molecule will be in excited vibrational states. However, for even the simplest polyatomic molecule the lowest vibrational state will be

TABLE 4

Level	Energy (meV)	T = 313K	T=573K	T = 673K
000	0.00	0.9076	0.7428	0.5526
01^10	82.75	0.0839	0.1943	0.2645
02^00	159.37	0.0025	0.0151	0.0357
02^20	165.54	0.0039	0.0254	0.0633
100	172.11	0.0016	0.0110	0.0286
03^10	239.59	0.0003	0.0041	0.0177
03^30	248.38	0.0002	0.0034	0.0154
11^10	257.50	0.0001	0.0027	0.0130
001	291.26	0.0000	0.0006	0.0037
04^00	317.00	0.0000	0.0003	0.0024
04^20	320.50	0.0000	0.0003	0.0020
12^00	331.20	0.0000	0.0002	0.0018
01^20	342.30	0.0000	0.0002	0.0016
200	348.00	0.0000	0.0002	0.0014
01^11	372.45	0.0000	0.0001	0.0012
05^10	394.50	0.0000	0.0001	0.0009
05^30	401.80	0.0000	0.0000	0.0007

The populations of the vibrational modes of CO_2 as a function of temperature

populated even at room temperature (Table 4). Indeed even when CO_2 condenses (T = 195K) 1.4% of the molecules are still vibrating.

The role of excited molecular targets has been recognised for many years in atmospheric studies with the definition of so called 'hot bands'. For example, the temperature dependence of photoabsorption cross sections has been the subject of several recent experiments using synchrotron radiation. Stark et al (1992) have reported high resolution absorption cross sections between 90 and 100 nm at temperatures of 20K, since it is only at these low temperatures that is possible to measure band oscillator strengths free of perturbation and overlap from higher excited states, such cross sections are then similar to those experienced under interstellar conditions.

In atmospheric studies it is important to remember that the terrestrial atmosphere has a complex temperature profile. The stratosphere is approximately 90K below that of the troposphere, therefore experiments performed at normal laboratory temperatures will have contributions from higher vibrational states which, if not allowed for, will lead to erroneous data being used in stratospheric UV flux models. Recently Brion et al (1993) have clearly

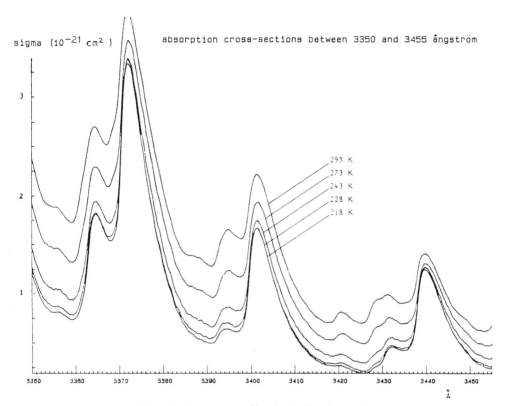

sigma $(10^{-21} \ cm^2)$ absorption cross-sections between 3350 and 3455 ångström

295 K
273 K
243 K
228 K
218 K

3350   3360   3370   3380   3390   3400   3410   3420   3430   3440   3450

Å

Figure 8. Temperature effect in the Huggins bands

demonstrated the importance of temperature effects in photoabsorption cross sections of ozone (figure 8). The strong UV absorbing Hartley band ($285 > \lambda > 225$ nm) is only slightly temperature dependent, but the Huggins band ($\lambda > 310$ nm) shows a strong temperature dependence, particularly in the regions of low absorption between the peaks where the photoabsorption cross section (σ_{pr}) may change by 300% between T = 218K and 295K. Such marked differences may also be apparent in other important stratospheric species, but at present the temperature dependency of σ_{pa} for many molecules remains unknown.

The role of excited molecules within target beams prepared for electron scattering has only recently been quantified [Johnstone et al (1993), Mason et al (1994)]. In a collision experiment performed in the laboratory the target is an ensemble of rotational states such that those measurements described as 'elastic' in the literature are in reality at best only vibrationally elastic (ie no change in vibrational quantum number). Many changes in rotational state may therefore be incurred without the knowledge of the observer because the resolution of modern electron spectrometers is insufficient to distinguish the rotational levels of any molecule except H_2 (and its isotopes).

When studying electron polyatomic molecule scattering the target beam is likely to have a significant proportion of vibrationally excited molecules. Clearly, any interpretation of such processes must allow for the electron-excited molecule interaction and not treat the process as if all the molecules were in their ground state. For example, the measured elastic scattering cross section will then comprise of two parts

$$\sigma_{elastic} = \sigma_{O \to O} + \sigma_{x \to x}$$

453

where $\sigma_{o \to o}$ is the true ground state elastic scattering cross section and $\sigma_{x \to x}$ is the cross section for elastic scattering from higher vibrational states. Should the latter be greater than $\sigma_{o \to o}$ then the presence of only a small percentage of vibrationally excited molecules will significantly alter σ elastic. Hence different experiments performed at different temperatures may lead to different elastic cross sections being determined and will lead to inaccurate comparisons with theory (which assumes all the molecules are in their lowest states of excitation).

Total cross section measurements [Buckman et al (1987), Ferch et al (1989)] suggested that electron scattering cross sections from vibrationally excited CO_2 in its first bending mode may be significantly higher than from the ground state. These results were tentatively explained by the electron dipole moment associated with the CO_2 010 bending vibration enhancing electron scattering cross section, it being known that the cross section for scattering of low electrons by polar molecules (eg H_2O) is larger. However, this hypothesis requires more accurate experiments to measure individual cross sections (elastic, inelastic, superelastic) and the extension to other targets whose initial state is polar or whose lowest lying vibrational states are linear. Johnstone et al (1993) reported measurements of high resolution electron scattering from vibrationally excited CO_2 and demonstrated that the elastic scattering cross section from vibrational states may be 15% higher than from the ground state. In contrast recent studies of the ionization of diatomic molecules has shown little vibronic dependence [Kulz et al (1995)]. Alternatively electron impact dissociative attachment cross sections may show an order of magnitude increases with increasing degrees of vibrational excitation [Christodoulides et al (1990)]. Therefore, it is necessary to study each excitation process in detail and commence state-to-state electron molecule collision experiments. The target beam may be prepared in specified vibrational states using a supersonic source and subsequent laser pumping,the products of the collision can then be analysed either by electron impact energy loss spectroscopy and laser induced fluorescence (LIF) or resonance enhanced multiphoton ionisation [REMPI].

Such experiments will have important consequences for our understanding of the spectroscopy of atmospheric molecules. For example, dissociative processes play a key role in molecular physics of the atmosphere, providing a route for the formation of reactive species. Population of a molecular excited state by electron impact at energy E_{1} followed by dissociation to form an electronically excited fragment species can be described by the following equations:

$$ABC(\upsilon) + e^- \ (E_i = E_1) \to ABC^* + e\text{-} \ [E_f = E_1 - \Delta E]$$
$$ABC^* \to AB^* + C$$

where AB* is either an excited (vibronic or electronic) state or ion and in some respects is analogous to the photodissociation process prominent in the atmosphere's ionosphere

$$ABC + h\upsilon \to AB^* + C$$

However, electron impact allows the population of electronic states that cannot be accessed by photoabsorption experiments but states that may be produced by chemical reactions, for example the molecular triplet states of ozone discussed above.

The initial vibronic state of ABC is crucial in determining the final state of distribution AB and C the amount of energy stored internally being enough to both shift the absorption spectrum and overcome the energy barriers to chemical or further photo chemical reactions.

It is therefore necessary to determine the temperature dependence of photon and electron impact cross sections and assess the importance of initial vibrational and electronic state energy.

SUMMARY

In this brief review the use of synchrotron radiation and electron impact to study the spectroscopy of molecules relevant to atmospheric studies has been discussed. The two methods are complimentary, with electron impact spectroscopy able to probe those electric dipole forbidden transitions as well as providing a cross check for optically derived cross sections. The two methods have been used to derive oscillator strengths (absorption cross sections) for ozone with good agreement with other measurements being found over a wide wavelength range. Electron impact spectroscopy may also be used to study infrared spectroscopy of molecular targets with enhanced vibrational cross sections being observed when resonances are formed. Finally, the role of internal excitation of the molecular targets is discussed and the need to perform state-to-state experiments revealed.

In conclusion, the combination of these two well established experimental techniques with the future possibility of laser preparation and analysis of the target molecule offers the prospect of producing new information on the spectroscopy and collision dynamics of atmospheric species that will be important in our understanding and modelling of the current environmental issues of ozone depletion, the greenhouse effect and, perhaps, acid rain. It therefore remains an exciting and challenging field of study that will take us into the new millennium.

ACKNOWLEDGEMENTS

NJM wishes to acknowledge the continued support of the Royal Society, JAD receipt of a Gassiot research studentship and JMG an EPSRC postgraduate studentship. We also wish to acknowledge our colleagues Dr G Marston, Dr R P Wayne, Dr M Siggel and Dr I C Walker without whose collaboration this work would not have been possible.

REFERENCES

Borowski P, Fulscher M, Malmqvist P A and Roos B O (1995) Chem. Phys. Lett **237** 195.

Brion J, Chaker A, Daumont D, Mabeet J and Parisse C (1993) Chem. Phys. Lett **213** 610.

Buckman S J, Elford M T and Newman D S (1987) J. Phys B**20** 5157.

Christodoulides A A, Christophorou L G and McCorkle D L (1987) Chem.Phys. Lett **139** 35.

Curtis M G and Walker I C (1989) J. Chem. Soc. Farad Trans 2 **85** 659.

Chan W F, Cooper G and Brion C E (1991) Phys. Rev. A. **44** 186.

Davies J A and Mason N J (1995) unpublished.

Ferch J, Masche C, Raith W and Wieman L (1989) Phys. Rev. A**40** 5407.

Furlan M, Hubin-Franskin M J, Delwiche J and Collin J E (1987) J Phys B**20** 6283

Gingell J M, Davies J A, Mason N J, Zhao H, Siggel M and Walker I C (1995) to be submitted to J.Phys.B.

Johnstone W M, Mason N J and Newell W R (1995) J.Phys.B. **26** L147.

Kulz M, Mortyna M, Keil B, Schellhaas and Bergmann K (1995) Z.Phys.D **33** 109.

Keenan G A, Walkjer I C and Dance D F (1982) J.Phys.B. **15** 2509.

Mason N J, Johnstone W M and Akther P (1994) Electron Collisions with Molecules, Clusters and Surfaces, ed H Ehrhardt and L A Morgan, p47, Plenum Press.

Stark G, Yoshino K, Smith P L, Ito K and Parkinson W H (1991) Astrophys. J. **369** 574.

Tanaka Y, Inn E C Y and Watanabse K (1953) J.Chem.Phys. **21** 1651.

TWO-PHOTON POLARIZATION FOURIER SPECTROSCOPY OF METASTABLE ATOMIC HYDROGEN

A J Duncan, Z A Sheikh and H Kleinpoppen
Atomic Physics Laboratory
University of Stirling
Stirling FK9 4LA
Scotland

INTRODUCTION

For many years the theoretical and experimental study of atomic hydrogen has been used to improve our understanding and extend our knowledge of the fundamental properties and behaviour of atoms. The states with principal quantum number n=2 are and have been of special interest and importance, in particular with regard to the determination of the fine structure constant and measurement of the Lamb shift. It was, of course, the observations of the Lamb shift in 1947 and 1950 by Lamb and Retherford [1] which, by demonstrating the nondegeneracy of the $2^2S_{1/2}$ and $2^2P_{1/2}$ states, confirmed that the $2^2S_{1/2}$ state would be metastable in experimentally realisable situations, and showed that it should be possible to observe the two-photon emission which is the main mode of decay of this state. However, Göppert-Mayer, in 1931, in a paper [2] which pioneered the field of multiphoton transitions, was the first to predict the possibility of the spontaneous two-photon decay process and, in 1940, Breit and Teller [3] applied this theory to the $2^2S_{1/2}$ - $1^2S_{1/2}$ transition in atomic hydrogen. Improved calculations of the characteristics of the two-photon decay process were carried out by Spitzer and Greenstein [4], Shapiro and Breit [5], Zon and Rapaport [6], Klarsfeld [7] and Johnson [8]. Further refinements to the theory have been made for example by Goldman and Drake [9], Parpia and Johnson [10], Tung et al [11], Florescu [12], Costescu [13] and Drake [14]. A comprehensive review concerning the metastability of atomic hydrogen up to 1969 was given by Novick [15] in which he emphasised the various controversies with regard to the metastability or otherwise of the 2S state during the first part of this century. There is a continuing interest in the theoretical aspects of the subject as exemplified by the work of Mu and Crasemann [16], Wu and Li [17] and Tong, Li, Kissel and Pratt [18]. In addition, Stancil and Copeland [19] have recently predicted interesting spectral features and a significant increase in the two-photon decay rate in the presence of a strong magnetic field.

Selected Topics on Electron Physics
Edited by Campbell and Kleinpoppen, Plenum Press, New York, 1996

Although the existence of the Lamb shift indicated that it should be possible to observe the two-photon decay process, the process is second-order with a comparatively low transition probability and it was not until 1965 that Lipeles, Novick and Tolk [20] made a successful measurement in singly ionized helium, further results concerning the spectrum being obtained by Artura, Tolk and Novick [21] in 1969. Subsequently, the two-photon decay of hydrogenlike argon and sulphur ions was observed by Schmeider and Marrus [22] and Marrus and Schmeider [23] in a beam-foil-type experiment. In a similar experiment Cocke et al [24] measured the lifetime of the $2^2S_{1/2}$ state of hydrogenlike fluorine and oxygen and, about the same time Prior [25] measured the lifetime of the $2^2S_{1/2}$ state of singly ionized helium using an ion-trapping technique, while Kocher et al [26] and later Hinds et al [27] did the same using a decay-in-flight method. In 1983, Gould and Marrus [28] carried out a further investigation of the two-photon decay process in hydrogenlike argon and derived a value for the Lamb shift by observing the quenching of the $2^2S_{1/2}$ state in an electric field. Other closely related experiments of interest have been performed, for example that of Bannett and Freund [29] in which K-shell hole states in molybdenum were filled by 2S-1S transitions with the accompanying emission of two X-ray photons, that of Braunlich et al [30] in which singly stimulated emission from the $2^2S_{1/2}$ state of atomic hydrogen was observed and that of Hippler [31] involving the observation of two-photon bremsstrahlung. Interesting experimental observations which test various relativistic and non-relativisitic calculations have recently been made by Ilakovac et al. for two-photon decay processes in hydrogenic xenon [32], silver and hafnium ions [33]. There has been considerable interest, too, shown in the properties of the two photons produced in the process of parametric down conversion in which an incident (pump) photon can be considered to be split into two lower frequency (signal and idler) photons to form a highly correlated photon pair [34-38].

Because of the long lifetime of the $2^2S_{1/2}$ state, and the problems involved in producing a source of sufficient intensity, the experimental observation of the two-photon decay process in atomic hydrogen itself is very difficult. However, in 1975 the observation was carried out successfully at Stirling by O'Connell et al [39] while almost simultaneously Kruger and Oed [40] at Tubingen published their results. These first experiments on atomic hydrogen concentrated on measurement of the lifetime of the metastable state as well as the angular correlation and spectral distribution of the two photons emitted in the decay process. However, considerable interest also attaches to the measurement of the polarization correlation of the two photons both as a sensitive test of the theory of the two-photon decay process itself and as a test of Bell's inequality [41] which allows a quantitative distinction to be made between, on the one hand, the predictions of quantum mechanics and, on the other, the predictions of local realistic or hidden variables theories. At Stirling, measurements of the linear and circular polarization have been carried out and Bell's inequality put to the test [42-46]. Professor Farago, well known among other things for his work with Drake and Wijngaarden [47-49] on metastable atomic hydrogen, took a great interest in this work at Stirling. His encouragement and constructive comments through the years was always much appreciated by the authors.

As mentioned above, Artura, Tolk and Novick [21] made a measurement of the spectral distribution of the two photons emitted in the decay of metastable singly ionized helium. Kruger and Oed [40] also made some attempt in their experiment to measure the spectral distribution of metastable atomic hydrogen. Both methods relied on the use of filters and only approximate estimates of the form of the spectral distribution could be made from the experimental results. In fact, any method using filters in these experiments is difficult given the low signal strength and the need accurately to calibrate the filters used. The present article describes an attempt to measure the spectral distribution of the two photons emitted in the decay of metastable atomic hydrogen by a novel Fourier transform spectroscopic technique which makes use of the subtle interplay between the entangled polarization properties and the spectral properties of the two-photon radiation.

CHARACTERISTICS OF THE TWO-PHOTON DECAY

For metastable atomic hydrogen [H(2S)], electric dipole and electric quadrupole transitions from the $2S_{1/2}$ to $1S_{1/2}$ state are forbidden. As a result the H(2S) state has a long lifetime of about 1/8 second and decays primarily by the simultaneous emission of two photons as indicated in Fig 1.

Contributions to the decay of the $2S_{1/2}$ state are also possible by:

(i)   single photon magnetic dipole transitions which only become significant, however, when relativistic effects are important, for example in high Z hydrogenic ions. For hydrogen itself the lifetime of this type of transition is about 4×10^5 seconds [10] and can be neglected for most purposes;

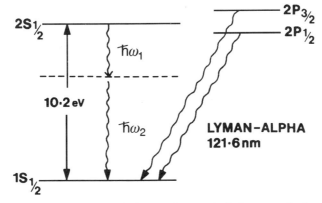

Figure 1.   Energy level diagram for atomic hydrogen neglecting hyperfine structure (not to scale).

(ii)   a cascade, involving the sequential emission of two photons through the $2P_{1/2}$ state which, because of the Lamb shift, lies slightly below the $2S_{1/2}$ state in energy. For hydrogen the associated lifetime is about 5×10^9 seconds [5] and hence this process can also be effectively neglected;

(iii)   other types of two-photon transition. For example, the $2S_{1/2}$ state could decay by emitting two quadrupole photons [50] but the effect of such processes is negligible for the present purposes.

It is also worth noting that since the electric dipole operator mediating the two-photon decay process is diagonal in the electronic and nuclear spin, fine and hyperfine structure play no part in determining the decay process [2,3] or the properties,

459

particularly the polarization correlation properties, of the two photons emitted.

The theoretically predicted [2,3,4] spectral distribution $A(\omega)$ of the photons emitted in the two-photon decay process is shown in Fig 2, where $A(\omega)d\omega$ represents the probability of detection of one photon of the pair in the range $d\omega$ in the neighbourhood of angular frequency ω. It can be seen from Fig 2 that each photon of the pair can be detected with any frequency (energy) between 1.55×10^{16} rad s^{-1} (10.2 eV) and 0 rad s^{-1} (0 eV) subject to the requirement that the sum of the frequencies (energies) is given by $\omega_o = 1.55 \times 10^{16}$ rad s^{-1} (10.2 eV). The spectral distribution curve has a maximum at a frequency (energy) of 0.775×10^{16} rad s^{-1} (5.1 eV) corresponding to a wavelength of 243 nm. Experimentally this fact is important since it allows measurements of photon characteristics to be made in air using simple optical components.

For photon pairs for which the recoil momentum of the atom is less than the uncertainty in momentum resulting from localisation, the frequency component of the state vector representing the two-photon radiation can be written in the form [51]

$$|\psi> \; = \int a(\omega)|\omega>_1|\omega_o\text{-}\omega>_2 d\omega \qquad (1)$$

where $|\omega>_1$ and $|\omega_o\text{-}\omega>_2$ are the state vectors representing photons with the complementary frequencies ω and $\omega_o\text{-}\omega$, and $|a(\omega)|^2 = A(\omega)$.

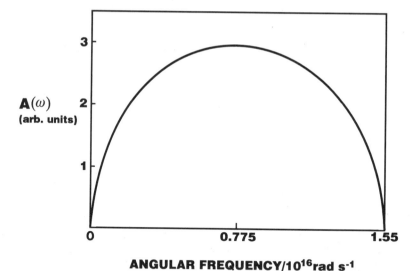

Figure 2. Predicted two-photon spectral distribution for the two-photon decay of metastable atomic hydrogen.

The polarization properties of the two photons are also well known [42,45]. Of particular interest is the situation where the two photons are detected in diametrically opposite directions, say the $\pm z$ directions. On the basis of consideration of conservation of angular momentum and parity it can be shown [52] that the polarization component of the two-photon state vector takes the form

$$|\psi> \ = \frac{1}{\sqrt{2}} \ (|x>_1|x>_2 + \ |y>_1|y>_2) \qquad (2)$$

where $|x>_1$ represents a photon on the right-hand-side (+z) say, of the source polarized in the x direction, $|x>_2$ a photon on the left hand side (-z) say, of the source also polarized in the x direction with corresponding definitions for $|y>_1$ and $|y>_2$.

It is interesting to note that both the frequency and polarization components of the state vectors above are in what is now commonly referred to as an "entangled" form. Before any detection event takes place neither the polarization nor frequency (energy) of a single photon is defined or, indeed, can be assigned any meaning. The entangled state vector represents the properties of the photon pair not single photons. However, if, for example, in this case where an arrangement is made to detect photons in diametrically opposite directions, a detection of a photon of particular polarization and frequency (energy) is made on one side of the source then the polarization and frequency (energy) characteristics of the photon on the other side are then known with certainty. In other words, in a sense which has proved heuristically useful, the properties of the photon on the right (say) can be considered to be determined by the measurement which is chosen to be made on the left.

PRINCIPLE OF THE METHOD

The basic experimental arrangement for the analysis and detection of the two-photon radiation is shown in Fig 3. The two photons emitted by the source are detected in

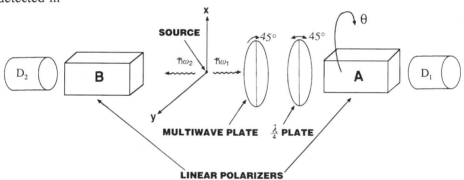

Figure 3. Schematic diagram of the experimental arrangement using two linear polarizers A and B and two detectors D_1 and D_2. The transmission axis of polarizer A is rotated through angles θ with respect to the x axis while that of polarizer B is fixed parallel to the x axis. The fast axis of the multiwave plate is set at 45° to the x axis while the fast axis of the quarter-wave plate is set at ±45° relative to the x axis to analyse circular polarization.

coincidence by the detectors D_1 and D_2. A linear polarizer with its transmission axis orientated in the x direction is placed on the left hand side of the source ensuring that upon detection of a photon on the left the complementary photon on the right, before being detected itself, can to all intents and purposes be regarded as also polarized in the x direction. In the absence of any wavelength filter on the left the frequency of the photon on the right is indeterminate, and the radiation on the right can be considered either to consist of a sequence of single frequency photons with a spectral distribution determined by the spectrum of the source or, alternatively and equivalently, as a sequence of minimum wavepackets with the spectrum and coherence length of each wavepacket determined by the spectral characteristics of the two-photon source.

If this radiation on the right is now passed through a uniaxial birefringent multiwave plate with its axis at an angle of 45° to the x axis as shown in Fig 3, its state of polarization will be changed and it will also be depolarized to an extent depending on the thickness d of the multiwave plate. The state of polarization of the radiation emerging from the multiwave plate may be monitored by measuring the Stokes parameters P_1, P_2 and P_3 of the radiation on the right detected in coincidence with the radiation on the left. The Stokes parameters are defined as [53],

$$P_1 = \frac{I(0°) - I(90°)}{I(0°) + I(90°)}, \quad P_2 = \frac{I(45°) - I(-45°)}{I(45°) + I(-45°)}, \quad P_3 = \frac{I(RHC) - I(LHC)}{I(RHC) + I(LHC)} \tag{3}$$

where $I(0°)$ is the strength of the coincidence signal when the transmission axis of polarizer A is set at an angle $\theta = 0°$ to the x axis with corresponding definitions for $I(90°)$, $I(45°)$ and $I(-45°)$. Similarly $I(RHC)$ and $I(LHC)$ refer to the strength of the coincidence signal with the achromatic quarter-wave plate shown in Fig 3 in place orientated with its axis at $\pm 45°$ to the x axis so as to detect, respectively, right-handed-circularly (RHC) polarized light or left-hand-circularly (LHC) polarized light.

If the retardation of the multiwave plate of thickness d is ϕ then

$$\phi = \frac{(n_e - n_o)\omega d}{c} \tag{4}$$

where n_e and n_o are respectively the extraordinary and ordinary refractive indices of the material of the multiwave plate. It is easy to show [54] then that for monochromatic radiation incident upon the multiwave plate the Stokes parameters of the emerging radiation are given by,

$$P_1 = \cos\phi, \quad P_2 = 0, \quad P_3 = -\sin\phi. \tag{5}$$

However, in the present case, averaged over the spectral distribution $A(\omega)$, the expected values for P_1, P_2 and P_3 of the emerging radiation are,

$$P_1 = \int_0^{\omega_o} \cos\phi\, A(\omega)d\omega \left/ \int_0^{\omega_o} A(\omega)d\omega \right.$$

$$P_2 = 0 \tag{6}$$

$$P_3 = -\int_0^{\omega_o} \sin\phi\, A(\omega)d\omega \left/ \int_0^{\omega_o} A(\omega)d\omega \right.$$

Examination of the above expressions reveals that, if the birefringence $(n_e - n_o)$ were frequency independent, P_1 and P_3 would be precisely the Fourier cosine and Fourier sine transforms of the spectral distribution $A(\omega)$ with the quantity $(n_e - n_o)d/c$ acting as the "time" variable.

However the birefringence is, in practice, frequency dependent and it is more convenient to write P_1, say, in the form

$$P_1(t) = \int_0^\infty g(\omega) \cos [n(\omega)\omega t]\, d\omega \tag{7}$$

where

$$g(\omega) = A(\omega) \Big/ \int_0^\infty A(\omega)\, d\omega$$

$$n(\omega) = (n_e - n_o), \quad t = d/c$$

Since $P_1(t)$ is a real even function of t_1 and $\omega \geq 0$ the Fourier transform of $P_1(t)$ can be written as,

$$F(\omega') = \frac{1}{\pi} \int_{-\infty}^\infty P_1(t) \cos\omega t\, dt \tag{8}$$

and substituting for $P_1(t)$ gives

$$F(\omega') = \frac{1}{\pi} \int_0^\infty g(\omega) \int_{-\infty}^\infty \cos [n(\omega)\omega t] \cos \omega' t\, dt\, d\omega$$

$$= \frac{1}{4\pi} \int_0^\infty g(\omega) \int_{-\infty}^\infty [e^{i(n\omega+\omega')t} + e^{i(n\omega-\omega')t}$$

$$+ e^{-i(n\omega-\omega')t} + e^{-i(n\omega+\omega')t}]\, dt\, d\omega \tag{9}$$

It follows that [55]

$$F(\omega') = \int_0^\infty g(\omega) \left[\sum_i \left\{ \frac{\delta(\omega-\omega_{oi}^+)}{\left|\dfrac{dn(\omega)\omega}{d\omega}\right|_{\omega_{oi}^+}} + \frac{\delta(\omega-\omega_{oi}^-)}{\left|\dfrac{dn(\omega)\omega}{d\omega}\right|_{\omega_{oi}^-}} \right\} \right] d\omega \tag{10}$$

where ω_{oi}^+ are the values of ω for which $n(\omega)\omega + \omega' = 0$ and ω_{oi}^- are the values of ω for which $n(\omega) - \omega' = 0$

As shown in Fig 4, the quantity $n(\omega)\omega$ is a monotomically increasing function of ω, so that for $\omega \geq 0$, there are no zeros of $n(\omega)\omega + \omega'$ and only one zero of $n(\omega)\omega - \omega'$. Hence, in this special case

$$F(\omega') = \int_0^\infty g(\omega) \frac{\delta(\omega-\omega_0)}{\left|\dfrac{dn(\omega)\omega}{d\omega}\right|_{\omega_0}}\, d\omega \tag{11}$$

where ω_0 is the value of ω for which $n(\omega)\omega - \omega' = 0$. Clearly the value of ω_0 depends on ω' in a way that can be determined graphically making use of Fig. 4, ie ω_0 is a function of ω'.

It follows that the relationship between the theoretical spectral distribution $g(\omega)$ and the "effective" spectral distribution $F(\omega')$ is

$$F(\omega') = \cfrac{g(\omega_0)}{\left| \cfrac{dn(\omega)\omega}{d\omega} \right|_{\omega_0}} \quad \text{with } \omega_0 = \omega_0(\omega') \qquad (12)$$

Hence given $g(\omega)$ and the dependence of the birefringence $n(\omega)$ on the angular frequency ω, $F(\omega')$ is easily calculated and the predicted variation of the Stokes parameters $P_1(t)$ and $P_3(t)$ found from

$$P_3(t) = -\int_0^\infty F(\omega') \sin\omega't \, d\omega', \qquad P_1(t) = \int_0^\infty F(\omega') \cos\omega't \, d\omega' \qquad (13)$$

Conversely, given the measured quantities $P_1(t)$ and $P_3(t)$, $F(\omega')$ and hence $g(\omega)$ may be found by taking, respectively the inverse Fourier cosine and sine transforms of these quantities. Alternatively $F(\omega')$ and $g(\omega)$ may be calculated by taking the complex Fourier transform of the combination $P_1 - iP_3$ which, interestingly, forms a representation in the complex plane of the Stokes vector corresponding to the polarization state of the radiation.

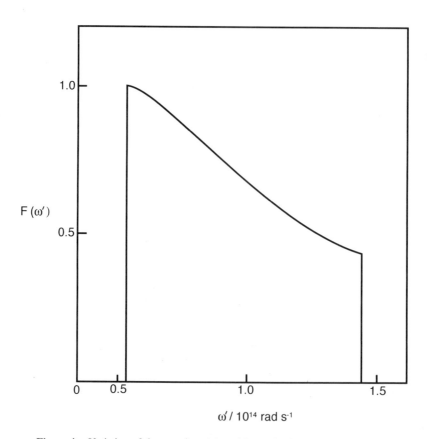

Figure 4. Variation of the quantity $n(\omega)\omega$ with angular frequency ω for quartz

APPARATUS

The metastable hydrogen source and the general experimental procedures have been described in detail elsewhere [42]. In summary, a 1-keV beam of metastable atomic deuterium, of density about $10^4 cm^{-3}$, is produced by charge exchange, in caesium vapour, of deuterons extracted from a radio-frequency ion source. Deuterium is used rather than hydrogen since a more stable beam with less collision induced noise is produced in this way. After emerging from the charge exchange cell the beam passes through collimating apertures and then the observation region into the monitor region at the end of the apparatus where it is quenched in an electric field, the resulting Lyman-α signal being used to normalize measurements taken over a long period. The two-photon radiation emitted by a small portion of the beam, after passing through 5 mm thick fused silica vacuum windows one on either side of the beam, is collected and collimated by two 50 mm diameter fused silica lenses each with a focal length of 43 nm at a wavelength of 243 nm placed diametrically on either side of the beam about 50 mm from its centre. The lenses produce an image of the source on the two photomultiplier cathodes each of which is 53 cm from the centre of the beam. The pulses from the photomultipliers are fed to a conventional coincidence circuit [39] to produce a time correlated spectrum which is displayed on a multichannel analyser.

The linear polarizers consist of twelve 2 mm thick fused silica plates optically polished flat to 2λ at 243 nm and set nearly at Brewster's angle. The transmission efficiencies M and m for light polarized parallel to and perpendicular to the transmission axis of the polarizers are M = 0.936±0.008 and m = 0.032±0.001 at a wavelength of 243 nm. The efficiencies M and m have a weak wavelength dependence. More details of the characteristics of the linear polarizers are given elsewhere [46].

The achromatic quarter-wave plates (Halle-Nachfl) consist of a combination of two double-plates, one of crystal quartz and one of magnesium fluoride. The plates of diameter 19.5 mm are cut parallel to the optic axis and polished flat to an accuracy of $\lambda/10$. The retardation produced by these plates shows some dependence on wavelength which has been described previously [46]. It is, of course, necessary to use such achromatic plates because of the continuum nature of the radiation from the source which emits at all wavelengths between 121.6 nm and infinity. However, in practice, because of absorption in oxygen there is an effective short wavelength cut-off at about 185 nm which, in turn, implies a long wavelength cut-off at the complementary wavelength of 355 nm. Hence only photons with wavelengths in the range 185 nm to 355 nm can contribute to the coincidence signal.

The "multiwave" plates themselves, placed one at a time on the right hand side of the source, are, in practice, chosen to be a series of zero-order half-wave plates at wavelengths of 200 nm, 243 nm, 300 nm and 486 nm. These plates consist of two flat pieces of crystal quartz of slightly different thicknesses cut parallel to the optic axis and placed in contact with their optic axes perpendicular to give "effective" thicknesses d = 7.69 μm, 10.84 μm, 14.56 μm and 26.27 μm deduced from the known birefringence properties of quartz [56]. Additional effective thicknesses of d = 3.15 μm and 18.53 μm are obtained by placing the 200 nm and 243 nm plates in series with their optic axes respectively at right angles and parallel. An effective thickness of d = 37.11 μm is obtained by placing the 243 nm and 486 nm plates in series with their axes parallel.

465

EXPERIMENTAL RESULTS AND COMPARISON WITH THEORY

The theoretical predictions for P_1 and P_3 are calculated from equation (13), and the theoretical spectral distribution shown in Fig 2 cut off above at an angular frequency of 1.02×10^{16} rad s^{-1} (185 nm) and below at the complementary frequency of 5.31×10^{15} rad s^{-1} (355 nm). The value of $F(\omega')$, deduced from the procedure outlined previously, is shown in Fig. 5, and the theoretical values for P_1, and P_3 calculated therefrom by Fourier transformation are shown in Fig 6. The "total" polarization $P = \sqrt{P_1^2 + P_2^2 + P_3^2}$ is shown in Fig 7, assuming that $P_2 = 0$.

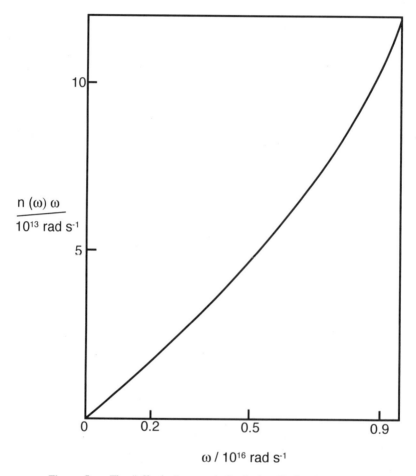

Figure 5. The "effective" spectral distribution F(ω') taking into account the frequency dependence of the birefringence of quartz.

In practice, in order to compare the measured values of the Stokes parameter with the theoretically predicted values, allowance must be made for the imperfection of polarizers A and B. Since the polarizers are essentially identical, this correction is accomplished by dividing the measured values of the Stokes parameters by the polarization $\Pi = (M-m)/(M+m)$ of each polarizer, ie by Π^2. The experimental results thus modified are shown in comparison with the theoretical predictions in Fig. 6. From equation (6), and also from considerations of symmetry, $P_2 = 0$ and it was confirmed that, within the limits of experimental error this was true for a representative sample of multiwave plates.

466

Taking account of the experimental errors it can be seen from Fig 6 that there is excellent agreement between the measurements and theoretical predictions.

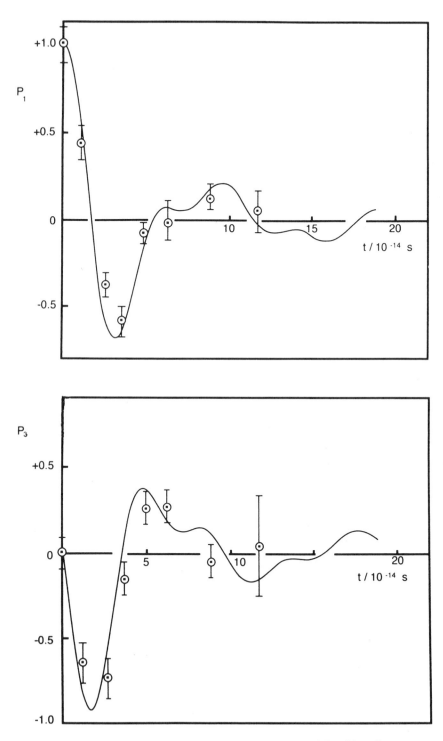

Figure 6. Variation of the Stokes parameters P_1 and P_3 with t=d/c where d is the effective thickness of the multiwave plate.

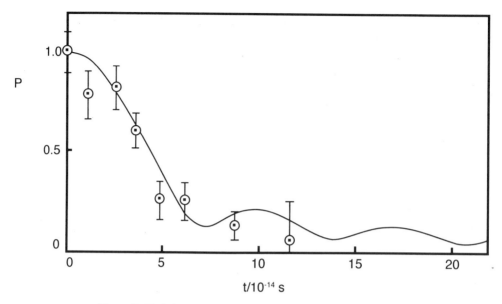

Figure 7. Variation of the "total" polarization P with t=d/c where d is the thickness of the multiwave plate.

If anything the results for P_1 are better than for P_3 possibly due to the use of imperfect quarter wave plates in the analysis of circular polarization

CONCLUSIONS

It has been shown that the insertion of a range of multiwave plates with varying thickness in one arm of what can be regarded as a two-photon polarization spectrometer, results in a series of measurements which constitute, in effect, sample points of the Fourier transform of the effective spectral distribution of the source obtained by taking into account the frequency dependence of the multiwave plates.

Assuming the correctness of the theoretical form for the spectral distribution of the source it is seen that the experimental measurements agree well with the Fourier transform of the effective spectral distribution and confirm, within the limits of experimental error, the well known theoretical predicted spectral distribution shown in Fig 2.

The inverse problem of determining the form of the spectral distribution from the experimental measurements is straightforward in principle but presents some difficulty in the present case. The measured points are not sufficient in number, not regularly spaced and do not extend to large enough values of the thickness of the multiwave plates to allow anything but a crude estimate of the spectral distribution to be made. Even making a reasonable interpolation it can only be concluded that the results are characteristic of a bandwidth limited signal of the expected bandwith and centre frequency, but no details of any structure within the bandwidth can be deduced. However the method is sound and has recently been applied to the construction of a Fourier transform spectometer with no moving parts capable of analysing radiation from a single pulse [57].

468

REFERENCES

1. W.E. Lamb and R.C. Retherford, *Phys. Rev.* **72**, 241 (1947); **79**, 549 (1950).
2. M. Göppert-Mayer, *Ann. Phys.* (N.Y.) **9**, 273 (1931).
3. G. Breit and E. Teller, *Astrophys. J.* **91**, 215 (1940).
4. L. Spitzer and J.L. Greenstein, *Astrophys. J.* **114**, 407 (1951).
5. J. Shapriro and G Breit, *Phys. Rev.* **113**, 179 (1959).
6. B.A. Zon and L.P. Rapaport, *JETP Lett.* **7**, 52 (1968).
7. S. Klarsfeld, *Phys. Lett.* **30A**, 382 (1969).
8. W.R. Johnson, *Phys. Rev. Lett.* **29**, 1123 (1972).
9. S.P. Goldman and G.W.F. Drake, *Phys. Rev. A* **24**, 183 (1981).
10. F.A. Parpia and W.R. Johnson, *Phys. Rev. A* **26**, 1142 (1982).
11. J.H. Tung, X.M. Ye, G.J. Salamo, and F.T. Chan, *Phys. Rev. A* **30**, 1175 (1984).
12. V. Florescu, *Phys. Rev. A* **30**, 2441 (1984).
13. A. Costescu, I. Brandus, and N. Mezincescu, *J. Phys. B* **18**, L11 (1985).
14. G.W.F. Drake, *Phys. Rev. A* **34**, 2871 (1986).
15. R. Novick, in *Physics of One and Two-Electron Atoms* (Edited by R Bopp and H Kleinpoppen) pp 296-325, North Holland, Amsterdam (1969)
16. X. Mu and B. Crasemann, *Phys. Rev. A.* **38** 4585 (1988).
17. Y.J. Wu and J.M. Li, *J. Phys. B: At. Molec. Opt. Phys.* **21**, 1509 (1988).
18. X.M. Tong, J.M. Li, L. Kissell and R.H. Pratt, *Phys. Rev. A.* **42**, 1442 (1990).
19. P.C. Stancil and G.E. Copeland, *Phys. Rev. A.* **46**, 132 (1992).
20. M. Lipeles, R. Novick, and N. Tolk, *Phys. Rev. Lett.* **15**, 690, 815 (1965).
21. C.J. Artura, N. Tolk, and R. Novick, *Astrophys. J.* **157**, L181 (1969).
22. R.W. Schmeider and R. Marrus, *Phys. Rev. Lett.* **25**, 1692 (1970).
23. R. Marrus and R.W. Schmeider, *Phys. Rev. A.* **5**, 1160 (1972).
24. C.L. Cocke. B. Curnette, J.R. Macdonald, J.A. Bednar, and R. Marrus, *Phys. Rev. A.* **9**, 2242 (1974).
25. M.H. Prior, *Phys. Rev. Lett.* **29**, 611 (1972).
26. C.A. Kocher, J.E. Clendenin, and R. Novick, *Phys. Rev. Lett.* **29**, 615 (1972).
27. E.A. Hinds, J.E. Clendenin, and R. Novick, *Phys. Rev. A.* **17**, 670 (1978).
28. H. Gould and R. Marrus, *Phys. Rev. A.* **28**, 2001 (1983).
29. Y. Bannett and I. Freund, *Phys. Rev. Lett.* **49**, 539 (1982).
30. P. Braunlich, R. Hall, and P. Lambropoulos, *Phys. Rev. A.* **5**, 1013 (1972).
31. R. Hippler, *Phys. Rev. Lett.* **66**, 2197 (1991).
32. K. Ilakovac, J. Tudoric-Ghemo, B. Busic and V. Horvat, *Phys. Rev. Lett.* **56**, 2469 (1986).
33. K. Ilakovac, V. Horvat, Z. Krecak, G. Jerbic-Zorc, N. Ilakovac and T. Bokulic, *Phys. Rev. A.* **48**, 516 (1993).
34. D.C. Burnham and D.L. Weinberg, *Phys. Rev. Lett.* **25**, 84 (1973).
35. C.K. Hong, Z.Y. Ou and L. Mandel, *Phys. Rev. Lett.* **59**, 2044 (1987).
36. A.M. Steinberg, P.G. Kwiat and R.Y. Chiao, *Phys. Rev. Lett.* **68**, 2421 (1992).
37. Z.Y. Ou, X.Y. Zou, L.J. Wang and L. Mandel, *Phys. Rev. Lett.* **65**, 321 (1990).
38. J. Brendel, E. Mohler and W. Martienssen, *Europhys. Lett.* **20**, 575 (1992).
39. D. O'Connell, K.J. Kollath, A.J. Duncan, and H. Kleinpoppen, *J. Phys. B* **8**, L214 (1975).
40. H. Kruger and A. Oed, *Phys. Lett.* **54A**, 251 (1975).
41. J.S. Bell, *Physics (N.Y.)* **1**, 195 (1964).
42. W. Perrie, A.J. Duncan, H.J. Beyer, and H. Kleinpoppen, *Phys. Rev. Lett.* **54**, 1790, 2647(E) (1985).

43. T. Haji-Hassan, A.J. Duncan, W. Perrie, H.J. Beyer and H. Kleinpoppen, *Phys. Lett.* **123A**, 110 (1987).
44. T. Haji-Hassan, Ph.D. Thesis, University of Stirling, Stirling, Scotland (1987)
45. T. Haji-Hassan, A.J. Duncan, W. Perrie, H. Kleinpoppen and E. Merzbacher, *Phys. Rev. Lett.* **62**, 237 (1989).
46. T. Haji-Hassan, A.J. Duncan, W. Perrie, H. Kleinpoppen and E. Merzbacher, *J. Phys. B: At. Mol. Opt. Phys.* **24**, 5035 (1991).
47. G.W.F. Drake, P.S. Farago and A. Van Wijngaarden, *Phys. Rev. A* **11**, 1621 (1975).
48. A. Van Wijngaarden, G.W.F Drake and P.S. Farago, *Phys. Rev. Lett.* **33**, 4 (1974)
49. A. Van Wijngaarden, E. Goh, G.W.F. Drake and P.S. Farago, *J. Phys. B: Atom. Molec. Phys.* **9**, 2017 (1976).
50. C.K Au, *Phys. Rev A.* **14**, 531 (1976).
51. Z.Y. Ou and L. Mandel, *Phys. Rev. Lett.* **61**, 54 (1988).
52. A.J. Duncan, in *Progress in atomic spectroscopy, part D* (edited by H.J. Beyer and H Kleinpoppen), pp 447-505, Plenum Press, New York (1987).
53. M. Born and E. Wolf, *Principles of Optics*, Pergammon Press, Oxford (1975).
54. Z.A. Sheikh, Ph.D. Thesis, University of Stirling, Stirling (1993).
55. A. Messiah, *Quantum Mechanics* Vol. I, p 469, North-Holland, Amsterdam (1965).
56. G. Kaye and T. Laby, *Tables of Physical and Chemical Constants*, Longmans, London (1968).
57. M.J. Padgett, A.R. Harvey, A.J. Duncan and W. Sibbett, Applied Optics **33**, 6035 (1994).

Index

Series Publications

Below is a chronological listing of all the published volumes in the *Physics of Atoms and Molecules* series.

RECENT STUDIES IN ATOMIC AND MOLECULAR PROCESSES
Edited by Arthur E. Kingston

QUANTUM MECHANICS VERSUS LOCAL REALISM: The Einstein-Podolsky-Rosen Paradox
Edited by Franco Selleri

ZERO-RANGE POTENTIALS AND THEIR APPLICATIONS IN ATOMIC PHYSICS
Yu. N. Demkov and V. N. Ostrovskii

COHERENCE IN ATOMIC COLLISION PHYSICS
Edited by H. J. Beyer, K. Blum, and R. Hippler

ELECTRON–MOLECULE SCATTERING AND PHOTOIONIZATION
Edited by P. G. Burke and J. B. West

ATOMIC SPECTRA AND COLLISIONS IN EXTERNAL FIELDS
Edited by K. T. Taylor, M. H. Nayfeh, and C. W. Clark

ATOMIC PHOTOEFFECT
M. Ya. Amusia

MOLECULAR PROCESSES IN SPACE
Edited by Tsutomu Watanabe, Isao Shimamura, Mikio Shimizu, and Yukikazu Itikawa

THE HANLE EFFECT AND LEVEL CROSSING SPECTROSCOPY
Edited by Giovanni Moruzzi and Franco Strumia

ATOMS AND LIGHT: INTERACTIONS
John N. Dodd

POLARIZATION BREMSSTRAHLUNG
Edited by V. N. Tsytovich and I. M. Ojringel

INTRODUCTION TO THE THEORY OF LASER–ATOM INTERACTIONS
(Second Edition)
Marvin H. Mittleman

ELECTRON COLLISIONS WITH MOLECULES, CLUSTERS, AND SURFACES
Edited by H. Ehrhardt and L. A. Morgan

THEORY OF ELECTRON–ATOM COLLISIONS, Part 1: Potential Scattering
Philip G. Burke and Charles J. Joachain

POLARIZED ELECTRON/POLARIZED PHOTON PHYSICS
Edited by H. Kleinpoppen and W. R. Newell

INTRODUCTION TO THE THEORY OF X-RAY AND ELECTRONIC
SPECTRA OF FREE ATOMS
Romas Karazija

VUV AND SOFT X-RAY PHOTOIONIZATION
Edited by Uwe Becker and David A. Shirley

DENSITY MATRIX THEORY AND APPLICATIONS (Second Edition)
Karl Blum

SELECTED TOPICS ON ELECTRON PHYSICS
Edited by D. Murray Campbell and Hans Kleinpoppen